3D and 4D Printing in Biomedical Applications

3D and 4D Printing in Biomedical Applications

Process Engineering and Additive Manufacturing

Edited by
Mohammed Maniruzzaman

WILEY-VCH

Editor

Dr. Mohammed Maniruzzaman
University of Sussex
School of Life Sciences
BN1 9QG Brighton
United Kingdom

Cover
3D Printer © iStock.com/AzmanL
Human Anatomy © courtesy of Luciano Paulino Silva & created by Ella Maru Studio

Library of Congress Card No.:
applied for

British Library Cataloguing-in-Publication Data
A catalogue record for this book is available from the British Library.

Bibliographic information published by the Deutsche Nationalbibliothek
The Deutsche Nationalbibliothek lists this publication in the Deutsche Nationalbibliografie; detailed bibliographic data are available on the Internet at <http://dnb.d-nb.de>.

Print ISBN: 978-3-527-34443-7
ePDF ISBN: 978-3-527-81367-4
ePub ISBN: 978-3-527-81369-8
oBook ISBN: 978-3-527-81370-4

Typesetting SPi Global, Chennai, India
Printing and Binding C.O.S. Printers Pte Ltd, Singapore
Printed on acid-free paper

10 9 8 7 6 5 4 3 2 1

The bond between father and daughter
happens instantly, starting right at birth.
When a father first lays eyes on his little girl,
He loves her more than anything on this earth.
(H. Twining)

To my wonderful little daughter
Shahrooz Myreen Zaman

Contents

Preface

Three-dimensional (3D) printing is a method of additive manufacturing, which involves materials, such as polymers or metals, deposited in sequential layers to produce 3D objects, i.e. medical devices. The convergence of additive manufacturing and suitable printing materials is of significant aptitude for the advancement of personalized products (e.g. biomaterials and pharmaceutical dosage forms). Geometric shapes as well as the visual effects of materials currently used in additive manufacturing play essential roles toward the smooth fabrication of the resultant objects. In most of the cases, especially in biomedical/pharmaceutical applications, functions of structures are surprisingly limited by the complexity of the manageable shapes. Besides, traditional processing techniques such as ink jet printing or molding suffer from the failure of meeting the growing needs because of both the difficulty and the associated cost. Up to date, 3D printing has been utilized as an attractive option because of its supreme flexibility and versatility in producing complex objects. However, the application and practical potential of 3D printing is at some extent limited because of its speed. Therefore, it is projected that moving from the stepwise layer-by-layer process, which is typical in 3D printing, to a continuous process may significantly accelerate the practical potential of printing technology. Four-dimensional (4D) printing can encompass a wide range of disciplines such as materials science, bioengineering, and chemistry/chemical engineering and has the true potential to emerge as the next-generation additive manufacturing technique. By utilizing stimuli-responsive (also known as shape memory) materials and existing 3D printing strategies, 4D printing aims to create dynamic 3D patterned structures that are capable of transforming from one shape to another, right off the print bed under various stimuli (e.g. temperature).

It is also not surprising that the continued interest in 3D/4D printing technologies and their applications has been supplemented by a wealth of publications. Within this context, *3D and 4D Printing in Biomedical Applications: Process Engineering and Additive Manufacturing* has been developed to provide a professional source on 3D and 4D printing technology in the biomedical and pharmaceutical fields. The process optimization, innovative process engineering, platform technology, i.e. behind world's first Food and Drug Administration (FDA) approved 3D printed medicine, materials and processes for bioprinting, novel and smart excipients used to fabricate 3D products, potential use of various biomaterials such as stimuli-responsive materials for the fabrication of

4D printed products, the state of the art and limitations that exist in the current 3D printing modality, and regulatory expectations are critically surveyed in this book.

Timely, and written by leading international experts from both academia and industry, *3D and 4D Printing in Biomedical Applications: Process Engineering and Additive Manufacturing* serves as an important professional resource and will help readers (e.g. pharmaceutical/biomedical scientists, researchers, and postgraduate students) develop a deep understanding of key aspects of 3D and 4D printing of medical and pharmaceutical products as well as fundamental challenges and advances associated with their development.

As the editor, I wish to acknowledge and thank all of the authors for their valuable contributions and insight without which this book would not have been possible. It is through their persistent and collective efforts that such a comprehensive and insightful book was created, and it is hoped that this book will aid in the continued advancement of 3D and 4D printing and their emergent applications across the industries.

Dr. Mohammed Maniruzzaman
Lecturer in Pharmaceutics and Drug Delivery
University of Sussex, UK

1

3D/4D Printing in Additive Manufacturing: Process Engineering and Novel Excipients

Christian Muehlenfeld[1] *and Simon A. Roberts*[2]

[1] *Ashland Industries Deutschland GmbH, Paul-Thomas-Straße 56, 40599 Düsseldorf, Germany*
[2] *Ashland Specialties UK Ltd., Vale Industrial Estate, Stourport Road, Kidderminster, Worcestershire, DY11 7QU, UK*

1.1 Introduction

In recent years, additive manufacturing, which is more colloquially referred to as three-dimensional (3D) printing, has seen high-impact implementation in manufacturing applications in areas such as aeronautics, robotics, electronics, industrial goods, and even the food industry. These wide-ranging applications have resulted in a change in focus for biomedical research [1]. 3D printing is a generic term that describes various methods of constructing objects in a layer-by-layer manner. Although the birth of 3D printing dates back to 1984, when Charles Hull invented the first stereolithographic printer, 3D printing started to increasingly change the way in which manufacturing was performed from the year 2000 onward.

This chapter will introduce the basic concepts of 3D and 4D printing technologies as they pertain to biomedical applications. In particular, 4D printing (printing of objects with the ability to change over time) has a strong potential for biomedical applications. Patient-specific products such as medical devices, tissue constructs (including muscle structures, bone, and ear tissue), and, eventually, artificial organs may be fabricated using 4D printing [2–6].

1.2 The Process of 3D and 4D Printing Technology

3D printing typically begins with a computer-aided design (CAD) file that describes the geometry and size of the objects to be printed. The object is sliced into a series of digital cross-sectional layers that are then fabricated by the 3D printer. This process can use many different types of materials such as thermoplastic polymers, powders, metals, and ultraviolet (UV) curable resins.

Four-dimensional (4D) printing is defined as printing of 3D objects with the ability to change the form or function under the influence of external stimuli over time [7, 8]. A schematic of printing dimensions is shown in Figure 1.1.

3D and 4D Printing in Biomedical Applications: Process Engineering and Additive Manufacturing, First Edition. Edited by Mohammed Maniruzzaman.

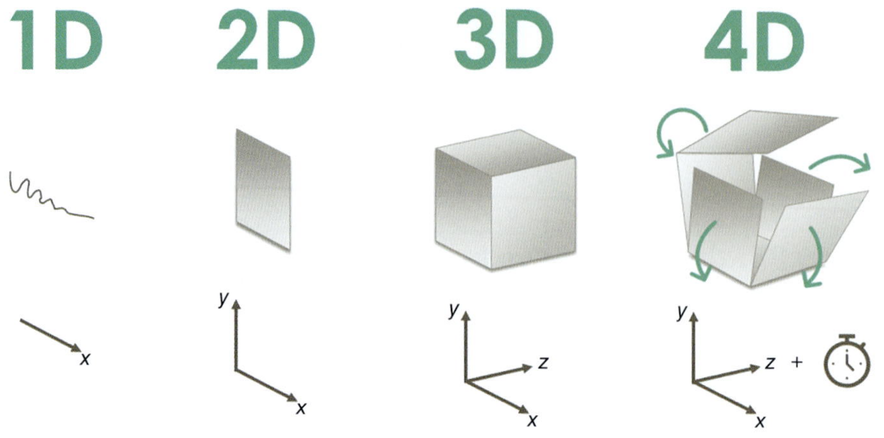

Figure 1.1 Schematic of 1D, 2D, 3D, and 4D printing dimensions. In a 4D system, a 3D printed object undergoes time-dependent deformations when exposed to various stimuli.

The essential difference between 4D printing and 3D printing is the addition of smart design, or responsive materials, that results in a time-dependent deformation of the object. In order to achieve this goal, the printed material needs to self-transform in form or function when exposed to an external stimulus such as osmotic pressure, heat, current, ultraviolet light, or another energy source [9]. Incorporating these additional functions poses major challenges to the design process because 4D printed structures must be preprogrammed in detail, based on the transforming mechanism of controllable smart materials that incorporate the requested material deformations. Because most 3D printing materials are designed only to produce rigid, static objects, the choice of materials for 4D printing is significant.

1.3 3D/4D Printing for Biomedical Applications

3D and 4D printing technologies have the potential for great impact in biomedical applications. 3D printing allows printing of biomaterials as well as living cells to build complex tissues and organs, whereas 4D bioprinting is an extension of the process that adds additional value. Different approaches can be used for 4D printing of biomaterials. The first approach strictly follows the original concept of 4D printing, in which a substrate material folds into a predefined 3D configuration upon stimulus. The printed cell or tissue material is incorporated within the device during printing and subsequently follows the folding of the substrate as it forms into a desired shape postimplantation.

The second approach is based on the maturation of engineered tissue constructs after printing and could be considered as a kind of *in vivo* 4D bioprinting. A 3D printed polymer medical device is implanted first and then accommodates the growth of tissue or organ over the postsurgical period.

1.4 Smart or Responsive Materials for 4D Biomedical Printing

The 3D and 4D printing technologies are classified mainly based on the types of materials used. The selection of materials has a direct influence on mechanical or thermal properties, as well as the transformation stimuli of the finished objects. Although the major difference between 3D and 4D printing is in the materials, the processes used to fabricate printed objects are the same. It should be pointed out that 4D printing is still in its early development stage. Herein, some example applications are presented to demonstrate its potential.

Although numerous materials are available for 3D printing, currently, limited stimuli-responsive biomaterials are available for 4D printing. At present, researchers are focused on the development of various, novel, smart materials; however, not every smart material can be 3D printed. The most common materials used in 4D printing are biocompatible materials such as hydrogels and polymers. Table 1.1 lists some examples of smart biomaterials intended for biomedical applications based on their stimulus responsiveness. Some of them have already been used for 4D printing, but it is unclear whether others of these materials can be used in 3D/4D printing in the future. The mechanisms facilitating 4D temporal shape transformation of 3D printed materials for biomedical applications range from temperature responsiveness, magnetic field responsiveness, and light responsiveness to humidity responsiveness.

A simple mechanism facilitating 4D shape transformation of 3D printed materials is the shape memory properties of thermoresponsive materials. Poly(*N*-isopropylacrylamide) (pNIPAM) hydrogels are well-known examples, in which the transformation principle is based on the wettability and solubility alteration of the thermoresponsive hydrogel following a change in temperature. Figure 1.2A shows an example of a photo-crosslinked, acrylic acid-functionalized pNIPAM (pNIPAM-AAc) in combination with polypropylene fumarate, where the pNIPAM-AAc component is transformed to a hydrophobic state showing shape transformation after increasing the temperature above 36 °C [10]. Zarek et al. [11] presented a strategy to capitalize on a series of medical imaging modalities to construct a printable shape memory endoluminal device, exemplified by a 4D printed tracheal stent made from methacrylated poly(ε-caprolactone) (PCL) that can be deformed into a temporary shape, inserted in the body, and then deployed back into its permanent shape with a local increase in temperature. Huang et al. [12] used biodegradable poly(L-lactic acid) (PLA) surgical staples as an alternative to biodegradable sutures in minimally invasive surgery for wound closure. Those staples are used in a stretched form and show a self-tightening function upon heating to slightly above body temperature (about 45 °C, which is within the glass transition temperature range of PLA) (Figure 1.2B). Another example based on the concept of temperature responsiveness is a poly(vinyl alcohol) (PVA)–poly(ethylene glycol) (PEG) double-network hydrogel, which was able to transform back from a stabilized helix structure after 15 seconds of immersion in hot water (90 °C), causing molten crystalline domains of PVA and

Table 1.1 Examples of smart or responsive materils suitable for biomedical purposes.

Stimulus	Material type or name	Composition and remarks	Print process	References
Temperature	pNIPAM-AAc	Poly(*N*-isopropylacrylamide-co-acrylic acid) (pNIPAM-AAc), polypropylene fumarate (PPF), iron oxide (Fe_2O_3) nanoparticles	—	[10]
	Methacrylated polycaprolactone	Poly(ε-caprolactone) (PCL) dimethylacrylate, 2,4,6-trimethylbenzoyl-diphenylphosphineoxide (TPO) as photoinitiator, vitamin E to prevent premature cross-linking, Toner Yellow 3GP	SLA (Freeform pico 2 SLA digital light processing printer)	[11]
	PLA surgical staples	Poly(L-lactic acid) (PLA)	Not mentioned	[12]
	PVA/PEG hydrogel	Poly(vinyl alcohol) (PVA)–poly(ethyleneglycol) (PEG) double-network hydrogel	—	[13]
	Soybean-oil-epoxidized acrylate liquid resin	Soybean-oil-epoxidized acrylate contains three major fatty acid residues (stearic, oleic, and linoleic acid) with pendant alkane groups that may freeze and benefit shape fixing at −18 °C.	SLA (modified Solidoodle® 3D printer platform)	[14]
Magnetic field	PEGDA/PHEMA soft microrobot	PEG acrylate (PEGDA), iron (II, III) oxide (Fe_3O_4); 2-hydroxyethyl methacrylate (PHEMA) layer	—	[15]
	Macroporous ferrogel	Peptides containing the arginine–glycine–aspartic acid (RGD) amino acid sequence, sodium alginate, Fe_3O_4 nanoparticles	—	[16]
Light	Optogenetic muscle ring-powered biobots	PEG acrylate (PEGDA) photosensitive resin	SLA (SLA 250/50; 3D systems)	[5]
	PHEMA hydrogel	Cross-linked PHEMA, functionalized with azobenzene groups	—	[17]
Humidity	PCAD@AG	PEG-conjugated azobenzene derivative (PCAD) and agarose (AG)	—	[18]
	$CSE_{0.3}$	Cellulose stearoyl ester with low degree of substitution (DS = 0.3)	—	[19]
Osmotic pressure	PEG hydrogel	Photo-crosslinkable PEG with 1-[4-(2-hydroxy-ethoxy)-phenyl]-2-hydroxy-2-methyl-1-propane-1-one (Irgacure 2959) photoinitiator	—	[20]
	Vinyl caprolactam/PE hydrogel	Vinyl caprolactam, polyethylene, epoxy diacrylate oligomer, Irgacure 819	Stratasys Connex 500 Multi-Material 3D Printer	[21]

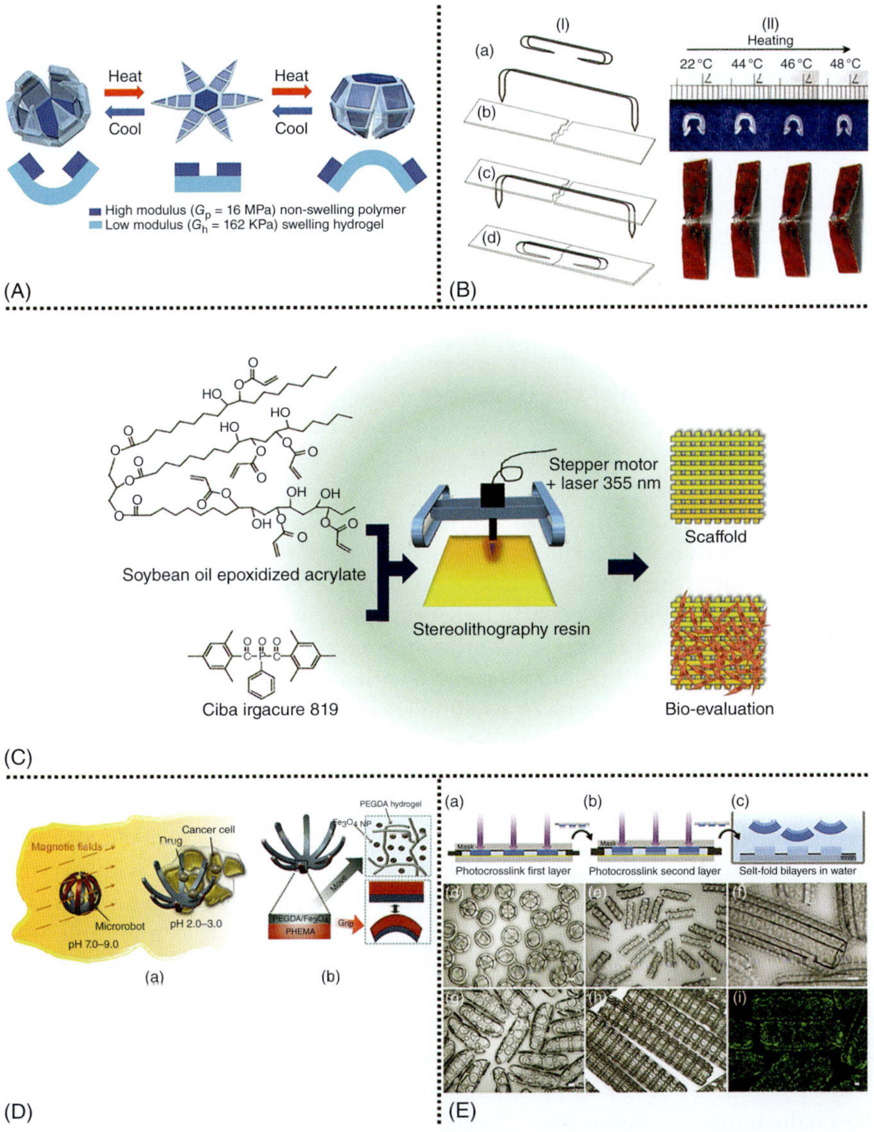

Figure 1.2 (A) Schematic diagram illustrating the reversible self-folding of soft microgrippers in response to temperature. Source: Breger et al. 2015 [10]. Reproduced with permission of ACS. (B) Self-tightening of a PLA staple. (I) Concept. (a) Original shape of a staple; (b) after programming; (c) after being fired into tissue; (d) after heating for tightening; (II) experimental result (Insorb® staple). Top: shrinking of staple upon heating; bottom: tightening of staple upon heating to bring two pieces of tissue closer. Source: Huang et al. 2013 [12]. Reproduced with permission of Elsevier. (C) Schematic of soybean-oil-epoxidized acrylate fabrication process from raw material through resin fabrication and application. Source: Miao et al. 2016 [14]. https://creativecommons.org/licenses/by/3.0/. (D) Schematic diagram illustrating the proposed soft microrobot, which can move freely by magnetic fields. Trapping and releasing of drug microbeads at the destination target by folding and unfolding motions is triggered by different pH values. Source: Hao et al. 2016 [15]. Reproduced with permission of IOP Publishing. (E) Schematic diagram illustrating the osmotic-pressure-driven deformation. Side view schematic of the three basic PEG bilayer photo-crosslinking steps (a–c) and examples of self-folded hydrogel geometries (d–i). (Source: Jamal et al. 2013 [20]. Reproduced with permission of John Wiley & Sons.)

thus transform back to a straight line [13]. Miao et al. [14] used the concept of thermoresponsiveness for biomedical scaffolds fabricated using a stereolithography (SLA) printer. Polymerized epoxidized soybean oil acrylate was used because of its thermoresponsive properties and glass transition temperature of approximately 20 °C, which could revert to its original shape at approximately 37 °C (Figure 1.2C).

Hydrogels containing magnetic particles, or ferrogels, are examples of magnetic-field-responsive materials. Figure 1.2D shows a 3D printed magnetic-field-responsive soft microrobot made of a poly(ethylene glycol) acrylate (PEGDA) and 2-hydroxyethyl methacrylate (PHEMA) hydrogel bilayer structure containing iron oxide particles (Fe_3O_4), which can move under an external magnetic field to the target site and release an encapsulated drug, triggered by a change in pH [15]. Another example is an alginate-based scaffold driving the outward movement of water from the internal pores under the influence of a magnetic field, thus triggering the release of cells or biological agents [16].

Light-responsive materials may convert their shape based on photoisomerization and photodegradation in the polymer chain. These mechanisms have been applied in artificial muscle biobots, where stereolithographic 3D printing was used to fabricate ring and strip injection molds and biobot skeletons from a PEGDA photosensitive resin [5]. Another example of light responsiveness is the use of cross-linked PHEMA functionalized with azobenzene groups, where light irradiation modifies the degree of swelling [17].

Humidity-responsive material uses include the humidity-induced bending of PEG-conjugated azobenzene derivatives with agarose (PCAD@AG) films [18], or cellulose-based materials [19].

An example of osmotic-pressure-driven hydrogels using intrinsic swelling characteristics was demonstrated using photo-crosslinkable PEG with varying molecular weights [20]. Printed as bilayered constructs with 1-[4-(2-hydroxyethoxy)-phenyl]-2-hydroxy-2-methyl-1-propane-1-one as the photoinitiator, differences in the swelling behavior of the hydrogel layers result in a shape transformation to form micropatterned structures (Figure 1.2E). This principle was adapted by adding a non-swelling but flexible material as the second layer to form joints between rigid linear structures [21].

However, all of these applications have been tested in biomechanically non-challenging environments. Therefore, direct biomedical application is currently restricted by material limitations and the complex host environment of the targeted tissue(s). Accordingly, not all stimuli may be applicable for use in biomedical applications. Although humidity responsiveness is widely present in nature, application of this stimulus could be restricted because of the limitations of humidity or osmotic pressure that can be applied to the constructs used for biomedical purposes. Taken together, these examples demonstrate that novel excipients and excipient combinations can be used to induce temporal shape transformation for 4D printing; however, performance has not been tested in biomechanically challenging environments. Thus, follow-up studies employing and characterizing these introduced concepts and, furthermore, using medical grade materials are necessary and important.

1.5 Classification of 3D and 4D Printing Technologies

Although a broad variety of technologies have been developed for industrial fabrication of 3D structures, there are only few major technologies used for biomedical printing. These include extrusion-based (fused filament), droplet-based (using chemical agents/binders), and laser-based systems (sintering/melting) to print the material. Each technique differs in the manner in which layers are built and printing materials are used (Figure 1.3). Furthermore, each of these shows certain process characteristics that might be preferable for different applications. The advantages and disadvantages associated with each approach can be demonstrated by comparing the dimensional accuracy, mechanical properties, surface roughness, build speed, and materials cost, across multiple 3D printing platforms [22]. A comprehensive summary of each technology is given in the following sections.

1.5.1 Fused Filament Fabrication (FFF) – Extrusion-Based Systems

Fused filament fabrication (FFF) is an extrusion-based printing technology, also known by the trademark name Fused Deposition Modeling™ (FDM)[23]. The FFF systems use solid filaments that are heated above the melting temperature of the material and then the extruded melt is deposited using a Cartesian coordinate robot in a continuous flow to build up a 3D printed part in a layer-by-layer manner. When each layer in the *xy* plane is finished, the platform (*z* axis) is lowered and the procedure is repeated (Figure 1.4). This process continues until the whole part is complete. Because of thermal fusion, the material bonds with the layer beneath and solidifies, thus forming a permanent bonding of the two layers. To improve the interlayer bonding, the entire process is performed in a closed chamber maintaining a constant temperature [24].

Multiple printheads can be accommodated in FFF devices, allowing the use of different materials within a single 3D printed object. If necessary, a second printhead is used to provide a temporary support substrate for complex structures with an overhang, offset, or cavity. This additional material prevents the

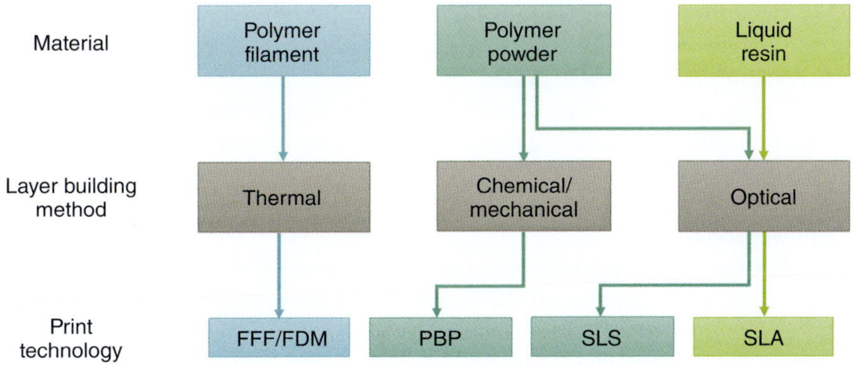

Figure 1.3 Overview of material types used with specific layer building methods in 3D/4D printing.

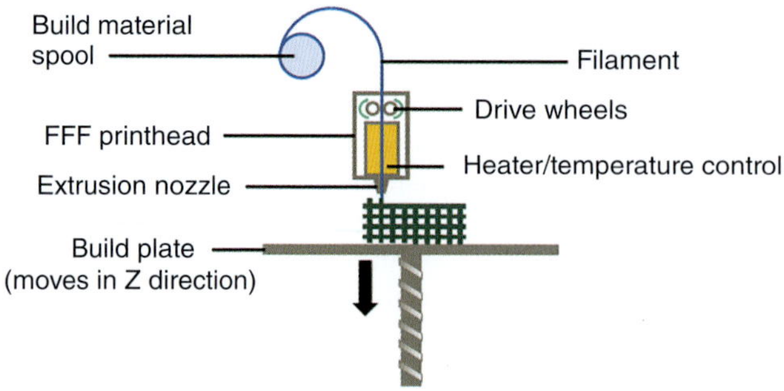

Figure 1.4 Schematic of FFF process.

component part from collapsing during the building process. The support material itself can be easily removed after the building process by breaking it off or dissolving it in a warm water bath.

The FFF approach allows fabrication of structures with controllable pore size and porosity by changing the material deposition amount, the spacing between the material paths, and the height interval (z axis). The most important material selection criteria for FFF materials are heat transfer characteristics and rheology because the FFF approach requires processable prepolymers as the building materials (filaments).

A major benefit of FFF printing is that the polymer filaments can be manufactured with hot melt extrusion (HME). This means that the knowledge and acceptance of HME-manufactured materials is already assured. However, the FFF process usually requires tight specifications for the filaments. Melocchi et al. [25] pointed out the need for homogeneous filaments with a minimum length of 25 cm, circular cross section, and consistent diameter as well as diameter tolerances (1.75 ± 0.05 mm) for filaments made of hydroxypropylcellulose (Klucel™ LF, Ashland). Undersized filament diameter led to the formation of air bubbles within the printed material and an oversized one resulted in clogging of the tip [25]. The diameter of the extruded filaments depends not only on the diameter of the extrusion die but also on the relaxation of the polymer and the speed of the conveying belt (Figure 1.5). Although both diameter and variance along the length of the filament matter, consistency is more important than exactly reaching 1.75 mm in diameter.

Suitable polymeric materials for FFF printer are thermoplastic and become molten at reasonably low heating temperatures (usually lower than 250 °C). They solidify fast enough (sufficiently high glass transition temperature) so that they hold their shape when hardened. Furthermore, the materials possess specific mechanical properties.

To predict the mechanical behavior of these materials, it is critical to understand the material properties of the raw material as well as the effect that the process parameters of FFF have on those mechanical properties [26]. There are various options for processing parameters such as layer thickness, build orientation, raster angle, raster width, and raster-to-raster air gap, all of which can significantly affect the mechanical properties and performance of the

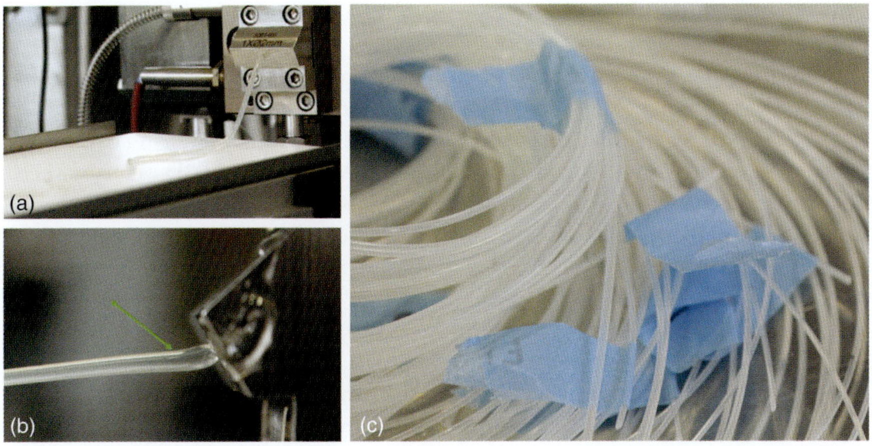

Figure 1.5 (a) Hot melt extrusion of hydroxypropylcellulose (HPC) filaments suitable for FFF. (b) Polymer relaxation (die swell) of HPC after leaving the extrusion die. (c) HPC filaments varying in diameter (<1.8 mm).

Table 1.2 Variables that can affect mechanical properties of printed materials.

Process variables	Design variables	Material variables
Build orientation	Printer model	Rheological properties
Layer thickness	Process type	Density of the unprocessed material
Raster angle	Extrusion nozzle diameter	Thermal properties of polymer and other ingredients
Raster width	Software	Formulation; miscibility and concentration of the components
Raster-to-raster air gap		Filament diameter
Fill pattern		Uniformity of filament diameter
Material deposition speed/amount		
Extrusion temperature		
Chamber temperature		
Environmental conditions (e.g. humidity)		

FFF material [27]. An overview of variables that might affect the mechanical properties of printed materials is given in Table 1.2.

Originally, acrylonitrile butadiene styrene (ABS) served as the feedstock material, delivered as fibers from spools. However, the range of materials that can be processed effectively is increasing, including new materials and polymer blends in the filament form. Other material options include polycarbonate (PC), polypropylene (PP), polyphenylsulfone (PPSF), polyglycolic acid

(PGA), poly(L-glutamic acid) (PLGA) and PCL, polydioxanone (PDO), and poly(ether-ether-ketone) (PEEK).

For biomedical purposes, biodegradable materials have been frequently used to replace metallic implants for internal fixation. For instance, PGA- or PLA-based screws and pins have been widely used for orthopedic surgery, offering the advantage of being resorbable [28]. With the use and development of 3D printing technologies such as FFF rising, more complex shapes are possible to print.

PLGA has been previously used with FFF to create scaffolds [29–31]; however, the comparably high glass transition temperature of 40–60 °C presents challenges to the extrusion process because a higher extrusion temperature is required to create the right material flow properties for extrusion from the nozzle and for fusion of the layers [30, 31].

Another polymer that has been widely used to fabricate bioresorbable scaffolds for bone tissue engineering applications is PCL. In contrast to PLGA, it has a low melting temperature of approximately 60 °C, low glass transition temperature of −60 °C, and high thermal stability [23, 32], although still being biodegradable by hydrolysis [33, 34]. Apart from bone tissue engineering, PCL can also be used for the preparation of implantable devices, such as drug-loaded implants [35], or long-lasting implantable intrauterine systems for birth control [36].

Other melt-extrudable polymer-based medical devices have incorporated PEEK because of its cell biocompatibility and desirable mechanical properties such as the elastic modulus being comparable to that of cortical bone, which results in reduced stress shielding after implantation [37]. Although more often processed using selective laser sintering (SLS)[38], it can also be formed with FFF, although it is quite challenging to process because of its very high melting temperature[39, 40].

Overall, the main advantages of the FFF process are that it does not require toxic or organic solvents, and the use of filaments allow for continuous low-cost production, with high flexibility in handling and processing of materials. Despite these advantages, the FFF process includes restrictions with regard to the material properties of the feedstock filament material necessary to feed it through the rollers and nozzle. Any changes in the properties of the material require considerable effort to recalibrate the feeding parameters. Additionally, parts manufactured by the FFF technique show some dimensional inaccuracy compared to other additive manufacturing techniques such as SLS because of the variety of conflicting and interacting process parameters that affect dimensional accuracy [41].

1.5.2 Powder Bed Printing (PBP) – Droplet-Based Systems

Powder bed printing (PBP) was developed at the Massachusetts Institute of Technology [42] and utilizes a liquid binder, delivered by an inkjet printhead, to build objects layer-by-layer from a bed of powder. The process begins by evenly spreading a layer of powder onto the build plate. The layer typically has a thickness of 200 μm and consists of powder with a particle size range of 50–100 μm [43]. The inkjet printhead then deposits droplets of the liquid binder solution onto the powder surface. The powder is solidified by the binder solution in the shape of the

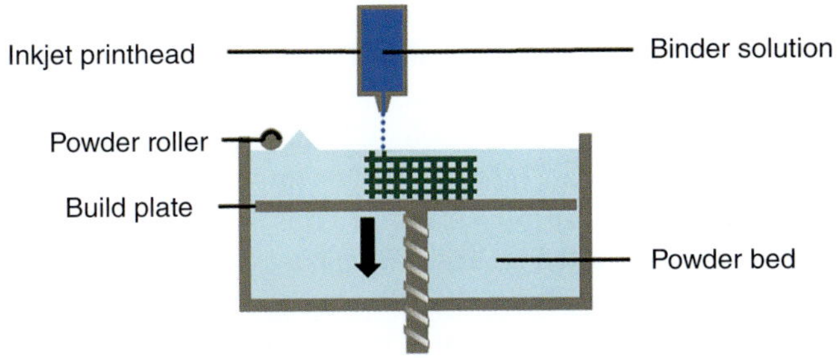

Figure 1.6 Schematic illustration of 3D printing by PBP.

two-dimensional cross section of that layer. The build plate is then lowered to the depth of the next layer and a new even layer of powder is spread across the surface in readiness for the printing of the new cross section. The process continues until the complete object is constructed within the powder bed. Overhanging structures and pores within the object are supported by the unbound powder during the printing process. Once complete, the object is removed from the surrounding unbound powder, including the removal of unbound powder from cavities and pores within the finished structure. The ease with which the object can be depowdered depends on the complexity of the design and may sometimes require the use of an air gun. The liquid binding of the powder tends to result in porous structures, which are sometimes sintered in order to improve the surface finish and mechanical strength [44, 45]. The process is illustrated in Figure 1.6.

The concepts behind inkjet printing were first described by Lord Rayleigh in the late nineteenth century [46, 47]. Development of these concepts has resulted in devices that can deliver either a continuous stream of droplets, known as continuous ink jetting (CIJ), or a drop-on-demand (DOD) ejection of droplets. The CIJ process releases a continuous stream of charged droplets, which are directed by electrostatic plates into the powder bed, or deviated into a waste recirculation line. On the other hand, the DOD process only dispenses the binder liquid droplets when required by the printing process, thus making it less wasteful than the CIJ process. Additionally, DOD is more precise than CIJ, with the ability to control droplet volume within a range of 1–300 pl [48, 49] at delivery frequencies of up to 10 000 Hz [50]. Production of droplets within a DOD printhead can be achieved by thermal or piezoelectrc methods. Thermal printheads consist of a thin film resistor that heats up rapidly when an electric pulse is applied. A superheated vapor bubble is formed, which expands and ejects a droplet from the print nozzle. Subsequent collapse of the bubble creates a partial vacuum, into which fresh binder solution is pulled [51]. Temperatures as high as 300 °C can be reached at the resistor surface, but the exposure time is in the order of milliseconds, and only a small fraction of the liquid, approximately 0.5% by volume, is heated, thus minimizing the potential effect of degradation of any thermally labile components [52]. Droplet formation within a piezoelectric printhead is a result of pressure waves that are induced within the liquid when a voltage is applied to the surrounding piezoelectric transducer causing it to deform. The liquid reservoir then refills once the piezoelectric material regains its original shape. In contrast to thermal inkjet printheads, the piezoelectric process is thermally constant and can be carried out at room temperature or in a localized cold environment [53].

A wide variety of powders, commonly utilized in medical applications, have been used in PBP in combination with suitable binders. It has been said that any combination of a powdered material with a binder that has low enough viscosity to form droplets could be used [54]. The physical and chemical properties of the binder solution need to be controlled in the PBP process. For successful delivery from the printheads, the viscosity of the binder solution needs to be in the range of 5–20 Pa s and the surface tension in the range of 35–40 mJ N^{-1} [51]. If an organic solvent is used in the binder solution, care must be taken to ensure the printhead is compatible with the solvent, as some organic solvents can dissolve the polymers used in most printheads [55]. The powder must have sufficient flow to enable it to be spread evenly to the thickness required for each layer. It must also be able to be removed easily from within the cavities and pores of the finished object. A number of synthetic polymers, including PCL, poly(lactide-*co*-glycolide), and PLA, have been used with organic solvent-based binders [56–58]. Natural polymers such as starch, dextran, and gelatin have been used in combination with water as a binder [44, 59]. Calcium phosphate-based bioceramics have also been used in PBP biomedical applications with acid- and solvent-based binder solutions [60]. There is also the ability to incorporate additional components within the powder bed or binder solution to expand the versatility of the PBP process. The additional components could be active pharmaceutical ingredients for drug delivery [61, 62] or biological agents such as peptides, proteins, polysaccharides, DNA plasmids, or cells [55].

1.5.3 Stereolithographic (SLA) Printing – Resin-Based Systems

SLA is a well-established 3D printing technology. In simple terms, it uses UV or visible light to solidify liquid, photocurable polymer resins. Objects are built up through sequential illumination of thin layers of resin, either by tracing a pattern with a laser beam or projecting the pattern with a digital projector, which solidifies the illuminated resin. Construction of a 3D object can be achieved in two ways. The object can be built from the bottom-up by illumination of the layer pattern on the upper surface of a bath of resin. Once the layer is complete, the build platform is lowered by the depth of the next layer and a blade is swept across the surface to apply a smooth layer of fresh resin. The new layer of resin is then illuminated with the next layer pattern. The alternative approach is to build the object from the top-down. In this case, the resin is placed in a bath with a base made of a UV–vis transparent material, such as polyethylene terephthalate (PET) [63]. The layer pattern is illuminated through the base window, onto the lower surface of the resin. The solidified layer is then raised upward by one layer thickness, liquid resin fills the space below the solid layer, and is illuminated with the next layer pattern. Schematic representations of these two approaches are illustrated in Figure 1.7.

In both cases, the depth of curing of each layer is slightly larger than the step movement of the build platform in the z-direction. Unreacted functional groups in the solid layer can then polymerize with the exposed resin in the new layer,

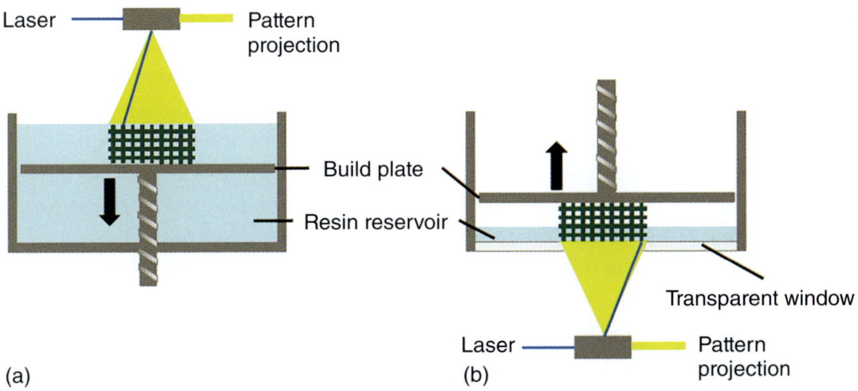

Figure 1.7 Schematic illustration of SLA process: (a) bottom-up process and (b) top-down process.

thus ensuring good adherence between each layer of the build. At the end of the process, when all layers have been completed, excess resin is drained and washed off. The completed object is often exposed to a final curing step using UV light to ensure full conversion of unreacted functional groups in the polymer.

The top-down process offers some advantages over the bottom-up process, but there are also some disadvantages. The consumption of resin is significantly lower in the top-down process because of not having to immerse the entire object in the resin. The time taken for the resin surface to settle and be swept by a blade is not required in the top-down process. Additionally, the top-down process exhibits greater control of layer thickness. This is a result of forming completely flat layers at the bottom of the resin tank. There is also no exposure to the air, which occurs in the bottom-up process, and can lead to oxygen inhibition of the polymerization reaction. However, because of the photopolymerization of each layer occurring in contact with the transparent window, there is the possibility of adherence between the solidifying layer and the window. Consequently, as the build platform rises upward, it could result in damage to the newly formed layer. Another potential drawback occurs during the manufacture of larger, heavier objects, where separation between the build platform and the object may occur, or weak sections may break. Addition of supports for weak sections can alleviate this problem.

The layer pattern can be transmitted to the surface of the resin by tracing the pattern with a single laser beam, known as scanning lithography or direct writing. Alternatively, it can be achieved by projecting the entire pattern onto the surface of the resin using a digital mask generator such as a digital micromirror device (DMD), known as projection-based SLA or the dynamic mask method. Projection-based SLA processes are less expensive than the direct writing process because of not requiring an expensive laser system. Projection of the entire layer pattern onto the resin surface enables a complete layer of resin to be cured simultaneously. The resultant build time is significantly reduced compared to direct writing, as it is dependent only on layer thickness and required exposure time, rather than also size in the xy-plane and the number of structures being built [64].

A combination of the scanning and projection methods has been developed by Emami and coworkers [63, 65] described as scanning–projection-based SLA. The projected image is continually updated as it is scanned over the surface of the layer being built. This allows larger objects to be constructed at higher resolution.

The resolution of the object being built is governed by a combination of the diameter of the laser beam or pixel size of the projection device and the curing depth of each layer. The depth of cure for each resin is determined by the amount of energy applied. This can be controlled by adjusting the power of the light source and the length of time the light is applied onto the resin. The complex kinetics of the photopolymerization curing process has been described in detail [66]. However, a simpler, semiempirical model based on the Beer–Lambert law has been described [67]. Equation (1.1) relates the cure depth in microns (C_d) to the dose of light irradiation, E (mJ cm^{-2}).

$$C_d = D_p \ln\left(\frac{E}{E_c}\right) \tag{1.1}$$

Determining the cure depth, or layer thickness, and plotting it against the applied radiation produce what has been called a working curve, which can then be used to determine the correct settings for the SLA process. SLA resins are characterized by a critical energy, E_c, and a penetration depth, D_p. Penetration depth, D_p, is directly related to the extinction coefficient in the Beer–Lambert equation. When the applied dose of irradiation, E, is greater than the critical energy, E_c, the resin solidifies from the surface. The concentrations of photoinitiator, dissolved oxygen, and other inhibiting species will all affect the value of E_c.

To ensure the effective layer-to-layer bonding, the value of C_d should be slightly higher than the layer thickness. This, however, results in additional curing in the preceding layer, potentially resulting in polymerization of voids within the design of the object being fabricated. For porous structures, such as scaffolds for tissue engineering, the effect of polymerization in the voids could be significant. Accurate control of the polymerization process can minimize the effect of filling voids with solidified resin. This can be achieved by reducing the light penetration depth, D_p, through the use of higher concentrations of photoinitiator, or the inclusion of nonreactive components such as dyes or UV absorbers [68]. The negative impact of reducing the value of D_p is that the build time is increased.

Possible limitations of SLA for biomedical applications include the small number of biocompatible resins that are suitable for the process and the complexities associated with using more than a single resin in the construction of the finished object. The number of suitable, photocurable resins is increasing and examples include poly(propylene fumarate) [69], poly(ε-caprolactone-*co*-trimethylene carbonate) [70], poly(D,L-lactide) [71], PCL [72], and PEG [73–76]. Resins with reported 4D properties based on soybean-oil-epoxized acrylate have also been reported by Miao et al. [14]. Overcoming the limitation of incorporating more than one resin into an object is more difficult. Techniques have been developed in which sequential polymerization and rinsing steps allow multiple resins to be built into each layer [73, 77]. A simple, automated method to switch between different resins would greatly expand the potential of the SLA technique.

1.5.4 Selective Laser Sintering (SLS) Printing – Laser-Based Systems

SLS was introduced soon after the SLA technique, but it primarily employs semicrystalline, particulate thermoplastic prepolymer as the building material. The technique involves a bed of tightly compacted powdered particles that is preheated close to the melting transition temperature. A laser beam is traced over the surface of the powder bed to bind the powder particles together. During the printing process, the laser draws a specific pattern onto the surface of the powder bed. After finishing the sintering of the first layer, the building platform is lowered by 100–200 μm and fresh powder is spread by a roller to build a new layer on top of the previous one. The resulting 3D printed object is built layer-by-layer and is finally recovered from underneath the powder bed. An illustration of the SLS process is shown in Figure 1.8.

Localized thermal sintering of particles is achieved by additional energy input from a high-power carbon dioxide (CO_2) laser, which scans the surface of the powder bed in a specific pattern and selectively melts the powder. The entire fabrication chamber is sealed and maintained at a temperature just below the melting point of the plastic powder. Thus, heat from the laser needs only to elevate the temperature slightly to cause sintering, greatly speeding up the process.

Especially for the selection of (new) laser sinter materials, an understanding of the different SLS subprocesses is essential. Different material properties and process parameters may affect the structural and mechanical properties of printed objects [78], and it is important to find a balance between effective sintering and avoiding polymer degradation that comes with overheating due to laser power and energy density [79]. SLS process parameters include part bed temperature, feed bed temperature, powder layer thickness, laser power, scan spacing, number of scans, time between layers, roller speed, build size, and heating/cooling rates, and these parameters are set differently according to powder properties and requirements of the application to achieve an optimum quality [80].

The basic material developed for SLS technology is a freely poured (loose) or slightly compacted polymeric powder, typically with a particle size in the range

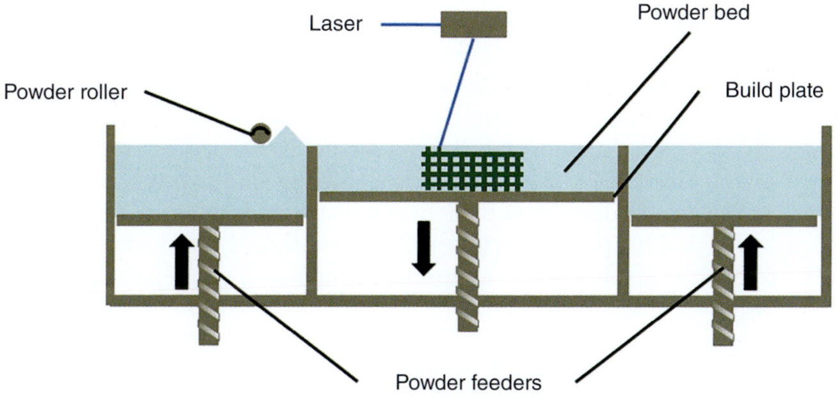

Figure 1.8 Schematic of SLS process using a CO_2 laser to locally fuse powder particles on the surface of a preheated build chamber that is incrementally lowered as the process progresses.

of several microns to several hundred microns [81]. The application of powders aims to apply a smooth, dense, and uniform powder layer when displaced by a roller or other spreading mechanism. After each finished layer, the new powder layer needs to be heated as fast as possible over its crystallization temperature to reach a fully melted state necessary for complete bonding between powder particles and to avoid cooling of the already finished layer, which can potentially cause shrinking and deformation of the sintered layer. Critical material attributes for deposition and laser sintering of the powder include powder density, particle shape, particle size distribution, and flowability [82, 83].

The most commonly used materials are powdered forms of plastics, ceramics, and metal alloys that require high temperatures and high-energy lasers to be sintered. These harsh printing conditions are the reason that the use of SLS printing in the medical field has been limited to medical instruments and implants, for example [38], or drug delivery devices where the drug was included after the printing process to circumvent the problem that the energy input of the high-power laser may degrade components if they are used as the starting material [84]. SLS polymer feedstock materials tend to be thermoplastic polymers that are either (semi)-crystalline polymers such as polyamide (PA, nylon) [85], poly(ABS), PEEK [38], and polyether block amide (PEBA) [86] or amorphous materials such as PC, and more recently polystyrene (PS), showing a different thermal behavior. The selection criteria of the semicrystalline polymers primarily include a broad process temperature window between the melting (T_m) and recrystallization temperature, a narrow melt transition, and a high melting enthalpy to minimize unwanted sintering [82], whereas the selection criteria for amorphous materials are somewhat different. Amorphous polymers tend to yield weaker, more porous structures than the semicrystalline polymer powders [87], and they do not undergo the significant dimensional contraction associated with crystallization as the process temperature is reduced [88]. Overall, glass transition temperature (T_g) and melting temperature (T_m) play important roles in the selection of process parameters and directly affect the mechanical properties of the SLS components. Other thermal properties such as specific heat capacity and thermal conductivity of polymers have great influence on the fabrication process as well [80].

Using SLS technology allows for complex 3D structures to be printed without the use of (organic) solvents. Moreover, as long as the material is in powdered form and can fuse but will not decompose under the laser beam, it can be used with SLS [89]. This opened the way for many biomedical applications ranging from the use of PEEK to produce non-resorbable implants for tissue engineering [38] and the production of bioactive implants and tissue scaffolds using composites of high-density polyethylene (HDPE) reinforced with hydroxyapatite (HA)[90], to biodegradable substitutes for tissue engineering to repair or replace damaged tissues such as PCL-[89]and poly-(L-lactic) acid (PLLA)[91]-fabricated scaffolds.

The potential disadvantages of SLS technology are the poor surface and dimensional accuracy and materials that sometimes require post-processing treatments that are considered critical for complex and controlled 3D printed structures.

1.6 Conclusions and Perspectives

Although the field of 4D printing in biomedical applications is just starting to emerge, the available pioneering examples demonstrate the possibility to incorporate the time scale into 3D printing to achieve transformation of printed objects along prescribed paths. It is necessary to develop new stimuli-responsive smart materials that, in addition to excellent biocompatibility, also possess appropriate rheological properties to ensure printability, appropriate mechanical properties, and stabilization mechanisms for the desired application and offer an effective interplay between cellular viability, function, and dynamic modulation of the printed objects. Merging these smart biomaterials within innovative technologies, 4D bioprinting is expected to become the next big thing to create transformable objects in biomedical applications, eventually mimicking the complex, dynamic deformation of native tissues such as the pumping behavior of the heart and the peristaltic movements of the gastrointestinal tract, among others.

References

1 Barnatt, C. (2013). *3D Printing: The Next Industrial Revolution*. CreateSpace Independent Publishing Platform.
2 Ionita, C.N., Mokin, M., Varble, N. et al. (eds.) (2014). Challenges and limitations of patient-specific vascular phantom fabrication using 3D Polyjet printing. In: *SPIE Medical Imaging*. SPIE.
3 Ozbolat, I.T. (2015). Bioprinting scale-up tissue and organ constructs for transplantation. *Trends in Biotechnology* 33 (7): 395–400.
4 Choong, C., Maleksaeedi, S., Eng, H. et al. (2017). 4D printing of high performance shape memory polymer using stereolithography. *Materials & Design* 126: 219–225.
5 Raman, R., Cvetkovic, C., Uzel, S.G.M. et al. (2016). Optogenetic skeletal muscle-powered adaptive biological machines. *Proceedings of the National Academy of Sciences* 113 (13): 3497–3502.
6 Morgan, N.B. (2004). Medical shape memory alloy applications – the market and its products. *Materials Science and Engineering: A* 378 (1): 16–23.
7 Tibbits, S. (2013). The emergence of "4D printing". https://www.ted.com/talks/skylar_tibbits_the_emergence_of_4d_printing/transcript?language=en (accessed 12 January 2018).
8 Gladman, A.S., Matsumoto, E.A., Nuzzo, R.G. et al. (2016). Biomimetic 4D printing. *Nature Materials* 15 (4): 413–418.
9 Campbell, T.A., Tibbits, S., and Garrett, B. (2014). The Next Wave: 4D Printing Programming the Material World. Technical report. Atlantic Council, USA.
10 Breger, J.C., Yoon, C., Xiao, R. et al. (2015). Self-folding thermo-magnetically responsive soft microgrippers. *ACS Applied Materials & Interfaces* 7 (5): 3398–3405.

11 Zarek, M., Mansour, N., Shapira, S., and Cohn, D. (2017). 4D printing of shape memory-based personalized endoluminal medical devices. *Macromolecular Rapid Communications* 38 (2): 1600628.
12 Huang, W.M., Song, C.L., Fu, Y.Q. et al. (2013). Shaping tissue with shape memory materials. *Advanced Drug Delivery Reviews* 65 (4): 515–535.
13 Li, G., Zhang, H., Fortin, D. et al. (2015). Poly(vinyl alcohol)–poly(ethylene glycol) double-network hydrogel: a general approach to shape memory and self-healing functionalities. *Langmuir* 31 (42): 11709–11716.
14 Miao, S., Zhu, W., Castro, N.J. et al. (2016). 4D printing smart biomedical scaffolds with novel soybean oil epoxidized acrylate. *Science Reports* 627226.
15 Hao, L., Gwangjun, G., Seong Yong, K. et al. (2016). Magnetic actuated pH-responsive hydrogel-based soft micro-robot for targeted drug delivery. *Smart Materials and Structures* 25 (2): 027001.
16 Zhao, X., Kim, J., Cezar, C.A. et al. (2011). Active scaffolds for on-demand drug and cell delivery. *Proceedings of the National Academy of Sciences* 108 (1): 67–72.
17 Unger, K., Salzmann, P., Masciullo, C. et al. (2017). Novel light-responsive biocompatible hydrogels produced by initiated chemical vapor deposition. *ACS Applied Materials & Interfaces* 9 (20): 17408–17416.
18 Zhang, L., Liang, H., Jacob, J., and Naumov, P. (2015). Photogated humidity-driven motility. *Nature Communications* 6: 7429.
19 Zhang, K., Geissler, A., Standhardt, M. et al. (2015). Moisture-responsive films of cellulose stearoyl esters showing reversible shape transitions. *Scientific Reports* 511011.
20 Jamal, M., Kadam, S.S., Xiao, R. et al. (2013). Bio-origami hydrogel scaffolds composed of photocrosslinked PEG bilayers. *Advanced Healthcare Materials* 2 (8): 1142–1150.
21 Raviv, D., Zhao, W., McKnelly, C. et al. (2014). Active printed materials for complex self-evolving deformations. *Scientific Reports* 4: 7422.
22 Kim, G.D. and Oh, Y.T. (2008). A benchmark study on rapid prototyping processes and machines: quantitative comparisons of mechanical properties, accuracy, roughness, speed, and material cost. *Proceedings of the Institution of Mechanical Engineers, Part B: Journal of Engineering Manufacture* 222 (2): 201–215.
23 Gibson, I., Rosen, D., and Stucker, B. (2015). *Additive Manufacturing Technologies – 3D Printing, Rapid Prototyping, and Direct Digital Manufacturing*. New York: Springer-Verlag.
24 Sun, Q., Rizvi, G.M., Bellehumeur, C.T., and Gu, P. (2008). Effect of processing conditions on the bonding quality of FDM polymer filaments. *Rapid Prototyping Journal* 14 (2): 72–80.
25 Melocchi, A., Parietti, F., Loreti, G. et al. (2015). 3D printing by fused deposition modeling (FDM) of a swellable/erodible capsular device for oral pulsatile release of drugs. *Journal of Drug Delivery Science and Technology* 30: 360–367.
26 Novakova-Marcincinova, L. and Novak-Marcincin, J. (2012). Testing of materials for rapid prototyping fused deposition modelling technology.

International Journal of Mechanical, Aerospace, Industrial, Mechatronic and Manufacturing Engineering 6 (10): 2082–2085.

27 Sood, A.K., Ohdar, R.K., and Mahapatra, S.S. (2010). Parametric appraisal of mechanical property of fused deposition modelling processed parts. *Materials & Design* 31 (1): 287–295.

28 Agrawal, C.M., Niederauer, G.G., and Athanasiou, K.A. (1995). Fabrication and characterization of PLA–PGA orthopedic implants. *Tissue Engineering* 1 (3): 241–252.

29 Kim, J.Y. and Cho, D.-W. (2009). Blended PCL/PLGA scaffold fabrication using multi-head deposition system. *Microelectronic Engineering* 86 (4): 1447–1450.

30 Park, S.H., Park, D.S., Shin, J.W. et al. (2012). Scaffolds for bone tissue engineering fabricated from two different materials by the rapid prototyping technique: PCL versus PLGA. *Journal of Materials Science: Materials in Medicine* 23 (11): 2671–2678.

31 Kim, J., McBride, S., Tellis, B. et al. (2012). Rapid-prototyped PLGA/β-TCP/hydroxyapatite nanocomposite scaffolds in a rabbit femoral defect model. *Biofabrication* 4 (2): 025003.

32 Hutmacher, D.W., Schantz, T., Zein, I. et al. (2001). Mechanical properties and cell cultural response of polycaprolactone scaffolds designed and fabricated via fused deposition modeling. *Journal of Biomedical Materials Research* 55 (2): 203–216.

33 Piskin, E., Nimet, B., Egri, S., and Isoglu, I.A. (2007). Electrospun matrices made of poly(α-hydroxy acids) for medical use. *Nanomedicine* 2 (4): 441–457.

34 Pitt, C.G., Chasalow, F.I., Hibionada, Y.M. et al. (1981). Aliphatic polyesters. I. The degradation of poly(ε-caprolactone) in vivo. *Journal of Applied Polymer Science* 26 (11): 3779–3787.

35 Kempin, W., Franz, C., Koster, L.-C. et al. (2017). Assessment of different polymers and drug loads for fused deposition modeling of drug loaded implants. *European Journal of Pharmaceutics and Biopharmaceutics* 115: 84–93.

36 Holländer, J., Genina, N., Jukarainen, H. et al. (2016). Three-dimensional printed PCL-based implantable prototypes of medical devices for controlled drug delivery. *Journal of Pharmaceutical Sciences* 105 (9): 2665–2676.

37 Sobieraj, M.C. and Rimnac, C.M. (2012). Chapter 5 – Fracture, fatigue, and notch behavior of PEEK A2. In: *PEEK Biomaterials Handbook* (ed. S.M. Kurtz), 61–73. Oxford: William Andrew Publishing.

38 Schmidt, M., Pohle, D., and Rechtenwald, T. (2007). Selective Laser Sintering of PEEK. *CIRP Annals – Manufacturing Technology* 56 (1): 205–208.

39 Vaezi, M. and Yang, S. (2015). Extrusion-based additive manufacturing of PEEK for biomedical applications. *Virtual and Physical Prototyping* 10 (3): 123–135.

40 Valentan, B., Kadivnik, Ž., Brajlih, T. et al. (2013). Processing poly(ether etherketone) an a 3D printer for thermoplastic modelling. *Materiali in Tehnologije* 47 (6): 715–721.

41 Mohamed, O.A., Masood, S.H., and Bhowmik, J.L. (2015). Optimization of fused deposition modeling process parameters: a review of current research and future prospects. *Advances in Manufacturing* 3 (1): 42–53.
42 Sachs, E.M., Haggerty, J.S., Cima, M.J., and Williams, P.A. (1993). Three-dimensional printing techniques. US Patent 5, 204,055, 20 April 1993.
43 Pfister, A., Landers, R., Laib, A. et al. (2004). Biofunctional rapid prototyping for tissue-engineering applications: 3D bioplotting versus 3D printing. *Journal of Polymer Science Part A: Polymer Chemistry* 42 (3): 624–638.
44 Seitz, H., Rieder, W., Irsen, S. et al. (2005). Three-dimensional printing of porous ceramic scaffolds for bone tissue engineering. *Journal of Biomedicals Materials Research Part B Applied Biomaterials* 74 (2): 782–788.
45 Lee, J.W., Kim, J.Y., and Cho, D.W. (2010). Solid free-form fabrication technology and its application to bone tissue engineering. *International Journal of Stem Cells* 3 (2): 85–95.
46 Rayleigh, L. (1878). The influence of electricity on colliding water drops. *Proceedings of the Royal Society of London* 28: 406–409.
47 Rayleigh, L. (1879). On the capillary phenomena of jets. *Proceedings of the Royal Society of London* 29: 71–97.
48 Sekitani, T., Noguchi, Y., Zschieschang, U. et al. (2008). Organic transistors manufactured using inkjet technology with subfemtoliter accuracy. *Proceedings of the National Academy of Sciences of the United States of America* 105 (13): 4976–4980.
49 Singh, M., Haverinen, H.M., Dhagat, P., and Jabbour, G.E. (2010). Inkjet printing – process and its applications. *Advanced Materials* 22 (6): 673–685.
50 Demirci, U. and Montesano, G. (2007). Single cell epitaxy by acoustic picolitre droplets. *Lab on a Chip* 7 (9): 1139–1145.
51 Shirazi, S.F.S., Gharehkhani, S., Mehrali, M. et al. (2015). A review on powder-based additive manufacturing for tissue engineering: selective laser sintering and inkjet 3D printing. *Science and Technology of Advanced Materials* 16 (3): 033502.
52 Alomari, M., Mohamed, F.H., Basit, A.W., and Gaisford, S. (2015). Personalised dosing: printing a dose of one's own medicine. *International Journal of Pharmaceutics* 494 (2): 568–577.
53 Sumerel, J., Lewis, J., Doraiswamy, A. et al. (2006). Piezoelectric ink jet processing of materials for medical and biological applications. *Biotechnology Journal* 1 (9): 976–987.
54 Guo, N. and Leu, M.C. (2013). Additive manufacturing: technology, applications and research needs. *Frontiers of Mechanical Engineering* 8 (3): 215–243.
55 Chia, H.N. and Wu, B.M. (2015). Recent advances in 3D printing of biomaterials. *Journal of Biological Engineering* 9 (1): 4.
56 Wu, B.M., Borland, S.W., Giordano, R.A. et al. (1996). Solid free-form fabrication of drug delivery devices. *Journal of Controlled Release* 40 (1): 77–87.
57 Kim, S.S., Utsunomiya, H., Koski, J.A. et al. (1998). Survival and function of hepatocytes on a novel three-dimensional synthetic biodegradable polymer scaffold with an intrinsic network of channels. *Annals of Surgery* 228 (1): 8–13.

58 Zeltinger, J., Sherwood, J.K., Graham, D.A. et al. (2001). Effect of pore size and void fraction on cellular adhesion, proliferation, and matrix deposition. *Tissue Engineering* 7 (5): 557–572.

59 Lam, C.X.F., Mo, X.M., Teoh, S.-H., and Hutmacher, D. (2002). Scaffold development using 3D printing with a starch-based polymer. *Materials Science and Engineering C: Materials for Biological Applications* 20 (1/2): 49–56.

60 Brunello, G., Sivolella, S., Meneghello, R. et al. (2016). Powder-based 3D printing for bone tissue engineering. *Biotechnology Advances* 34 (5): 740–753.

61 Acosta-Vélez, G. and Wu, B.M. (2016). 3D pharming: direct printing of personalized pharmaceutical tablets. *Polymer Sciences* 2 (1:3).

62 Alhnan, M.A., Okwuosa, T.C., Sadia, M. et al. (2016). Emergence of 3D printed dosage forms: opportunities and challenges. *Pharmaceutical Research* 33 (8): 1817–1832.

63 Emami, M.M., Barazandeh, F., and Yaghmaie, F. (2014). Scanning-projection based stereolithography: method and structure. *Sensors and Actuators A: Physical* 218 (Supplement C): 116–124.

64 Melchels, F.P.W., Feijen, J., and Grijpma, D.W. (2010). A review on stereolithography and its applications in biomedical engineering. *Biomaterials* 31 (24): 6121–6130.

65 Emami, M.M., Barazandeh, F., and Yaghmaie, F. (2015). An analytical model for scanning-projection based stereolithography. *Journal of Materials Processing Technology* 219 (Supplement C): 17–27.

66 Andrzejewska, E. (2001). Photopolymerization kinetics of multifunctional monomers. *Progress in Polymer Science* 26 (4): 605–665.

67 Jacobs, P.F. (1993). *Rapid Prototyping and Manufacturing: Fundamentals of StereoLithography*. Dearborn, MI: Society of Manufacturing Engineers.

68 Heller, C., Schwentenwein, M., Russmueller, G. et al. (2009). Vinyl esters: low cytotoxicity monomers for the fabrication of biocompatible 3D scaffolds by lithography based additive manufacturing. *Journal of Polymer Science Part A: Polymer Chemistry* 47 (24). 6941–6954.

69 Cooke, M.N., Fisher, J.P., Dean, D. et al. (2003). Use of stereolithography to manufacture critical-sized 3D biodegradable scaffolds for bone ingrowth. *Journal of Biomedial Materials Research Part B Applied Biomaterials* 64 (2): 65–69.

70 Matsuda, T. and Mizutani, M. (2002). Liquid acrylate-endcapped biodegradable poly(epsilon-caprolactone-co-trimethylene carbonate). II. Computer-aided stereolithographicmicroarchitectural surface photoconstructs. *Journal of Biomedial Materials Research* 62 (3): 395–403.

71 Melchels, F.P., Feijen, J., and Grijpma, D.W. (2009). A poly(DL-lactide) resin for the preparation of tissue engineering scaffolds by stereolithography. *Biomaterials* 30 (23–24): 3801–3809.

72 Elomaa, L., Teixeira, S., Hakala, R. et al. (2011). Preparation of poly(epsilon-caprolactone)-based tissue engineering scaffolds by stereolithography. *Acta Biomaterialia* 7 (11): 3850–3856.

73 Arcaute, K., Mann, B.K., and Wicker, R.B. (2006). Stereolithography of three-dimensional bioactive poly(ethylene glycol) constructs with encapsulated cells. *Annals of Biomedical Engineering* 34 (9): 1429–1441.

74 Dhariwala, B., Hunt, E., and Boland, T. (2004). Rapid prototyping of tissue-engineering constructs, using photopolymerizable hydrogels and stereolithography. *Tissue Engineering* 10 (9–10): 1316–1322.

75 Lu, Y., Mapili, G., Suhali, G. et al. (2006). A digital micro-mirror device-based system for the microfabrication of complex, spatially patterned tissue engineering scaffolds. *Journal of Biomedial Materials Research Part A* 77 (2): 396–405.

76 Arcaute, K., Mann, B., and Wicker, R. (2010). Stereolithography of spatially controlled multi-material bioactive poly(ethylene glycol) scaffolds. *Acta Biomaterialia* 6 (3): 1047–1054.

77 Mapili, G., Lu, Y., Chen, S., and Roy, K. (2005). Laser-layered microfabrication of spatially patterned functionalized tissue-engineering scaffolds. *Journal of Biomedial Materials Research Part B Applied Biomaterials* 75 (2): 414–424.

78 Kolan, K.C.R., Leu, M.C., Hilmas, G.E., and Velez, M. (2012). Effect of material, process parameters, and simulated body fluids on mechanical properties of 13-93 bioactive glass porous constructs made by selective laser sintering. *Journal of the Mechanical Behavior of Biomedical Materials* 13 (Supplement C): 14–24.

79 Jain, P.K., Pandey, P.M., and Rao, P.V.M. (2008). Effect of delay time on part strength in selective laser sintering. *The International Journal of Advanced Manufacturing Technology* 43 (1): 117.

80 Gibson, I. and Shi, D. (1997). Material properties and fabrication parameters in selective laser sintering process. *Rapid Prototyping Journal* 3 (4): 129–136.

81 Gusarov, A.V., Laoui, T., Froyen, L., and Titov, V.I. (2003). Contact thermal conductivity of a powder bed in selective laser sintering. *International Journal of Heat and Mass Transfer* 46 (6): 1103–1109.

82 Dupin, S., Lame, O., Barrès, C., and Charmeau, J.-Y. (2012). Microstructural origin of physical and mechanical properties of polyamide 12 processed by laser sintering. *European Polymer Journal* 48 (9): 1611–1621.

83 Van der Schueren, B. and Kruth, J.P. (1995). Powder deposition in selective metal powder sintering. *Rapid Prototyping Journal* 1 (3): 23–31.

84 Leong, K.F., Chua, C.K., Gui, W.S., and V. (2006). Building porous biopolymeric microstructures for controlled drug delivery devices using selective laser sintering. *The International Journal of Advanced Manufacturing Technology* 31 (5): 483–489.

85 Berry, E., Brown, J.M., Connell, M. et al. Preliminary experience with medical applications of rapid prototyping by selective laser sintering. *Medical Engineering and Physics* 19 (1): 90–96.

86 Bourell, D.L., Watt, T.J., Leigh, D.K., and Fulcher, B. (2014). Performance limitations in polymer laser sintering. *Physics Procedia* 56 (Supplement C): 147–156.

87 Kruth, J.P., Levy, G., Klocke, F., and Childs, T.H.C. (2007). Consolidation phenomena in laser and powder-bed based layered manufacturing. *CIRP Annals* 56 (2): 730–759.

88 Yan, C., Shi, Y., and Hao, L. (2011). Investigation into the differences in the selective laser sintering between amorphous and semi-crystalline polymers. *International Polymer Processing* 26 (4): 416–423.

89 Williams, J.M., Adewunmi, A., Schek, R.M. et al. (2005). Bone tissue engineering using polycaprolactone scaffolds fabricated via selective laser sintering. *Biomaterials* 26 (23): 4817–4827.
90 Hao, L., Savalani, M.M., Zhang, Y. et al. (2006). Selective laser sintering of hydroxyapatite reinforced polyethylene composites for bioactive implants and tissue scaffold development. *Proceedings of the Institution of Mechanical Engineers, Part H: Journal of Engineering in Medicine* 220 (4): 521–531.
91 Tan, K.H., Chua, C.K., Leong, K.F. et al. (2005). Selective laser sintering of biocompatible polymers for applications in tissue engineering. *Biomedical Materials and Engineering* 15 (1–2): 113–124.

2

3D and 4D Printing Technologies: Innovative Process Engineering and Smart Additive Manufacturing

Deck Tan, Ali Nokhodchi, and Mohammed Maniruzzaman

University of Sussex, School of Life Sciences, Department of Pharmacy/Chemistry, John Maynard Smith (JMS) Building, Falmer, Brighton, BN1 9QJUK

2.1 Introduction

3D printing, also known as additive manufacturing or rapid prototyping, is a manufacturing process that is used to create a three-dimensional solid object from a 3D digital model. It was first invented by Charles Hull in the early 1980s and has gained much interest since [1]. It is a quick and effective method to create physical objects from digital designs [2]. The digital 3D models are normally generated using a computer-aided design (CAD) software or obtained from 3D scanners that capture images and distance information of real objects and then transfer the data to a computer [3]. 3D printing has been used by the manufacturing industry for decades, primarily for making prototypes of products due to its cost-effective manufacturing process. 3D printing has also allowed large flexibility in the design of an object. Complicated structures can be easily produced by 3D printing, which would be impossible or expensive to be produced with conventional manufacturing methods [4]. There has been increasing interest to use 3D printers in the pharmaceutical and medical field. In the medical field, 3D printing technologies have been used for the production of personalized medicines, oral dosage forms, medical devices, and tissue engineering applications [3]. In general, a 3D printed object is made of many thin layers of a material, laying on top of each other. There are several 3D printing technologies currently available in the market. The name of the technology is usually related to the technique involved in the formation of the 3D object.

2.2 Types of 3D Printing Technologies

2.2.1 Stereolithographic 3D Printing (SLA)

The stereolithographic (SLA) 3D printing technique uses light-sensitive polymers (also known as photopolymers) as the starting material. It uses a photopolymerization process, which is the curing of photosensitive materials, to produce a

3D and 4D Printing in Biomedical Applications: Process Engineering and Additive Manufacturing, First Edition. Edited by Mohammed Maniruzzaman.

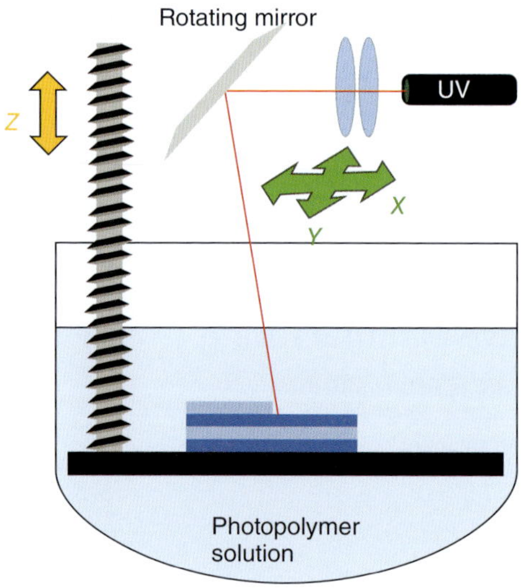

Figure 2.1 Mechanism of an SLA 3D printer. Source: Alhnan et al. 2016 [6]. Reproduced with permission of Springer Nature.

3D object [5]. This printing technique will produce support structures that are used to connect the different parts of an object when building so that the object will not collapse during the printing process. The photopolymers are normally cured using ultraviolet (UV) light, or by using the digital light-projecting (DLP) technique [6]. The SLA technique uses a digital mirroring device to cause a chemical reaction in the photopolymer. The photopolymers exposed to the light will undergo gelation [7, 8]. The photopolymerization process repeats itself to form many layers of the material until the desired 3D object is built. In 3D printing, a first layer printed usually solidifies quickly before another layer is laid on top of it. In SLA 3D printing, each of the layers printed bond together when the unreacted functional groups on the solidified structure in the first layer polymerize with the illuminated resin in the layer above it [9]. Figure 2.1 illustrates the working mechanism of an SLA 3D printer.

When a 3D object is built from an SLA 3D printing technique, further steps to improve the mechanical properties of the 3D printed object are required. The steps also include polishing and removing the unwanted support structures attached to the object [5]. This 3D printing technology is very efficient in producing objects that require high level of accuracy and resolution. It is capable of building objects up to several cubic centimeters with a resolution of approximately 0.2 μm [10].

The main drawback of this method is that the resins used are potential carcinogens and can be a potential health hazard. Another disadvantage is that the resins are photosensitive, which means they might have problems with long-term stability. SLA printing can also be very time-consuming because of its slow printing speed of about 1–3 $cm\,h^{-1}$ [6].

2.2.2 Powder-Based 3D Printing

Powder-based 3D printing uses a powder bed system or a powder jetting mechanism (Figure 2.2) to distribute layers of powder for building a 3D object.

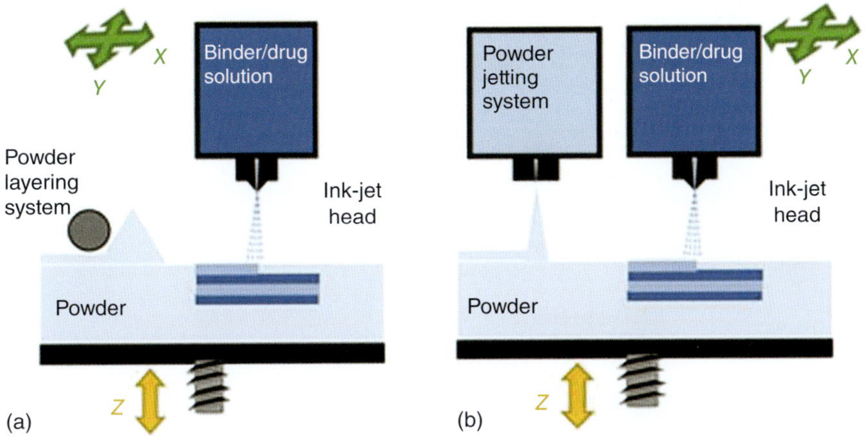

Figure 2.2 Mechanism of powder-based 3D printing: (a) powder bed system; (b) powder jetting system. Source: Alhnan et al. 2016 [6]. Reproduced with permission of Springer Nature.

The layers of powder can be joined by applying drops of a liquid binder from an inkjet or piezoelectric printer head [11]. Alternatively, binder solutions can also be jetted onto the powder layers [12]. The printed 3D object must go through an additional drying step to remove residual solvents and improve the physical resistance of the printed product. Excess powder accumulated during the printing process needs to be cleaned from the finished object as well. Therefore, the powder-based 3D printers can only be used in specialized laboratories that can handle excess powder and significant wastage [6]. The powder-based 3D printing technology has been widely used as a method to 3D print a metal object. However, researchers in the pharmaceutical industry have also shown interest in utilizing the power-based 3D printing technology in fabricating delayed release tablets with complex release profiles such as immediate-extended, dual pulsatile, and zero-order release profiles. This is because this technique has the ability to construct complex formulation design, even for those with loose powders in the inner regions.

The disadvantage of using the powder-based 3D printing technique is that the printed object could be brittle and have poor mechanical strength because of the high porosity of the structures [13, 14]. Apart from that, this 3D printing technique can only produce an object with poor resolution as the resolution is limited by the thickness of the powder layer [9].

2.2.3 Selective Laser Sintering (SLS)

A selective laser sintering (SLS) 3D printer also uses powdered material as the starting material and is commonly used to produce metal objects. Other materials that can be used in an SLS printing technique include polyamides (PAs), polystyrene, and polycarbonates [15]. During the printing process, a layer of powder is laid onto the printing bed. A laser is used to draw the shape of the printed object in the powder, which then liquefies and fuses the powder together. Then, a new layer of powder is deposited again and the laser is used again to draw

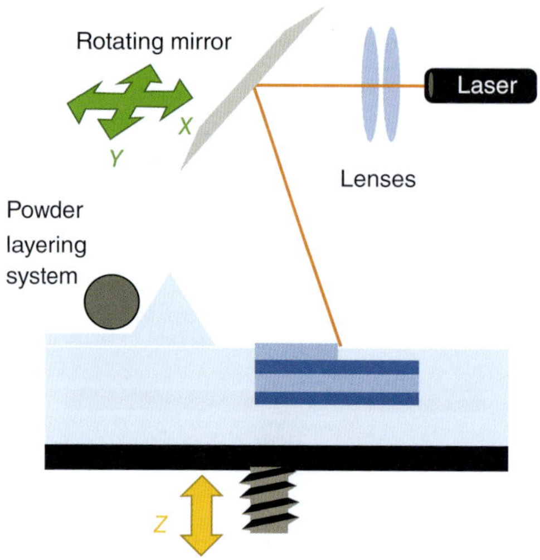

Figure 2.3 Mechanism of selective laser sintering (SLS). Source: Alhnan et al. 2016 [6]. Reproduced with permission of Springer Nature.

the shape of the object in the new layer of powder. The printing process repeats itself layer by layer until the whole object is formed [2]. The working mechanism of an SLS printer is depicted in Figure 2.3. When the printing process is finished, post-processing is required to remove the support structure. The sintered part of the powder will form the final object and the unsintered part will be the supporting material [16]. The SLS printing technique has been used to create objects made of metal, plastic, and ceramics. It is possible to create very detailed and delicate structures with this printing technique [2].

The SLS 3D printing method has not been used in pharmaceutical applications despite its ability to produce detailed and delicate objects. The main concern is the high energy from the laser beam used, which could possibly degrade the drug and pharmaceutical excipients [6].

2.2.4 Fused Deposition Modeling (FDM)

Fused deposition modeling (FDM) is the most commonly used 3D printing technique mainly because of its low cost of printing. This method is also known as fused filament fabrication (FFF) and was first commercialized in 1991 [15]. Thermoplastic filaments are most commonly used as the starting material for this printing technique. The filaments are melted and then extruded from a printer head that is able to move in the x- and y-directions. A 3D object is formed by many layers of melted thermoplastic. Each layer fuses and bonds to the layers below. The layers of melted thermoplastic solidify almost immediately after extruded from the printhead. As one layer cools down and solidifies, another layer of melted filament will be deposited on top of it until the printing of the object is finished. Some FDM printers have enhanced features such as having multiple printheads for printing in different materials within an object. One of the most famous FDM 3D printer manufactures is Makerbot from the United States. It has produced a dual-extruder FDM machine, which is called the Makerbot Replicator 2X (Figure 2.4).

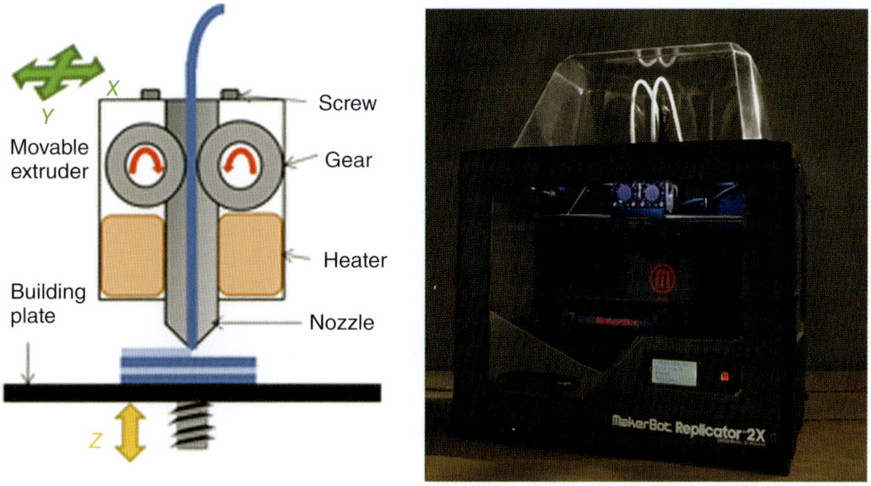

Figure 2.4 Mechanism of fused deposition modeling (FDM) and Makerbot Replicator 2X, a dual-extruder FDM 3D printer. Source: Alhnan et al. 2016 [6]. Reproduced with permission of Springer Nature.

FDM 3D printers are so common nowadays that they can be easily found in homes for personal use. The main advantage of FDM machines is that they are very user-friendly. It is also very economical and cost-effective in terms of producing items with complex designs. 3D printed objects can have very good mechanical properties. Some FDM printed 3D objects can even have similar mechanical properties when compared to the objects manufactured using the traditional injection molding or machining when the same thermoplastics are used [1]. This makes 3D printing a more cost-effective way to produce some good-quality products. Recently, thermoplastic polymers such as polyvinyl alcohol (PVA) have also been used as drug carriers in the pharmaceutical industry [6].

The main disadvantage of the FDM printing technique is the high temperature required to melt the thermoplastic filament. The high temperature may easily destroy a significant number of pharmaceutical excipients and active drugs [17].

2.2.5 Semisolid Extrusion (EXT) 3D Printing

The semisolid extrusion 3D printing method uses semisolids as the starting material. The semisolid material is extruded through a syringe-based toolhead. The semisolids are usually in the form of gels and pastes. The viscosity of the semisolids can be controlled by varying amount of polymers and solvents mixed together. This printing method does not require high temperatures like the FDM printing method, which makes it a preferred technology for bioprinting. However, the printing material needs to be in the form of gels or pastes. The printing mechanism of a semisolid extrusion 3D printer is shown in Figure 2.5.

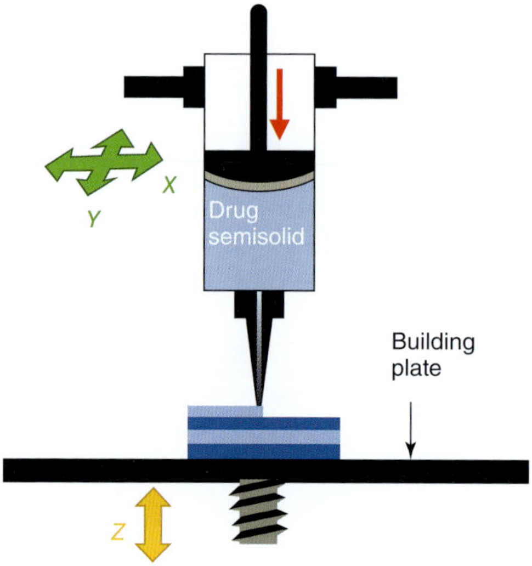

Figure 2.5 Mechanism of semisolid extrusion (EXT) 3D printing. Source: Alhnan et al. 2016 [6]. Reproduced with permission of Springer Nature.

The main disadvantage of this printing technique is that the printed objects have low resolution because of the size of the printing head, which is around 0.4–0.8 mm. Apart from that, the object may have the possibility to collapse during the printing process if the printed layers did not harden sufficiently. As the material extruded needs to be gel-like, there is a possibility for the printed object to shrink or deform [6]. An example of a semisolid extrusion 3D printer that can be found in the market is the EnvisionTEC's 3D Bioplotter by a German 3D printing company. This is a 3D bioprinter that has been used by many medical researchers and manufacturers for the fabrication of cells and organs. Similar to other 3D printers, the 3D bioplotter requires a 3D digital model or a CAD model as an input. For instance, the 3D bioplotter can take a 3D digital file from a patient's computed tomography (CT) and then convert it into a physical 3D scaffold [18].

2.2.6 Thermal Inkjet Printing

One other 3D printing technology that is commonly used is the thermal inkjet printing technology. The thermal inkjet printing technology is the technology that is being used in some desktop printers for printing documents. In thermal inkjet printing, tiny droplets of ink (or other materials used) are ejected from a heated printing head and deposited onto a substrate. This is a noncontact printing method and it usually uses heat to eject the ink droplets. The heat in the printhead creates small air bubbles that will collapse. This will then create pressure pulses that eject small droplets of ink. Figure 2.6 shows the mechanism of the printing head of a thermal inkjet printer. Apart from thermal, there are other inkjet printing methods such as electromagnetic or piezoelectric. The size of the ink droplet can be varied by adjusting the applied temperature gradient, pulse frequency, and ink viscosity. The size of the droplet can range from 10 to 150 pl [20].

Researchers in the medical field such as tissue engineering and regenerative medicine have shown great interest in using the thermal inkjet printing method.

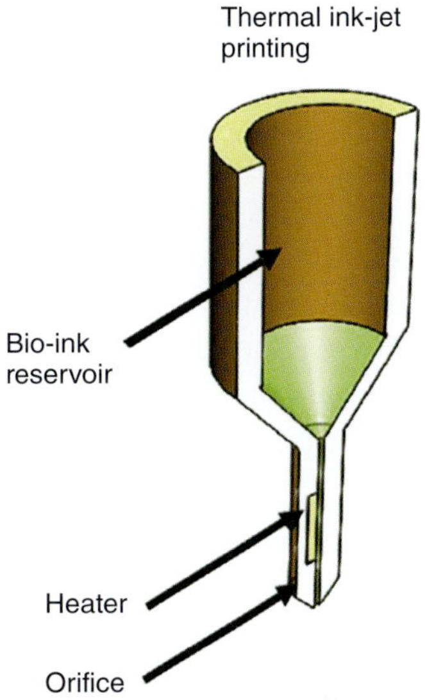

Figure 2.6 Schematic diagram of thermal inkjet printing. Source: Munaz et al. 2016 [19]. Reproduced with permission of Elsevier.

This is because this technique has high precision, control, and versatility. This printing technique also has benign effect on mammalian cells. This technology has already been used to print simple tissues and organs [20]. An example of 3D printer that uses the inkjet technology is the Jet Fusion 3D printer by the famous inkjet printers' manufacturer, Hewlett Packard (HP). The HP Jet Fusion 3D printer uses the thermal inkjet array technology that is also being used in their traditional 2D printers. It first lays a thin layer of powder and then uses the thermal inkjet array to introduce chemicals that can fuse the powder material together. This process is repeated layer-by-layer until the 3D object is finished [21].

2.3 FDM 3D Printing Technology

The FDM 3D printing can be considered the most low-cost and easily accessible technology among all 3D printing techniques. This technique is based on the deposition of successive layers of thermoplastic materials following the softening/melting of the raw materials [22]. Nowadays, FDM printers can also be found in homes for personal use. These types of 3D printers are also called the desktop 3D printers, as they are as easy to use as a traditional 2D printer is. It is also possible to build your own FDM 3D printer. The first two open source DIY 3D printers are RepRap and Fab@Home. The aim of these DIY 3D printers is to make the 3D printing technology more affordable and accessible when 3D printing was first introduced. One of the most popular RepRap 3D printer designs is the Prusa i3.

There is a wide range of materials that can be used in a FDM 3D printer. The materials can range from thermoplastics (such as polylactic acid, PLA and acrylonitrile butadiene styrene, ABS) to engineering materials (such as PA, thermoplastic polyurethane (TPU), and polyethylene terephthalate glycol (PETG))

Table 2.1 Material properties of ABS and PLA.

Properties	ABS	PLA
Tensile strength (MPa)	27	37
Elongation (%)	3.5–50	6
Flexural modulus (GPa)	2.1–7.6	4
Density (g cm^{-3})	1.0–1.4	1.3
Melting point (°C)	N/A (amorphous)	173
Biodegradable	No	Yes, under correct conditions
Glass transition temperature (°C)	105	60

Source: Giang 2017 [24]. Reproduced with permission of Elsevier.

and high-performance thermoplastics (such as poly(ether-ether-ketone) (PEEK) and polyether imide (PEI)). The two most common thermoplastics used are PLA and ABS [23]. ABS is a type of plastic that is mostly used in the injection molding industry and has been used to make LEGO, electronic housings, and automotive bumper parts. On the other hand, PLA is biodegradable and is made from renewable sources such as corn starch or sugarcane. Hence, it is classified as a bioplastic and is used for many applications ranging from plastic cups to medical implants [24].

Table 2.1 compares the main properties of ABS and PLA.

ABS has a higher flexural strength, better elongation, and higher heat resistance than PLA. Hence, ABS is often preferred as it has better mechanical properties. PLA can quickly lose its structural integrity and can begin to droop and deform, particularly if under load, as it approaches 60 °C. As PLA is a bioplastic, it is biodegradable, provided the conditions are favorable. PLA can degrade within 50 days in general atmospheric conditions and 48 months in water. ABS is not biodegradable but it is recyclable. ABS is best suited for applications because of its strength, ductility, machinability, and thermal stability. PLA is suitable where aesthetics are important [24].

Most of the time, post-processing is required for FDM 3D printed parts as the print layers of the finished object are often visible. Acetone is often used to smooth ABS so that it has a glossy finish. ABS can also be machined or easily sanded after printing.

In general, FDM is the most cost-effective way to produce custom thermoplastic parts and prototypes. The lead times of FDM are short because of the high availability of the technology. A wide range of thermoplastic materials is available for FDM printing. These materials include those that are suitable for both prototyping and some noncommercial functional applications. However, FDM has the lowest dimensional accuracy and resolution compared to other 3D printing technologies, so it is not suitable for parts with intricate details. All FDM printed parts have visible layer lines, so post-processing is required for a smooth finish.

2.3.1 FDM 3D Printing Applications in Unit Dose Fabrications and Medical Implants

3D printing has been widely used in the medical field because of its customizability and the ability to create complex shapes with precision. FDM 3D printing technique, in particular, has attracted much attention on medical research because it is easily accessible, less expensive, and versatile. The characteristics of FDM 3D printing show a good potential for the fabrication of single-unit dosage forms [22]. Traditional pharmaceutical processes, such as tablet compression, have very limited process capability and manufacturing flexibility as these processes are slightly outdated. Therefore, there has been increasing research efforts toward the production of personalized solid oral formulations by utilizing the FDM 3D printing method. The 3D printing technology has competitive advantages in terms of producing complex products and personalized products, hence creating opportunities for improved safety, efficacy, and accessibility of medicines [25].

As FDM 3D printing has been considered as an alternative method to conventional tableting technique, researchers have been investigating the potential of this technology to incorporate drug molecules using commercially available PVA filaments. However, there are some limitations such as limited drug loading, the use of non-pharmaceutical grade ingredients, high temperature, and basic tablet designs [26].

Research has also been carried out to create "polypills" formulation, which can incorporate multiple drugs, by using the 3D printing technology. With 3D printing, pills can be designed to house several drugs, all with different release times [27]. Different release profiles of the drug can be obtained by changing the exposed surface area or the geometry of the dosage form and by using more than one kind of carrier material with different drug release characteristics [28, 29]. A 3D printed "polypill" that contains three different drugs has already been developed for patients with diabetes and hypertension [30].

One of the major drawbacks in the pharmaceutical application of FDM 3D printing is the limited number of printable materials currently available. Most FDM printers use thermoplastic polymers. This is because materials are required to be highly thermoplastic in FDM printers so that they can be effectively extruded from the nozzle head. However, these thermoplastic materials may not be pharmaceutically approved or they might not be ideal for optimizing dosage form performance of poorly soluble compounds. Currently, most oral solids were printed in PVA, PLA, and polycaprolactone (PCL) [27]. Polymer blend system has also been used as a formulation strategy, which can be used to control the disintegration and drug release of FDM printed solid dispersion. This approach opens the possibility of using pharmaceutically approved polymers in FDM printing of oral solid dosage forms. It is shown that polymer blends possess excellent printability and are suitable for processing using a commercially available FDM 3D printer [27].

In most cases, FDM printing requires high printing temperature in order to melt the printing materials. This can easily cause degradation of the 3D printed drugs and polymers. Therefore, lower temperatures have been used for 3D printing when printing of drugs is made from different polymer mixtures.

However, lower temperatures can possibly cause blockages in the nozzle of the printer head because of increased viscosity of the semisolid drug polymer mixture [26].

3D printers have also been extensively used in the medical field for medical modeling, surgical planning, and production of custom plates, screws, and surgical guides. With the current 3D printing technology available, it is now possible to produce custom-made medical products and equipment. 3D printing is also considered relatively cost-effective when compared with the traditional manufacturing method for medical products, especially when produced in smaller quantities [1]. Surgeons can use 3D printers to produce a 3D printed medical model to educate the patients and plan the surgical approach. The surgical model may help to decrease surgical time and minimize surgical complications. 3D printed models for orthopedic and craniofacial applications can also be used intraoperatively as a surgical guide [31].

With 3D printing, it is possible to create even customized prosthetic limbs and surgical implants with complex geometries within a short period of time. This can be normally done by translating the X-ray, magnetic resonance imaging (MRI), or CT scans of the patients into digital 3D print files. This approach has been applied to produce dental, spinal, and hip implant. Apart from that, 3D printing has been extensively used for the production of customized hearing aids. This is because everyone's ear canal differs in shape and 3D printing has allowed customized hearing aids to produce them efficiently and cost-effectively [1].

FDM 3D printing technology has a promising future in the medical applications because of its low cost, availability, and ease of use. FDM has also been utilized in tissue engineering and regenerative medicine. It is anticipated that 3D printing can be applied in the bioprinting of complex organs in the future. There is also a possibility of *in situ* printing, which means the implants or organs are printed directly in the human body during operation. Currently, *in situ* bioprinting is used for repairing external organs, such as skin [1].

2.4 Hot Melt Extrusion Technique to Produce 3D Printing Polymeric Filaments

Hot melt extrusion (HME) is the process of forming a new material by heating and melting a raw material, most of the time polymers, and then forcing the melted material through a die under controlled conditions. This process is usually used to produce polymer products of uniform shape and density. The main component of the HME process is the extruder, which is a barrel containing one or two rotating screws that transport materials down the barrel. The processes that occur in HME includes feeding of the raw materials, melting and mixing of the materials, extrusion of the mixed materials through a die, and further downstream processing such as cooling, cutting, or collecting the products [32]. Figure 2.7 shows a schematic diagram that explains the hot melt extrusion process.

HME has been applied to the medical and health care industry more recently where it is used to manufacture medical devices and to mix active pharmaceutical ingredients (APIs). This is because HME can enhance the bioavailability and solubility of the APIs. When producing materials that are made of multiple

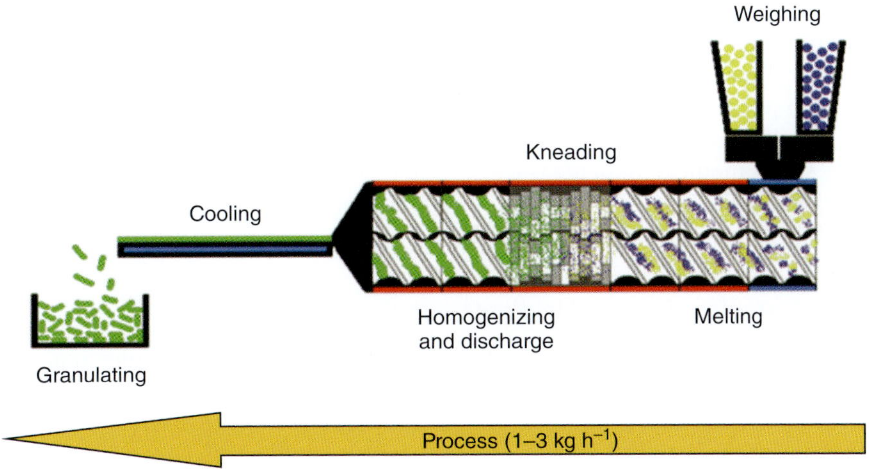

Figure 2.7 Hot melt extrusion process [33]. Source: https://creativecommons.org/licenses/by/3.0/

formulation ingredients, twin screw extruders are preferred as it provides a better mixing in order to produce a homogeneous solid containing finely dispersed API particles.

HME has also the main production method for the filaments needed for FDM 3D printing. There has been increasing research interest in the fabrication of different types of filaments from different materials because of the increasing popularity of using 3D printers as an alternative manufacturing method to the traditional ones. There have been attempts to produce filaments from materials such as PVA, XT copolyester, polyethylene terephthalate, nylon, and TPU in order to obtain filaments with different physical and mechanical properties. However, there is still very limited amount of filaments that are made of pharmaceutical grade materials suitable for FDM printing. By using HME, it is possible to produce filaments from polymers that are insoluble (ethylcellulose), promptly soluble (polyethylene oxide), enteric soluble (hydroxypropyl methyl cellulose acetate succinate), and swellable/erodible (hydrophilic cellulose derivative, PVA). These filaments can potentially be used to 3D print capsules and coating layers for immediate or modified release. They can also be used to print any other type of dosage forms when added with active ingredients [22]. There has been research in using the HME method to fabricate theophylline-loaded filaments. The drug-loaded filaments were used to 3D print theophylline-loaded tablets. This filament was produced by processing a physical mixture of the drug and polymer through HME. Therefore, it is now possible to bridge FDM 3D printing with HME in an attempt to increase the range of polymers that can be adapted with FDM [26].

2.5 Smart Medical Implants Integrated with Sensors

Integration of sensors with medical implants has gained much research attention within recent years. The purpose of integrating implants with sensors is to be able to immediately detect and monitor the response of the human body in all conditions. This can play an important role in the detection of life-threatening diseases. Therefore, medical implants integrated with sensors are also called smart implantable devices. In general, smart medical implants are able to improve the functionality and longevity of the implantable devices. The function of the smart medical implants is to provide real-time information of any part of the human body that is of interest. This information is very useful as it can help to further improve the quality of life [34].

2.5.1 Examples of Medical Implants with Sensors

Some orthopedic implants, especially for artificial joints used for joint replacement surgery, are currently integrated with sensors. This is because there has been an increasing number of complications related to bacterial biofilm infections at the artificial joint prostheses, which could cause chronic infection and result in aseptic loosening of the prostheses. By integrating sensors to these implants, it is then possible to detect and monitor the response of the human body to the implants after the surgery is performed. This could also help to monitor the progress of the patients' recovery. One possible application of the biosensors is that it can be used as a chimeric protein switch with the prostheses. The chimeric protein switch can detect the predevelopment of any antigen or antibody in the body after implantation. When an unusual activity is detected, the enzymatic function can then be activated through this switch. For instance, the chimeric protein switch and a single-walled carbon nanotube (SWNT) bio-actuator can be integrated to a prosthesis to create a smart medical implant system. This system can detect the development of bacteria in the bacteria-infected area of the prosthesis by sensing the viscosity change in this area. When the viscosity changes, a voltage is produced by the switch that can give a displacement to SWNT bio-actuator, which will then release the medicine stored in the reservoir within the prosthesis [34].

Another smart medical implant that has attracted much research attention is the needle-type continuous glucose-monitoring (CGM) system. There could be some subcutaneous implant containing sensors that allow the CGM in a human body. This type of glucose-monitoring system detects the glucose level of human body through the fluid in the subcutaneous tissue. This allows us to monitor our glucose level under any conditions and determine how our glucose level reacts to different activities that we carry out throughout our daily lives. Most of these subcutaneous glucose-sensing systems are composed of a glucose sensor with an external electrical connector and a glucose monitor [35]. The sensing electrode of the glucose sensor can be made of a Teflon-coated platinum–iridium wire. The exposed part of the electrode can be coated with cellulose acetate, on which glucose oxidase is immobilized. The part of the electrode that acts as a reference can be an Ag/AgCl electrode. This could be made from a wrapped silver wire treated with HCl to form AgCl on the surface [35].

The sensing electrodes of a CGM-sensing system are normally of microsize or nanosize. Therefore, the single-walled carbon nanotube (SWCNT) and multiwalled carbon nanotube (MWCNT) have been a very popular material for making electrodes because of their ultrasmall size. Apart from that, the carbon nanotubes (CNTs) have properties such as high strength, large flexibility, excellent chemical stability, and high surface area [34]. They also have a high electrocatalytic effect and fast electron transfer rate. Because of the electrochemical properties of CNTs, it has recently gained much research interest in the application for electrochemical sensors and biosensors [36].

An example where CNTs have been used is the production of a glucose biosensor using the carbon nanotube–nanoelectrode ensembles (CNT–NEEs). A sensitive glucose biosensor that is made of CNT–NEEs can selectively detect glucose even in different types of biological fluids such as saliva, sweat, urine, and serum. This means that the biosensor is able to perform a selective electrochemical analysis of glucose even in the presence of interferents by avoiding the generation of an overlapping signal caused by the interferers. As a result, the design and fabrication of the sensor can be further simplified and miniaturized because there is no need for the sensor to first eliminate the interference [36].

The pressure in the body is one of the most vital signs of the human body. Some examples of the most measured pressures in the human body include pressure in the blood circulatory system, urinary bladder, muscle compartments, joints, and brain [37]. However, the traditional methods of taking pressure measurements do not provide any real-time information or the exact pressure in the organ of interest at that particular moment. Therefore, much research work is being done for the development of implantable pressure sensors. Implantable sensors have been made possible through the development of microelectromechanical systems (MEMS) as they can miniaturize the design of the implants when MEMS-based sensors are used.

Most pressure sensors convert pressure to the deflection of a mechanical element, most of the time a diaphragm deformation. The deflection of the diaphragm because of the change in pressure can be categorized into piezoresistive based and piezoelectric based. The piezoresistive pressure sensors that have a silicone membrane are commonly used for *in vivo* monitoring of blood pressure. In a piezoresistive-based pressure sensor, the resistivity changes as the pressure changes. The piezoresistors are mounted on a diaphragm, and the diaphragm deflects when there is a change in pressure. For thin diaphragms and small deflections, the resistance change is linear with the pressure. As for a piezoelectric-based sensor, an electric charge is induced when there is a mechanical strain applied. The diaphragm of a piezoelectric sensor is commonly made of polyvinylidene fluoride (PVDF). When there are compression and stress on the membrane, the electric charge will be produced and is recorded using the electrodes coated on either side of the diaphragm. Polymer-based piezoelectric devices are commonly used for implantable applications because of their mechanical flexibility. PVDF has the highest piezoelectricity among all piezoelectric polymers. PVDF has been widely used for the fabrication of sensitive sensors because of its high electromechanical coupling coefficient and its semicrystalline nature. Apart from that, PVDF is also an ideal biomaterial because it is nontoxic, inert, water resistant, biocompatible, environment friendly [35].

In general, the development of smart medical implants integrated with sensors is to provide an indication of a person's health or disease progression. As most measurement methods for vital signs are currently being widely used to only provide information at a single point of time, there is a need for the development of implantable sensors. The advantage of having implantable sensors is to be able to obtain information continuously, accurately, and conveniently. However, there are still many challenges that need to be overcome for these smart medical implants. Biocompatibility, surgical placement, and patient comfort are the main concerns while developing smart medical implants. One of the main challenges faced is the power source needed for the operation of the sensors implanted, as this will directly affect the longevity of the implant. At the moment, most implantable sensors are being powered by an external battery source, which would be limited for the lifetime of the implant. Another challenge is the transmission of the data measured from the sensor to a reading device. The challenge is that there will be attenuation or interference of signals while traveling through the organs, which could affect the reliability and accuracy of the measurements obtained.

2.6 4D Printing and Future Perspectives

2.6.1 4D Printing and Its Transition in Material Fabrication

In recent years, much work has been done to discover more materials that can be compatible with 3D printing technologies. Smart materials have attracted much research interest, resulting in the development of 4D printing from the original 3D printing technology. The idea of 4D printing was first introduced in 2013 by a research group in MIT. The working principle of this technology originates from the use of smart materials in 3D printing [38]. Hence, 4D printing can be defined as the printing of 3D objects that can evolve in a predefined manner when influenced by external stimuli. This means that the 3D printed object can transform as well as change their geometry and function when stimulated [39]. For the printed object to transform, the starting material used needs to be programmable and responsive to various external stimuli such as heat, pH, light, or magnetic field. These materials that are able to transform as a response to external stimuli are commonly called smart materials [40].

In general, there are two main requirements for 4D printing. First, the materials need to be responsive to external stimuli to make the printed objects programmable. Second, there must be the presence of external stimuli such as heat, pH, light irradiation, or water that can act as a driving force for the transformation to occur. Only when these two requirements are fulfilled, the 4D printed object can be activated to achieve its full functionality [41].

Origami is one of the ideas that have inspired the evolution of 4D printing. Researchers have successfully explored this type of traditional art to provide some innovative solutions in terms of compacting large objects into a small volume of space [39]. Therefore, one of the biggest advantages of 4D printing is that the design of a complex or large structure can be simplified while being printed. The printed object can then be programmed and activated to transform into its final complex form. The ability of an object to be programmed and be transformed later on is similar to an active origami, where an origami object is able to self-fold or self-unfold. Reducing the complexity of the object while printing will help reduce the printing time and material waste, hence resulting in a more cost-effective fabrication method [38].

2.6.2 Shape Memory or Stimuli-Responsive Mechanism of 4D Printing

Materials that are used in 4D printing are called smart materials. These materials can be programmed and then activated to change their shape or configuration when stimulated by the surrounding factors [42]. Some examples of the smart materials are shape memory polymers (SMPs), light-activated polymers, and shape memory alloys (SMAs). 4D printing has also been particularly useful in producing biomedical products, organs, or tissues through bioprinting. The examples of biocompatible materials that have been used in the 4D printing of medical applications are lipids, responsive hydrogels, and polymers. In particular, SMPs have been widely used in any 4D printing applications. SMPs can be defined as materials that carry the shape memory effect (SME), which is the ability to inelastically deform and then recover to their original shape when there is the presence of external stimuli [43]. SMPs have biocompatibility and biodegradability properties, and these properties can be altered by changing the physical and chemical properties of the SMPs [39]. The advantages of using SMPs are the high strain recovery, low density, low cost, shapes can be easily programmed, and good controllability over recovery temperature [39]. The ability for SMPs to reshape has been successfully used in biomedical applications to produce sutures and stents that are very useful in minimally invasive surgery. SMPs have also found their way into the field of robotics as they have been widely used to produce sensors and actuators to make robots [44]. An example of SMPs used in 4D printing is the printing of active composites. In the active composites, SMP fibers are precisely prescribed so that the printed product can be activated to change to a complex of the product. For instance, SMP fibers could have different glass transition temperatures to produce multishape active composites. Some SMPs are made to have different responding rates in order to produce sequential self-folding structures (Figure 2.8) [45]. SMPs that are used to produce soft composites can be reinforced using glassy polymer fibers. In this case, the glassy polymer fibers reinforcing the elastomeric matrix of the SMP act as a switch to affect the memory behavior of the composite [46].

Hydrogels have been used in 4D printing, particularly for biomedical applications as they have biocompatibility properties. The high water content and soft consistency of hydrogels make it very similar to natural living tissues as well as providing their biocompatibility properties [47]. Hydrogels that are used in biomedical applications include peptide hydrogels, natural polymeric hydrogels, and synthetic polymeric hydrogels [41]. These hydrogels are physiologically responsive, which means it can change as a result of the changing external environment. These hydrogels are also termed smart hydrogels. Smart hydrogels have unique features such as shape-memory, self-healing, and controllable sol–gel transition [39]. Some of the factors that can affect the change in smart hydrogels include pH, ionic strength, temperature, light, and electric and

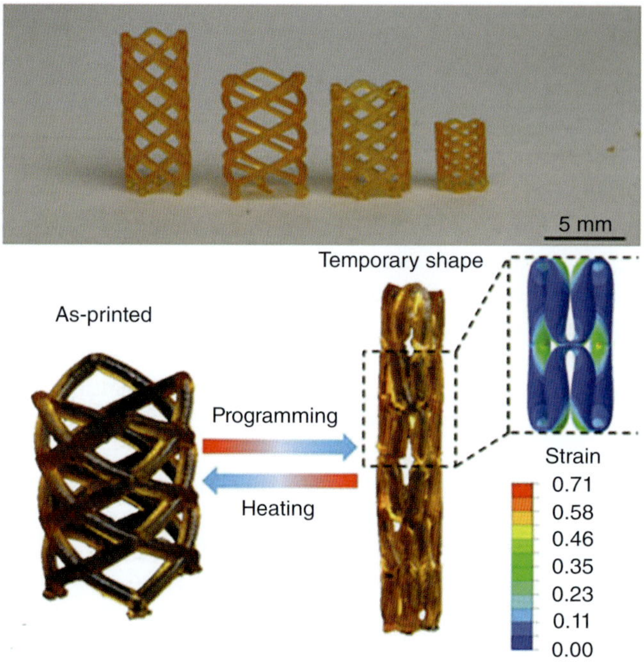

Figure 2.8 4D printed stent [45]. Source: https://creativecommons.org/licenses/by/4.0/

magnetic field [47]. Some examples of hydrogel-based 4D printing are the production complex self-evolving structures and 4D printed active valves using thermally sensitive hydrogels. Anisotropic hydrogel composites with cellulose fibrils have also been used in 4D bioprinting [45]. These synthetic polymeric hydrogels are used in the printing of scaffolds that can self-fold or self-unfold for tissue engineering applications. These hydrogels are suitable for the use of printing scaffolds as their porosity and interconnectivity allow the supply of gas and nutrition to cells in the scaffolds [48].

2.6.3 Factors Affecting 4D Printing

2.6.3.1 Humidity-Responsive Materials

The humidity of the environment is one of the factors that can be used to activate the change in the structure of a printed object. An example of a material that can respond to the presence of water is hydrogel. Hydrogel is an extremely hydrophilic polymer material. It can geometrically fold, curl, expand, or perform other shape changes that have been programmed in the presence of water [49]. The work done by Raviv et al. [49] has successfully demonstrated the concept of self-evolving complex structures, in which the 3D printed active hydrogel structures change its geometry when activated through the presence of water. Three different structures were created to demonstrate the ability of the structures to linearly expand, ring stretch, and fold over time when immersed in water.

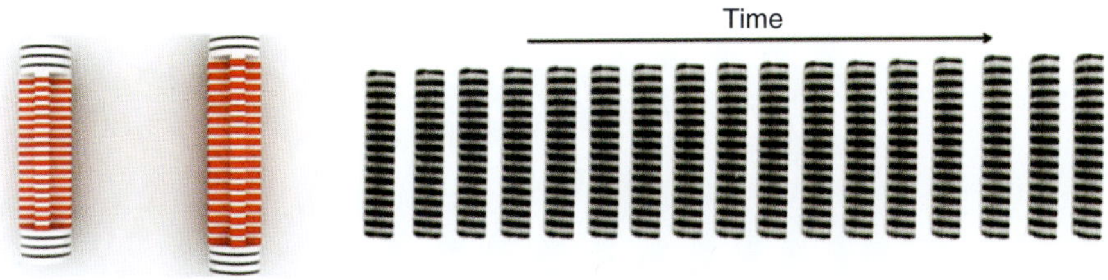

Figure 2.9 Linear stretching primitive. Assembly of rigid disks with expanding materials in the middle. Adjusting the ratio of expanding materials to the rigid disks can control the length of stretching. Linear expansion in the presence of water over time [49]. Source: https://creativecommons.org/licenses/by/4.0/

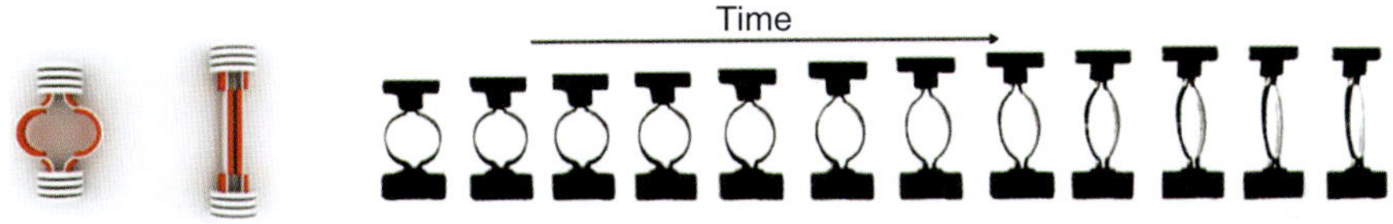

Figure 2.10 Ring stretching. Expansion of a ring shape into a bar. Stretching length can be adjusted by controlling the radius of the ring [49]. Source: https://creativecommons.org/licenses/by/4.0/

To demonstrate linear stretching (Figure 2.9), the structure is made of an alternating layer of rigid materials and expandable materials in a circular disk shape. The rigid part is made of rigid plastic, whereas the expandable part is made of very hydrophilic UV curable hydrogel with low cross-link density. The cross-linking in the polymer is able to change the water-soluble properties of the hydrogel to water-swellable. In the presence of water, the hydrogel will absorb the water and expand to up to two times the original volume. The ratio of the expanding material to the rigid portions can be altered to control the percentage of linear expansion of the joint.

For ring stretching (Figure 2.10), two rings (an inner ring and an outer ring) were printed using two different materials in two layers of each ring. In the presence of water, the inner rings expand and force the rings to deform into a bar shape that is longer than their original formation.

To demonstrate the ability of a structure to fold (Figure 2.11), a structure that is made of bars and disks is used. A bar is made of a strip of expanding material printed over a strip of rigid material. By switching the material order, where the rigid material is on top of the expanding material, the object will fold in the other direction. To achieve a certain final angle, rigid disks are placed in between the bars. The disks in the center act as stoppers that will stop the folding action when they hit one another. The spacing and diameters of the disks can be changed to determine the final folding angle.

2.6.3.2 Temperatures

Another factor that can be used as a stimulus to the smart materials is the change in temperature. Printed active composites (PACs) can be programmed to respond

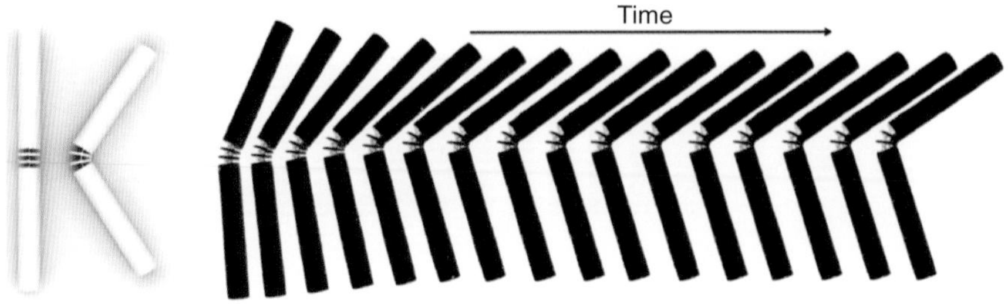

Figure 2.11 Folding action of the structure [49]. Source: https://creativecommons.org/licenses/by/4.0/

as a result of temperature change. The way that they can be programmed is by deforming the object in order to obtain the desired temporary shape. The process of programming includes heating up the object, physically deforming, and cooling the object or drawing the object at a low temperature (which is called cold drawing). Through the programming process, the permanent shape of the object is stored while the object is in its temporary shape. By heating up the object above a certain transition temperature, the object will absorb the thermal energy and is able to restore to its original permanent shape that has been stored through the programming process [50]. In general, thermoresponsive materials possess a characteristic transition temperature, such as the glass transition temperature [43]. One way to produce a PAC that can be thermomechanically programmed is by printing SMPs in an elastomeric matrix. The SMP fibers allow the fixing a temporary shape of a printed object and then recovering the object to its permanent shape in response to temperature change. The final shape can be designed based on the theoretical understanding of the thermomechanical shape memory behavior of the composites and their constituents [42].

The concept of self-assembling origami has been demonstrated in the work done by Ge et al. [42]. Here, flat polymer sheets are printed, and the sides of the polymers are connected by hinges consisting of PACs that can be thermomechanically programmed. Once the hinges are programmed, it will fold the flat sheet into the desired final shape (Figure 2.12).

Another example of an object that can respond to temperature change is through a directly printed 3D box. The 3D box was unfolded and made flat by heating to a higher temperature, by applying a mechanical load, and then cooled down to a lower temperature. When the unfolded structure is heated again, the structure reforms to its original shape, forming a 3D box again. It has also been seen that the SMPs can be activated by heat to 3D print on ceramics (Figure 2.13), an artificial flower bud that can blossom upon heating (Figure 2.14). The flower is made of three layers, each with SMPs that have different transition temperatures. As a result, each layer of the flower will be able to unfold sequentially when being heated, imitating the notion of a natural blossom. The ability for an object to respond according to the temperature change can also be useful for medical applications, especially for minimally invasive surgery. This is because a temporary shape of a medical device, which can be much smaller, can be inserted into the human body through just a small

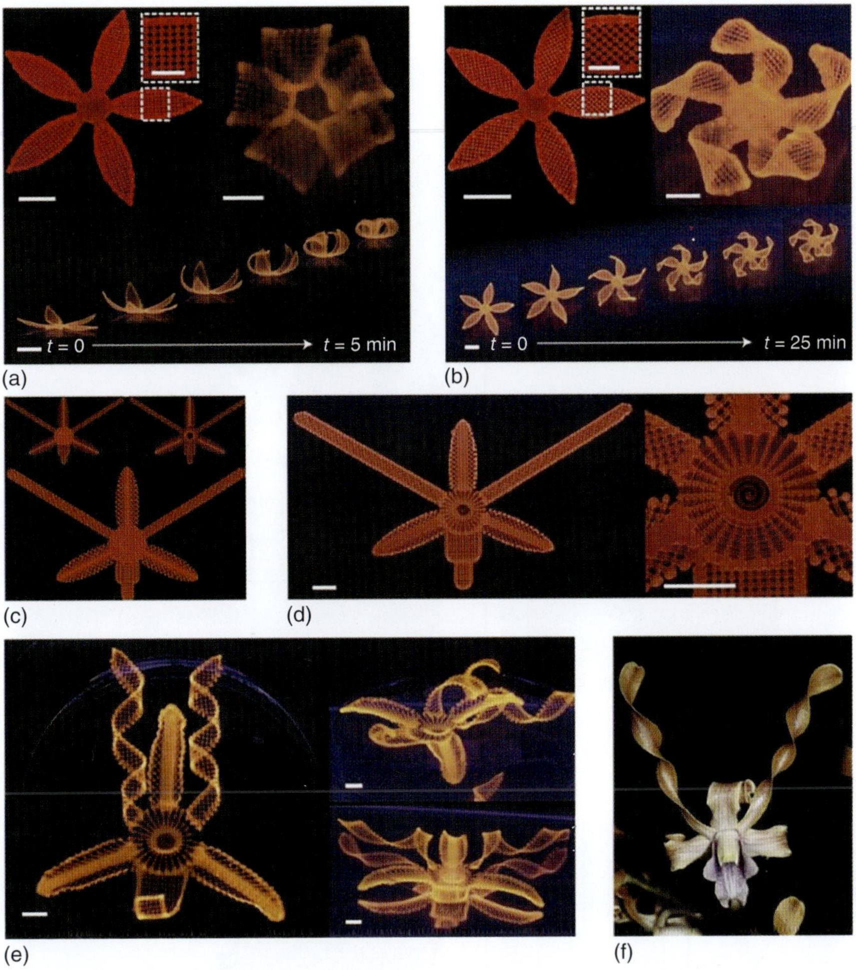

Figure 2.12 Biomimetic 4D printing. Source: Lee et al. 2017 [40]. Reproduced with permission of Elsevier.

cut in the body. Thermal energy can then be used to activate the medical device within the body to transform to its desired shape and function. However, the source of thermal energy required remains a challenge [51].

2.6.3.3 Electronic and Magnetic Stimuli

The electric or magnetic energy has also been used to cause a change in the shape of SMP-printed objects. For the material to respond to the electric field, the SMPs need to be incorporated with highly dielectric susceptible components such as carbon nanotubes or liquid-crystalline phases [51]. Apart from that, the incorporation of magnetosensitive materials in SMPs such as the superparamagnetic magnetite nanoparticles can cause the SMPs to be activated through the electromagnetic field. Inductive heating can occur in this type of SMP, which means the electromagnetic energy, from an external high-frequency electromagnetic field,

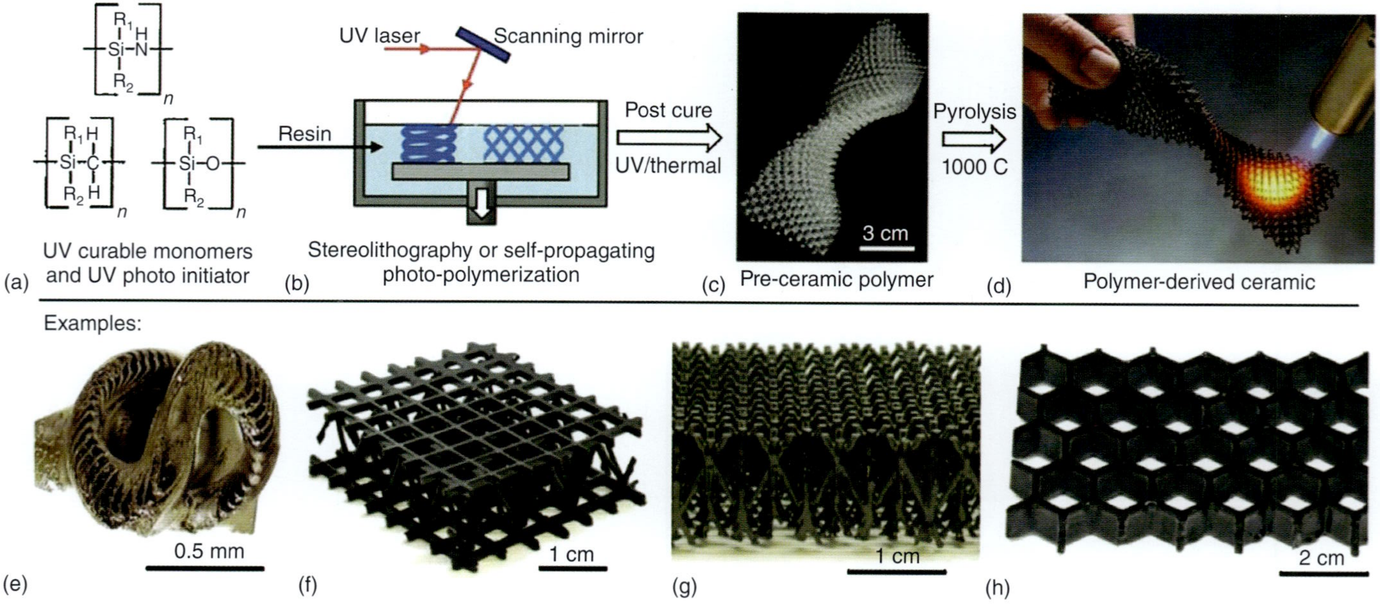

Figure 2.13 Additive manufacturing of polymer-derived ceramics. Source: Lee et al. 2017 [40]. Reproduced with permission of Elsevier.

Figure 2.14 4D printed flower that self-opens upon heat stimulation. Source: Lee et al. 2017 [40]. Reproduced with permission of Elsevier.

can be transformed to heat energy in the composite [51]. For the programming of a magnetosensitive material, the permanent shape of the printed object is first deformed at a temperature above the glass transition temperature of the composite and then let to cool down to form a desired temporary shape. The transformation of the object can then be activated electromagnetically, by an external electromagnetic field.

Another type of material that can respond to the electric field is the dielectric elastomer. The dielectric elastomer is a type of electroactive polymer, and it is suitable to be used to manufacture soft robotics as it is also a type of soft material. This type of polymer can be used in 3D printing to fabricate dielectric elastomer actuators (DEA), which are used in soft robotics. In the presence of electrical energy, the structure of DEA will be transformed, generating a significant amount of strain on the structure [52]. The 3D printing of DEA is a preferred manufacturing method for DEA as the conventional methods are more labor-intensive [39].

2.6.3.4 Light

Another type of smart material is the light-activated polymer, and it has been used to make an active origami that can be activated by light [53]. This type of active origami can be called the photo-origami and is activated through the localized photoinduced stress relaxation in the material. This type of light activation is also called the photochemical approach [53]. The photo-origami starts with a flat, unstressed polymer sheet containing photo-absorbing molecules. The polymer sheet is then being programmed by first straining the surface. When light is irradiated onto the strained surface of the polymer sheet, the photo-absorbing molecules in the polymer absorb the light. Radicals are produced photochemically in the polymers during the light absorption process and will cleave and reform polymer chains resulting in stress relaxation. When the load is removed and the light irradiation is stopped, the stress in the polymer network will redistribute itself. As a result, the polymer will then change its shape to form a 3D box in order to achieve a mechanical equilibrium [53].

2.6.4 Future Perspectives of 4D Printing

4D printing is still a relatively new technology and has significant potential for expansion. It opens up new fields for application in which a structure can be activated for self-assembly, reconfiguration, and replication through environmental free energies. The main advantage of 4D printing is that the printed object can be significantly small in volume, resulting in a reduction for storage space. As the transformation of the object can be activated after being printed, the 4D printed structure can be very small or flat in volume during its initial stage. This is also particularly useful when building a highly complicated object as it is not required to directly print the complicated 3D shape of the object through the printing process. Simple components or lower dimension shape can be 3D printed first using smart materials. The printed parts can then self-assemble or transform to achieve the final complex shape with the desired performance. This is able to reduce the manufacturing cost as printing in more simple shapes is relatively easier and faster. In general, 4D printing has great potential to be applied to many fields. It can be applied in producing biomedical devices, security, fabrication of precise patterned surfaced for optics, electronic devices, structures with multidirectional properties, and soft actuators [43].

In the medical field, 4D bioprinting has been used in the manufacturing of drugs that are programmable. For instance, the 4D printing technology has allowed precise drug delivery at targeted tumor cells in a programmable manner. This novel manufacturing technique could also be a promising solution to unaddressed medical needs, such as tissue regeneration, and may find widespread application in the biomedical field [48]. Previously, the 3D printed medical objects are static and lack stimuli responsiveness, which restrict the structural and functional transformation to perform any function-like natural tissues and organs. With 4D printing, it is now possible to create objects that can better mimic the functions of the natural living tissues or organs as the objects can now be programmed to respond in the same way [41].

There are still many challenges ahead of 4D printing. These challenges include technological limitation, material limitation, and design limitation [39]. In terms of technological limitation, there are not many 3D printing technologies that are able to create objects made of multiple materials. In 4D printing, multiple materials are required to be embedded into a single 3D structure in most cases. Therefore, more work still needs to be done on multimaterial printing technology.

2.7 Regulatory Aspects

As 3D printing has been recently gained much interest in the medical field, there has been an increasing number of medical devices and products that have been produced using the 3D printing technology. All medical devices need to first pass a set of standards and regulations before they can be sold to the public. Therefore, there is no exception for the 3D printed medical products. According to the U.S. Food and Drug Administration (FDA), medical devices can be classified into three classes (i.e. class I, II, or III) based on the risk of the device and the level of control that are necessary to assure safety and efficacy [54]. The devices classified as class I have less risk and the devices classified as class III have the highest risk. Hence, class I devices have the least regulatory control, class II devices have more stringent controls, and class III are subjected to the most complete and stringent controls. Most implantable devices are categorized as class III devices, and they are required to go through a full Premarket Approval (PMA) procedure. The PMA has the toughest regulations, and sufficient scientific data are required to prove the safety and effectiveness of the medical device. Apart from the technical information of the device, the nonclinical laboratory studies such as immunology and biocompatibility are also required in the PMA [35]. An alternative pathway for class III devices to pass the FDA approval and reaching to the public is through the Humanitarian Use Device (HUD) program. This program does not require efficacy data of the device and the decision from FDA will be given within 45 days. The HUD pathway was created with the intention to encourage inventors to develop potential devices to fight rare and life-threatening diseases. This is because the conventional PMA path is a very costly and time-consuming process as sufficient efficacy data are required. However, there must be a comprehensive description of the device and documentation to prove the rarity of the disease that the device is intended for [55].

As 3D printed medical devices are customizable and may have more complexities, there are some new regulatory controls introduced specifically for 3D printed medical devices. The control of a 3D printed medical device starts at the design stage. The FDA suggests a design control model for the development of any medical devices through 3D printing. The design control model is a system that allows the identification of potential design flaws, the creation of multiple design concepts, as well as the verification and validation of design efficacy through repeated design reviews. The design reviews are to be carried out in all the five different stages during the designing process. The five stages include user needs, design inputs, design processes, design outputs, and validation.

After the design stage, the next control is the printing process. FDA Quality System (QS) Regulation/Medical Device Good Manufacturing Practices should be implemented throughout the whole 3D printing process. The first step to be considered is the raw materials being used. The source of the material should be evaluated with appropriate quality control to ensure homogeneous and traceable manufacturing substrate. One unique manufacturing process that can be done by 3D printing but not in other conventional manufacturing method is the recycling of manufacturing materials. There are concerns that recycling may introduce potential for material contamination, the performance of a device made of recycled materials may be affected, and the material traceability with added complexities. Apart from that, the biocompatibility of the materials used as a medical device needs to be evaluated. The materials need to comply with the standards set by the International Organization Standards (ISO) and the biocompatible materials that can be used in the manufacturing of medical devices are collectively termed ISO 10993.

In the process of 3D printing, some parameters that can be changed could significantly change the physical properties of the printed product. These parameters include printing speed, deposition velocity, humidity within the build environment, etc. Therefore, FDA requires all printing parameters and processes to be documented to decrease the risk of build inaccuracies. Variations in the final printed device can still occur despite appropriate quality controls throughout the printing process. Therefore, post-manufacture quality assurance needs to be carried out in order to ensure the quality of the final product. The method of post-manufacturing is entirely dependent on the 3D printing technique used, and there is still no standard guidance for it. Some examples of the post-manufacture quality assurance that have been used are surface laser scanning, micro-CT, and various printer-monitoring strategies. Cleaning of the final 3D printed devices is also necessary and needs to be demonstrated to be sufficient for regulatory approval. Depending on the 3D printing technique used, cleaning methods may differ. Most cleaning methods include removal of support materials or removal of residual monomers. For certain 3D printing techniques, additional finishing steps are required. If finishing techniques have been employed, it is important to validate that the finishing processes do not alter the overall structure or the mechanical properties of the finished device [55]. There is a possibility that the 3D printing process will alter the properties of the materials through cross-linking and melting. Therefore, degradation testing and studies need to be carried out for the 3D printed device, especially for implants made of biodegradable materials. Sterilization of the finished product is essential to minimize the risk of infection with implantable medical devices. Sterility assurance level (SAL) is the probability of a single unit being nonsterile after it has been sterilized. 3D printed medical implants require the same validation of SAL as any other medical implant and the results must be presented to the FDA as part of the marketing application.

2.8 Conclusions

Up to date, 3D printing has been utilized as an emerging and attractive technology option because of its supreme flexibility and versatility in producing complex objects with finer possible details. However, the application and practical potential of 3D printing is at some extent limited because of its speed and time taken for the fabrication of the 3D objects. Therefore, it is projected that moving from the stepwise layer-by-layer process, which is typical in 3D printing, to a continuous process may significantly accelerate the practical potential of printing technology. Four-dimensional (4D) printing that generally relies on introducing stimulant into a printed 2D structure can make that fabrication process slightly faster. Therefore, 4D printing has the true potential to emerge as the next-generation additive manufacturing techniques. Similar to other medical and pharmaceutical products, all 3D printed products (medical/ pharmaceutical) need to first pass a set of standards and regulations before they can be sold to the public. Therefore, US FDA has issued specific guidelines that need to be adhered to in order to mass customize and commercially exploit any suitable 3D printed medical product.

References

1 Ventola, C.L. (2014). Medical applications for 3D printing: current and projected uses. *Pharmacy and Therapeutics* 39 (10): 704–711.
2 Hoy, M.B. (2013). 3D printing: making things at the library. *Medical Reference Services Quarterly* 32 (1): 94–99.

3 Goyanes, A., Det-Amornrat, U., Wang, J. et al. (2016). 3D scanning and 3D printing as innovative technologies for fabricating personalized topical drug delivery systems. *Journal of Controlled Release* 234 (Supplement C): 41–48.

4 Melnikova, R., Ehrmann, A., and Finsterbusch, K. (2014). 3D printing of textile-based structures by fused deposition modelling (FDM) with different polymer materials. *IOP Conference Series: Materials Science and Engineering* 62: 012018.

5 Melchels, F.P.W., Feijen, J., and Grijpma, D.W. (2010). A review on stereolithography and its applications in biomedical engineering. *Biomaterials* 31 (24): 6121–6130.

6 Alhnan, M.A., Okwuosa, T.C., Sadia, M. et al. (2016). Emergence of 3D printed dosage forms: opportunities and challenges (in eng). *Pharmaceutical Research* 33 (8): 1817–1832.

7 Liska, R., Schuster, M., Inführ, R. et al. (2007). Photopolymers for rapid prototyping. *Journal of Coatings Technology and Research* 4 (4): 505–510.

8 Melchels, F.P.W., Feijen, J., and Grijpma, D.W. (2009). A poly(D,L-lactide) resin for the preparation of tissue engineering scaffolds by stereolithography. *Biomaterials* 30 (23): 3801–3809.

9 McMains, S. (2005). Layered manufacturing technologies. *Communications of the ACM* 48 (6): 50–56.

10 Suzuki, M., Sawa, T., Terada, Y. et al. (2015). Fabrication of microneedles precisely imitating mosquito's proboscis by nanoscale tree dimensional laser lithography and its characterization. *2015 Transducers – 2015 18th International Conference on Solid-State Sensors, Actuators and Microsystems (TRANSDUCERS)*, pp. 121–124.

11 Pati, F., Shim, J.-H., Lee, J.-S., and Cho, D.-W. (2013). 3D printing of cell-laden constructs for heterogeneous tissue regeneration. *Manufacturing Letters* 1 (1): 49–53.

12 Seitz, H., Rieder, W., Irsen, S. et al. (2005). Three-dimensional printing of porous ceramic scaffolds for bone tissue engineering. *Journal of Biomedical Materials Research Part B: Applied Biomaterials* 74B (2): 782–788.

13 Katstra, W.E., Palazzolo, R.D., Rowe, C.W. et al. (2000). Oral dosage forms fabricated by Three Dimensional Printing™. *Journal of Controlled Release* 66 (1): 1–9.

14 Yu, D.-G., Shen, X.-X., Branford-White, C. et al. (2009). Novel oral fast-disintegrating drug delivery devices with predefined inner structure fabricated by three-dimensional printing. *Journal of Pharmacy and Pharmacology* 61 (3): 323–329.

15 Wendel, B., Rietzel, D., Kühnlein, F. et al. (2008). Additive processing of polymers. *Macromolecular Materials and Engineering* 293 (10): 799–809.

16 Van Hooreweder, B., Moens, D., Boonen, R. et al. (2013). On the difference in material structure and fatigue properties of nylon specimens produced by injection molding and selective laser sintering. *Polymer Testing* 32 (5): 972–981.

17 Goyanes, A., Buanz, A.B.M., Hatton, G.B. et al. (2015). 3D printing of modified-release aminosalicylate (4-ASA and 5-ASA) tablets. *European Journal of Pharmaceutics and Biopharmaceutics* 89 (Supplement C): 157–162.

18 Sher, D. (2015). The top 15 bioprinters. https://3dprintingindustry.com/news/top-10-bioprinters-55699/ (accessed 06 July 2018).
19 Munaz, A., Vadivelu, R.K., John, J.S. et al. (2016). Three-dimensional printing of biological matters. *Journal of Science: Advanced Materials and Devices* 1 (1): 1–17.
20 Cui, X., Boland, T., D'Lima, D.D., and Lotz, M.K. (2012). Thermal inkjet printing in tissue engineering and regenerative medicine. *Recent Patents on Drug Delivery & Formulation* 6 (2): 149–155.
21 Krassentein, E. (2014). Why HP's 3D printing technology is amazing, what the future holds, and where their competitors stand. https://3dprint.com/22033/hp-3d-printing-technology-hpq/ (accessed 06 July 2018).
22 Melocchi, A., Parietti, F., Maroni, A. et al. (2016). Hot-melt extruded filaments based on pharmaceutical grade polymers for 3D printing by fused deposition modeling. *International Journal of Pharmaceutics* 509 (1–2): 255–263.
23 Varotsis, A.B. (2017). Common FDM materials. https://www.3dhubs.com/knowledge-base/introduction-fdm-3d-printing?action= (accessed 06 July 2018).
24 Giang, K. (2017). PLA vs ABS: what's the difference? https://www.3dhubs.com/knowledge-base/pla-vs-abs-whats-difference (accessed 06 July 2018).
25 Norman, J., Madurawe, R.D., Moore, C.M.V. et al. (2017). A new chapter in pharmaceutical manufacturing: 3D-printed drug products. *Advanced Drug Delivery Reviews* 108 (Supplement C): 39–50.
26 Pietrzak, K., Isreb, A., and Alhnan, M.A. (2015). A flexible-dose dispenser for immediate and extended release 3D printed tablets. *European Journal of Pharmaceutics and Biopharmaceutics* 96 (Supplement C): 380–387.
27 Alhijjaj, M., Belton, P., and Qi, S. (2016). An investigation into the use of polymer blends to improve the printability of and regulate drug release from pharmaceutical solid dispersions prepared via fused deposition modeling (FDM) 3D printing. *European Journal of Pharmaceutics and Biopharmaceutics* 108 (Supplement C): 111–125.
28 Goyanes, A., Robles Martinez, P., Buanz, A. et al. (2015). Effect of geometry on drug release from 3D printed tablets. *International Journal of Pharmaceutics* 494 (2): 657–663.
29 Goyanes, A., Wang, J., Buanz, A. et al. (2015). 3D printing of medicines: engineering novel oral devices with unique design and drug release characteristics. *Molecular Pharmaceutics* 12 (11): 4077–4084.
30 Khaled, S.A., Burley, J.C., Alexander, M.R. et al. (2015). 3D printing of five-in-one dose combination polypill with defined immediate and sustained release profiles. *Journal of Controlled Release* 217 (Supplement C): 308–314.
31 Kaye, R., Goldstein, T., Zeltsman, D. et al. (2016). Three dimensional printing: a review on the utility within medicine and otolaryngology. *International Journal of Pediatric Otorhinolaryngology* 89 (Supplement C): 145–148.
32 Patel, A., Sahu, D., Dashora, A., and Garg, R. (2013). A review of hot melt extrusion technique. *International Journal of Innovative Research in Science, Engineering and Technology* 2012: 1–9.

33 Maniruzzaman, M., Boateng, J.S., Snowden, M.J., and Douroumis, D. (2012). A review of hot-melt extrusion: process technology to pharmaceutical products. *ISRN Pharmaceutics* 2012: 1–9. 436763.

34 Ponmozhi, J., Frias, C., Marques, T., and Frazão, O. (2012). Smart sensors/actuators for biomedical applications: review. *Measurement* 45 (7): 1675–1688.

35 Zhang, J. and Hoshino, K. (2014). Chapter 7 – Implantable sensors. In: *Molecular Sensors and Nanodevices*, 415–465. Oxford: William Andrew Publishing.

36 Lin, Y., Lu, F., Tu, Y., and Ren, Z. (2004). Glucose biosensors based on carbon nanotube nanoelectrode ensembles. *Nano Letters* 4 (2): 191–195.

37 Clausen, I. and Glott, T. (2014). Development of clinically relevant implantable pressure sensors: perspectives and challenges. *Sensors* 14 (9): 17686.

38 Momeni, F., Mehdi Hassani.N, S.M., Liu, X., and Ni, J. (2017). A review of 4D printing. *Materials & Design* 122: 42–79.

39 Khoo, Z.X., Teoh, J.E.M., Liu, Y. et al. (2015). 3D printing of smart materials: a review on recent progresses in 4D printing. *Virtual and Physical Prototyping* 10 (3): 103–122.

40 Lee, J.-Y., An, J., and Chua, C.K. (2017). Fundamentals and applications of 3D printing for novel materials. *Applied Materials Today* 7: 120–133.

41 Jadhav, R.G. and Das, A.K. (2017). Chapter 10 – Four dimensional printing in healthcare A2. In: *3D Printing in Medicine* (ed. M. Kalaskar Deepak), 207–218. Woodhead Publishing.

42 Ge, Q., Conner, K.D., Qi, H.J., and Martin, L.D. (2014). Active origami by 4D printing. *Smart Materials and Structures* 23 (9): 094007.

43 Shin, D.-G., Kim, T.-H., and Kim, D.-E. (2017). Review of 4D printing materials and their properties. *International Journal of Precision Engineering and Manufacturing-Green Technology* 4 (3): 349–357.

44 Choong, Y.Y.C., Maleksaeedi, S., Eng, H. et al. (2017). 4D printing of high performance shape memory polymer using stereolithography. *Materials & Design* 126: 219–225.

45 Ge, Q., Sakhaei, A.H., Lee, H. et al. (2016). Multimaterial 4D printing with tailorable shape memory polymers. *Scientific Reports* 6: 31110.

46 Ge, Q., Qi, H.J., and Dunn, M.L. (2013). Active materials by four-dimension printing. *Applied Physics Letters* 103 (13): 131901.

47 Peppas, N.A., Bures, P., Leobandung, W., and Ichikawa, H. (2000). Hydrogels in pharmaceutical formulations. *European Journal of Pharmaceutics and Biopharmaceutics* 50 (1): 27–46.

48 Gao, B., Yang, Q., Zhao, X. et al. (2016). 4D bioprinting for biomedical applications. *Trends in Biotechnology* 746–756.

49 Raviv, D., Zhao, W., McKnelly, C. et al. (2014). Active printed materials for complex self-evolving deformations. *Scientific Reports* 4: 7422.

50 Lendlein, A. and Kelch, S. (2002). Shape-memory polymers. *Angewandte Chemie International Edition* 41 (12): 2034–2057.

51 Schmidt, A.M. (2006). Electromagnetic activation of shape memory polymer networks containing magnetic nanoparticles. *Macromolecular Rapid Communications* 27 (14): 1168–1172.

52 Rossiter, J., Walters, P., and Stoimenov, B. (2009). Printing 3D dielectric elastomer actuators for soft robotics. In: *SPIE Smart Structures and Materials + Nondestructive Evaluation and Health Monitoring*, vol. 7287 (ed. Y. Bar-Cohen and T. Wallmersperger), 10. SPIE.

53 Ryu, J., D'Amato, M., Cui, X. et al. (2012). Photo-origami – bending and folding polymers with light. *Applied Physics Letters* 100 (16): 161908.

54 FDA (2014). Classify your medical device. https://www.fda.gov/MedicalDevices/DeviceRegulationandGuidance/Overview/ClassifyYourDevice/ucm2005371.htm (accessed 06 July 2018).

55 Morrison, R.J., Kashlan, K.N., Flanangan, C.L. et al. (2015). Regulatory considerations in the design and manufacturing of implantable 3D-printed medical devices. *Clinical and Translational Science* 8 (5): 594–600.

3

3D Printing: A Case of ZipDose® Technology – World's First 3D Printing Platform to Obtain FDA Approval for a Pharmaceutical Product

Thomas G. West[1] *and Thomas J. Bradbury*[2]

[1] *Aprecia Pharmaceuticals, LLC, Intellectual Property, 89 Twin Rivers Drive, East Windsor, NJ 08520, USA*
[2] *Aprecia Pharmaceuticals, LLC, Development Engineering, 89 Twin Rivers Drive, East Windsor, NJ 08520, USA*

3.1 Introduction

3D printing of pharmaceuticals has been the subject of a number of research publications and patents from academia and small companies alike over the last two decades. However, despite this substantial interest, an exemplar for commercial deployment in a regulated context remained absent until only recently. This situation effectively changed on 31 July 2015 when the US Food and Drug Administration (FDA) approved the New Drug Application (NDA) for SPRITAM® (levetiracetam), marking the first regulatory approval of a pharmaceutical product that uses 3D printing in its manufacturing process. It is noteworthy as the first product approval based on the ZipDose® formulation technology from Aprecia Pharmaceuticals Company (Aprecia), as well as the company's first product approval generally. This milestone represents the culmination of several years of work in proprietary machine and product development, as well as in regulatory positioning of a new manufacturing technology. More broadly, SPRITAM's approval punctuates a resurgence of general research interest over the most recent decade regarding the application of 3D printing and other variations on additive manufacturing techniques to pharmaceuticals and life sciences.

This chapter will endeavor to provide substantial description of the technology development leading to SPRITAM's approval, while also placing this milestone into historical context.

3.2 Terminology

A set of terms for additive manufacturing has been established jointly by the International Standards Organization (ISO) and American Society for Testing and Materials (ASTM) International under the standard ISO/ASTM 52900 [1]. However, in this diverse area of technology, there is no single standard set of nomenclature that works unambiguously across disciplines and applications.

3D and 4D Printing in Biomedical Applications: Process Engineering and Additive Manufacturing, First Edition. Edited by Mohammed Maniruzzaman.

Accordingly, this section will point out some aspects of naming that may benefit from further refinement, and this chapter as a whole will make use of additional nonstandard terminology where needed. Under ISO/ASTM, both "additive manufacturing" and "3D printing" are used as category names that capture somewhat different aspects of layer-by-layer manufacturing techniques, and the terms are often used without distinction between them in nontechnical publications.

Additive manufacturing effectively conveys the aspect of joining (adding) material during the process of forming parts, in contrast with techniques that remove material (subtractive) or reshape material (formative) from simpler standardized shapes. Unfortunately, in the pharmaceutical context, older processes such as coating of tablets or beads, filling of capsules, or multistage compression of tablets can be considered both additive and layered in nature. Therefore, the intended meaning of additive manufacturing must also imply *planar* laminar manufacturing in most instances, with a greater degree of control on the spatial placement of material during its addition.

3D printing under ISO/ASTM is directed to fabrication via deposition of material using a printhead, nozzle, or other technology. This definition is a more generalized use of one of the names originally associated with the powder and liquid process of the Massachusetts Institute of Technology (M.I.T.), as noted in Section 3.3, which entails inkjet printing of droplets onto thin layers of powder. For this more specific process category, ISO/ASTM provides the term *binder jetting*. In this phrasing, the powder aspect is unstated and must therefore be taken as implicit. Although adding specificity in regard to the jetting of droplets, the new definition can cause confusion in the pharmaceutical context because the term "binder" may apply to a binding agent (ingredient) present either in a powder blend or in a liquid solution, and in the latter case often invokes the image of binder solutions that are delivered quite differently (sprayed) onto powder for wet granulation. This is a sharp contrast to the instant usage for designed planar additive manufacturing. Moreover, a solid or nonvolatile binding agent need not always be included in the printing fluid. In this sense, the M.I.T. version of 3D printing is perhaps better described as liquid jetting onto powder.

For clarity, in this chapter, the phrasing "powder–liquid 3D printing" will be used to help delineate the manufacturing process that operates via jetting of liquid droplets onto thin layers of powder, in sequential vertical arrangement layer by layer, to build objects in a progressive additive manner from bottom to top. This phrasing will be shortened to "3D printing" during more detailed process discussions in which the two required forming materials are already evident based on the context.

3.3 Historical Context for This Form of 3D Printing

As per the terminology discussion above, in modern parlance, the term "3D printing" is used as an umbrella term for a wide range of additive, layer-by-layer manufacturing techniques. However, this broad usage was not always the case. The genesis of that phrasing is intertwined with the advent of a particular form of 3D printing that carefully combines powder and liquid as the material inputs in order to make the final parts. And of particular relevance to this chapter, the lineage of this specific type of 3D printing more closely contextualizes the work done by Aprecia for approval of SPRITAM using its own form of powder–liquid 3D printing.

Initially described under the alternative names "three-dimensional printing," "3DP," and "3D printing," the powder and liquid version of 3D printing was originally invented by researchers at M.I.T. in the late 1980s and is based on inkjet printing of liquid droplets onto thin layers of powder [2, 3]. M.I.T.'s seminal work on this manufacturing process touched multiple industries largely because of the diversity of powders that could be handled. In regard to life sciences, the work included exploration of materials and architecture appropriate for the fabrication of implantable tissue scaffolds and drug delivery devices exhibiting complex release patterns over time [4–6]. Although M.I.T. initially filed trademark applications for the names three-dimensional printing, 3DP, and 3D printing, in the end, those terms were not registered as trademarks and over time became subject to more generic usage and new efforts at standardization.

In the mid-1990s, M.I.T.'s process was licensed for medical applications to a start-up company named Therics, Inc., which also sponsored continued research at M.I.T. Therics branded its own version of the powder–liquid process under the name TheriForm, a registered trademark that later lapsed. An extended description of the TheriForm process is beyond the scope of this chapter, so the reader is directed to a prior text [7] for a discussion of the process and proof-of-concept work for controlled release applications, and to a separate example showing a low-dose fast melt formulation [8]. The process has also been rediscovered in recent reviews [9, 10]. Even so, the present authors had the privilege of working on the TheriForm systems at Therics before the formation of Aprecia, so it is useful to make some distinctions. Briefly, most of the Therics machine designs shared several features. The first aspect is a bed-to-bed design. In this regard, the machines had two adjacent walled assemblies ("beds") in a fixed position against one another, each with a vertically movable piston therein to control the depth. The first of these beds would be used with the piston in a low position and serve as a "feed bed" to house and supply powder for each layer. The second of these beds would begin with an empty chamber and the piston in a raised position, allowing it to drop incrementally to receive thin layers of powder in succession, layer-by-layer, during the printing process. A roller assembly would move across from feed bed to build bed in order to spread each layer of powder, taking turns in alternating series with droplet deposition by the printhead. The printhead could be either continuous jet or drop-on-demand in design, with the majority of work conducted using microvalve-based drop-on-demand printheads. The printheads were designed to move across the build bed in a reciprocating manner, line by line, until the entire printable area of the build bed had been addressed. On the largest machines, the printhead included up to 32 jets in order to subdivide the printable area. Although the TheriForm version of 3D printing demonstrated substantial utility in various forms of drug delivery at a small scale, the overall machine designs were still regarded as too slow for centralized production of pharmaceutical products. Consequently, Therics as a business would ultimately discontinue work on pharmaceuticals to focus exclusively on direct manufacturing of medical devices requiring lower volumes at higher price points, with all pharmaceutical rights reverting to M.I.T. Hence, in 2003 and 2004, Therics would obtain some of the first FDA clearances for implantable medical devices that were made using direct powder–liquid 3D printing in a centralized manufacturing process [11, 12]. These products were cleared under the 510(k) premarket notification process by FDA's Center for Devices and Radiological Health (CDRH) and comprised synthetic bone graft substitutes based on hydroxyapatite and β-tricalcium phosphate, respectively.

Aprecia Pharmaceuticals Company was founded in 2003 with the explicit goal of achieving pharmaceutically relevant production rates for the unique types of dosage forms enabled by 3D printing, becoming the new exclusive licensee of M.I.T. for pharmaceutical applications of powder–liquid 3D printing, including the pharmaceutical work from Therics. Initial efforts focused on new machine designs for improved throughput, and the company strategically selected the "fast melt" style of dosage forms as its first multiproduct platform application, branding that category of formulations as ZipDose technology. A high-dose cardiometabolic drug was pursued for early experiments at scale on its proprietary equipment, providing the first successful demonstration of 1000 mg drug loading in a ZipDose unit capable of disintegrating in seconds and also generating positive stability and comparative bioavailability/bioequivalence (BA/BE) results. This effort provided a valuable foundation as the company later made a strategic switch to favor central nervous system (CNS) drugs, with levetiracetam as the lead compound having a top dose of 1000 mg, eventually reaching approval as SPRITAM. It is in this overarching historical context that the development of SPRITAM has taken place.

3.4 ZipDose® Technology

ZipDose technology is the brand name of Aprecia's *formulation* technology related to orodispersible dosage forms. The name does not refer to the manufacturing process or machinery and is instead directed to the resulting family of formulations themselves. ZipDose technology is a platform in the sense that the formulation approach has applicability to a broad swathe of compounds, including pharmaceuticals and nutraceuticals. SPRITAM is the first example of a commercial product made using ZipDose technology.

A main point of differentiation for ZipDose formulations is that they are able to accommodate large amounts of their target compound while still retaining the ability to disintegrate in the mouth within seconds. Thus far, dose loading of 1000 mg has been shown commercially. This combination of features requires knowledge of the target compound's attributes, appropriate choice of excipients, and deliberate selection of printing parameters. The goal is to create a threshold level of interparticle binding adequate for handling strength as a unitary structure while also maximizing the ability to redisperse quickly when the structure is reintroduced to liquid. Rather than relying on unusual or novel excipients, the technology makes creative use of well-established materials that are compendial, generally recognized as safe (GRAS), or otherwise approved for use under specific regulations. Functionally, excipients are chosen to support initial wetting and binding during manufacturing, to aid rewetting and disintegration of the dosage form upon administration, and to provide any required taste masking while in the oral cavity. The technology is compatible with well-known taste-masking approaches, including direct masking by flavors and sweeteners, physical barriers via coating or encapsulation, or chemical complexation using ion exchange resins, cyclodextrins, or the like. Because large amounts of material can be accommodated, there is greater capacity to deploy these taste-masking techniques while still maintaining fast disintegration functionality. In general terms, the mass of the active ingredient and any required taste-masking ingredients comprise the total payload that can be accommodated in each unit. Thus, the highest drug loads are achieved with active ingredients that can be directly masked or require little allocation of payload to masking. SPRITAM is one such example.

Products designed to disintegrate directly in the mouth are not new, but the ability to do so quickly while also delivering high doses has remained absent in the marketplace. Previous techniques to make orodispersible dosage forms include freeze-drying and soft compression. Freeze-dried forms are made by dispensing a solution or suspension into empty blister cavities followed by rapid freezing and sublimation of the moisture, leaving behind highly porous dosage forms. In contrast, soft compressed forms have limited porosity and make use of superdisintegrants and lower compression forces in order to attain more rapid disintegration than conventional compressed tablets. In the United States, the dosage forms made from these processes are generally categorized as orally disintegrating tablets (ODTs). These formulations were designed most often for delivery without water, and so both the total mass and the drug loading within them tends to be limited. FDA has provided a guidance document for the ODT dosage form designation based on the intended in-mouth disintegration using only saliva and without use of additional liquid. The guidance requires disintegration in less than 30 seconds when tested using the USP disintegration apparatus or suitable alternative and recommends that the total mass of each ODT be below about 500 mg [13].

Figure 3.1 provides a graphical representation of the current prescription ODT products in the US market, showing the available strengths on the x-axis. The y-axis provides the proprietary name of each product, with any generic products grouped simply under the name of the active ingredient. The data were obtained from the electronic version of FDA's database/publication, *Approved Drug Products and Therapeutic Equivalence Evaluation*, commonly known as the Orange Book [14]. This resource lists the drug products approved by FDA on the basis of safety and effectiveness. The listed products include marketed brand and generic prescription drug products, some over-the-counter (OTC) products, and some products discontinued for reasons unrelated to safety or efficacy.

As shown on the graph, the top product strength is 275 mg for the fixed-dose combination of carbidopa and levodopa, which is now available generically but was originally offered as the brand product PARCOPA® (UCB, Inc.). After that, the products with the next highest strengths are 200 mg of lamotrigine in LAMICTAL® ODT (GlaxoSmithkline LLC) and its generics, and 200 mg

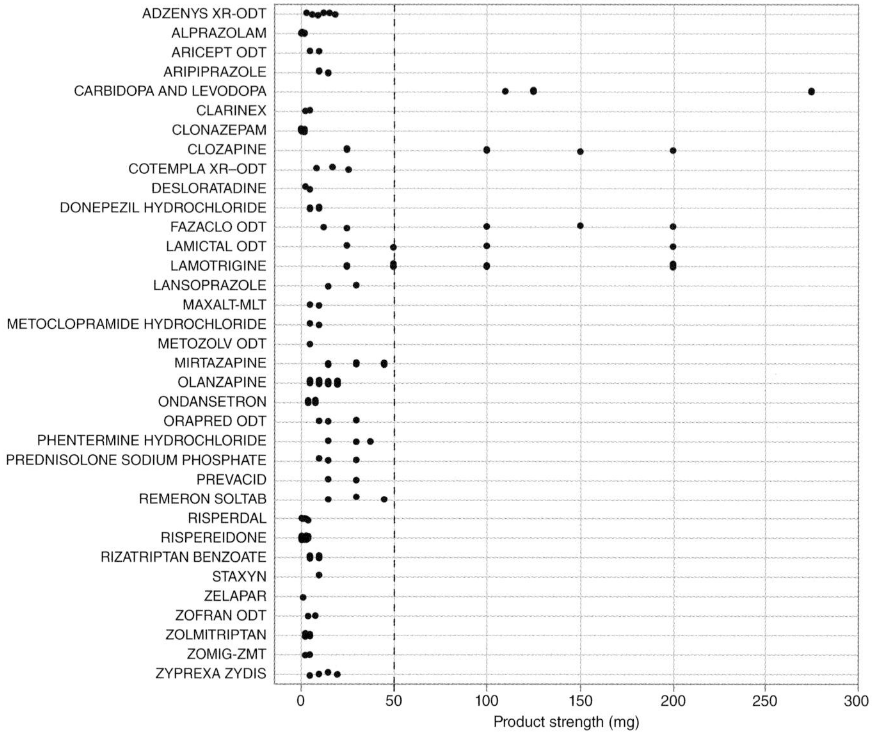

Figure 3.1 Dose loading of active ingredients for prescription ODTs in the US market.

of clozapine in FAZACLO® ODT (Jazz Pharmaceuticals) and its generics. Apart from these drugs, there are no ODTs listed in the Orange Book with strengths above 50 mg, including the OTC and the discontinued ODTs (not shown). Given that a major motivation of the ODT form is to benefit patients who may have difficulty swallowing tablets intact and that drugs requiring higher doses will tend to have larger physical size, the reality is that many of the most difficult-to-swallow solid oral dosage forms remain unaddressed by prior technologies. These unaddressed products are a primary target for ZipDose technology.

Figure 3.2 compares the drug loading and total mass per unit of two of the largest ODTs to those of SPRITAM and other example ZipDose formulations from R&D and prototyping work at Aprecia. The ZipDose examples represent compounds from BCS Classes I, II, and III, and one example of an intensely bitter compound that required physical encapsulation in a hydrophobic material ahead of the 3D printing step in order to achieve a desirable level of taste masking. Overall, the figure illustrates the ability of ZipDose formulations to deliver higher amounts of drug and, in many instances, higher concentrations of drug, when compared to these ODTs.

Consistent with its positioning primarily for high-dose applications, ZipDose technology was specifically designed to create formulations that differ from ODTs and that are most responsive to a small sip of liquid (about 10–15 ml)

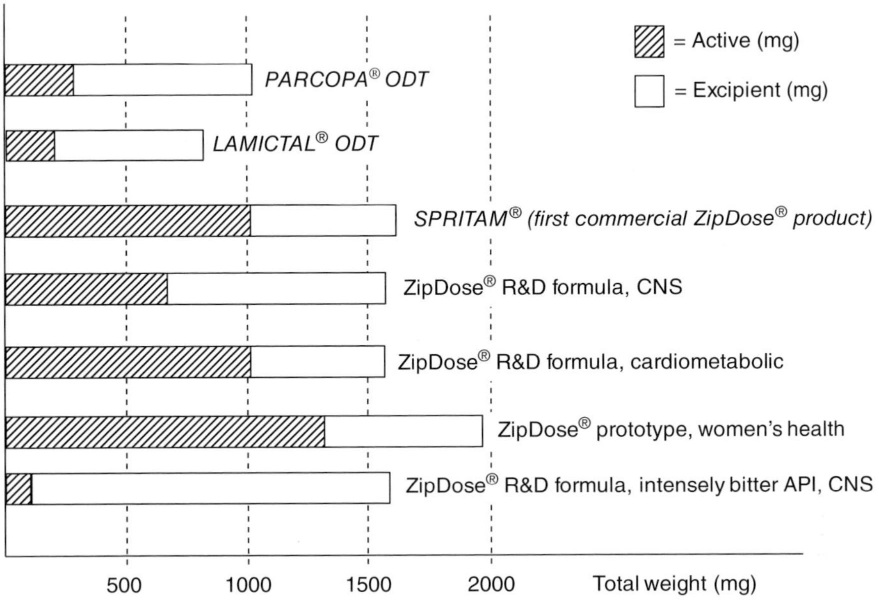

Figure 3.2 Illustration of dose loading and total mass in ZipDose units as compared to the largest prescription ODTs in the USA.

as the primary mode of administration. This approach to administration was based on the recognition that the amount of saliva present can vary from dose to dose in a given patient, and more so across a wide population of patients. In addition, the variation itself becomes more noticeable in high-strength products with larger physical sizes. For a total dry mass of perhaps 500 mg, saliva alone may not always be adequate to initiate rapid disintegration *consistently*, such that a patient may observe slower disintegration or a feeling of dryness in the mouth. The residence time in the mouth may be longer, which is less desirable for quick and convenient dosing. In contrast, administration of a ZipDose formulation with a small sip of liquid always ensures adequate fluid to trigger fast disintegration of the entire unit, regardless of the amount of saliva available in a patient and irrespective of different sizes and strengths of a given product. For established material sets (drug plus excipients), the physical size of an individual ZipDose unit can be scaled considerably while maintaining rapid disperse ability in response to a sip of liquid. Conversely, the technology can also be adapted to drugs with lower strengths, including certain formulations designed for administration without liquid similar to ODTs, provided that the total mass is sufficiently low. In this way, the knowledge base for ZipDose technology can be used to emulate the attributes of ODTs if desired.

The fast disintegration of ZipDose formulations is a function of both the ingredient selection and the interconnected porous network within the ZipDose dosage forms. In contrast with the compressed dosage forms, the available porosity on the surface and throughout the ZipDose dosage form can more quickly make use of the supplemental liquid provided when administering with a sip. The liquid wicks quickly through the porous network of the dosage form, inducing bulk disintegration of the interparticle connections and rapid loss of structural integrity. Excipient selection impacts the speed with which the pores may be wetted, and the rate at which interconnections dissolve or disengage. The deliberate selection of printing parameters in the 3D printing process enables the final unitary form to retain a significant amount of "powder-like" character, including a substantial amount of the void space from the bulk powder layers that are spread during the forming steps of the manufacturing process. In this aspect, 3D printing is truly enabling – whereas other pharmaceutical applications demand a higher degree of binding and densification, for fast-melt dosage forms "less is more." Hence, the choice for location and extent of binding of the powder throughout a ZipDose part design is discriminating. In addition, the 3D printing process enables versatility in terms of materials. A diverse range of powders that flow and spread well can be processed into dosage forms. Active ingredient is most often incorporated as part of the powder blend in order to achieve high dosing. However, if dosing needs are low enough, for certain active ingredients the printing fluid may be used. With these two avenues for incorporating the drug, there are multiple options for creating fixed-dose combination products.

In effect, each part design can be considered as a three-dimensional arrangement of voxels that are designed to be printed or unprinted. Very detailed patterns of droplet placement may be created, providing local regions of greater or lesser saturation as needed for the part design. The regional density of printed voxels versus unprinted voxels determines the extent of binding with a given material set. Once a design is deployed in practice, the physical starting point of the structure is a thin layer of dry powder, and consequently, it begins with a significant amount of void space between the particles of powder before the limited selective wetting from printing. As subsequent layers of material are added vertically atop the preceding layers, the wetted zones bind together horizontally and vertically, eventually defining the entire 3D part when all layers of the design are complete.

Finished ZipDose dosage forms are generally packaged in unit-dose blisters designed with crush-resistant cavities and peel-only lidding. This package design provides mechanical protection for the porous dosage form, which is not suited to bulking in bottles or highly energetic transfer motions. For prescription products, the blister design also includes child-resistant and senior-friendly functionality. No special moisture barrier is required for the packaging except as may be required by a specific active ingredient.

3.5 3D Printing Machines and Pharmaceutical Process Design

3.5.1 Overview

As noted above, Aprecia's version of 3D printing uses two types of material inputs, a powder and a liquid, and is descended from the technology that originated at M.I.T. As with other layered manufacturing techniques, there are certain additional requirements when transitioning from applications in rapid prototyping (i.e. appearance prototypes) to direct manufacturing of functional parts. For oral pharmaceuticals, the needs of direct manufacturing involve requirements for compliance with current Good Manufacturing Practices (cGMPs) as the finished parts are intended for ingestion by patients. From a machine design perspective, these standards invoke considerations related to cleanability and selection of materials of construction for direct and indirect product contact, in addition to controls for ensuring consistent operation and adequate batch documentation. The equipment must not alter the safety, identity, strength, quality, or purity of the drug product beyond the official or other established requirements [15].

Implicit in a machine design approach is a decision on manufacturing strategy. Although various decentralized approaches continue to be discussed for layered manufacturing technologies, the present approach implemented for regulatory entrée of SPRITAM was one of centralized manufacturing. In this way, and perhaps counterintuitively for many observers, this deployment of technology is via direct analogy to manufacturing of traditional pharmaceutical products. This framework casts 3D printing in the role of a new unit operation, and more specifically as a new way of forming unit doses from bulk quantities of pharmaceutically suitable starting materials. Thus, the apparatus for 3D printing itself is part of a larger manufacturing process around it that covers all steps to make the drug product, spanning from raw materials procurement through packaging of finished dosage forms. Regulatory review of the complete manufacturing process is conducted in the context of (and to the extent of) a specific marketing authorization application for a specific drug product. Such an application need not require changes to the regulatory pathway simply because 3D printing is included as part of the process. On the contrary, as the present case illustrates, and as is described further in later sections, for the centralized manufacturing strategy described above marketing authorization may be sought entirely within the existing regulatory paradigm. Current and former FDA staff members have expressed this view more generally, illustrating the accommodation of additive manufacturing technologies within both the existing medical device regulations [16] and the existing drug product regulations [17] using the same kind of science- and risk-based approaches as are used for development, review, and approval of products manufactured by more traditional means.

To succeed with a centralized approach for pharmaceuticals, machine innovations were required. For this model, the equipment and process would need to be scaled significantly to achieve a level of throughput sufficient for specialized pharmaceutical processing. Although not a replacement for conventional high-speed tableting, the requirement was for a high enough production rate to make ZipDose products that are functionally differentiated from both conventional tablets and ODTs made by other specialized processes such as lyophilization or soft compression. In a 3D printing sense, this task is a transition from making tens of parts at a time using prototyping machines to making thousands of parts at a time on newly designed production equipment. Moreover, it is a reversal from the highly customized and often larger and externally complex shapes demanded for appearance prototypes and other applications. For pharmaceuticals, the parts are smaller, identical, and externally simple tablet-like shapes that need to be produced in far greater quantities.

Accordingly, novel and proprietary 3D printing machines were designed and built for Aprecia's purposes. Aprecia's versions of 3D printing machines, equipment assemblies, or processes using them are the subject of several issued patents in the USA [18–22], with additional patents pending in the USA and related patents issued or pending abroad. Because powder–liquid 3D printing can handle a variety of materials beyond those used with pharmaceuticals (e.g. ceramics and metals), these machine innovations have potential applicability to other industries as well.

Expanding on the historical context discussed earlier, it is useful to clarify the significance of these innovations. Although production rates in the tens of thousands of pills per hour have been attributed by recent publications to the equipment once made by Therics [23], those figures are often cited without adequate context. Those projected rates represented the fastest possible throughput of the largest model of TheriForm machine in and of itself (absent post-processing) and when making the smallest pills, sized for approximately 4 mg of drug loading per unit as a low percentage of a small total mass per unit. The rates did not include time for downstream processing or transfers between equipment, including a separate step for feeding and packaging. By comparison, the proprietary manufacturing systems used by Aprecia provide a factor of roughly 10× to 30× higher throughput versus the legacy Therics systems if each were tasked with making parts of substantially the same size, design, and composition. This state-of-the-art equipment is still specialized when compared to conventional high-speed tableting, and currently, it is best suited to differentiated premium products for which the functionality enabled by 3D printing justifies the additional cost of manufacturing parts in a layer-by-layer manner.

3.5.2 Generalized Process in the Pharmaceutical Context

The overall process flow follows that of traditional solid oral dose pharmaceutical manufacturing in many respects, with the exception of 3D printing substituted for tableting in order to carry out the formation of individual unit doses. A generalized schematic of this process is shown in Figure 3.3.

Powder preparation techniques have perhaps the most similar elements. Generally speaking, the powder used for 3D printing is not a single constituent, but rather it is a blend of multiple ingredients in powder form. This aspect matches conceptually with the normal practice for making traditional compression tablets that incorporate a combination of active ingredient and excipients to form the tablet core. Powder blends may be prepared via dry blending, dry granulation, wet granulation, or other known techniques and include milling/screening as needed to remove agglomerates. In contrast, the qualitative and quantitative composition of the powder blend for 3D printing will differ from tableting because (i) the binding mechanism of the forming step is different and (ii) the final part to be made is structurally and functionally different – in the case of ZipDose formulations, the forms are porous and uncompressed.

Similarly, the preparation of the printing fluid resembles other solution preparation steps used in traditional pharmaceutical processing, e.g. the preparation of

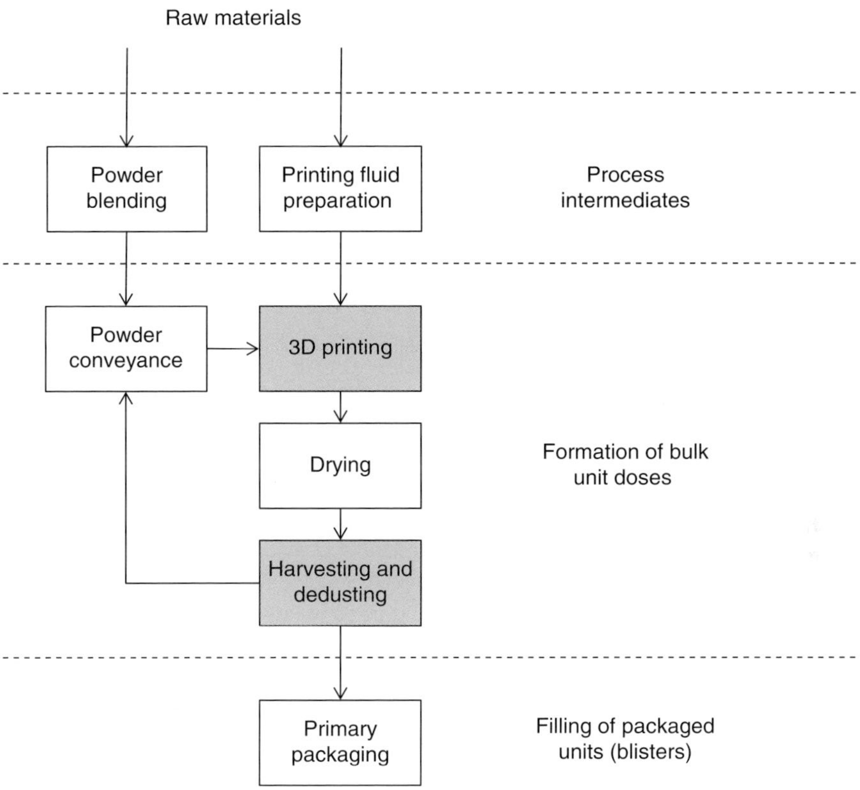

Figure 3.3 Simplified process schematic for pharmaceutical manufacturing using 3D printing as the forming step.

binding solutions used as process intermediates for wet granulation of powders ahead of tableting, or the preparation of oral solutions and syrups that are bottled as finished liquid dosage forms. Once again, the distinctions for 3D printing are mainly compositional. In particular, the printing fluid must exhibit suitable attributes for consistent jetting of liquid droplets from the printheads, and the overall interaction of printing fluid and powder blend must result in the desired attributes for a ZipDose finished dosage form.

Combining powder and liquid via 3D printing is the most unique step. This form of 3D printing assembles pharmaceutical products layer by layer in a manner somewhat similar to inkjet printing of documents but uses accepted pharmaceutical powders and liquids as the analogue to paper and ink, respectively. Unlike paper, the powder is not preformed and must be deposited and spread into smooth, thin layers at the proper rate during the process. During pharmaceutical 3D printing, groups of unit doses are formed by a repeated sequence of spreading thin layers of powder and depositing ("printing") preprogrammed patterns of printing fluid onto selected regions of each layer. The process begins by spreading the first set of dry powder layers on empty build plates to serve as a foundation for the parts to be built, followed by alternating steps of spreading powder and printing liquid to build the parts. The spreading and printing steps are repeated

Figure 3.4 Photo of dosage forms after printing and drying (contrast enhanced).

several times for each group of unit doses, with each new powder layer spread on top of any preceding layers. In the document analogy, this arrangement of layers is like the stacked pages in a paper document and can include different printed patterns on each page. This sequence continues until the full height of the unit doses has been assembled based on the specific predefined instructions for the product and strength. At the conclusion of the 3D printing step, each build plate contains two-dimensional arrays of wet unit doses that are spaced apart from one another, each surrounded and supported by dry unprinted powder. The build plates and their contents (wet unit doses and surrounding powder) are then transferred for subsequent processing.

A drying step is conducted to solidify the wet unit doses at controlled temperature, relative humidity, and time that are determined for each product based on the formula and size/strength. An example photo of the dried printed surface for a cardiometabolic compound is shown in Figure 3.4. Contrast has been enhanced in this image, which shows darker printed regions and lighter unprinted regions. In principle, this drying step is roughly comparable to drying of wet granulated materials, but with a much more specific spatial arrangement of powder and liquid that delineates wet and dry zones within planar beds of material. During drying of 3D printed material, the layered arrangement of wetted powder finishes binding together horizontally and vertically in the printed zones, yielding "3D" unitary forms. In contrast, the surrounding powder remains unbound and is separated from the unit doses after drying ("harvesting") and fed back into the powder input stream for 3DP ("recirculation").

In the harvesting step, plates of printed and dried material are placed in a covered custom assembly with a rotary shaker, which causes the loose unprinted powder to separate from the solid printed parts aided by light rubbing and impact between the parts. The plates used for transferring parts from 3D printing through harvesting include perforations, which allow the unprinted powder to pass through for collection and recirculation back to the 3D printing step. A dedusting step is then performed on the harvested parts using filtered compressed air (air knife). Figure 3.5 shows a photo of the same dosage units from Figure 3.4 after the harvesting and dedusting steps have been performed to remove the loose powder. The units are stored in bulk on trays in a single-layer arrangement ahead of blister packaging. Because of the unique handling needs

Figure 3.5 Photo of dosage forms after harvesting.

of the units, they are fed into the blister cavities using vacuum pick-and-place capability.

Groups of unit doses that are 3D printed concurrently will move through the printing, drying, and harvesting steps together. Collectively, those steps are referred to as the "build cycle" along with the resulting grouping of parts. A batch of product consists of a continuous series of several build cycles that are carried out using the same starting materials (i.e. the powder blend and printing fluid that are process intermediates). During each batch, the unprinted powder from build cycles is continuously recirculated in conjunction with fresh feed of powder until insufficient powder remains for the 3DP step. Once this recirculated powder becomes available, it is redeployed in its entirety and the amount of fresh feed powder is adjusted to make up the remainder. Dense-phase pneumatic conveying is used to transport powder from fresh feed and recirculated sources for use in the 3D printing step. To help ensure physical uniformity, all powder is passed through a cone mill before introduction to the 3D printing machine. This step removes any aggregates that might result from recirculation or consolidation during powder holding, handling, or conveyance. Screen size and impeller designs for the cone mill are selected based on formulation requirements. The amount of fresh feed powder is approximately the same as the amount of powder leaving the process as finished dosage forms, ensuring mass balance. Overall, the movement of materials through the process steps of each build cycle may be characterized as intermittent and incremental in nature, step to step, but the overall batch framework is a continuous series of these build cycles.

3.5.3 Exemplary 3DP Machine Designs

As described earlier, new machine designs were needed to improve the throughput and support a centralized manufacturing approach for specialized value-added pharmaceutical products. In reviewing prior powder–liquid 3D printers, several opportunities for improvement were observed when developing Aprecia's proprietary equipment.

In terms of time efficiency, many prototyping systems use only a single working space within which the powder spreading and liquid printing steps must take turns – each must wait for the other to complete its respective operation and any associated movement in order to avoid spatially obstructing the other apparatus.

As adding powder and adding liquid are the fundamental repetitive operations for forming parts, it is desirable to maximize the proportion of the total print job time during which each one of these operations may be conducted. Similarly, recognizing that a complete manufacturing process requires time to offload parts in between execution of each print job, a longer print job leads to higher productivity of forming parts relative to real (clock) time. This improvement holds true as long as downstream post-processing steps do not become rate limiting and are able to keep pace with the 3D printer. These types of changes lead to longer periods of continuous forming of parts and are complementary with dimensional scaling via larger machine dimensions and replicated machine elements.

Accordingly, Aprecia's designs have a deliberate separation of the spatial zones in which powder spreading and liquid printing occur. Rather than moving a powder spreader and liquid printer over a single common forming area repeatedly, the powder and the liquid stations are mounted in semifixed positions, spatially separated from one another, and a plurality of build modules move underneath them to receive their respective material contributions in succession. An illustration of this separation is shown in Figure 3.6. This image shows a series of build modules that move left to right, providing a near-continuous working surface for the powder spreading and liquid printing stations. At the left of this image, the build surface is lowered vertically by a z-axis stage within each build module to create a shallow planar depression for receiving the next incremental layer of powder. As a build module moves under the powder feeder and roller, the planar depression captures powder, and the new top surface of powder is leveled to provide a smooth substrate. When each build module moves under the printhead, a preprogrammed pattern of liquid is selectively deposited to define the architecture of each respective layer of the part. In this configuration, the stations for powder spreading and liquid printing carry out their functions uninterrupted for nearly the entire job. To eliminate the need for reciprocating movements of the printhead (e.g. orthogonal to the direction of build module movement), the printhead design incorporates adequate stacking and interlacing of component linear jetting assemblies to provide sufficient nozzle count and aggregate print density to span the width of each build module. In this arrangement, each individual nozzle of the printhead addresses a single unique line of voxels through the overall pattern of parts to be made.

With improved time utilization of each powder station and each liquid station, the focus shifts to ensure that adequate working surface is provided by the number and configuration of build modules. Accordingly, the number of build modules and their movement must be designed to prevent any interruption in availability of the build surface that receives material beneath the powder station and the liquid station. Accordingly, the design implementation used by Aprecia defines a horizontal circuitous path for build module movement, with a section of the circuit placed beneath the powder station and the liquid station. Thus, each build module performs multiple passes to repeatedly receive material from each station. When a single powder station and single liquid station are present, each lap around the circuit corresponds with the formation of a single constituent layer of the parts. Subject to space constraints, the design

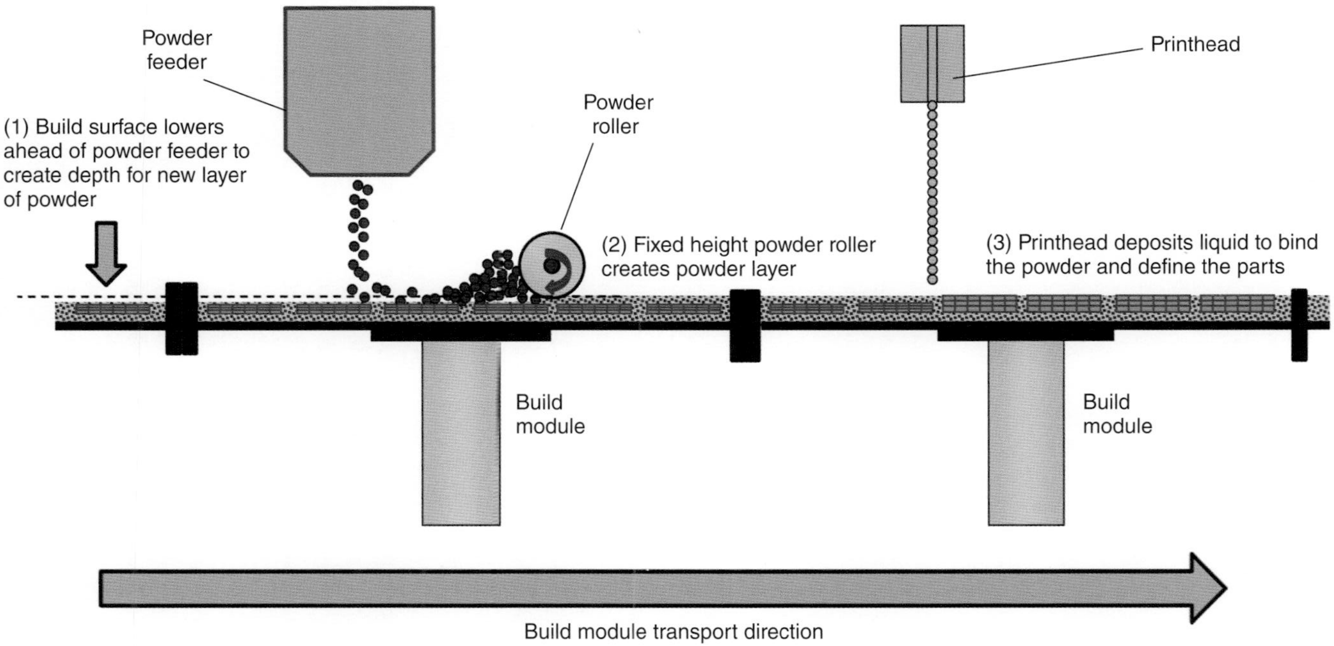

Figure 3.6 Side view schematic of Aprecia's 3D printing process.

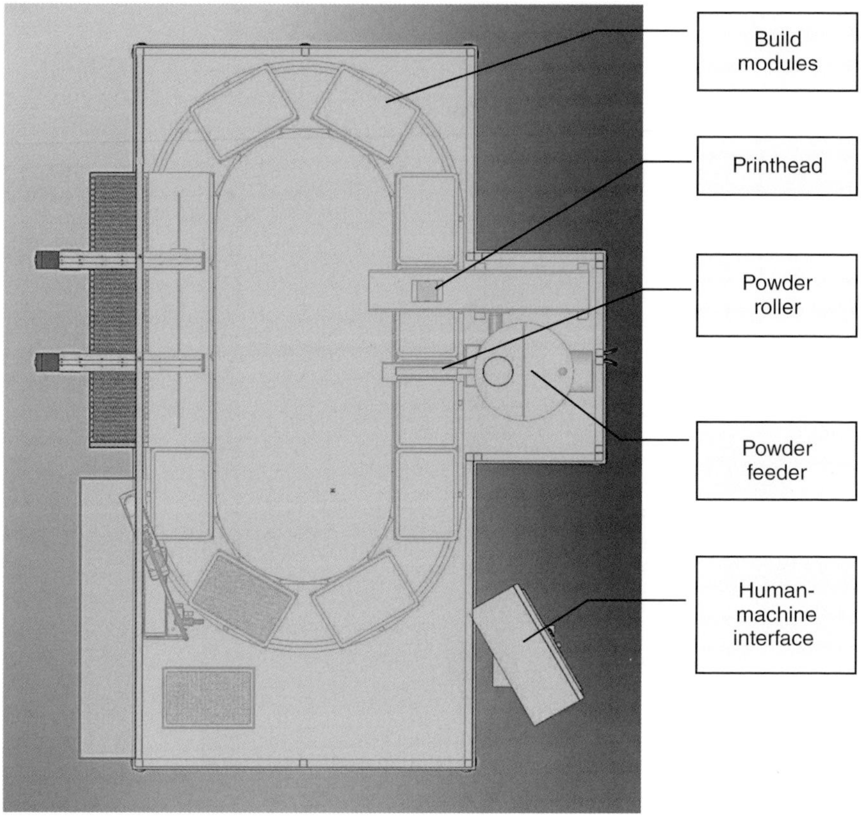

Figure 3.7 Top view schematic of Aprecia's "M0" 3D printing machine.

can incorporate additional spatially separated stations around the circuit in order to form multiple constituent layers of the parts per lap. For certain implementations of the circuit design, at the start of each print job, the build modules can accelerate to reach a constant operating speed that may be used through all the layer formation until deceleration when the job is complete.

Aprecia refers to machines with these design elements as its M-Series of machines. The first of these machines is identified as M0 (pronounced "em zero"). This machine was designed and built initially as an engineering prototype, and later refined and qualified to support operation in a full cGMP production environment. It was used in the majority of the original development work on SPRITAM, including registration, validation, and commercial launch activities. A top view schematic of M0 is provided in Figure 3.7.

The build module path for M0 consists of two parallel linear segments with half circles at each end to enclose an oblong path. As viewed from above, the build modules move in a counterclockwise manner. Twelve build modules fill the entire path, showing angular separation along the circular sections and becoming fully abutted to one another in the linear sections. A single powder station

Figure 3.8 Photo of Aprecia's "M0" 3D printing machine.

and liquid station are arranged along the linear segment on the right. This configuration illustrates a design decision to perform the essential material-forming steps on M0 according to a Cartesian coordinate system and associated physics. As such, this approach avoids the complication of adjusting the designed voxel pattern for parts based on their radial positioning (i.e. array position within a build module relative to center of rotation) because of the difference in tangential velocity both within-part and part-to-part beneath each printhead nozzle. An offloading station is positioned along the linear segment on the left, capable of removing the contents of three build modules at a time after job completion. An enclosure surrounds M0 to help protect the parts during the open forming process and the safety of operators while a print job is running. A photo of M0 is shown as Figure 3.8.

A second machine design, identified as M1, has been completed, and a first embodiment of this design has been built, deployed, and qualified for cGMP operations. This design incorporates wider build modules of more interlocked shape, improved configurations of powder and liquid stations, and increased automation for loading and offloading, resulting in an aggregate throughput of roughly twice that of the initial M0-based process. At the time of this writing, the M1-based process is poised to begin a process validation campaign in order to complete requirements for commercial manufacturing of SPRITAM on the new equipment. Additional machine configurations are contemplated for both small-scale development and manufacturing use.

3.6 Development of SPRITAM®

3.6.1 Product Concept and Need

SPRITAM was conceived to be a novel easy-to-swallow dosage form containing the active pharmaceutical ingredient levetiracetam. It was designed to be a unitary solid dosage form with a porous structure that is made via a powder–liquid 3D printing process. This version of 3D printing binds the powders together without compression, maintaining much of the regional void space that exists between the powder particles in bulk. The primary goals were for the dosage form to provide adequate handling strength while enabling disintegration directly in the mouth within seconds when taken with a small sip of liquid. For this active ingredient, the new dosage form and mode of administration were expected to offer similar rate and extent of absorption as existing products, allowing the same strengths and dosages to be used.

The choice of molecule for this application is deliberate. Levetiracetam is a widely prescribed off-patent medication used in the treatment of multiple forms of epilepsy and was first approved for use in the USA in 1999 under the brand name KEPPRA (UCB, Inc.) via NDA 21035 [24]. The dosing requirements for levetiracetam are among the highest for anti-epilepsy drugs. Moreover, epilepsy diagnoses are most frequent among pediatric and geriatric patients, for whom swallowing of large tablets and capsules is more likely to cause difficulty.

Conventional film-coated compressed tablets of levetiracetam are available in fixed strengths of 250, 500, 750, and 1000 mg and are designed to be swallowed intact but otherwise to provide an immediate release (IR) profile of the drug. A patient treated with levetiracetam in this form must take it twice a day, every day for extended periods of their life in order to help ensure seizure control either alone or in conjunction with other epilepsy medications. Because of the large amount of drug that must be accommodated in each unit, the tablets are predisposed to have large dimensions, which can make them more difficult to swallow. For example, KEPPRA tablets have an elongated shape with lengths of 13, 16, 18, and 19 mm for the respective dosing strengths. These sizes all exceed 8 mm, a threshold at which patient complaints because of size may become more frequent, as noted in a recent FDA guidance related to generics [25].

Alternative dosage forms of levetiracetam have been available over the past decade. Extended release tablets are available that can reduce the frequency of dosing to once daily in a swallow tablet format. However, the number of tablets required is still two per day, and the tablets are still quite large because of the dosing strengths of 500 and 750 mg and the use of specific excipients to effectuate the extended release profile necessary for less frequent dosing. Accordingly, extended release tablets do not address difficulties related to swallowing large tablets intact. In contrast, oral solutions of levetiracetam are also available, primarily for pediatric use. Because the drug is predissolved in a liquid form, these oral solutions provide for ease of swallowing, but they are supplied in bottles requiring measurement at each instance of dosing, twice daily. Hence, these types of liquid preparations can present a different set of limitations in terms of consistent accurate dosing, portability, and convenience.

In general terms, the effectiveness of any medication relies on its ability to be taken reliably as directed, often phrased in terms of patient adherence to (or compliance with) physicians' instructions. In regard to epilepsy, nonadherence remains a challenge, with some specific study estimates of 39% nonadherence for adults [26] and 58% nonadherence for children (first six months of treatment) [27]. The impact of uncontrolled epilepsy is substantial, with one analysis in the USA showing added costs of $12 258 for Medicaid patients and $14 582 for patients with private insurance [28]. Overall, the US Centers for Disease Control and Prevention (CDC) calculates an annual cost of $15.5 billion in medical costs and lost or reduced earnings and production because of epilepsy [29].

3.6.2 Regulatory Approach

At the start of SPRITAM development, the drug substance levetiracetam was already off-patent and available in both branded and generic versions as conventional IR tablets, XR tablets, and oral solutions. Conversely, no 3D printed drug product of any sort had been approved yet. Thus, an appropriate regulatory pathway needed to be identified, supported by the substantial clinical and regulatory history for levetiracetam.

In the USA, generic products are intended to be duplicates of the original brand products in most respects, exhibiting both pharmaceutical equivalence and BE, and are filed via the Abbreviated New Drug Application (ANDA) process under section 505(j) of the Federal Food Drug & Cosmetic Act (FD&C Act). As such, an ANDA allows for the marketing of a generic product with the same active ingredients, strength, route of administration, dosage form, performance characteristics, and labeling as the original brand product. Unlike a full NDA, in general, an ANDA is not required to include preclinical or clinical data to establish safety and efficacy. Instead a showing of BE is made by *in vivo*, *in vitro*, or other means in comparison with a previously approved innovator drug known as a reference-listed drug (RLD). Products approved under ANDAs are considered to be therapeutically equivalent to the RLD and therefore substitutable under the same conditions of use.

Based on the product concept and desired attributes for SPRITAM, the ultimate product was envisioned to represent a new dosage form in comparison with all previously approved forms of levetiracetam, including noteworthy differences in appearance on the shelf, composition, disintegration behavior, and instructions for administration. From a marketing perspective, the intent was for a distinct branded product that would avoid potential confusion with dosage forms that have dissimilar administration. Because of these anticipated differences, the ANDA filing route would be inappropriate. Similarly, a full standalone NDA would also be inappropriate, as it would unnecessarily repeat the original work by the innovator in regard to the drug substance as well as preclinical and clinical studies of the RLD.

In the USA, these types of innovations are pursued using a special type of NDA known as a 505(b)(2) NDA. These applications are similar to conventional NDAs in the sense that full reports of investigations of safety and efficacy are provided, but with the notable difference that at least some of the information required for approval comes from studies not conducted by the applicant and for which the applicant has not obtained a right of reference. This information may include published literature as well as FDA's prior findings of safety and efficacy on a RLD [30]. In this situation, sponsors are required to focus on generating appropriate bridging data to specifically address the differences of the new product candidate in comparison with the reference drug. Accordingly, the 505(b)(2) NDA route was selected for pursuing approval of SPRITAM. More recently, FDA has published a draft guidance to help applicants determine whether an ANDA or 505(b)(2) application is most appropriate for their programs [31].

3.6.3 Introduction of the Technology to FDA

As a background to the development of SPRITAM, Aprecia had multiple interactions with FDA to introduce 3D printing and ZipDose technology and to seek formally FDA's advice on specific development programs in accordance with customary meeting practices [32]. To provide a foundational scientific introduction to 3D printing, a seminar was held for FDA staff members in December 2008 at FDA's White Oak Campus. This seminar provided general information on 3D printing and its potential use in pharmaceutical applications and was coordinated through the Office of Compliance at CDER. Additionally, formal type C and type B interactions were conducted with the FDA's Division of Metabolism and Endocrine Drug Products and Office of New Drug Quality Assessment (ONDQA) in 2011 and 2012, respectively, in relation to a metabolic product candidate before the company's strategic focus moved to SPRITAM. For the SPRITAM program, type B meetings were conducted with FDA in 2013 and 2014, respectively. This cumulative dialogue with FDA provided the necessary advice at key stages of development to guide the program to successful submission of the NDA for SPRITAM in October 2014.

3.6.4 Target Product Profile

The target product profile (TPP) is a tool used to guide the overall development program based on its end goals. For drug development, this is expressed in terms of the desired labeling of the product candidate (i.e. prescribing information) and the associated studies that are needed to generate appropriate data in support of any label claims. A TPP may be included as part of the briefing document provided for formal meetings with FDA, and in this way, it can help focus the sponsor's drug development efforts and facilitate efficient and constructive feedback from the agency [33]. For the SPRITAM program, a TPP was defined in terms of anticipated changes to labeling versus the conventional tablet version of the RLD. Based on the SPRITAM product concept, these potential changes included the instructions for in-mouth administration with a sip of liquid, the listing and description of the dosage forms and how they are supplied, the qualitative composition, and the summary results from comparative BA and food effect studies. These in-man studies were proposed and discussed as part of formal meetings with FDA in the course of development.

3.6.5 Synopsis of Formulation and Clinical Development

Like traditional preparations, the design of ZipDose formulations begins with the consideration of the drug substance and its properties. Levetiracetam drug substance is a single enantiomer with chemical name (−)-(*S*)-α-ethyl-2-oxo-1-pyrrolidine acetamide. The compound is a nonhygroscopic crystalline powder with slight odor and bitter taste and is very soluble in water (104.0 g/100 ml). It has a molecular weight of 170.21 with the following structural formula:

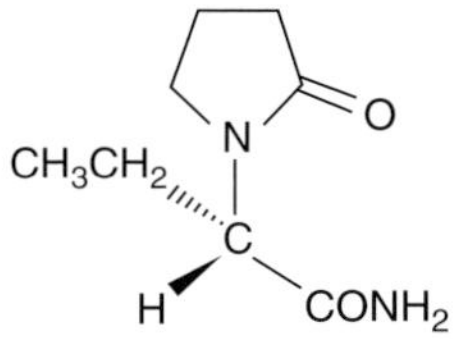

The oral BA of levetiracetam tablets is 100%, and the oral solution version of levetiracetam provides equivalent rate and extent of absorption to the tablet form [34]. As noted above, the desired product strengths for SPRITAM matched those of existing tablets: 250, 500, 750, and 1000 mg.

In considering the product concept, certain physical challenges were evident for the formulation. To achieve high drug loading in an uncompressed form of reasonable size and dimensions, a high weight percentage of the active ingredient would be needed in the final formulation. Accordingly, the drug substance would need to be incorporated in solid form as part of the powder blend for 3D printing. Given the high aqueous solubility of the drug and the preference for water-based fluids in pharmaceutical 3D printing, both the excipients and the product design itself (3D architecture) would need to be chosen carefully to preclude potential overbinding of the powder and to obtain the desired balance of adequate physical integrity with very rapid disintegration. Prior knowledge from a cardiometabolic drug candidate had shown that a drug loading of 1000 mg could be achieved with a water-soluble compound in a ZipDose format while still retaining the ability to disperse in seconds. The work on SPRITAM would extend this approach to a compound that is approximately three times more soluble.

Initial taste evaluation determined that the bitter aspect of the drug has a mild intensity that can be directly masked by inclusion of sweeteners and flavors, and a series of experiments were conducted to identify suitable flavor and sweetener combinations and levels. This work was performed in concert with range-finding experiments to maximize levetiracetam loading in the formulation while also achieving target quality attributes. Prototyping of taste-masked formulations was conducted on small- and intermediate-scale 3D printing machines before scaling up to Aprecia's M0 3D printing machine described in earlier sections. Statistically designed experiments were used to aid formula iteration through each scale of equipment, and select formulations were used to generate stability data to guide development. Ultimately, the bulk of development work was conducted on the M0 system. At M0 scale, design factors that were evaluated included print job (part dimensions, layer number, and layer thickness), foundation layer thickness and number, printhead height, powder feeder type, feed rate, powder spreading, product marking, powder circulation, harvesting and dedusting, drying conditions, and other variables specific to manufacturing by 3D printing.

An inherent aspect of the formulation work was establishment of the powder blend formula and printing fluid formula that serve as the two process inputs for the 3D printing step. Development of these two intermediates is interactive, but it is driven principally by the powder blend when the active ingredient is a component of the powder and when prior (drug-free) printing fluids are available as a starting point. For SPRITAM, the powder blend was designed for good flow to support feeding, spreading, and pneumatic conveyance as part of circulation of unprinted powder in the full-scale 3D printing process train. A direct dry blending process was adopted, and all constituents were selected to provide sufficient overlap in their respective particle size distributions to minimize the chance for segregation during processing. A majority of the final powder blend is levetiracetam by weight. Similarly, for the printing fluid, constituents were chosen to attain desirable ranges of surface tension and viscosity to support reliable drop formation and jetting from the printhead, with limited need for supplemental binding agents. Because the levetiracetam powder blend is largely water soluble, an ample level of binding can be activated in situ by depositing a predominantly aqueous fluid, as was selected for SPRITAM. Appropriately, the final part designs and printing parameters represent a fairly low aggregate level of saturation (print density), with localized regions of denser printing on the exterior surfaces of the dosage form. Final mass contribution from the printing fluid components is less than five percent of the total for the unit dose.

The final formulation for SPRITAM includes the labeled strength of levetiracetam and the inactive ingredients colloidal silicon dioxide, glycerin, mannitol, microcrystalline cellulose, polysorbate 20, povidone, sucralose, butylated hydroxyanisole, and natural and artificial spearmint flavor. These components are commonly used in oral solid and solution formulations. Each component is USP or NF grade with the exception of spearmint flavor, which has GRAS status and is used at very low levels. The spatial architecture, printing parameters, and quantitative composition of SPRITAM are proprietary; however, the use levels of each excipient are generally consistent with those outlined in FDA's Inactive Ingredient Database for approved products having the same route of administration. In the final designs, the SPRITAM formula is proportionally similar across all the four strengths of the product, and all strengths share the same formulations for their process intermediates (i.e. powder blend and printing fluid). A photo of the four strengths of SPRITAM is shown in Figure 3.9. The visible markings on the dosage forms are uniquely created in the 3D printing step as negative (unprinted) features. During harvesting and dedusting, the unprinted powder empties out, resulting in recessed surfaces that define the characters. In effect, this approach is a kind of noncontact debossing.

Eight registration batches of the final formulation were manufactured at intended commercial scale on the M0 system, including three batches each of the top and bottom strengths to bracket one batch each of the middle strengths. For context, these batches would be considered small by traditional pharmaceutical standards but large for 3D printing of identical parts. All batches conformed

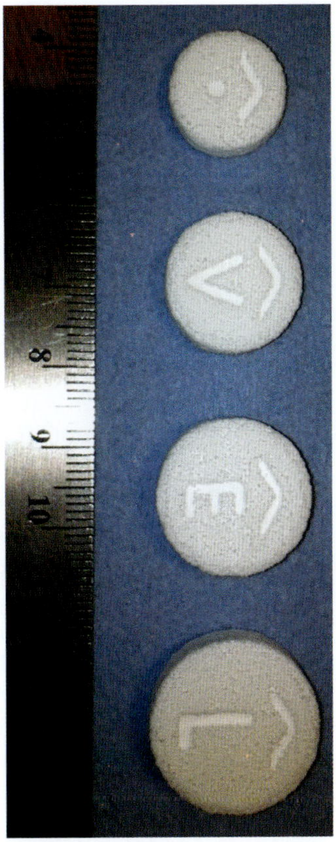

Figure 3.9 Photo showing the top view of all the four strengths of SPRITAM.

to a release specification that included tests for appearance, identity, assay, organic impurities, dissolution, uniformity of dosage units (content uniformity), disintegration, and residual solvents. The container closure design for the units included mechanical protection without the need for a special moisture barrier. Accordingly, all units were packaged on the intended commercial equipment into six-count, child-resistant unit-dose blister cards that can be separated into individual blisters via perforations. The blisters were comprised of crush-resistant PVC/Aclar® cavity material and peel-only aluminum and paper lidding. These batches generated primary stability data at long-term conditions of 25 °C/60%RH and accelerated conditions of 40 °C/75%RH. One 1000 mg batch was also used for in-man studies. These data sets formed the foundation of the NDA submission for SPRITAM.

To help establish a scientific bridge to the RLD, a pivotal study was conducted to compare the *in vivo* BA of 1000 mg SPRITAM (test product) with that of 1000 mg levetiracetam IR tablet (reference product). This study is the subject of its own publication [35]. Briefly, the study comprised a three-way crossover design in 33 healthy adult subjects using pharmacokinetic endpoints. Both the test and reference products were evaluated in the fasted state, and the test product was also evaluated in the fed state after consumption of FDA-compliant high-fat, high-calorie meal to establish food effect. Importantly, in order to support the preferred mode of administration using an unmeasured sip of liquid, the test product was given in that mode. As implemented, each subject was provided 30 ml of water to take a natural sip in order to disperse the dosage form intraorally before swallowing; the remaining water was then measured to infer the size of the sip taken (average of 13 ml). In contrast, the reference product was administered with 240 ml of water, consistent with general recommendations on BA/BE studies [36]. The results confirmed equivalent rate and extent of absorption of levetiracetam in the fasting state, traversing the differences in dosage form, composition, and method of administration. The test product showed similar safety and tolerability to the reference product, with no change observed in the oral mucosa. Subsequent to this pivotal study, a set of *in vitro* data were generated to support use of liquids other than water for administration. In addition, a pilot *in vivo* study was conducted to demonstrate that administration without any water, a condition contrary to the intended label instructions, would not adversely affect the BA of the drug. Biowaivers were obtained for the three lower strengths of SPRITAM based on the pivotal study results on the highest strength and the following factors across all strengths: proportionally similar formula, same manufacturing process, same dosage form, and similar dissolution in three media.

Based on these original data sets and reference to FDA's prior findings of safety and efficacy of the RLD, the NDA for SPRITAM was approved on 31 July 2015 within the standard review cycle. An NDA supplement later clarified the dosage form designation for the product (tablets, for oral suspension) and added a second mode of administration as an alternative to the preferred (primary) mode of dispersing directly in the mouth with a sip of liquid.

3.7 Conclusion

This chapter has provided an overview of the history and technology development leading to the approval of SPRITAM, the first product designed using Aprecia's proprietary ZipDose formulation technology and the first FDA-approved pharmaceutical product incorporating powder–liquid 3D printing as the centerpiece of the manufacturing process. More directly, this accomplishment serves as a foundation for the development of additional ZipDose products by Aprecia and its prospective partners to improve medicines and expand patient reach in the USA and eventually abroad. In the future, other platform applications of powder–liquid 3D printing are contemplated, including those for controlled-release dosage forms, and other deployments beyond the current example for centralized manufacturing.

Two decades of diverse research in 3D printing and its application to pharmaceuticals have set the stage. With the first FDA approval recently demonstrated and widespread interest in the broader technology area, in regard to 3D printing of pharmaceuticals one may now describe the present moment most aptly, in the words of Churchill, as perhaps the end of the beginning.

Acknowledgments

Novel product and process advancements do not occur in isolation. The technological achievements described in this chapter include meaningful contributions both large and small that are too numerous to list individually. Therefore, the authors would like to thank collectively all internal and external contributors to the SPRITAM development team and their forebears, as well as the investors of Aprecia, for their contributions and steadfast commitment in achieving this historic milestone.

References

1 ISO/ASTM52900-15 (2015). *Standard Terminology for Additive Manufacturing–General Principles–Terminology*. West Conshohocken, PA: ASTM International www.astm.org.

2 Sachs, E.M., Haggerty, J.S., Cima, M.J., and Williams, P.A. (1989). Three-dimensional printing techniques. US Patent 5, 204,055, filed 8 December 1989 and issued 20 April 1993.

3 Sachs, E., Cima, M., Williams, P. et al. (1992). Three-dimensional printing: rapid tooling and prototypes directly from a CAD model. *Journal of Engineering for Industry* 114: 481–488.

4 Cima, L.G. and Cima, M.J. (1993). Preparation of medical devices by solid freeform fabrication methods. US Patent 5, 490,962, filed 18 October 1993 and issued 13 February 1996.

5 Cima, L.G. and Cima M.J. (1994). Tissue regeneration matrices by solid freeform fabrication techniques. US Patent 5, 518,680, filed 23 February 1994 and issued 21 May 1996.

6 Cima, M.J., Sachs, E., Cima, L.G. et al. (1994). Computer-derived microstructures by 3D printing: bio- and structural materials. In: *Proceedings of the Solid Freeform Fabrication Symposium* (ed. H.L. Marcus, J.J. Beaman, J.W. Barlow, et al.), 181–190. Austin, TX: University of Texas at Austin.

7 Rowe, C.W., Wang, C.C., and Monkhouse, D.C. (2002). TheriForm technology. In: *Modified Release Drug Delivery Technology*, 1e (ed. M.J. Rathbone and J. Hadgraft), 77–87. Boca Raton: CRC Press.

8 Lee, K.J., Kang, A., Delfino, J.J. et al. (2003). Evaluation of critical formulation factors in the development of a rapidly dispersing captopril oral dosage form. *Drug Development and Industrial Pharmacy* 29 (9): 967–979.

9 Goole, J. and Amighi, K. (2016). 3D printing in pharmaceutics: a new tool for designing customized drug delivery systems. *International Journal of Pharmaceutics* 499: 376–394.

10 Hsiao, W.K., Lorber, B., Reitsamer, H., and Khinast, J. (2017). 3D printing of oral drugs: a new reality or hype? *Expert Opinion on Drug Delivery* 15: 1–4.

11 TheriRidge Block. Bone Graft Substitute, 510(k) Number K023998, decision date March 26, 2003.

12 TheriLink Bone Void Filler. 510(k) Number K040134, decision date March 9, 2004.
13 U.S. FDA (2008). Guidance for industry: orally disintegrating tablets. https://www.fda.gov/ucm/groups/fdagov-public/@fdagov-drugs-gen/documents/document/ucm070578.pdf (accessed 12 November 2017).
14 FDA (2018). Orange Book. Current through January 2018: https://www.accessdata.fda.gov/scripts/cder/ob/ (accessed 12 March 2018).
15 21 CFR 211 subpart D Equipment. https://www.gpo.gov/fdsys/pkg/CFR-2011-title21-vol4/xml/CFR-2011-title21-vol4-part211-subpartD.xml (accessed 26 November 2017).
16 Di Prima, M., Coburn, J., Hwang, D. et al. (2016). Additively manufactured medical products – the FDA perspective. *3D Printing in Medicine* 2 (1): 1–6.
17 Norman, J., Madurawe, R.D., Moore, C.M. et al. (2017). A new chapter in pharmaceutical manufacturing: 3D-printed drug products. *Advanced Drug Delivery Reviews* 108: 39–50.
18 Yoo, J., Bradbury, T.J., Bebb, T.J. et al. (2013). Three-dimensional printing system and equipment assembly. US Patent 8, 888,480, filed 3 September 2013 and issued 18 November 2014.
19 Yoo, J., Bradbury, T.J., Bebb, T.J. et al. (2016). Three-dimensional printing system and equipment assembly. US Patent 9, 517,591, filed 18 February 2016 and issued 13 December 2016.
20 Yoo, J., Bradbury, T.J., Bebb, T.J. et al. (2016). Three-dimensional printing system and equipment assembly. US Patent 9, 517,592, filed 22 March 2016 and issued 13 December 2016.
21 Yoo, J., Bradbury, T.J., Bebb, T.J. et al. (2014). Three-dimensional printing system and equipment assembly. US Patent 9, 610,735, filed 30 September 2014 and issued 4 April 2017.
22 Yoo, J., Bradbury, T.J., Bebb, T.J. et al. (2017). Three-dimensional printing system and equipment assembly. US Patent 9, 908,293, filed 2 February 2017 and issued 6 March 2018.
23 Gebhardt, A. and Hötter, J.A. (2016). *Additive Manufacturing, 3D Printing for Prototyping and Manufacturing*. Munich: Carl Hanser Verlag.
24 Drugs@FDA.com: FDA approved drug products. https://www.accessdata.fda.gov/scripts/cder/daf/index.cfm?event=overview.process&ApplNo=021035 (accessed 29 September 2017).
25 U.S. FDA (2015). Guidance for industry: size, shape, and other physical attributes of generic tablets and capsules. https://www.fda.gov/downloads/drugs/guidances/ucm377938.pdf (accessed 26 November 2017).
26 Davis, K.L., Candrilli, S.D., and Edin, H.M. (2008). Prevalence and cost of nonadherence with antiepileptic drugs in an adult managed care population. *Epilepsia* 49 (3): 446–454.
27 Modi, A.C., Rausch, J.R., and Glauser, T.A. (2011). Patterns of nonadherence to antiepileptic drug therapy in children with newly diagnosed epilepsy. *JAMA* 305 (16): 1669–1676.
28 Manjunath, R., Paradis, P.E., Parisé, H. et al. (2012). Burden of uncontrolled epilepsy in patients requiring an emergency room visit or hospitalization. *Neurology* 79 (18): 1908–1916.

29 U.S. CDC (2018). Rechecking moments ago. https://www.cdc.gov/epilepsy/basics/fast-facts.htm (accessed 13 July 2018).

30 U.S. FDA (1999). Guidance for industry (draft): applications covered by Section 505(b)(2). https://www.fda.gov/ucm/groups/fdagov-public/@fdagov-drugs-gen/documents/document/ucm079345.pdf (accessed 9 October 2017).

31 U.S. FDA (2017). Guidance for industry (draft): determining whether to submit and ANDA or a 505(b)(2) application. https://www.fda.gov/ucm/groups/fdagov-public/@fdagov-drugs-gen/documents/document/ucm579751.pdf (accessed 13 October 2017).

32 U.S. FDA (2009). Guidance for industry: formal meetings between the FDA and sponsors or applicants, rev 1. https://www.fda.gov/downloads/drugs/guidances/ucm153222.pdf (accessed 14 October 2017).

33 U.S. FDA (2007). Guidance for industry and review staff (draft): target product profile – a strategic development process tool. https://www.fda.gov/downloads/drugs/guidancecomplianceregulatoryinformation/guidances/ucm080593.pdf (accessed 21 October 2017).

34 Approved Labeling for KEPPRA Tablet, NDA 021035 (2017). https://www.accessdata.fda.gov/drugsatfda_docs/label/2017/021035s100,021505s040lbl.pdf (accessed 3 July 2018).

35 Boudriau, S., Hanzel, C., Massicotte, J. et al. (2016). Randomized comparative bioavailability of a novel three-dimensional printed fast-melt formulation of levetiracetam following the administration of a single 1000-mg dose to healthy human volunteers under fasting and fed conditions. *Drugs in R&D* 16 (2): 229–238.

36 U.S. FDA (2014). Guidance for industry (draft): bioavailability and bioequivalence studies submitted in NDAs or INDs–general considerations. https://www.fda.gov/ucm/groups/fdagov-public/@fdagov-drugs-gen/documents/document/ucm389370.pdf (accessed 27 November 2017).

4

Manufacturing of Biomaterials via a 3D Printing Platform

Patrick Thayer, Hector Martinez, and Erik Gatenholm

CELLINK LLC, Cambridge, MA, USA
CELLINK LLC, Blacksburg, VA, USA

4.1 Additive Manufacturing and Bioprinting

In recent years, advances in additive manufacturing, also known as three-dimensional (3D) printing, have attracted great attention as a "new industrial revolution" [1]. The 3D printing field as a whole is projected to make an impact in nearly every industry [2] because of both the flexibility of the technology and the continual decrease in the costs. The medical industry is not immune from this disruption as 3D printing will advance the field as a whole in new directions. At the moment, 3D printing has found a niche application as a critical tool in surgical preplanning [3]. In this application, a digital model of a region of interest can be generated through CT or MRI scans that can be used as a basis for a physical model fabricated through 3D printing [4, 5]. Other applications include surgical guides [6, 7] and fabrication of some implants [8, 9]. These implants are primarily structural in native, serving more of a structural or a space-filling role to facilitate the growth of a new tissue rather as a tissue engineering implant. Often, the materials that make up these implants are synthetic, such as titanium [8, 9] or thermoplastics [10]. Although 3D printing shows the potential in the fabrication of patient-specific implants, these traditional 3D printing systems are not compatible with biological materials that would be utilized to fabricate truly regenerative and cellularized implants. The convergence of 3D printing and tissue engineering can be thought as a new field known as 3D bioprinting. 3D bioprinting will enable the manufacturing of biological constructs that can not only serve as regenerative implants in the clinics but also as models of organ systems for applications such as drug screening or a mechanism to study cell behavior in three dimensions for research and cell expansion rather than the currently utilized 2D cell culture techniques.

The term "bioprinting" was popularized by Dr. Thomas Boland's group [11, 12] in 2004 with the development of early versions of bioprinter systems. Since the early 2010s, the field has experienced a renaissance driven by a wider adaption of bioprinting technology by researchers. From this growth, several types of bioprinters that utilize several different bioprinting modalities have

3D and 4D Printing in Biomedical Applications: Process Engineering and Additive Manufacturing,
First Edition. Edited by Mohammed Maniruzzaman.

become commercially available. These include laser-assisted bioprinting [13, 14], stereolithography bioprinting [15, 16], inkjet bioprinting [17, 18], and extrusion bioprinting [19]. Extrusion bioprinting is by far the most popular technology and its use has increased in recent years because of the decreased cost of entry, ease of use of the systems, the technology's broad application [19], and diversity of printable materials [20].

Biological tissues are composed of three components, the cells, the surrounding extracellular matrix (ECM), and the intracellular fluid [21], that provide a favorable environment suitable for cell proliferation and maintenance of phenotype. The recapitulation of the characteristics of the native ECM must be considered when developing biomaterials. However, the arrangement of these biomaterials within a construct to recapitulate the native tissue architecture is critical for the functionality of the resulting construct. Bioprinting can be utilized to precisely position these biomaterials and therefore cells within the construct. These printable biomaterials can be considered a specialized subclass known as bioinks [22]. Bioinks have two key characteristics that distinguish themselves from traditional 3D printing materials. First, bioinks must support cell viability during and after the printing process [23, 24]. Cellular compatibility greatly dictates the types of materials that can be considered bioinks. For example, the need for preservation of cell viability eliminates the use of thermoplastics such as polycaprolactone (PCL), poly(lactic-*co*-glycolic acid) (PLGA), and polyurethanes as bioinks, relegating their role to support the structures because of the need for significant heating for printing. However, although thermoplastics are not suitable as bioinks, these materials can be utilized for the fabrication of more traditional biomaterials via 3D printing or incorporated as support structures during the printing process for primary application in load-bearing tissues such as bone and cartilage.

Second, bioinks must retain their structural fidelity [25–27] both after printing and before cross-linking or self-assembly. Modulating the biomaterial composition to impart this shape fidelity while retaining the original characteristics of the base biomaterial can be a challenge [28]. Most hydrogel biomaterials were originally developed to be pipetted into a well plate or a petri dish and therefore often exhibit non-crosslinked mechanical strength similar to water or syrup. The resulting gel is mechanically softer because of its low network density. A common method of improving printability is the blending of these base biomaterials with thickening agents [29] or increasing their viscosity by adjusting their concentration [30, 31], average molecular weight, and extent of entanglement or self-assembly [23, 32]. Although changing the intrinsic properties of the biomaterial is one strategy for improving printability, the other option is the modulation of its external environment. Many mammalian protein and synthetic polymer-based hydrogels possess thermosensitivity [33, 34]. Intrinsic properties such as viscosity, extent of chain entanglement, extent of cross-linking [35, 36], self-assembly [37], and the extent of shear-thinning behavior can also change depending on the temperature [38]. To take advantage of these dynamic properties, the temperature of these bioinks should be controlled by the bioprinting systems to control bioink printability.

4.2 Bioinks

As discussed earlier, bioinks are a distinct subclass of biomaterials that have been developed to be both extrudable by a 3D bioprinter system while supporting cell proliferation and differentiation. The key challenge with the development of bioink is the balance between imparting printability on a traditional biomaterial while minimizing the impact to biological activity [28]. A biomaterial that possesses shear-thinning properties is a strong indicator that the material can be utilized as a bioink. Shear thinning allows a bioink to flow through the nozzle under an applied force but does not flow or collapse once the force is absent, allowing the bioink to maintain its shape [39]. A common approach for improving printability includes the blending of a low viscosity base biomaterial with a thickening agent that can be materials such as nanofibrillar cellulose [29, 40] or hyaluronic acid methacrylate (HAMA) [34] to impart shear-thinning characteristics. Other approaches include increasing the concentration of the base biomaterial; however, as the concentration increases, this can have an adverse effect on cell behavior because of the increased matrix network density [30, 41, 42]. Modulation of the thermal characteristics of the bioink can also be utilized to improve printability. Many bioinks such as those based on materials such as gelatin, collagen [43, 44], and decellularized ECM [45, 46] are sensitive to temperature and self-assemble or entangle under different thermal conditions.

4.2.1 Printability Control – Bioink Composition and Environmental Factors

Several factors and parameters can influence the resulting characteristics and properties of a bioprinted structure (Figure 4.1). These characteristics and properties of the resulting bioprinted structure can be quantified and utilized to evaluate the performance of the bioink within a 3D bioprinter system. These characteristics and properties as a whole can evaluate a bioink and be utilized to convey and compare the "printability" of the different bioinks [25]. Printability encompasses several measurable metrics that when put together can be used to evaluate if a given bioink can consistently and predictability form filaments during the printing process. Bioink printability can be influenced by many factors and parameters, both intrinsic and extrinsic, as detailed below.

Intrinsic factors that can influence printability include but are not limited to the following:

- Average molecular weight of the protein or polymer chains within the bioink
- Concentration of the bioink
- Strength of interactions and entanglements between the protein and polymer chains
- Presence of a thickening agent or other shear-thinning material
- Ability to self-assemble
- Ability to be cross-linked

Extrinsic factors that can influence printability include but are not limited to the following:

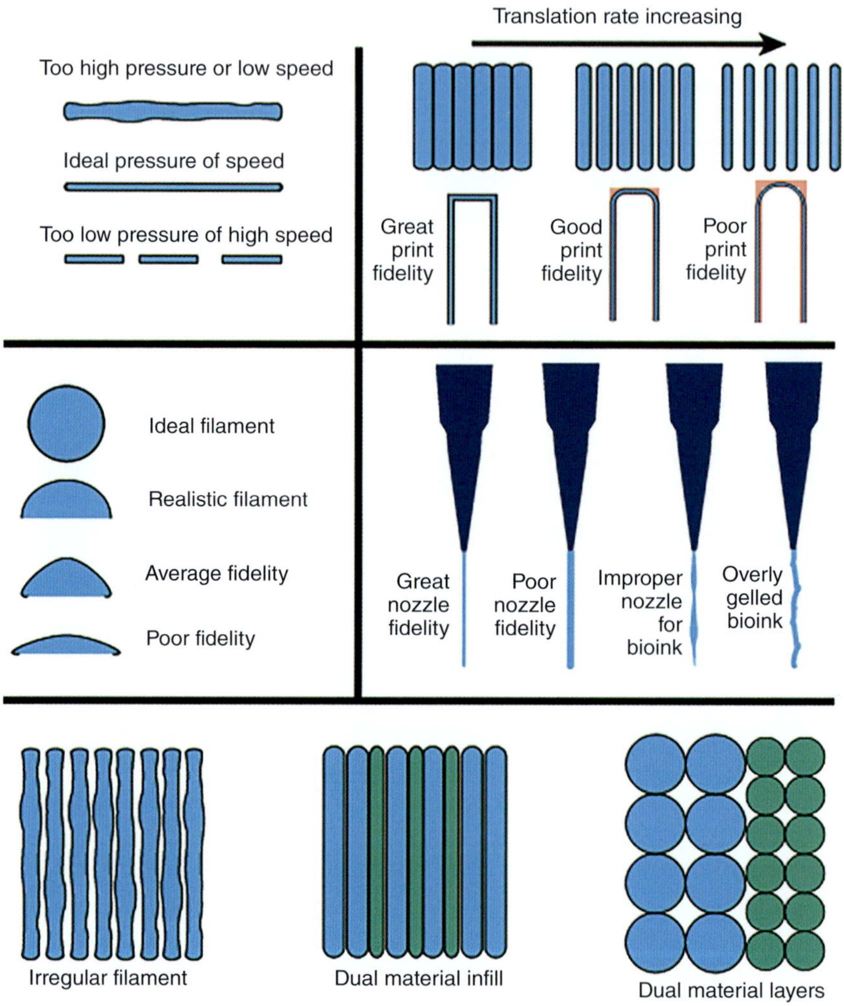

Figure 4.1 *Printability*. Several factors can influence the characteristics of a printed filament. These characteristics include the filament diameter, uniformity, overall shape, and behavior at directional changes and in proximity to other filaments. Translation rate and printing pressure are two factors that can have an effect on the resulting filament properties. For example, a too slow translation rate or high pressure can result in a nonuniform filament that is wider than anticipated. In contrast, a fast translation rate and low pressure can result in a noncontinuous filament. In particular, manipulation of the translation rate can be utilized to control filament diameter while keeping the pressure constant. However, at faster translation rates, the fidelity of the printed structure to printing commands can suffer. The physical and chemical characteristics of the bioinks can also have an effect on printability. Although ideally a deposited filament would have a cylindrical shape, in more realistic cases, a deposited filament will form a hemispherical shape on the surface. Bioinks with too low viscosity exhibit poor fidelity and are unstable. Additionally, the shape of the bioink filament immediately after extrusion through a nozzle can exhibit good or poor nozzle fidelity and nonuniformity in shape because of an improper nozzle or overly gelled or nonhomogeneous bioink. Understanding printability is important for minimizing the deposition of irregular filaments that can result in poor printed structures. It is a paramount that the printability of a bioink is well characterized to allow the fabrication of structures comprising of multiple bioinks that have distinct printing characteristics.

- Temperature of the bioink in the cartridge
- Temperature of the nozzle
- Nozzle diameter
- Nozzle shape such as tapered vs straight
- Temperature of the printbed
- Hydrophobicity of the printing surface
- Translation speed of the printhead
- Applied pressure or displacement on the bioink by the bioprinter system
- Layer height

The many factors and parameters that influence bioink printability necessitate that bioprinter systems have the ability to precisely control the environment before and after printing. Additionally, precise control over the mechanics of the system such as printhead positioning, motion, and nozzle properties is necessary to control and hone filament characteristics. These bioprinter systems should also exploit bioink intrinsic properties such as thermosensitivity and photo-crosslinkability to enhance the formation of and post-deposition maintenance of filament stability and stability during cell culture or implantation.

4.2.2 Mechanisms for Filament Formation and Stability

Bioprinter systems can apply external stimuli to modulate chain entanglement of polymer networks or to induce their photochemical cross-linking to impart stability (Figure 4.2). Bioinks can also consist of components that can be cross-linked through the addition of a binding solution that can catalyze cross-linking chemically or enzymatically. Classic examples of this mechanism are alginate-based biomaterials [47] and fibrin-based biomaterials [48], respectively. These materials possess low viscosities and stiffness that result in a poor stability of the filament after deposition but before cross-linking. This necessitates that cross-linking be initiated during the bioprinting process or immediately after to result in a stable construct. A bioprinter system can apply the cross-linking solution layer-by-layer or at other designated intervals to initiate bioink cross-linking. Alternatively, the bioprinter system and printbed could be modulated to bioprinting directly into the cross-linking solution [42].

Another mechanism for controlling printability and stability of bioinks is through the modulation of chain entanglements [49] and dynamics of network self-assembly. The dynamics of entanglements and network self-assembly is often influenced by the thermal environment of the bioink pre-printing and post-printing. Generally, physical entanglements can be reversible; however, self-assembly can be more difficult to reverse and may require the addition of agents that change pH and other environmental factors. For example, heating gelatin methacrylate above its gelatin point to disrupt the chain entanglements but still within a region of partial entanglements is necessary to ensure printability [34, 49]. Conversely, the cooling of a collagen-based bioink is necessary to delay irreversible self-assembly until deposition is performed onto a warmed printbed. The relationship between temperature and chain entanglement and reversible assembly is the basis for many sacrificial materials such as gelatin [50]

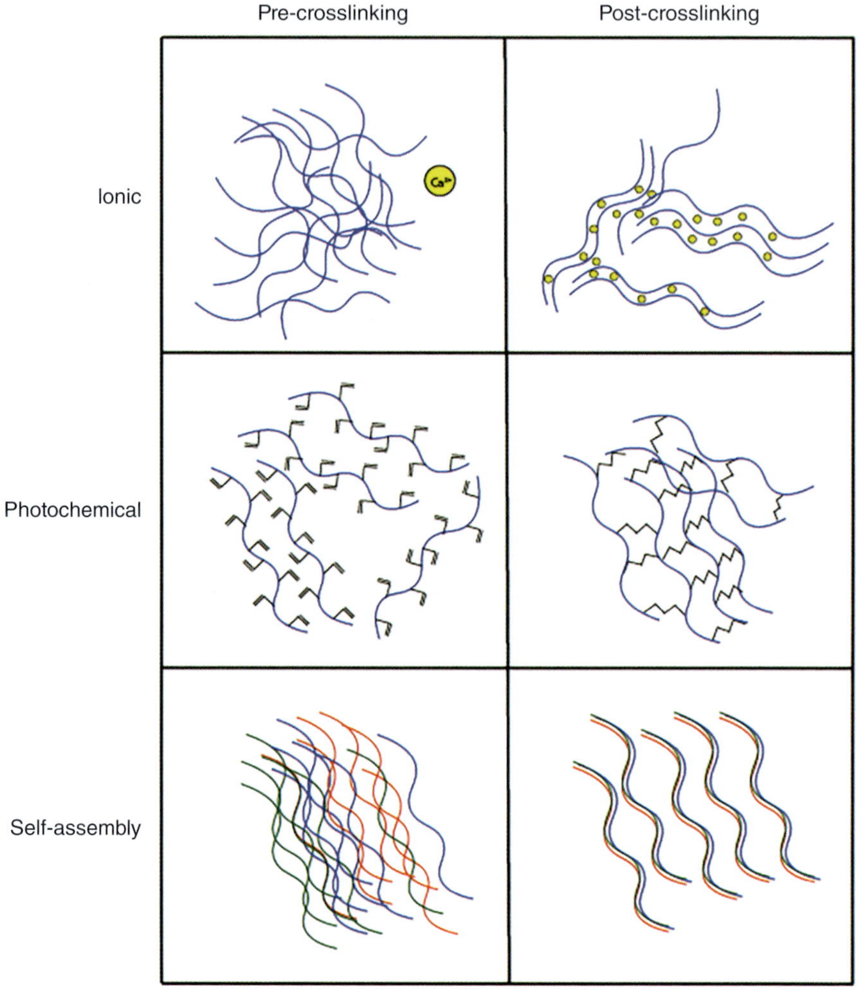

Figure 4.2 *Bioink cross-linking*. Bioinks can be cross-linked through several different modalities, including ionically, photochemically, and through intrinsic properties such as self-assembly. Cross-linking is necessary to impart stability on a deposited filament so that complex structures can be bioprinted and remain stable during culture and tissue development.

or Pluronics F-127 [51]. For example, a pluronics-based bioink can be bioprinted while heated to form a stable filament, and once deposited, the temperature of the structure can be lowered to cause the pluronics to liquefy and be washed away to result in channels. Conversely, these entangled or assembled network chains can be cross-linked through the activation of a photoinitiator [30, 52] to generate more stable structures that will not dissolve away. For some materials, such as thermoplastics including PCL [53], PLGA [54], and polyurethanes [55], it is necessary to expose the material to high heat to melt the polymer and allow their extrusion. The diversity of printable materials requires a bioprinter system to be flexible in its ability to control the environment and temperature of the bioinks and other materials that are utilized.

4.3 3D Bioprinting Systems

As discussed in previous sections, the materials that can be considered bioinks are extremely diverse. Therefore, bioprinter systems must have a wide range of capabilities to facilitate the utilization of many types of bioinks. However, no two bioinks will print alike. Several bioinks may require special printing conditions and protocols. There are several methodologies (Figure 4.3) that have been developed as "bottom-up" approaches for the bioprinting of scaffolds, including laser-assisted bioprinting [13, 14], extrusion bioprinting [19], and droplet-based approaches such as inkjet bioprinting [17, 18]. Extrusion-based techniques such as pneumatically or mechanically driven filament formation are suitable for the building up of large constructs. Similarly, droplet-based techniques such as inkjet and microsphere deposition can be utilized to build constructs without a fibrillar microarchitecture [56]. Despite the precision allowed by techniques

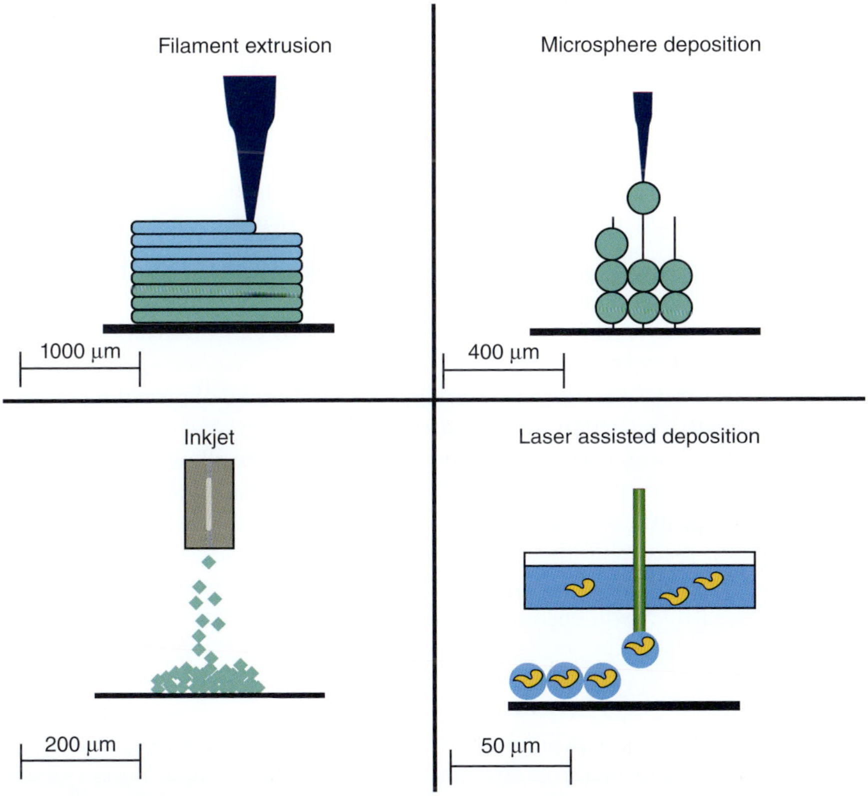

Figure 4.3 *Bioprinting modalities.* The several prevalent additive bioprinting modalities include filament extrusion-based techniques, droplet deposition-based techniques such as microsphere deposition and inkjet printing, and laser-assisted bioprinting.

such as laser-assisted deposition or laser-induced forward transfer [57, 58], the single-cell deposition enabled by the technique is not suitable for the generation of thick tissues.

4.3.1 Multifaceted Systems

Bioprinter systems that possess multifaceted printing modalities are necessary for the fabrication of complex structures, tissue constructs, and model tissues. Key characteristics and features of these multifaceted bioprinting systems are as follows:

- Multiple or interchangeable printheads that perform a wide range of printing modalities
- Capacity to cool bioinks down to 4 °C and heating thermoplastics up to 250 °C
- Capacity to cool the printbed surface down to 4 °C and heat the printbed surface up to 80 °C
- Ability to accommodate a wide range of print surface materials such as plastic, glass, and hydrogel and structures such as well plates, slides, dishes, and bioreactors
- Ability to apply diverse cross-linking modalities, whether ionic, light-based, or thermal to stabilize deposited structures
- Ability to maintain a clean chamber or sterile environment to allow the printing of cells and eliminate particulates in the final structure
- Ability to measure the characteristics of the resulting structure and evaluate the data for post-fabrication quality control
- Ability to adjust the printing parameters and the printing instructions based on monitoring of the ongoing print to improve the print.

Advanced bioprinting systems have been introduced to market as the technology continues to advance and its adaption by researchers continues to grow. The next sections will discuss the BIO X system developed by CELLINK, which fulfills many of these previously outlined requirements.

The BIO X system (Figure 4.4) was developed with the goal of providing unparalleled flexibility to the user. In contrast to systems that consist of a single printing modality, the BIO X was designed to be modular in nature to allow the use of multiple printing modalities. To accomplish this, the BIO X system is equipped with interchangeable printheads that allow the user to customize the bioprinter system toward their research rather than being restricted by a preset configuration. This allows the use of a wide range of bioinks. As the field continues to mature, systems like the BIO X will become ubiquitous in the field.

4.3.2 Major Components

The ability to configure the printer with different printing modalities (Figure 4.5) is necessary for the fabrication of complex structures. The modular design allows a wide range of printheads to be compatible with the system and allows yet-to-be designed printing modalities to be utilized in the system as they are developed. Additionally, the modular printheads can be designed to perform functions such

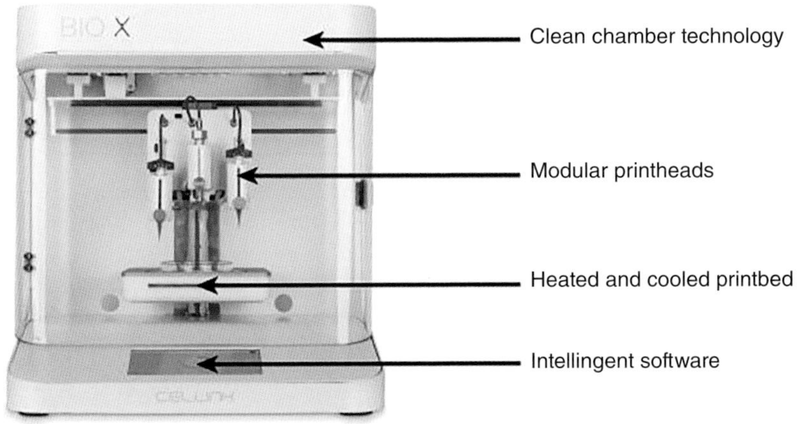

Figure 4.4 *Bio X diagram.* The Bio X bioprinter system contains several advanced features that facilitate the fabrication of multifaceted structures. Modular printheads allow flexibility in the types of bioinks that can be printed and the use of more precise printing parameters that allows better control over printability. The incorporation of printbed that can be either heated or cooled allows optimization of the post-deposition environment to better maintain filament fidelity or facilitating stabilization through cross-linking. The incorporation of dual-layer HEPA filters in conjugation with the closed front allows for the preservation of sterility and the maintenance of a clean chamber environment for bioprinting of cells on the benchtop or ensuring the minimization of particulates into the bioprinted construct. Finally, intelligent software that can adjust printing parameters as fabrication is occurring to result in better optimized constructs is necessary for quality control and better reproducibility between prints.

as cross-linking, imaging, or measurement of the printed structure. This ability allows more sensors and other devices to be incorporated in the printer to give it functionality beyond bioprinting.

4.3.3 Pneumatic Printhead

A pneumatic printhead can dispense a bioink from a cartridge utilizing pressurized air. Increasing or decreasing the applied pressure controls the dispensing rate. Precise control over the applied pressure is important for the maintenance of cell viability during the printing process. Too high of an applied pressure (in combination with a small nozzle diameter) can result in cell rupture or damage [59–61]. As discussed previously, the diameter of a deposited filament is a function of the nozzle diameter, translation speed, layer height, and bioink flow rate [62]. The relationship between the applied pressure and the flow rate is variable depending on the bioink chemistry [63] and its thermal environment [64]. Thorough characterization of the bioinks is necessary to calibrate the bioinks and the applied parameters for use with a pneumatic printhead.

4.3.4 Mechanical Displacement Printhead

Mechanical displacement can be considered a contrary bioprinting modality to pneumatic-driven dispensing. A mechanical displacement printhead functions

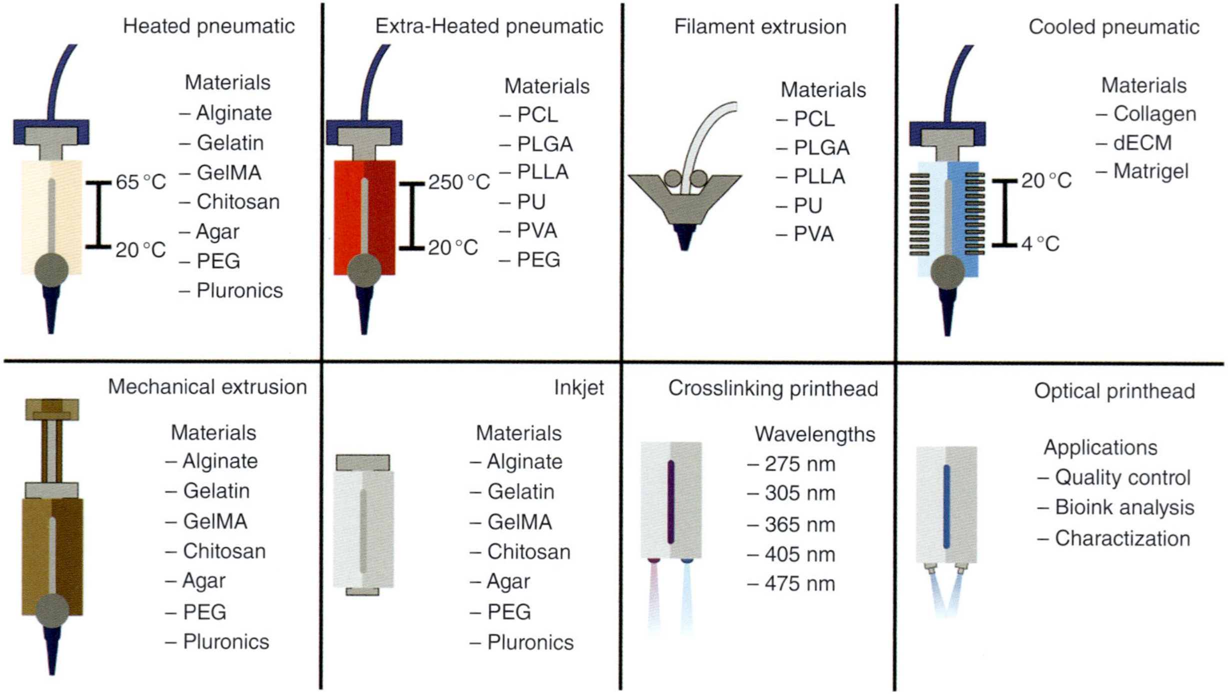

Figure 4.5 *Modular printheads*. Modular printheads allow the use of multiple printing techniques and diverse bioinks within a singular construct. Each printhead has advantages and disadvantages in its use to fabricate a construct. Heated pneumatic printheads allow the printing of thermosensitive bioinks such as gelatin, GelMA, agar, and pluronics and also allow the maintenance of bioink temperature at physiological levels to improve cell viability during printing. A printhead that can heat up to high temperatures such as 250 °C allows the printing of thermoplastics such as biocompatible materials such as PCL, PLGA, and PVA. Alternatively, one could utilize a printhead such as a heated filament extrusion head that mimics traditional 3D printing. A cooled pneumatic printhead allows the deposition of thermosensitive bioinks such as collagens and decellularized ECMs such as Matrigel that self-assemble at room temperature but can be printed at a low temperature. A mechanical extrusion-based printhead can be utilized to deposit very small volumes such as those necessary for droplets or fluid dispensing. An inkjet printhead may be utilized for the deposition of thin layers of bioinks. Finally, other printheads could be developed that do not print materials but aid in cross-linking of other materials or facilitate analysis of printed structures such as through optical imaging.

essentially as a syringe pump and therefore offers precise control over the flow rate of the bioink and therefore the total volume dispensed. This has several advantages compared to pneumatic dispensing such as more predictable filament diameter when coordinated with translation speed and correct nozzle characteristics, applicability of greater forces for the extrusion of highly viscous materials, and the ability to deposit smaller volumes. The deposition of small volumes facilitates the ability to place cells in precise patterns and arrangements, fabricate microspheres, and encapsulate cells in small matrix volumes. However, the mechanical displacement mechanism can inadvertently result in unintended high pressures on cells, which can adversely affect cell viability.

4.3.5 Inkjet Printhead

Inkjet bioprinting can utilize thermal [65] or piezoelectric [66] techniques to expel droplets and pattern them on a surface. Thermal inkjet printing has proven compatible with cells because of the minimization of heating and vibrations that may damage or kill cells [65, 66]. However, when compared to extrusion-based techniques, an inkjet bioprinting approach may result in a slower buildup of the construct and a loss of continuous internal organization as filaments are replaced by droplet agglomeration. However, the trade-off to this slow accumulation is the high spatial precision of the technique. Droplets are expelled from the inkjet nozzle with volumes between 10 and 100 pl, allowing the precise deposition of both cells and ECM [11, 18]. The combination of an inkjet printhead with other bioprinting techniques such as extrusion-based systems would facilitate the generation of hybrid constructs. For example, an extrusion-based technique may be useful for fabricating a construct with isotropic organization, whereas the inkjet technique can be utilized to deposit regions of specialized cells or coatings.

4.3.6 Heated and Cooled Printheads

As discussed in previous sections, many bioinks exhibit thermosensitivity in their viscosity, rate of cross-linking, or assembly, and therefore, their thermal environment must be precisely regulated. It is necessary that the ability to control thermal regulation of the bioinks be integrated within the printheads. Achieving desired bioprinted structure characteristics may require the maintenance of the bioink within the temperature range that is conducive for predictable filament formation. This is particularly important for pneumatic-driven extrusion where an applied air pressure drives the bioink flow rate rather than mechanical displacement. Improper temperature control can result in a different extrusion rate than expected because of changes in viscosity. These changes in viscosity may be caused by changes in the extent of chain entanglement or the extent of material self-assembly caused by the temperature. The BIO X system was developed to be compatible with printheads with either preset temperatures or controlled heating up to 250 °C or cooling down to 4 °C, for use with a wide range of bioinks. Many bioinks have a region where temperature plays a critical role in their properties such as gelatin-based materials or decellularized ECMs. Furthermore, some bioinks such as collagen-based materials or bioinks that contain growth factors can be particularly sensitive to temperatures, and a temperature fluctuation beyond physiological temperatures may lead to denaturation and comprise stability. The thermal control must extend down to the nozzle to minimize the temperature gradients that will form in the nozzle during the extrusion pressure. Steep temperature gradients may cause the nozzle clog because of the gelation or self-assembly of the bioink when the flow is slowed or stopped. This may occur during switching of printheads, during cross-linking, or pauses in printing where flow is interrupted.

4.3.7 High-Temperature Extruder

Biodegradable polyesters and other thermoplastics are widely utilized in the fabrication of medical devices and implants. These materials such as PCL, PLGA, and polylactic acid are favorable because of their biological inertness, degradability, ability to be customized, and mechanical strength. Although not traditionally considered a bioink because of non-cell-friendly printing conditions, these thermoplastics serve a critical role in many printed constructs. For example, deposited thermoplastics can act as rigid supports for softer bioprinted materials such as alginates or gelatins to extend use within load-bearing constructs. Additionally, these thermoplastics can form the basis of drug capsules or drug delivery systems. To enable the utilization of these materials in the BIO X bioprinter system, high-temperature extruder (HTE) printheads have been developed to apply sufficient temperature (up to 250 °C) to melt these materials and enable their extrusion. The melting of these materials is necessary to avoid the use of cytotoxic and volatile organic solvents such as dichloromethane or trifluoroethanol to solubilize the thermoplastic for printing.

4.3.8 Multimaterial Printhead

All the printhead technologies discussed in prior sections involve the bioprinting of a singular bioink. However, homogeneous filaments may not be ideal when fabricating a tissue construct. The native tissues exhibit heterogeneous microarchitecture as demonstrated by distinct regions of matrix chemistry and organization [67] that impart and facilitate distinct functionalities [68]. Gradual or sharp gradients in matrix composition, cell type, and organization can be found between these regions. Recapitulation of these regions is necessary for the successful fabrication of a bioprinted construct. One approach to recapitulate these native microarchitectures is through the deposition of multifaceted filaments or structures that possess changing chemical and mechanical properties along the filament. These gradient building blocks can be assembled and arranged into broader macrostructures through infill patterning and layering. Considerations must be made to facilitate the integration of these distinct regions to maintain stability and mechanical rigidity. For example, a blood vessel can be thought as a heterogeneous tissue that consists of three distinct layers with distinct matrix and cellular compositions [69, 70]. It is necessary that all the subsequent layers are mechanically integrated to withstand the stresses they may experience.

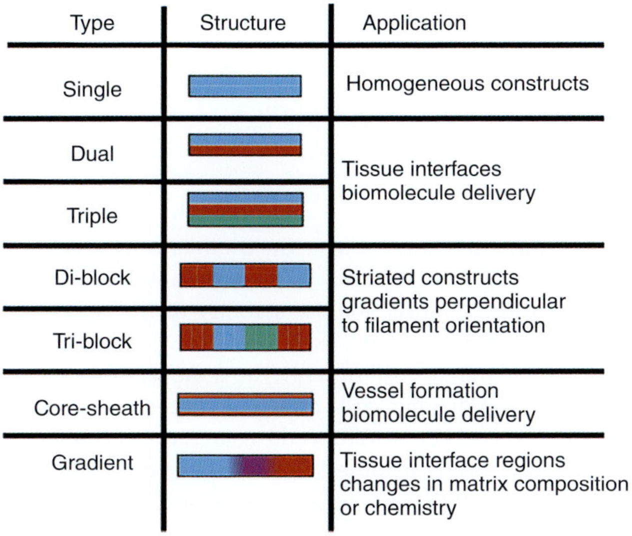

Figure 4.6 *Multimaterial filaments*. Specially designed printheads and nozzles permit the formation of filaments that can contain several types of bioinks in specific arrangements. These specialized nozzles can generate filaments more complicated than the single-material filaments, with structures such as dual or triple material, block structure arrangements, core sheath, or gradient arrangements. These multimaterial filaments can enhance constructs or enable gradients that can better recapitulate native tissue architecture.

Multifaceted filaments can be generated through printheads that layer bioinks together within the nozzle via laminar flow during extrusion [71]. This is a distinct departure from multiple printhead bioprinting in which several printheads are utilized to generate structurally heterogeneous materials through deposition of homogeneous filaments in distinct regions [72]. These multimaterial printheads can exhibit several conformations through distinct material feeding and nozzle structures to generate these multifaceted bioink filaments (Figure 4.6).

- A printhead that blends two bioinks together within the nozzle during extrusion through independent control over the flow rate of each component [73]. This could be used to generate gradients in matrix composition along the length of a single filament.
- A printhead that generates coaxially structured filaments consisting of two distinct materials as the core and sheath [74–76]. This could be used to generate filaments that can deliver growth factors or other molecules from the core. Additionally, through the use of a sacrificial material as the core, generate hollow tubes that could be used as blood vessels or other conduits.
- A printhead that generates dual-material filaments such as a filament with hemispheric differences in matrix compositions through layering of bioinks using laminar flow [71, 72]. This could be utilized to generate filaments that bridge two distinct regions in a construct or for barrier layers where side-specific cell behavior is desired.

- A printhead that provides stability mechanisms during the printing process such as the application of a binding solution or providing an ultraviolet (UV) source for cross-linking. This may be necessary for the deposition of bioinks that have a low viscosity and would otherwise not print [77].

4.3.9 Heated and Cooled Printbed

As discussed previously, environmental control over the bioinks is necessary to better control the characteristics of the resulting extruded filament. Thermal control over the printbed can directly influence the stability of the deposited filament by strengthening chain entanglement [78] or catalyzing biomolecule self-assembly [79] to stabilize the filament. Additionally, the removal of sacrificial components such as gelatin [50] or pluronics [80] can be facilitated by temperature changes. After a bioprinted construct is stabilized through cross-linking or self-assembly [79], the temperature of the printbed can be changed to cause the sacrificial components of the construct to become solubilized and removed.

4.3.10 Clean Chamber Technology

A printing environment that is free from particulates, bacteria, viruses, and fungal spores is necessary for the fabrication of constructs suitable for implantation *in vivo* in an animal model and ultimate clinical application. Although steps can be taken during the preparation of bioinks and the cells to ensure an elimination of these impurities, it is necessary to supplement this with a clean printing environment. However, additional engineering controls are necessary to ensure a clean environment for bioprinting. Although the quickest solution would be to place the bioprinter in a clean room or a biosafety cabinet, this approach is not ideal as it restricts the access to and the use of the bioprinters system. The integration of engineering controls into the bioprinter system may be a more favorable solution. Systems such as the BIO X incorporate several engineering controls. These include the following:

- Doors that enclose the chamber during the printing and sterilizing process to both protect the user and the construct.
- Integrated UV sources to sterilize the chamber before bioprinting.
- Incorporation of serial filtration units that can trap both large particulates and microscopic agents.
- Surfaces that can be cleaned with standard cleaning reagents.

4.3.11 Video-Capture Printhead and Sensors

A self-calibrating and autonomous bioprinter system necessitates the integration of video-capture devices along with sensors that can monitor and collect metrics of the bioink during the printing process and the resulting filament and construct characteristics. This is particularly important in bioink development where the reliable and precise monitoring of printability is necessary for the optimization of bioink composition [81]. One such approach for this integration is the development of specialized printheads that contain cameras for the visualization and capture of deposited filaments. Two types of camera printbeds have been developed and integrated within the BIO X system. The first type is a printhead containing a low-magnification camera, which is necessary for the visualization of the structure as a whole. The second type is a magnification printhead that would allow the visualization of the filament with high resolution. This high-resolution printhead can be utilized to analyze and quantify the specific details of a deposited filament such as diameter, uniformity, thickness, surface rough, and filament properties at or around changes in directions or intersections. This is necessary for validation of the applied printing parameters and to also rapidly screen and test bioink compositions. These data and images can be recorded for quality control of printed constructs but also be used to optimize further prints. This visual integration can be supplemented by data collected from strategic sensors placed throughout the bioprinter. Collected metrics such as temperature, flow rates, pressures, deposited mass, and volume can complement the visual data to provide the user a more complete picture of how the bioink is printing and the structure is being fabricated and maintain cell viability during the process [59]. This is necessary to ensure reproducibility in the bioprinted constructs between samples within a single study or across several studies. One can envision that a quality control system will play a critical role in the translation of bioprinting systems from the laboratory to clinical application.

4.3.12 Integrated Intelligence

Finally, the combination of user input and material validation in combination with the data collected by the bioprinter systems will allow the development of integrated intelligence. This integrated intelligence could automatically and rapidly correct and adjust the printing parameters without or with minimal user input. This is critical when managing the bioprinting of a large number of samples or when the desired filament diameter or other construct properties must be precise and whose visualization is beyond what the human eye can see. This integrated intelligence can be used in several ways to enhance the quality of printed constructs. It can provide feedback for better optimization of bioink composition. It can also be utilized to automatically adjust the G-code to change printhead translation speeds, temperatures, or pressures to improve the print and optimize the G-code for more rapid and optimally deposited filaments. Most importantly, it can recognize when a bioprinted construct is not within the acceptable error.

4.4 Applications

The previous two sections of this chapter discussed bioinks and their unique characteristics that necessitate environmental controls that are integrated into bioprinter systems to facilitate and modulate printability. This section will focus on what an advanced bioprinting system such as the BIO X can achieve with all these controls and features necessitated by the materials.

4.4.1 Internal Architecture

Bioprinting can permit precise control over the overall structure of a construct and its microarchitecture within those determined by infill patterns (Figure 4.7). In traditional 3D printing, the infill pattern geometry and density are directly correlated with the resulting mechanical characteristics [82–84]. In this case, the different infill patterns can be designed to enhance the mechanical characteristics

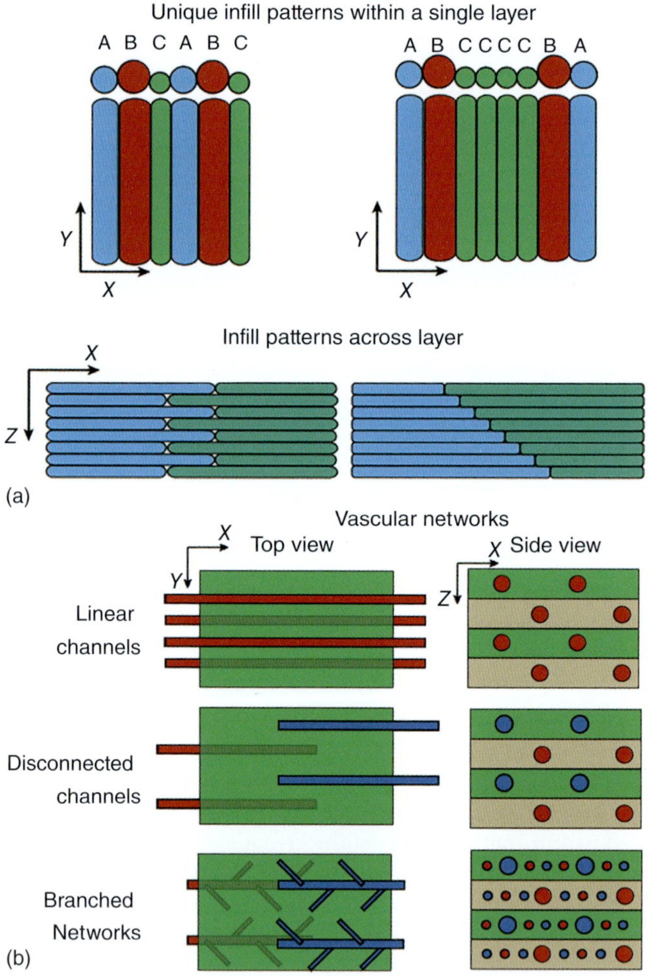

Figure 4.7 *Patterning (a) and vascular structures (b).* Bioprinting can be utilized to fabricate complicated structures that better mimic native tissue microarchitecture. Advanced bioprinter systems can fabricate constructs that can contain both patterned extracellular matrix and cells along with conduits that can guide the formation of a vascular network. The generation of these advanced microarchitectures within a bioprinted construct is necessary for the development of mature tissues. Additionally, the incorporation of sacrificial bioinks can be utilized to generate vascular networks that can either be used to perfuse the construct or act as a base for vasculogenesis or angiogenesis through continuous or noncontinuous channels and branched networks.

of the structure under different loads such as under compression, torsion, or isotropic strain [85, 86]. Additionally, the overall density of the infill pattern can be decreased to allow for the inclusion of void regions that function as interconnected pores that can allow better nutrient supply [87], absorb applied stresses [88], be filled with a secondary material such as a bioink containing cells [89, 90], or left empty to permit tissue infiltration after implantation [91]. The classic example is that the incorporation of various sized pores changes overall construct permeability and its resulting mechanical characteristics [92, 93]. However, the internal architecture can have a significant influence on construct mechanical strength and integrity as demonstrated in scaffolds fabricated from metals or thermoplastics [92, 93]. Applying these trends and observations to softer biological materials can be more difficult.

The ability to control the microarchitecture of a bioprinted construct has several advantages over traditional 3D cell culture techniques that are often limited by their homogeneous chemistry and structure. The most obvious advantage is that bioprinting could be utilized to enhance the resulting mechanical characteristics of a 3D construct. However, the microarchitecture that can be imparted from bioprinting is not as dramatic or long lasting as constructs that are composed of thermoplastics. For example, Zhang et al. demonstrated that while initial compressive modulus can be tailored through controlling the orientation and thickness of the deposited filaments, this result was not stable because of the cell-induced remodeling of the construct [94]. However, control over the internal microarchitecture may be more valuable in guiding cell behavior within these constructs [95]. In traditionally fabricated 3D cell culture constructs, such as a molded homogeneous hydrogel, the resulting construct microarchitecture is essentially nonorganized and relatively random. The network characteristics of the scaffold itself are dependent on the bioink utilized and the material's gelation or self-assembly mechanism. This network architecture can range from cross-linked biomacromolecules to self-assembled twisted chains to interconnected polymer networks comprising two distinct materials. If external stimuli are applied to the construct during gelatin such as mechanical strain [96], compression [97], or shear [98, 99], the resulting network characteristics and structure can be influenced because of induced network organization. There is growing evidence that the internal configuration of the bioprinted construct can have an effect on the incorporated cells [100]. For example, Costantini et al. recently demonstrated that the compartmentalization of the cells within a deposited filament guides their alignment and orientation, which resulted in the organization of the bioprinted construct to more closely mimic the microstructure of skeletal muscle [101]. Additionally, multilayering is necessary for the fabrication of multilayered constructs that contain gradients of several material types. A bioprinting system greatly simplifies the generation of these more complicated constructs.

Control over the internal microarchitecture of bioprinted constructs can be utilized for several applications, including:

- Spatial control of deposited materials through positioning such as multilayering [102] and patterning [51].

- Spatial control over the positioning [103] or layering of cell types [104] to generate tissue-mimetic patterns.
- Modulation of the construct infill characteristics to guide cell orientation, migration, and matrix organization.
- Deposition of dual-material filaments that contain guiding features to induce cell organization, directed migration, or differentiation.
- Deposition of regions composed of sacrificial or rapidly degradable materials to incorporate void regions or structures that can be utilized as conduits for cell infiltration, perfused to enhance cell viability, or for the generation of a vascular network [80].
- Fabrication of constructs derived from patient-specific cells, data, or scans such as tumor geometry [105], patient-specific tissue models [106], or defects [107].

The last two points will be expanded in the following sections as they utilize and build on many of the features outlined in the previous points.

4.4.2 Integrated Vascular Networks and Microstructure Patterning

Insufficient nutrient supply is a critical challenge in the development of engineered tissues. The recapitulation of dense vessel and capillary networks that are necessary to deliver these nutrients can be difficult to achieve [108, 109]. Although no fabrication technique at the moment can incorporate networks that mimic the complexity of an *in vivo* capillary bed at the scale necessary for tissue regeneration, they can begin to move toward that objective. Bioprinting can be used to template a construct with perfusable conduits that can enable nutrient delivery into the interior of the construct [51, 110, 111]. The rapid fabrication of conduit networks with tailorable width, length, and degree of branching allows one to experiment with different geometries and flow profiles. However, new vessel formation from these networks is necessary for the maturation of a bioprinted construct. Seeding of these networks with endothelial cells [112, 113] and application of relevant stimuli could guide new vessel formation and angiogenesis [114–116]. It may be necessary to pattern inductive bioinks to guide vessel outgrowth from these incorporated conduit networks [117]. For example, strategically depositing bioinks that contain releasable growth factors [118] may be used to form biochemical gradients that induce cell migration and tubule formation [119]. Alternatively, generation of regions that consist of distinct network densities or stiffness could guide cell infiltration and migration to generate this vascular network [120]. However, it has yet to be seen what is the ideal combination of prefabricated conduits, growth factors, and mechanical cues that will result in vascularization. Advanced bioprinting systems will enable researchers to begin to conduct these studies and better understand these systems and the role of stimulation.

Beyond the incorporation of conduit networks, advanced bioprinter systems can generate extensive gradients within a construct. The recapitulation of native tissue gradients in cell phenotype and ECM is a major challenge in the fabrication of implantable constructs [121–123]. Bioprinter systems offer a methodology to address this challenge by enabling the positioning of materials in precise locations to begin to assemble constructs consisting of multifaceted chemistries [119, 120, 124], mechanical characteristics, network architecture, and cell type [125, 126]. These gradients are critical in their functionality of many tissues within the body [121, 122]. Bioprinting can permit the positioning or patterning of cells to result in a template for tissue regeneration and growth [51]. This prepositioning to mimic the native tissue architecture could accelerate the regeneration of the tissue and its integrate with the body. Other critical structures and microarchitectures that may be bioprinted include nephron components such as kidney proximal tubule units [127, 128], liver lobes [46, 129, 130], bone-to-ligament transition regions [131], and lung alveoli mimetics such as the air–blood barrier [132]. These structures are of interest because their fabrication allows small-scale testing of larger organ building blocks. Although the combination of many of these smaller tissues could be an approach to regenerate an entire organ, they could also be used to begin to screen and test drug therapies and other treatments.

4.4.3 Personalized Medicine

In recent years, personalized medicine has become more prevalent as techniques for performing patient-specific tests have advanced [10, 133, 134]. Bioprinting will play a critical role in the advancement of personalized medicine by enabling the fabrication of patterned and organized constructs and prostheses [10]. Additionally, the ability to fabricate constructs that are comparable to patient tissues will allow more rapid and reliable testing of therapies. Through immediate applications such as organ-on-a-chip [135, 136], patterned constructs that contain conduits and multifaceted microarchitectures can begin to recapitulate the functionality of a native organ or tissue on a much smaller scale [127, 134, 135]. One can envision, with optimized inputs, that bioprinter systems can fabricate several dozens of miniature livers or kidneys [127] with the push of a button. These miniature livers and kidneys could contain patient-derived cells or matrix, essentially recreating the patients' native tissue in the lab. This would enable testing of different therapies or drugs, without risking the health of the patient through side effects such as drug toxicity [137]. Similarly, these bioprinted miniature organs could also contain cancerous regions that are derived from patient biopsy [138–140]. From these constructs, one could use chemotherapy or other cancer treatments to evaluate both the effect on the cancerous cells and on whether the therapy results in adverse side effects on the noncancerous cells.

Finally, personalized medicine can also include the fabrication of implantable tissue constructs that match patient-specific shape and size. Although this idea is not new, bioprinting allows the generation of a patterned microarchitecture and regenerative stimuli that enables a more rapid tissue generation and integration with the adjacent tissue [141–143]. Although the field of bioprinting has not reached maturation where the technique is clinically translatable, the potential impact can already be predicted. If one looks at the nascent implantation of traditional 3D printing in treatments, one can foresee where bioprinting may find an application. A review by Martelli et al. indicated that 71.5% of the use of traditional 3D printing in surgeries between 2005 and 2015 utilized the technique for surgical planning [144], whereas surgical guides (25.3%) and implants (9.5%) made up a substantially smaller application. It is noted that the use of the technique was dominated by skeletal applications such as a maxillofacial role in 50% of the cases and orthopedic role in 24.7% of the studies. However, when 3D printing was utilized to fabricate patient implants, the results were beneficial. In particular, 3D printing was invaluable in the surgical treatment of complicated deformities in areas such as the spine. Two cases presented by Mobbs et al. demonstrated how 3D printing was utilized for both the reproduction of the patient anatomy as a printed structure for surgical planning and in the fabrication of the prosthesis to treat the pathology [145]. In both these cases, 3D printing was critical in visualizing the complex patient pathology and providing a specific treatment for the ailment without the need to additional surgeries for the harvesting of autografts or a prolonged production time. Similarly, case studies presented by Liang et al. discussed the clinical outcome of the use of 3D printed pelvic endoprosthesis for reconstruction after resection of pelvic tumors [146]. 3D printing proved critical in the generation of endoprosthesis specific to the size and shape of the patient bone defect after resection and aided in osseointegration because of the incorporation of pores on the surface of the prostheses. However, these 3D printed prostheses were nonbioactive and fabricated from similar materials found in the existing prostheses. This is where bioprinting can make an impact in these applications and in other clinical treatments.

Recently, various research groups have incorporated bioactive components such as minerals [147–149] into these prostheses with the aim to improve osseointegration and therapeutic response. However, through bioprinting, implants can be fabricated that can be integrated into the body through tissue ingrowth and remodeled by the body. This integration can be achieved through the patterning of stimuli, matrix, and other cues provide a more favorable environment that the interface of the implant and the body for tissue ingrowth and integration. Ultimately, this can allow the fabrication of materials that integrate with and are replaced by native tissues rather than are tolerated. As discussed by Colasante et al., the marriage of 3D printing with tissue engineering promises to facilitate the engineering of functional constructs known as bioprostheses that incorporate biological cues into the structural construct to enhance clinical outcomes [150]. However, these bioprostheses at the moment are hybrid structures consisting of titanium or other orthopedic alloys and bioactive cues and stimuli that enhance the bioactivity and biocompatibility of the material. On the other extreme, bioprostheses fabricated completely out of natural materials have the potential to be completely integrated and replaced by native tissues.

Truly, 3D bioprinted constructs will result in a revolution in surgical treatment. Where damaged tissues may be excised and left to heal on its own, surgeons may infill the site with a construct that can initiate and support total healing (Figure 4.8). 3D bioprinted skin constructs will enable better repair of burn injuries. Bioprinted tendons and ligaments may augment healing of connective tissue injuries and laboratory-grown cartilage may reverse arthritis and eliminate

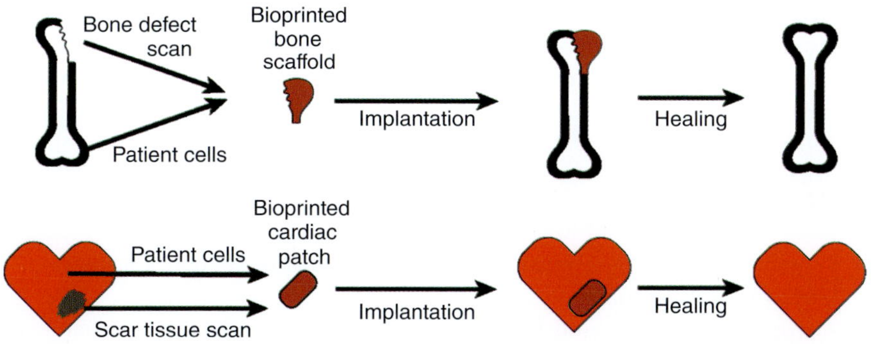

Figure 4.8 *Personalized medicine*. The maturation of the bioprinting field will permit the technology's use as a tool for the fabrication of constructs for personalized medicine applications. One such application may be the use of bioprinting to fabricate patient-specific constructs based on clinical data such as biologically active bone grafts or constructs that reverse scarring of cardiac tissue.

knee pain. The possibilities are endless and are enabled by the engineering of advanced bioprinter systems that can expatiate and enable the research.

4.5 Steps Necessary for Broader Application

A major challenge still exists in the scaling up of the bioprinting process to allow facility level control and management. Most bioprinter systems on the market must be controlled directly by the user. Often, laboratories have a single bioprinter system, utilized primarily for research and development. However, in the future, one can imagine a facility or research laboratory where dozens of bioprinters are controlled from a single console and are fabricating tissue models for various applications. Another group of bioprinters may be directed from the same console to fabricate customized pills and drug capsules for various patients. A third group of bioprinters may be located in the operating room in a hospital but be controlled by a group of tissue specialists in another room to fabricate various cartilage, bone, and muscle tissue grafts which are to be surgically implanted to heal a patient.

Several innovations must be developed to fulfill these goals.

- Systems must be integrated securely to the network to allow control of several printers as a unit or individually.
- An infrastructure must be developed so that the quality control of each individual bioprinter can be monitored to ensure the consistency between all bioprinters.
- Systems must be developed that fulfill Good Manufacturing Practice (GMP) and other guidelines to ensure clean manufacturing.
- Algorithms must be developed to convert a patient-specific scan to a printable construct with all the necessary internal features and tissue microstructures.
- Nondestructive methodologies to validate the internal structure of the bioprinted construct.

With the development of new innovations, the bioprinting field will continue to mature. Bioprinter systems will move from the research laboratory to being essential parts of clinics, pharmaceutical production, surgical suites, and implant fabrication.

References

1 Berman, B. (2012). 3-D printing: the new industrial revolution. *Business Horizons* 55: 155–162.
2 Kietzmann, J., Pitt, L., and Berthon, P. (2015). Disruptions, decisions, and destinations: enter the age of 3-D printing and additive manufacturing. *Business Horizons* 58: 209–215.
3 Rengier, F., Mehndiratta, A., von Tengg-Kobligk, H. et al. (2010). 3D printing based on imaging data: review of medical applications. *International Journal of Computer Assisted Radiology and Surgery* 5: 335–341.
4 Zein, N.N., Hanouneh, I.A., Bishop, P.D. et al. (2013). Three-dimensional print of a liver for preoperative planning in living donor liver transplantation. *Liver Transplantation* 19: 1304–1310.
5 Sodian, R., Weber, S., Markert, M. et al. (2008). Pediatric cardiac transplantation: three-dimensional printing of anatomic models for surgical planning of heart transplantation in patients with univentricular heart. *The Journal of Thoracic and Cardiovascular Surgery* 136: 1098–1099.
6 Sarment, D.P., Sukovic, P., and Clinthorne, N. (2003). Accuracy of implant placement with a stereolithographic surgical guide. *International Journal of Oral and Maxillofacial Implants* 18: 571–577.
7 Flügge, T.V., Nelson, K., Schmelzeisen, R., and Metzger, M.C. (2013). Three-dimensional plotting and printing of an implant drilling guide: simplifying guided implant surgery. *Journal of Oral and Maxillofacial Surgery* 71: 1340–1346.
8 Moradiellos, J., Amor, S., Córdoba, M. et al. (2017). Functional chest wall reconstruction with a biomechanical three-dimensionally printed implant. *The Annals of Thoracic Surgery* 103: e389–e391.
9 Park, E.-K., Lim, J.-Y., Yun, I.-S. et al. (2016). Cranioplasty enhanced by three-dimensional printing: custom-made three-dimensional-printed titanium implants for skull defects. *Journal of Craniofacial Surgery* 27: 943–949.
10 Morrison, R.J., Hollister, S.J., Niedner, M.F. et al. (2015). Mitigation of tracheobronchomalacia with 3D-printed personalized medical devices in pediatric patients. *Science Translational Medicine* 7: 285ra264.
11 Xu, T., Jin, J., Gregory, C. et al. (2005). Inkjet printing of viable mammalian cells. *Biomaterials* 26: 93–99.
12 Roth, E.A., Xu, T., Das, M. et al. (2004). Inkjet printing for high-throughput cell patterning. *Biomaterials* 25: 3707–3715.
13 Pagès, E., Rémy, M., Kériquel, V. et al. (2015). Creation of highly defined mesenchymal stem cell patterns in three dimensions by laser-assisted bioprinting. *Journal of Nanotechnology in Engineering and Medicine* 6: 021006.

14 Michael, S., Sorg, H., Peck, C.-T. et al. (2013). Tissue engineered skin substitutes created by laser-assisted bioprinting form skin-like structures in the dorsal skin fold chamber in mice. *PLoS ONE* 8: e57741.

15 Dhariwala, B., Hunt, E., and Boland, T. (2004). Rapid prototyping of tissue-engineering constructs, using photopolymerizable hydrogels and stereolithography. *Tissue Engineering* 10: 1316–1322.

16 Wang, Z., Abdulla, R., Parker, B. et al. (2015). A simple and high-resolution stereolithography-based 3D bioprinting system using visible light crosslinkable bioinks. *Biofabrication* 7: 045009.

17 Gao, G., Yonezawa, T., Hubbell, K. et al. (2015). Inkjet-bioprinted acrylated peptides and PEG hydrogel with human mesenchymal stem cells promote robust bone and cartilage formation with minimal printhead clogging. *Biotechnology Journal* 10: 1568–1577.

18 Xu, C., Chai, W., Huang, Y., and Markwald, R.R. (2012). Scaffold-free inkjet printing of three-dimensional zigzag cellular tubes. *Biotechnology and Bioengineering* 109: 3152–3160.

19 Ozbolat, I.T. and Hospodiuk, M. (2016). Current advances and future perspectives in extrusion-based bioprinting. *Biomaterials* 76: 321–343.

20 Chimene, D., Lennox, K.K., Kaunas, R.R., and Gaharwar, A.K. (2016). Advanced bioinks for 3D printing: a materials science perspective. *Annals of Biomedical Engineering* 44: 2090–2102.

21 Frantz, C., Stewart, K.M., and Weaver, V.M. (2010). The extracellular matrix at a glance. *Journal of Cell Science* 123: 4195–4200.

22 Hospodiuk, M., Dey, M., Sosnoski, D., and Ozbolat, I.T. (2017). The bioink: a comprehensive review on bioprintable materials. *Biotechnology Advances* 35 (2): 217–239.

23 Ouyang, L., Yao, R., Zhao, Y., and Sun, W. (2016). Effect of bioink properties on printability and cell viability for 3D bioplotting of embryonic stem cells. *Biofabrication* 8: 035020.

24 Zhao, Y., Li, Y., Mao, S. et al. (2015). The influence of printing parameters on cell survival rate and printability in microextrusion-based 3D cell printing technology. *Biofabrication* 7: 045002.

25 Kyle, S., Jessop, Z.M., Al-Sabah, A., and Whitaker, I.S. (2017). 'Printability' of candidate biomaterials for extrusion based 3D printing: state-of-the-art. *Advanced Healthcare Materials* 6: 1700264.

26 Ribeiro, A., Blokzijl, M.M., Levato, R. et al. (2017). Assessing bioink shape fidelity to aid material development in 3D bioprinting. *Biofabrication* 10 (1): 014102.

27 Rutz, A.L., Lewis, P.L., and Shah, R.N. (2017). Toward next-generation bioinks: tuning material properties pre- and post-printing to optimize cell viability. *MRS Bulletin* 42: 563–570.

28 Malda, J., Visser, J., Melchels, F.P. et al. (2013). 25th anniversary article: engineering hydrogels for biofabrication. *Advanced Materials* 25: 5011–5028.

29 Markstedt, K., Mantas, A., Tournier, I. et al. (2015). 3D bioprinting human chondrocytes with nanocellulose – alginate bioink for cartilage tissue engineering applications. *Biomacromolecules* 16: 1489–1496.

30 Nichol, J.W., Koshy, S.T., Bae, H. et al. (2010). Cell-laden microengineered gelatin methacrylate hydrogels. *Biomaterials* 31: 5536–5544.
31 Rezende, R.A., Bártolo, P.J., Mendes, A., and Filho, R.M. (2009). Rheological behavior of alginate solutions for biomanufacturing. *Journal of Applied Polymer Science* 113: 3866–3871.
32 Colosi, C., Shin, S.R., Manoharan, V. et al. (2016). Microfluidic bioprinting of heterogeneous 3D tissue constructs using low-viscosity bioink. *Advanced Materials* 28: 677–684.
33 Mouser, V.H., Abbadessa, A., Levato, R. et al. (2017). Development of a thermosensitive hama-containing bio-ink for the fabrication of composite cartilage repair constructs. *Biofabrication* 9: 015026.
34 Schuurman, W., Levett, P.A., Pot, M.W. et al. (2013). Gelatin-methacrylamide hydrogels as potential biomaterials for fabrication of tissue-engineered cartilage constructs. *Macromolecular Bioscience* 13: 551–561.
35 Melchels, F.P., Dhert, W.J., Hutmacher, D.W., and Malda, J. (2014). Development and characterisation of a new bioink for additive tissue manufacturing. *Journal of Materials Chemistry B* 2: 2282–2289.
36 Atabak Ghanizadeh, T., Miguel, A.H., Nicholas, R.L., and Wenmiao, S. (2015). Three-dimensional bioprinting of complex cell laden alginate hydrogel structures. *Biofabrication* 7: 045012.
37 Shafiee, A., McCune, M., Forgacs, G., and Kosztin, I. (2015). Post-deposition bioink self-assembly: a quantitative study. *Biofabrication* 7: 045005.
38 Yu, L. and Ding, J. (2008). Injectable hydrogels as unique biomedical materials. *Chemical Society Reviews* 37: 1473–1481.
39 Guvendiren, M., Lu, H.D., and Burdick, J.A. (2012). Shear-thinning hydrogels for biomedical applications. *Soft Matter* 8: 260–272.
40 Müller, M., Öztürk, E., Arlov, Ø. et al. (2017). Alginate sulfate – nanocellulose bioinks for cartilage bioprinting applications. *Annals of Biomedical Engineering* 45: 210–223.
41 Ouyang, L., Highley, C.B., Rodell, C.B. et al. (2016). 3D printing of shear-thinning hyaluronic acid hydrogels with secondary cross-linking. *ACS Biomaterials Science and Engineering* 2: 1743–1751.
42 Highley, C.B., Rodell, C.B., and Burdick, J.A. (2015). Direct 3D printing of shear-thinning hydrogels into self-healing hydrogels. *Advanced Materials* 27: 5075–5079.
43 Silver, F.H., Freeman, J.W., and Seehra, G.P. (2003). Collagen self-assembly and the development of tendon mechanical properties. *Journal of Biomechanics* 36: 1529–1553.
44 Leikin, S., Rau, D., and Parsegian, V. (1995). Temperature-favoured assembly of collagen is driven by hydrophilic not hydrophobic interactions. *Nature Structural & Molecular Biology* 2: 205–210.
45 Jang, J., Kim, T.G., Kim, B.S. et al. (2016). Tailoring mechanical properties of decellularized extracellular matrix bioink by vitamin B2-induced photo-crosslinking. *Acta Biomaterialia* 33: 88–95.
46 Pati, F., Jang, J., Ha, D.-H. et al. (2014). Printing three-dimensional tissue analogues with decellularized extracellular matrix bioink. *Nature Communications* 5: doi: https://doi.org/10.1038/ncomms4935.

47 Lee, K.Y., Rowley, J.A., Eiselt, P. et al. (2000). Controlling mechanical and swelling properties of alginate hydrogels independently by cross-linker type and cross-linking density. *Macromolecules* 33: 4291–4294.

48 Ahmed, T.A., Dare, E.V., and Hincke, M. (2008). Fibrin: a versatile scaffold for tissue engineering applications. *Tissue Engineering Part B: Reviews* 14: 199–215.

49 Landers, R., Hübner, U., Schmelzeisen, R., and Mülhaupt, R. (2002). Rapid prototyping of scaffolds derived from thermoreversible hydrogels and tailored for applications in tissue engineering. *Biomaterials* 23: 4437–4447.

50 Golden, A.P. and Tien, J. (2007). Fabrication of microfluidic hydrogels using molded gelatin as a sacrificial element. *Lab on a Chip* 7: 720–725.

51 Kolesky, D.B., Truby, R.L., Gladman, A.S. et al. (2014). 3D bioprinting of vascularized, heterogeneous cell-laden tissue constructs. *Advanced Materials* 26: 3124–3130.

52 Drzewiecki, K.E., Parmar, A.S., Gaudet, I.D. et al. (2014). Methacrylation induces rapid, temperature-dependent, reversible self-assembly of type-I collagen. *Langmuir* 30: 11204–11211.

53 Park, S., Kim, G., Jeon, Y.C. et al. (2009). 3D polycaprolactone scaffolds with controlled pore structure using a rapid prototyping system. *Journal of Materials Science Materials in Medicine* 20: 229–234.

54 Yu, D., Li, Q., Mu, X. et al. (2008). Bone regeneration of critical calvarial defect in goat model by PLGA/TCP/rhBMP-2 scaffolds prepared by low-temperature rapid-prototyping technology. *International Journal of Oral and Maxillofacial Surgery* 37: 929–934.

55 Xu, W., Wang, X., Yan, Y., and Zhang, R. (2008). Rapid prototyping of polyurethane for the creation of vascular systems. *Journal of Bioactive and Compatible Polymers* 23: 103–114.

56 Cui, X., Gao, G., Yonezawa, T., and Dai, G. (2014). Human cartilage tissue fabrication using three-dimensional inkjet printing technology. *Journal of Visualized Experiments* e51294.

57 Gudapati, H., Yan, J., Huang, Y., and Chrisey, D.B. (2014). Alginate gelation-induced cell death during laser-assisted cell printing. *Biofabrication* 6: 035022.

58 Fernández-Pradas, J.M., Florian, C., Caballero-Lucas, F. et al. (2017). Laser-induced forward transfer: propelling liquids with light. *Applied Surface Science* 418: 559–564.

59 Chang, R., Nam, J., and Sun, W. (2008). Effects of dispensing pressure and nozzle diameter on cell survival from solid freeform fabrication – based direct cell writing. *Tissue Engineering Part A* 14: 41–48.

60 Nair, K., Gandhi, M., Khalil, S. et al. (2009). Characterization of cell viability during bioprinting processes. *Biotechnology Journal* 4: 1168–1177.

61 Blaeser, A., Duarte Campos, D.F., Puster, U. et al. (2016). Controlling shear stress in 3D bioprinting is a key factor to balance printing resolution and stem cell integrity. *Advanced Healthcare Materials* 5: 326–333.

62 Song, S.J., Choi, J., Park, Y.D. et al. (2010). A three-dimensional bioprinting system for use with a hydrogel-based biomaterial and printing parameter characterization. *Artificial Organs* 34: 1044–1048.

63 Findley, W.N. and Davis, F.A. (2013). *Creep and Relaxation of Nonlinear Viscoelastic Materials*. Courier Corporation.

64 Morland, L.W. and Lee, E. (1960). Stress analysis for linear viscoelastic materials with temperature variation. *Transactions of the Society of Rheology* 4: 233–263.

65 Cui, X. and Boland, T. (2009). Human microvasculature fabrication using thermal inkjet printing technology. *Biomaterials* 30: 6221–6227.

66 Saunders, R.E., Gough, J.E., and Derby, B. (2008). Delivery of human fibroblast cells by piezoelectric drop-on-demand inkjet printing. *Biomaterials* 29: 193–203.

67 Reddi, A., Gay, R., Gay, S., and Miller, E.J. (1977). Transitions in collagen types during matrix-induced cartilage, bone, and bone marrow formation. *Proceedings of the National Academy of Sciences of the United States of America* 74: 5589–5592.

68 Moffat, K.L., Sun, W.-H.S., Pena, P.E. et al. (2008). Characterization of the structure – function relationship at the ligament-to-bone interface. *Proceedings of the National Academy of Sciences of the United States of America* 105: 7947–7952.

69 Tan, W. and Desai, T.A. (2005). Microscale multilayer cocultures for biomimetic blood vessels. *Journal of Biomedical Materials Research Part A* 72: 146–160.

70 Kakisis, J.D., Liapis, C.D., Breuer, C., and Sumpio, B.E. (2005). Artificial blood vessel: the holy grail of peripheral vascular surgery. *Journal of Vascular Surgery* 41: 349–354.

71 Hardin, J.O., Ober, T.J., Valentine, A.D., and Lewis, J.A. (2015). Microfluidic printheads for multimaterial 3D printing of viscoelastic inks. *Advanced Materials* 27: 3279–3284.

72 Liu, W., Zhang, Y.S., Heinrich, M.A. et al. (2017). Rapid continuous multimaterial extrusion bioprinting. *Advanced Materials* 29: 1604630.

73 Bakarich, S.E., Gorkin, R., Gately, R. et al. (2017). 3D printing of tough hydrogel composites with spatially varying materials properties. *Additive Manufacturing* 14: 24–30.

74 Mistry, P., Aied, A., Alexander, M. et al. (2017). Bioprinting using mechanically robust core – shell cell-laden hydrogel strands. *Macromolecular Bioscience* 17: 1600472.

75 Luo, Y., Lode, A., Wu, C. et al. (2015). Alginate/nanohydroxyapatite scaffolds with designed core/shell structures fabricated by 3D plotting and in situ mineralization for bone tissue engineering. *ACS Applied Materials & Interfaces* 7: 6541–6549.

76 Dai, X., Liu, L., Ouyang, J. et al. (2017). Coaxial 3D bioprinting of self-assembled multicellular heterogeneous tumor fibers. *Scientific Reports* 7: doi: https://doi.org/10.1038/s41598-017-01581-y.

77 Ouyang, L., Highley, C.B., Sun, W., and Burdick, J.A. (2017). A generalizable strategy for the 3D bioprinting of hydrogels from nonviscous photo-crosslinkable inks. *Advanced Materials* 29: 1604983.

78 Okawa, Y., Komuro, W., Kobayashi, H., and Ohno, T. (1997). Rheological study on gelatin gelation. *The Imaging Science Journal* 45: 197–200.

79 Gillette, B.M., Jensen, J.A., Tang, B. et al. (2008). In situ collagen assembly for integrating microfabricated three-dimensional cell-seeded matrices. *Nature Materials* 7: 636–640.
80 Bertassoni, L.E., Cecconi, M., Manoharan, V. et al. (2014). Hydrogel bioprinted microchannel networks for vascularization of tissue engineering constructs. *Lab on a Chip* 14: 2202–2211.
81 Lee, J.M. and Yeong, W.Y. (2015). A preliminary model of time-pressure dispensing system for bioprinting based on printing and material parameters: this paper reports a method to predict and control the width of hydrogel filament for bioprinting applications. *Virtual and Physical Prototyping* 10: 3–8.
82 Fernandez-Vicente, M., Calle, W., Ferrandiz, S., and Conejero, A. (2016). Effect of infill parameters on tensile mechanical behavior in desktop 3D printing. *3D Printing and Additive Manufacturing* 3: 183–192.
83 Tymrak, B.M., Kreiger, M., and Pearce, J.M. (2014). Mechanical properties of components fabricated with open-source 3-D printers under realistic environmental conditions. *Materials and Design* 58: 242–246.
84 Lanzotti, A., Grasso, M., Staiano, G., and Martorelli, M. (2015). The impact of process parameters on mechanical properties of parts fabricated in pla with an open-source 3-D printer. *Rapid Prototyping Journal* 21: 604–617.
85 Farzadi, A., Solati-Hashjin, M., Asadi-Eydivand, M., and Osman, N.A.A. (2014). Effect of layer thickness and printing orientation on mechanical properties and dimensional accuracy of 3D printed porous samples for bone tissue engineering. *PLoS ONE* 9: e108252.
86 Torrado, A.R. and Roberson, D.A. (2016). Failure analysis and anisotropy evaluation of 3D-printed tensile test specimens of different geometries and print raster patterns. *Journal of Failure Analysis and Prevention* 16: 154–164.
87 Mohanty, S., Larsen, L.B., Trifol, J. et al. (2015). Fabrication of scalable and structured tissue engineering scaffolds using water dissolvable sacrificial 3D printed moulds. *Materials Science and Engineering C* 55: 569–578.
88 Tsouknidas, A., Pantazopoulos, M., Katsoulis, I. et al. (2016). Impact absorption capacity of 3D-printed components fabricated by fused deposition modelling. *Materials and Design* 102: 41–44.
89 Wu, J., Clausen, A., and Sigmund, O. (2017). Minimum compliance topology optimization of shell – infill composites for additive manufacturing. *Computer Methods in Applied Mechanics and Engineering* 326: 358–375.
90 Clausen, A., Andreassen, E., and Sigmund, O. (2017). Topology optimization of 3D shell structures with porous infill. *Acta Mechanica Sinica* 33: 778–791.
91 Temple, J.P., Hutton, D.L., Hung, B.P. et al. (2014). Engineering anatomically shaped vascularized bone grafts with hascs and 3D-printed pcl scaffolds. *Journal of Biomedical Materials Research Part A* 102: 4317–4325.
92 Li, J., De Wijn, J., Van Blitterswijk, C., and De Groot, K. (2010). The effect of scaffold architecture on properties of direct 3D fiber deposition of porous Ti6Al4V for orthopedic implants. *Journal of Biomedical Materials Research Part A* 92: 33–42.

93 Malinauskas, M., Skliutas, E., Jonušauskas, L. et al. (2015). Tailoring bulk mechanical properties of 3D printed objects of polylactic acid varying internal micro-architecture. In: *SPIE Optics+ Optoelectronics: International Society for Optics and Photonics*, vol. 9505, 1–14.

94 Zhang, T., Yan, K.C., Ouyang, L., and Sun, W. (2013). Mechanical characterization of bioprinted in vitro soft tissue models. *Biofabrication* 5: 045010.

95 Hogrebe, N.J., Reinhardt, J.W., and Gooch, K.J. (2017). Biomaterial microarchitecture: a potent regulator of individual cell behavior and multicellular organization. *Journal of Biomedical Materials Research Part A* 105: 640–661.

96 Heher, P., Maleiner, B., Prüller, J. et al. (2015). A novel bioreactor for the generation of highly aligned 3D skeletal muscle-like constructs through orientation of fibrin via application of static strain. *Acta Biomaterialia* 24: 251–265.

97 Braziulis, E., Diezi, M., Biedermann, T. et al. (2012). Modified plastic compression of collagen hydrogels provides an ideal matrix for clinically applicable skin substitutes. *Tissue Engineering Part C: Methods* 18: 464–474.

98 Wall, B.D., Diegelmann, S.R., Zhang, S. et al. (2011). Aligned macroscopic domains of optoelectronic nanostructures prepared via shear-flow assembly of peptide hydrogels. *Advanced Materials* 23: 5009–5014.

99 Aubin, H., Nichol, J.W., Hutson, C.B. et al. (2010). Directed 3D cell alignment and elongation in microengineered hydrogels. *Biomaterials* 31: 6941–6951.

100 Zhu, W., Qu, X., Zhu, J. et al. (2017). Direct 3D bioprinting of prevascularized tissue constructs with complex microarchitecture. *Biomaterials* 124: 106–115.

101 Costantini, M., Testa, S., Mozetic, P. et al. (2017). Microfluidic-enhanced 3D bioprinting of aligned myoblast-laden hydrogels leads to functionally organized myofibers in vitro and in vivo. *Biomaterials* 131: 98–110.

102 Chen, S. and Zhang, K. (2016). *Layer-by-Layer Bioprinting of Stem Cells for Retinal Tissue Regeneration*. San Diego, CA: University of California.

103 Hong, S., Song, S.-J., Lee, J.Y. et al. (2013). Cellular behavior in micropatterned hydrogels by bioprinting system depended on the cell types and cellular interaction. *Journal of Bioscience and Bioengineering* 116: 224–230.

104 Lee, V., Singh, G., Trasatti, J.P. et al. (2013). Design and fabrication of human skin by three-dimensional bioprinting. *Tissue Engineering Part C: Methods* 20: 473–484.

105 Zhang, Y., Ge, H.-W., Li, N.-C. et al. (2016). Evaluation of three-dimensional printing for laparoscopic partial nephrectomy of renal tumors: a preliminary report. *World Journal of Urology* 34: 533–537.

106 Silberstein, J.L., Maddox, M.M., Dorsey, P. et al. (2014). Physical models of renal malignancies using standard cross-sectional imaging and 3-dimensional printers: a pilot study. *Urology* 84: 268–273.

107 Olivieri, L.J., Krieger, A., Loke, Y.-H. et al. (2015). Three-dimensional printing of intracardiac defects from three-dimensional echocardiographic images: feasibility and relative accuracy. *Journal of the American Society of Echocardiography* 28: 392–397.

108 Novosel, E.C., Kleinhans, C., and Kluger, P.J. (2011). Vascularization is the key challenge in tissue engineering. *Advanced Drug Delivery Reviews* 63: 300–311.

109 Bae, H., Puranik, A.S., Gauvin, R. et al. (2012). Building vascular networks. *Science Translational Medicine* 4: 160ps123.

110 Kolesky, D.B., Homan, K.A., Skylar-Scott, M.A., and Lewis, J.A. (2016). Three-dimensional bioprinting of thick vascularized tissues. *Proceedings of the National Academy of Sciences of the United States of America* 113: 3179–3184.

111 Richards, D., Jia, J., Yost, M. et al. (2017). 3D bioprinting for vascularized tissue fabrication. *Annals of Biomedical Engineering* 45: 132–147.

112 Visconti, R.P., Kasyanov, V., Gentile, C. et al. (2010). Towards organ printing: engineering an intra-organ branched vascular tree. *Expert Opinion on Biological Therapy* 10: 409–420.

113 Lee, V.K., Kim, D.Y., Ngo, H. et al. (2014). Creating perfused functional vascular channels using 3D bio-printing technology. *Biomaterials* 35: 8092–8102.

114 Wragg, J.W., Durant, S., McGettrick, H.M. et al. (2014). Shear stress regulated gene expression and angiogenesis in vascular endothelium. *Microcirculation* 21: 290–300.

115 Kaully, T., Kaufman-Francis, K., Lesman, A., and Levenberg, S. (2009). Vascularization – the conduit to viable engineered tissues. *Tissue Engineering Part B: Reviews* 15: 159–169.

116 Sun, J., Jamilpour, N., Wang, F.-Y., and Wong, P.K. (2014). Geometric control of capillary architecture via cell-matrix mechanical interactions. *Biomaterials* 35: 3273–3280.

117 Newman, A.C., Nakatsu, M.N., Chou, W. et al. (2011). The requirement for fibroblasts in angiogenesis: fibroblast-derived matrix proteins are essential for endothelial cell lumen formation. *Molecular Biology of the Cell* 22: 3791–3800.

118 Poldervaart, M.T., Gremmels, H., van Deventer, K. et al. (2014). Prolonged presence of vegf promotes vascularization in 3D bioprinted scaffolds with defined architecture. *Journal of Controlled Release* 184: 58–66.

119 Chen, R.R., Silva, E.A., Yuen, W.W., and Mooney, D.J. (2007). Spatio – temporal VEGF and PDGF delivery patterns blood vessel formation and maturation. *Pharmaceutical Research* 24: 258–264.

120 Isenberg, B.C., DiMilla, P.A., Walker, M. et al. (2009). Vascular smooth muscle cell durotaxis depends on substrate stiffness gradient strength. *Biophysical Journal* 97: 1313–1322.

121 Sant, S., Hancock, M.J., Donnelly, J.P. et al. (2010). Biomimetic gradient hydrogels for tissue engineering. *The Canadian Journal of Chemical Engineering* 88: 899–911.

122 Seidi, A., Ramalingam, M., Elloumi-Hannachi, I. et al. (2011). Gradient biomaterials for soft-to-hard interface tissue engineering. *Acta Biomaterialia* 7: 1441–1451.

123 Singh, M., Berkland, C., and Detamore, M.S. (2008). Strategies and applications for incorporating physical and chemical signal gradients in tissue engineering. *Tissue Engineering Part B: Reviews* 14: 341–366.

124 He, J., Du, Y., Villa-Uribe, J.L. et al. (2010). Rapid generation of biologically relevant hydrogels containing long-range chemical gradients. *Advanced Functional Materials* 20: 131–137.

125 Tseng, P. and Di Carlo, D. (2014). Substrates with patterned extracellular matrix and subcellular stiffness gradients reveal local biomechanical responses. *Advanced Materials* 26: 1242–1247.

126 Spalazzi, J.P., Doty, S.B., Moffat, K.L. et al. (2006). Development of controlled matrix heterogeneity on a triphasic scaffold for orthopedic interface tissue engineering. *Tissue Engineering* 12: 3497–3508.

127 Jang, K.-J., Mehr, A.P., Hamilton, G.A. et al. (2013). Human kidney proximal tubule-on-a-chip for drug transport and nephrotoxicity assessment. *Integrative Biology* 5: 1119–1129.

128 Homan, K.A., Kolesky, D.B., Skylar-Scott, M.A. et al. (2016). Bioprinting of 3D convoluted renal proximal tubules on perfusable chips. *Scientific Reports* 6: doi: https://doi.org/10.1038/srep34845.

129 Ma, X., Qu, X., Zhu, W. et al. (2016). Deterministically patterned biomimetic human iPSC-derived hepatic model via rapid 3D bioprinting. *Proceedings of the National Academy of Sciences of the United States of America* 113: 2206–2211.

130 Byoung Soo, K., Hyeonji, K., Ge, G. et al. (2017). Decellularized extracellular matrix: a step towards the next generation source for bioink manufacturing. *Biofabrication* 9: 034104.

131 Smith, L., Xia, Y., Galatz, L.M. et al. (2012). Tissue-engineering strategies for the tendon/ligament-to-bone insertion. *Connective Tissue Research* 53: 95–105.

132 Horváth, L., Umehara, Y., Jud, C. et al. (2015). Engineering an in vitro air-blood barrier by 3D bioprinting. *Scientific Reports* 5: 7974.

133 Ventola, C.L. (2014). Medical applications for 3D printing: current and projected uses. *Pharmacy and Therapeutics* 39: 704.

134 Radenkovic, D., Solouk, A., and Seifalian, A. (2016). Personalized development of human organs using 3D printing technology. *Medical Hypotheses* 87: 30–33.

135 Lee, J.B. and Sung, J.H. (2013). Organ-on-a-chip technology and microfluidic whole-body models for pharmacokinetic drug toxicity screening. *Biotechnology Journal* 8: 1258–1266.

136 Bhise, N.S., Ribas, J., Manoharan, V. et al. (2014). Organ-on-a-chip platforms for studying drug delivery systems. *Journal of Controlled Release* 190: 82–93.

137 Esch, M., King, T., and Shuler, M. (2011). The role of body-on-a-chip devices in drug and toxicity studies. *Annual Review of Biomedical Engineering* 13: 55–72.

138 Villasante, A. and Vunjak-Novakovic, G. (2015). Tissue-engineered models of human tumors for cancer research. *Expert Opinion on Drug Discovery* 10: 257–268.

139 Gu, L. and Mooney, D.J. (2016). Biomaterials and emerging anticancer therapeutics: engineering the microenvironment. *Nature Reviews Cancer* 16: 56–66.

140 Kwakwa, K.A., Vanderburgh, J.P., Guelcher, S.A., and Sterling, J.A. (2017). Engineering 3D models of tumors and bone to understand tumor-induced bone disease and improve treatments. *Current Osteoporosis Reports* 15: 247–254.

141 Theodoropoulos, J.S., De Croos, J.A., Park, S.S. et al. (2011). Integration of tissue-engineered cartilage with host cartilage: an in vitro model. *Clinical Orthopaedics and Related Research* 469: 2785.

142 Shandalov, Y., Egozi, D., Koffler, J. et al. (2014). An engineered muscle flap for reconstruction of large soft tissue defects. *Proceedings of the National Academy of Sciences of the United States of America* 111: 6010–6015.

143 Baranski, J.D., Chaturvedi, R.R., Stevens, K.R. et al. (2013). Geometric control of vascular networks to enhance engineered tissue integration and function. *Proceedings of the National Academy of Sciences of the United States of America* 110: 7586–7591.

144 Martelli, N., Serrano, C., van den Brink, H. et al. (2016). Advantages and disadvantages of 3-dimensional printing in surgery: a systematic review. *Surgery* 159: 1485–1500.

145 Mobbs, R.J., Coughlan, M., Thompson, R. et al. (2017). The utility of 3D printing for surgical planning and patient-specific implant design for complex spinal pathologies: case report. *Journal of Neurosurgery: Spine* 26: 513–518.

146 Liang, H., Ji, T., Zhang, Y. et al. (2017). Reconstruction with 3D-printed pelvic endoprostheses after resection of a pelvic tumour. *Bone and Joint Journal* 99-B: 267–275.

147 Huifeng, S., Xiurong, K., An, L. et al. (2017). Bone regeneration in 3D printing bioactive ceramic scaffolds with improved tissue/material interface pore architecture in thin-wall bone defect. *Biofabrication* 9: 025003.

148 Inzana, J.A., Olvera, D., Fuller, S.M. et al. (2014). 3D printing of composite calcium phosphate and collagen scaffolds for bone regeneration. *Biomaterials* 35: 4026–4034.

149 Castro, N.J., Patel, R., and Zhang, L.G. (2015). Design of a novel 3D printed bioactive nanocomposite scaffold for improved osteochondral regeneration. *Cellular and Molecular Bioengineering* 8: 416–432.

150 Colasante, C., Sanford, Z., Garfein, E., and Tepper, O. (2016). Current trends in 3D printing, bioprosthetics, and tissue engineering in plastic and reconstructive surgery. *Current Surgery Reports* 4: 6.

5

Bioscaffolding: A New Innovative Fabrication Process

Rania Abdelgaber[1], David Kilian[1], and Hendrik Fiehn[2]

[1] *Technische Universität Dresden, Centre for Translational Bone, Joint and Soft Tissue Research, Medical Faculty Carl Gustav Carus, Fetscherstrasse 74, 01307, Dresden, Germany*
[2] *GeSiM mbH, Radeberg, Germany*

5.1 Introduction: From Bioscaffolding to Bioprinting

For the past two decades, tissue engineering (TE) has been an emerging interdisciplinary field in biomedical science, which aims to the development of biological substitutes that can replace pathological or degenerated tissues to maintain physiological function. TE combines the use of technologies acquired from mechanical engineering, material science, cell biology, and physical chemistry [1, 2]. As demonstrated in early approaches [3], in order to create a new tissue, three general approaches are employed: (i) the replacement of diseased cells by isolated cells or cell substitutes to provide the required function, (ii) the delivery of signaling molecules and growth factors to targeted locations, and (iii) the development of tissue-engineered biomaterial constructs that can mimic the complexity of native tissues, possibly in combination with seeded (autologous) cells. These constructs are referred to as scaffolds. Scaffolds can be defined as structures with controlled pore size and interconnectivity, which are able to support cells and formation of tissues [4]. In this context of TE, the term "Biofabrication" emerged.

Biofabrication refers to the automated generation of constructs that can closely mimic the complex heterogenous nature and geometry of tissues and organs more than the current regenerative therapies do. The community recently updated and completed the definition of biofabrication to "the automated generation of biologically functional products with structural organization from living cells, bioactive molecules, biomaterials, cell aggregates such as microtissues, or hybrid cell–material constructs, through bioprinting or bioassembly and subsequent tissue maturation processes"[5]. Furthermore, it is essential to differentiate between the two terms bioprinting and bioassembly as they are often misused. Based on this, bioprinting was defined as the use of computer-aided transfer processes for patterning and assembling living and nonliving materials with a prescribed 2D and 3D organization to produce bioengineered structures. Bioassembly was referred to as the fabrication of hierarchical constructs with a prescribed 2D or 3D organization through automated assembly of preformed

3D and 4D Printing in Biomedical Applications: Process Engineering and Additive Manufacturing,
First Edition. Edited by Mohammed Maniruzzaman.

cell-containing fabrication units generated via a cell-driven self-organization or through the preparation of hybrid cell–material building blocks aiming for a close simulation of natural tissue organization/development.

The human body shows a very limited ability to autoregenerate a damaged tissue or organs resulting from medical disorders or tissue dysfunctions [6]. With the aid of 3D printing as an additive manufacturing technology, fabrication of 3D constructs with high complexity and patient specificity became possible. Considering these technological advances for the field of TE, one could clearly see that such techniques are able to improve the quality of life in various cases of human disease. Therefore, a variety of tissues and organs might be potentially repaired or replaced one day using tissue-engineered constructs harboring therapeutic cells for reconstructive therapies [7]. For example, in cases of severe skin burns and scarring, 3D construction of skin grafts containing differentiated stem cells could be able to attain the structural and functional similarity to natural skin and prevent scar formation [8]. The 3D printing technology comprises a set of tools that work together to finally obtain the desired 3D printed construct. These tools include a computer, a 3D modeling software (computer-aided design (CAD), generated from clinical imaging date such as computed tomography (CT) or magnetic resonance imaging (MRI) scan images), a hardware including machine equipment, and the printing materials. In order to obtain the correct shapes of a tissue, data from CT are captured and processed by a CAD software, which is then translated to the printer to produce the desired 3D construct (Figure 5.1) [9].

The printability of the biomaterials is dependent on the manufacturing technique and the rheological and mechanical properties of the initial material. The composition and final form of the material differ significantly depending on the printing technique used. In the past, a number of conventional fabrication techniques were used to create porous scaffolds such as gas foaming, phase separation, fiber bonding, molding, particulate leaching, membrane lamination, and particulate leaching [10]. Within a decade, technological advances emerged for 3D printing applications. The most commonly used 3D printing technologies to produce scaffolds for TE include stereolithography, selective laser sintering, fused deposition modeling, and finally 3D plotting/bioprinting [11].

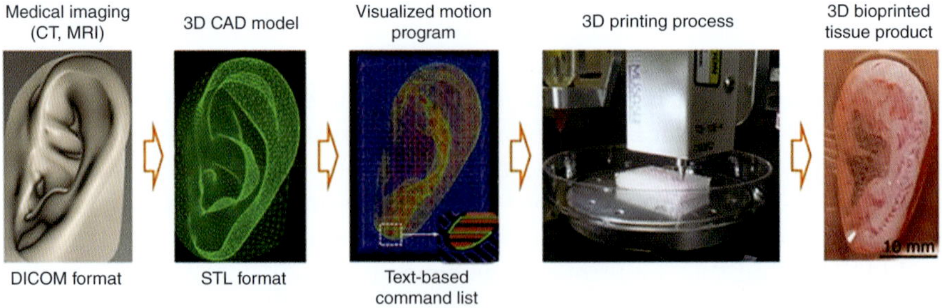

Figure 5.1 Automated printing process of a 3D structure of a human ear-mimicking target tissue obtained from a CAD model developed from medical image data. Source: Kang et al. 2016 [9]. Reproduced with permission of Springer Nature.

This chapter will give an overview about the possibilities of mainly extrusion-based manufacturing technologies to create scaffolds for 3D tissue models and techniques to approach a simulated natively heterogenous tissue structure.

5.2 Scaffolding

5.2.1 Properties of Scaffolds

Scaffolds are designed with the aim of mimicking the *in vivo* microenvironment of the cells where cell–cell and cell–ECM (extracellular matrix) interactions occur, triggering cellular behavior and tissue function. Thus, it is crucial to maintain material properties suitable for cell responses and crosstalk with the material and neighboring cells. When designing a scaffold, it is important to consider three architectural levels to ensure proper nutrient and oxygen delivery as well as natural cell–matrix interaction. Macroarchitecture describes the three-dimensional geometry of the scaffolds that has the ability to promote cell proliferation and differentiation, designed to mimic the native ECM of the cells in an optimal way. It represents the overall geometry of the scaffold including patient-specific design and structural features. Microarchitecture on the other hand describes the tissue architecture (shape, structural distribution, porosity, and pore size), whereas nanoarchitecture reflects the respective surface properties (e.g. triggering cell attachment, migration, and proliferation) [12].

The macro- and microstructural properties were proven to have an impact on cell growth, cell propagation, and reorganization; shape remodeling of the tissue; and even preserving its intrinsic phenotypes. However, a number of aspects should also be taken into consideration to determine the suitability of a scaffold for TE:

Biocompatibility: Scaffolds should support the growth and differentiation of encapsulated cells as well as their attachment and migration in both *in vitro* culture conditions and after implantation. Moreover, the materials used must be biocompatible not only with cellular components of the scaffold but also with the endogenous host cells.

Pore size and pore geometry: Porosity is one of the essential and most critical players in the geometry of scaffolds. Ideally, high porosity with high pore network interconnectivity of the 3D scaffolds is vital for oxygen, nutrients, and waste exchange through the scaffold [13].

The optimized pore size of scaffolds can determine the regenerative ability of certain tissues by creating a specific suitable microenvironment for the cells [14].

Moreover, pore geometry has also shown to be influential on cell performance and function. A study was conducted showing how pore geometry manipulation can affect ovarian follicle survival and spreading. With this variation in the architecture of a 3D printed scaffold, these findings could help to develop an *in vivo* functional ovarian implant later [15].

Biomimicking native structure: For enhancing cellular attachment, scaffolds are often modified with cell-adhesive ligands that provide bioactive sites for cells influencing their morphology and alignment. Scaffolds also serve as reservoirs or delivery systems encapsulating growth factors or proteins that can be released in a controlled manner while retaining their bioactivity [16].

Biodegradability: Scaffolds are designed in a way to be biodegradable to allow for the cells to function and produce their own ECM and for the host cells to replace and remodel this implanted tissue-engineered construct finally. Therefore, the degradation by-products of these implants must be noncytotoxic and easily excreted from the system [17].

Surface characteristics: Large surface area-to-volume ratio is an important aspect for harboring cells in sufficient density and thus influencing cell attachment and migration. Moreover, with mechanotransduction pathway cells are able to detect the stiffness of the surrounding environment, tension, and compression and therefore interact accordingly with the matrix [18].

Mechanical properties: Scaffolds must provide the sufficient strength to maintain their shape after extrusion and during *in vitro* cell culture. The degraded scaffold must also retain certain mechanical strength against physiological stresses *in vivo* [14]. Another important aspect to keep in consideration is to ensure the similarity of the intrinsic mechanical properties to the host tissue. Moreover, studies have highlighted the importance of mechanosensitivity because the cells' adhesive properties and differentiation capacity are influenced directly by the stiffness of the biomaterial [16].

Despite all the efforts directed toward mimicking ECM constitution and function using advanced scaffolding approaches, the use of hydrogel scaffolds in TE remains a significant challenge. This might be because of a number of limitations such as (i) poor mechanical properties of hydrogels limiting their usage in soft, non-load-bearing tissues and (ii) difficulty in attaining complex microvasculature in engineered scaffolds, which negatively influence the viability and function of cells because of lack of nutrient and oxygen transport [19].

5.2.2 Bioprinters vs Common 3D Printers: Approaches for Extrusion of Polymers

3D printing in mechanical and civil engineering has been established and available for 20 years now. The main purpose is to provide mechanically exact models at a certain stiffness. Manufacturers of related instruments usually stick to very specific materials (polymers, metals, resins, and paper) with fixed properties. The specifications of those conventional machines can be achieved only using these materials.

Requirements for bioprinting are quite different: (Cell) biology needs very specific polymers and bioinks to provide a physiological but well-oriented environment for cells and interface to cells. On the other hand, mechanical accuracy is not such a critical factor, at least compared to high-precision engineering.

Manufacturers of state-of-the-art bioprinters offer a range of various tools for the automated extrusion of very different material and for the preparation of these materials for printing, which is also possible within the machine, respectively. Furthermore, containers and cartridges must be easily accessible to the user for the following reasons:

1. Most researchers prepare their own bioinks/polymer solutions and do not use commercially available materials. They must be able to feed granules, powders, or pastes.
2. Addition of living cells to the bioink must be done rapidly and right before printing in order to keep the cells alive during the printing process.

Traditional extrusion approaches such as FDM (fused deposition modeling) usually rely on filaments (coiled wires). It is usually not applicable for bioprinting because for this technique the users are limited to commercially available filaments and the required temperature range is harmful to cells.

Due to the typical properties of bioinks with low polymer content, a 3D printed scaffold often consists of more than one material. The instrument must be able to print several materials of very different mechanical properties into one object, e.g. a biocompatible thermoplastic polymer along with a soft hydrogel. From a technical perspective, the machine must be able to synchronize tools with different extrusion techniques and different dispensing needles, see Section 5.3.2 for more details.

5.2.3 Comparing Cell Seeding Techniques to 3D Bioprinting or Cell-Laden Hydrogels

3D printing as an additive manufacturing technology now surpasses conventional scaffolds in a variety of properties including more precise control over scaffold architecture, which influence biological response and ensure more uniform and reproducible structures with control over pore size, geometry, and interconnectivity [18].

5.2.3.1 From Printing to Bioprinting

Bioink is the term mainly used for hydrogels that enable to include cells within the printing process. Bioinks therefore offer the basis of 3D printed constructs that can be implanted in damaged tissues later to allow for controlled cell-based regeneration. In this context, the bioink properties become a critical aspect in TE because these properties should account for a suitable biological environment for cells, along with finding optimum printing conditions that result in stable structures, with maintaining cellular integrity [20]. In order to fulfill these requirements, investigations have been made on the printability of hydrogel by assessing the ink rheology and consistency, as well as the question how these parameters would influence printability and its correlation with mechanical properties.

Therein, the importance of viscosity of the printed material must be highlighted. While adjusting the viscosity of the biomaterial, one has to keep in mind that the biomaterial should be of low viscosity to allow for easy extrusion and avoid nozzle clogging but also not liquid enough to cause deformation and strand collapse after printing. Optimal materials are those that have low viscosity upon printing to be extrudable along with the ability of fast gelation during stacking of hydrogel layers. One of the materials that are used to optimize viscosity is gelatin, as it is known to adjust the viscosity of other materials such as alginate to improve its printability as well as maintain and fix the shape of the printed structure. Therefore, viscosity is an important parameter without which poor printing quality or even failure may occur [21].

Limitations of Cell-Seeded Scaffolds

1. Lack of spatial and temporal control that results in irregular cell distribution throughout the scaffold, which will consequently influence cell migration and penetration.
2. Difficulty to optimize biomaterial concentration to cell density ratio because cell density might influence viability and tissue function.
3. Restriction with engineering complex tissues with different materials and multiple cell types that would be more feasible and controllable with bioprinting than cell seeding of printed scaffolds.

Fortunately, with the advancements of bioprinting technologies, many of the limitations of cell seeding have been overcome. This is because of the homogeneity of the final paste loaded with cells, which is deposited in a spatially and temporally controlled manner. However, with the complexity of this technology, some challenges also arise.

Challenges of 3D Bioprinting Some important considerations also reflect the complexity of 3D bioprinting processes compared to scaffold printing. This includes material choice because different cell types favor survival and metabolic activity in different biomaterials. The number of cytocompatible materials for cell encapsulation is limited [22]; therefore, suitable materials will provide appropriate cell–matrix interactions, maintaining their viability and interaction with each other. Other aspects include inclusion of growth and differentiation factors in printed scaffolds and technical challenges comprising construction of scaffolds. Additionally, cell-encapsulated materials are frequently exposed to chemical cross-linking agents or photoactivatable cross-linkers using UV light [23] before printing for extended periods of time, possibly causing cellular damage. During deposition, the mechanical stress caused by printing itself can result in serious cell damage and loss of function by cell extension or shear stress [24]. Although cell seeding scaffolds have the drawback of uneven cell distribution and commonly low cell density as mentioned by Niklason and Langer stating that "...gels seeded with cells have been limited by the fact that the resultant cell densities per unit volume that can be achieved are much lower than those observed *in vivo*..." [25], it would still cause less distress to the cells during processing.

5.2.3.2 Approaches of Stabilizing Printed Constructs

The key challenge for fabricating a functional scaffold with real volumetric dimensions ($>1\,cm^3$) is to develop a material that is printable with an ability to keep a stable geometrical structure allowing for proliferation and differentiation of seeded cells later. Therein, various stabilization techniques have been described in the literature, such as double cross-linking by covalent bond formation, thiol–ene dimerization, physical hydrogels, coreshell printing in which strands are stabilized by the mechanical support of the material, and finally printing into self-healing gels [23]. Other examples include chemical cross-linking of collagen by carbodiimide EDC [26], cross-linking of calcium phosphate cement (CPC) scaffolds by setting in humidity [27], or *in situ* cross-linking of bioinks such as gelatin methacryloyl (GelMA) and polyethylene glycol diacrylate (PEGDA) by UV [28]. However, in most of these strategies, cells cannot be included in the material during fabrication. Therefore, other stabilization approaches are being discussed in Section 5.3.1 that allow inclusion of cells during the printing process.

5.2.4 Examples/Applications of Cell-Seeded Scaffolds

As mentioned earlier, the 3D printed scaffolds are developed to provide structural support for cells and induce new tissue formation. Two main strategies that are widely used for developing scaffolds are (i) seeding cells onto prefabricated scaffolds and (ii) premixing cells and hydrogels before scaffold fabrication. Therein, with cell seeding process, harsher conditions can be employed. This involves using a variety of hydrophilic/hydrophobic precursors, solvents, and reagents. However, the final printed scaffold must ensure cytocompatibility for the seeded cells.

Earlier approaches in cell seeding 3D printed constructs also included a study that aimed to design a 3D construct that mimics cartilaginous tissue structure and function. The printed strands in the scaffold seeded with cells would then act as a template for tissue growth. Further, for determining the flow and distribution of nutrients and waste products, the design criteria of the scaffold was optimized for the development of functional tissues [29].

Further applications of cell-seeded scaffolds employed 3D plotting of concentrated gelatin/alginate scaffolds in which the synergistic effect of the composite enabled mechanical stability of the scaffold as well as cell adhesion and proliferation of seeded human bone-marrow-derived mesenchymal stem cells (MSCs) [30].

Fabrication of collagen scaffolds by 3D printing is known to be challenging because of its low viscosity. In a more recent study, Lode et al. were able to fabricate 3D printed scaffolds using highly viscous collagen, which formed stable strands that maintained their shape after plotting. These scaffolds also supported the growth of the seeded mesenchymal stromal cells [26].

5.2.5 Data Processing of 3D CAD Data for Bioscaffolds

Currently, 3D bioscaffolds are mostly printed to study the cell growth or for performing the basic material and viability tests. Usually, basic 3D objects such as cylinders and cones are sufficient for these experiments. Bioprinters shall have entry CAD functions to meet these basic requirements.

More clinically related applications such as scaffold printing for bone replacement require patient-derived freeform models. Nowadays, the data processing chain is complemented by standard software packages. DICOM (Digital Imaging and Communications in Medicine) data from medical imaging devices (CT and MRI) provide input data for selecting regions of interest, e.g. bone fracture areas. Then, these data will be transferred into STL data for 3D printing.

The commonly used STL format (abbreviation arising from the manufacturing technique of stereolithography) is a surface description language based on triangularization. For 3D printing, STL formats are usually sectioned into horizontal slices by the operation software of the 3D printer. Infills will be created by the printer software. This step is typically not connected to the read-in CAD model. Afterward, the machine-specific printing data, e.g. G-code, can be created.

The infill data for bioscaffolds are more challenging than infill data for 3D bulk objects: Porosity, shape, size, and orientation are important parameters to ensure cell seeding efficiency, cell viability, and matrix resemblance. Multimaterial scaffolds representing very different material properties may require individual layer heights and strand sizes for each material: Technically, thermoplastic melts can be printed with much thinner strands than soft hydrogels.

Therefore, infill design is a real challenge for bioprinters. Pattern and strand design can be controlled according to the user's requirements. The final model may not compensate failures and data inconsistencies and extruded strands may collapse during printing. A typical freeware software for the processing of STL data often does not meet these requirements.

It should be considered that the popular STL format typically does not show material-specific information but rather simply shows geometrical data. However, multimaterial bioscaffolds require machine data for each printing tool. There are two options for data processing:

1) The design for a multimaterial scaffold can be generated individually using the instrument-specific software. A dedicated material can be assigned to each strand and each layer. This way of design generation does not require an external STL model. The dimensions for all materials can be changed easily.
2) The multimaterial design based on CAD data requires an individual STL model for each material. The machine software must be able to read-in and merge/overlap these STL models. The 3MF (3D manufacturing format) – an alternative option to STL format – directly supports CAD data for individual materials/printing tools.

5.3 Bioprinted Scaffolds

5.3.1 Bioinks

As introduced in the first part of this chapter, recent developments enable the transition from hydrogels and printable materials described for bioscaffolding to a more specified approach of bioprinting that includes encapsulated cells inside those materials already during the fabrication process. This enables the achievement of a spatially defined distribution of different cell types inside the scaffold design overcoming the limitations of conventional seeding methods and improving homogeneous and effective seeding. The field defines the term "bioink" as a biomaterial or material combination, often applied as a hydrogel, that can be processed by additive manufacturing and enables the encapsulation of cells or biologically active factors before processing [31].

To a large extent, hydrogels are used here because of their high water content and their structural resemblance of the native ECM [2]. Bioinks require specific properties superior to the characteristics of common biomaterials and hydrogels that might be applicable merely for the production of conventional scaffolds. Especially, the stabilization of bioprinted scaffolds, hence the cross-linking scheme, presents a more critical challenge for method development. Although the majority of materials introduced for pure bioscaffolds are cross-linked via chemical reaction after fabrication, such as collagen via carbodiimide or glutaraldehyde treatment, or as other polymers via photopolymerization by an intensive UV treatment, those procedures bare a risk of harming the included cells. However, the shear-thinning properties need to be maintained for the hydrogels, being extrudable without deliquescing of the strands. Furthermore, the temperature and extrusion pressure need to be selected in a physiological range for included cells. Because those manufacturing procedures then often interfere with the printability of the material and particularly an appropriate stability for volumetric constructs, research came up with different strategies of stabilizing 3D bioprinted structures [32].

Strategies of internal stabilization combine different components within one bioink to ensure cell viability, printability, and shape fidelity while increasing the viscosity. The addition of – often naturally derived – thickeners increases the viscosity of the polymer-based hydrogels, preventing extruded strands from deliquescing before the actual cross-linking, ionically, chemically, or via photopolymerization. Although a high overall polymer content will lead to damage of encapsulated cells [2], developing bioinks so far has always been a matter of compromise between material stability and attractiveness for cells. One early approach for the generation of such tissue-mimicking structures was blending sodium alginate with nanofibrillated cellulose (NFC) by Markstedt et al. in 2015 to increase viscosity and produce cartilage-like structures [33]. Chondrocytes were successfully included in the manufacturing process and showed cell viability of 70% after printing. This fibrous insoluble supportive component remained inside the bioink after extrusion, cross-linking, and during incubation. Alginate as a biocompatible material, yet without providing adhesive sites for cells and growth factors, is a popular basis for bioinks. In other approaches, additional components were used to ensure high viscosity while printing and were potentially released later after alginate cross-linking, to some extent or entirely. To create 3D constructs in a clinically relevant size, methyl cellulose was applied as such a stabilizing material to support alginate before ionic cross-linking mediated by calcium ions. Schütz et al. presented this approach for the bioprinting of human mesenchymal stem cell (hMSC) that could be differentiated into adipogenic direction to create 3D models of fat tissues (Figure 5.2) [34]. Another example for this internal way of stabilization was provided by Müller et al., applying pluronic F127 as an additional stabilizer for a bioink based on acrylated hyaluronic acid, to print cartilage structures from primary bovine chondrocytes [35]. Those temporarily stabilizing components have another advantage of creating micropores after release, which could support cell adhesion and migration inside the construct. To increase the attractiveness of alginate-based systems, studies suggested the modification

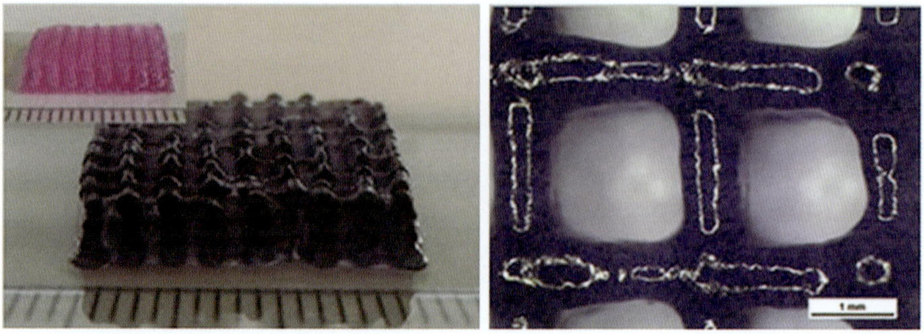

Figure 5.2 3D plotted scaffolds in clinically relevant dimensions with metabolically active (MTT staining, dark violet) hMSC, encapsulated in alginate–methyl cellulose bioink, compared to an empty non-cell-laden scaffold (upper left), bar = 1 mm. Source: Schütz et al. 2017 [34]. Reproduced with permission of John Wiley & Sons.

of the polymers with arginyl-glycyl-aspartic acid (RGD) motifs mediating cell adhesion and potential proliferation [36].

Another biocompatible naturally derived biopolymer that was introduced as an attractive basis for bioinks was agarose [37], which, in this approach, was stabilized by dispensing in a ultrahydrophobic high-density fluid of fluorocarbon. With agarose utilization, an additional thermoresponsive gelation behavior offered a different type of stabilization mechanism. Molecules that present a structural or molecular similarity to the native ECM in the human body are preferred. Early approaches used gelatin as degraded collagen to provide encapsulated cells with adhesive properties and with cleavable sites for degradation and migration [38]. By adding actual collagen to agarose solution, smooth muscle cells were printed by Köpf et al. [39]. Later, bioprinting of collagen type I hydrogels mixed with a biocompatible polyphenol cross-linking solution (2% tannic acid) was presented to process human adipose-tissue-derived stem cells, resulting in a significantly increased cell viability, adhesion, and metabolic activity compared to alginate-based approaches, enabling manufacturing of constructs in a core–shell manner [40]. Other ECM-mimicking studies suggested the use of decellularized matrix as a bioink to maintain a native environment [41].

Additional components mediating stem cell differentiation and providing a high affinity to bind biologically active factors can be included. In an approach presented by Ahlfeld and coworkers, a nanoclay was added to an alginate–methyl cellulose blend maintaining its ability to produce variable constructs in clinically relevant size, improving cell viability and offering high potential as a drug-delivering ingredient (Figure 5.3) [42].

Those naturally derived biopolymers provide a high biocompatibility while most of the components originating from plant or bacterial species lack consistency, reproducibility, and usually adhesive sites for mammalian cells. In other approaches, synthetic or chemically modified compounds and bioinks ensure a more controllable gelation and degradation behavior and therefore might trigger cell behavior to a more predictable way.

The majority of thermoplastics such as polycaprolactone (PCL) or polylactic-*co*-glycolic acid (PLGA) show a suitable printing behavior but are not applicable

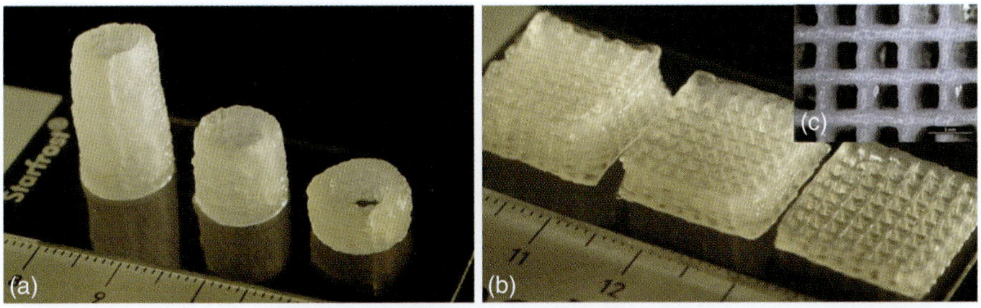

Figure 5.3 Alginate–methyl cellulose–laponite blend as a versatile bioink for the fabrication of constructs of varying shape and of clinically relevant dimensions (a). Cubic scaffolds of 30, 20, and 10 strand layers (b), with open macropores, bar = 2 mm (c). Source: Ahlfeld et al. 2017 [27]. Reproduced with permission of Springer Nature.

for the processing of encapsulated cells because of their melting temperatures above 60 °C. As a consequence, they are applied as a construct-stabilizing phase in addition to the cell-laden bioink (see Section 5.3.2) or can be used merely for cell seeding after extrusion. However, synthetic thiol-functional linear poly(glycidol) was applied for bioprinting. This material could be UV polymerized in a biocompatible time frame, forming a stable network and presenting high (80%) viability of encapsulated and extruded primary hMSC [43]. To combine photoinitiated and thermally mediated cross-linking, chemically modified blends of poly(*N*-isopropylacrylamide)-grafted hyaluronan and methacrylated hyaluronan were applied for the encapsulation and cultivation of bovine chondrocytes [44].

5.3.2 Tools for Multimaterial Printing

Multimaterial printing describes the capability to print 3D objects from materials with very different properties.

The GeSiM BioScaffold printer (Figure 5.4) operates up to four (different) printing tools on separate *z*-axes. The system offers the following extrusion methods:

- Pneumatic extrusion from disposable plastic cartridges, with a maximum pressure of 600 kPa. Shell heaters for the cartridges are available for a temperature range up to 190 °C. An optional Peltier chiller reaches a temperature close to 0 °C. The cartridge volume is 10 ml for temperature-controlled printing and 30 ml for printing at room temperature, respectively.
- A micropipetting unit applies liquid samples to individual layers of a bioscaffold during printing. This provides the option of separating cell handling and 3D printing but requires the cells to adhere to the surface of the extruded scaffold strands. This unit allows to process volumes in microliter range. Therefore, it supports the use of rare and expensive substances/cell lines in the fabrication process.
- The pneumatic coreshell extruder is an elegant tool to bridge the gap between mechanical stiffness – required to build up real 3D objects – and cell-friendly

Figure 5.4 GeSiM BioScaffold Printer 3.2: Printhead with different tools for multimaterial scaffolds.

low viscosity. It combines two materials into one strand: A core lane will be surrounded by a shell lane. For more details, see Section 5.3.4.

- A high-power/high-temperature piston extruder for polymer melts provides pressure >100 bar and works for melts up to 250 °C. In combination with thin steel needles, it allows to extrude very thin strands. Two of these extruders fit to a special polymer mixing head to print polymer blends and gradients of polymer blends. The latter one is of importance to mimic the composition of complex tissue, such as bone.
- Melt electrospinning writing describes the high-voltage-induced extrusion of polymer melts or solved polymers. Strands are getting really thin, down to the range of 10–20 μm. An ongoing project will use this technology to mimic the eardrums.

For all tools and nozzle configurations, an automated tip alignment for *XYZ* coordinates exists for simultaneous printing of different materials into one 3D object.

5.3.3 Multimaterial Scaffold

Arising from the technological advancement of combining several cartridges and different materials in one fabrication process as possible with the GeSiM BioScaffolder 3.2, various applications of combining types of materials inside one construct can be found. Each tissue to be mimicked and each organ has their characteristic pattern, defining (internally varying) mechanical properties, factor gradients, and particularly cell type or cell distribution. In contrast to articular cartilage containing only one cell type (chondrocytes), bone consists of at least three different cell types. Most human tissues contain blood vessels, which already make it a more complex structure than a single-material system could achieve. Only articular cartilage, corneal structures of the eye, and the epidermal layer of the skin lack blood vessels and are only supplied with oxygen and nutrients via diffusion.

A gradient of mechanical properties and stiffness inside one scaffold could also be induced via specific cross-linking regimes of the same material, as shown for alginate [45]. Beyond the application of different factors in regions of a monomaterial scaffold to drive differentiation in several lineages, as it was presented for a PCL-based biphasic osteochondral scaffold [46], multimaterial options offer the opportunity to use materials with different properties specific for a cell type, region of interest, or release behavior. Therefore, Ahlfeld and coworkers suggested the coextrusion of an alginate/gellan gum blend along with extruded strands of printable pasty CPC, to fulfill the requirements of both mechanical and structural resemblance of bone tissue as well as provision of the respective growth factors [27]. One step further into the direction of bioprinting of bone and osteochondral tissue, the combination of this 3D-plotted bone-mimicking CPC with cell-laden alginate/methylcellulose strands inside volumetric stabilized constructs was shown [32] (Figure 5.5), bringing together the advantages of favorable mechanical properties for bone replacement and a defined cell distribution. Those approaches are rather unique because both coextruded materials would individually be sufficient to create real volumetric constructs in clinically relevant dimensions without the necessity of stabilizing each other. Earlier studies suggested the use of a stable 3D plotted structure of PCL [47] or PCL–PEG [48] supporting a cell-laden alginate gel of low viscosity. This ensured the volumetric shape fidelity of the construct because the pure hydrogels would not be stable before cross-linking. Therefore, besides the option of including several materials mimicking the actual structure, many approaches require the necessity of a stabilizing high-polymer material. Some approaches even used only one actually printed material for the raw robust grid structure or a surrounding cast of the scaffold to fill this construct up with a low-viscosity hydrogel presenting favorable characteristics for the survival of encapsulated

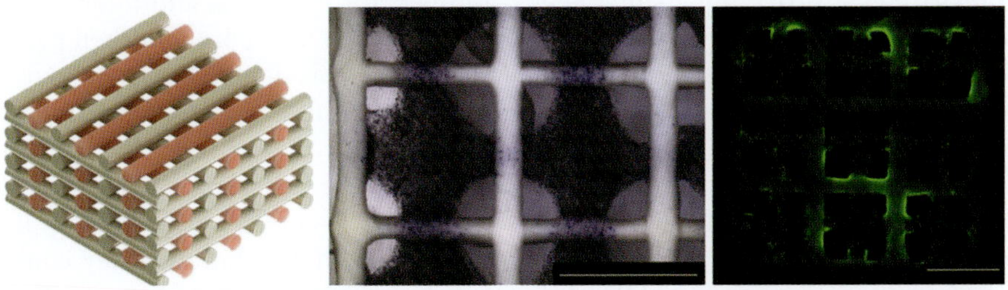

(a) (b) (c)

Figure 5.5 Multimaterial scaffold. Schematic scaffold design of alternating CPC (gray) and cell-laden hydrogel strands (red) (a). MTT 3-(4,5-dimethylthiazol-2-yl)-2,5-diphenyltetrazolium bromide staining for metabolically active cells inside the hydrogel strands, bar = 3 mm (b). Viability of encapsulated hMSC after 21 days of cultivation (>90%), viable cells labeled green (Calcein AM), dead cells labeled red (ethidium homodimer-1), green autofluorescence of CPC strands, bar = 3 mm (c). Source: Kilian et al. 2017 [32]. Reproduced with permission of Cambridge University Press.

cells. Kolesky et al. presented an innovative idea for the fabrication of vascularized 3D tissues combining techniques of bioprinting, stabilizing structures, sacrificial materials, and conventional cell seeding of molded hydrogel [49].

In another material- and technique-combining approach, mineralization of alginate hydrogels was achieved via alkaline phosphatase/alginate capsules containing osteoblast cells, mixed into a printable alginate hydrogel [50]. Later, other concepts also introduced the combination of different techniques of additive manufacturing creating and optimizing hybrid scaffolds. Therefore, phases of 3D plotted PCL strands and nanofibers of PLGA produced via electrospinning were combined in one multiscale scaffold to trigger biological response from MSC through mechanical and physical gradients inside the construct [51]. Even better resolution of polymer fibers could be achieved via melt electrospinning writing [52], a variation of the electrospinning technique raising more and more attention for biomaterial application.

5.3.4 Core–Shell Scaffolds

Technologies for scaffolding materials hold promising approaches toward therapeutic molecule delivery as well as TE with stem cells. Of these approaches, the core–shell design of scaffolds has been employed to attain these functions. In this concept, cells and therapeutic molecules can be encapsulated in different compartments of the strand in a coaxial manner to be delivered to its targeted site. Core–shell structures comprise an inner compartment, which is the "core" and an outer compartment (the "shell"), which surrounds the core completely. This design in structure offers a number of advantages, such as controllable and sequential release of signaling molecules and drugs loaded in the core and/or shell region. Cells can also benefit from being encapsulated in the core region by being protected in the inner part from the surrounding environment, allowing the shell to stabilize the less viscous core containing cells. The core–shell design can thus provide multifunctionality of a scaffold loaded with therapeutic molecules and cells in different compartments or different cells in different compartments, providing a suitable environment for cross talk and interaction through the core–shell interface. Perez et al. described how core and shell parts contribute to achieve cell encapsulation for TE as well as delivery of therapeutic molecules [53]. Figure 5.6 shows a schematic diagram illustrating fabrication of core–shell scaffolds.

In an attempt to design composite scaffolds for bone TE, Luo et al. fabricated alginate/nano-HAP (hydroxyapatite) composite core–shell scaffolds using 3D printing technique and *in situ* mineralization. With this technique, they could show how the presence of HAP enhanced the mechanical properties as well as cell attachment and differentiation of seeded cells with regard to pure alginate scaffolds (Figure 5.7) [55].

As mentioned above, growth factors and cells can be incorporated within the biomaterial before printing to ensure homogeneous distribution and localization within the coreshell struts. This was demonstrated by Akkineni et al. who showed that controlling shell thickness and composition of biomaterial resulted in different release behaviors of the loaded growth factors. Not only were they able

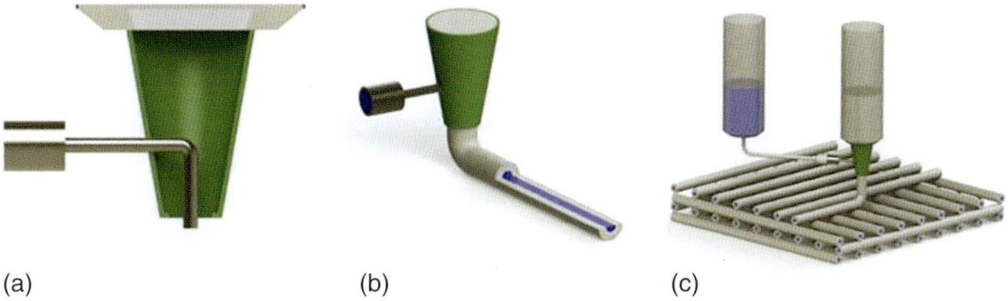

Figure 5.6 Schematic representation illustrating core–shell scaffold fabrication model. (a) Cross section of the coaxial needle. (b) Extruded core–shell strand. (c) Fabrication of a core–shell scaffold. Source: Akkineni et al. 2016 [54]. Reproduced with permission of IOP.

Figure 5.7 3D printed scaffolds of alginate/HAP core–shell scaffolds. (a) Scaffolds from left to right; mineralized alginate (pH 9.5), mixed alginate/HAP, and pure alginate. (b) Mineralized alginate with an inset showing core–shell morphology of cross-sectional view. (c) Mixed alginate/HAP. Scale bars: a, millimeters and b–e, 500 μm [55]. Source: Luo et al. 2015 [55]. Reproduced with permission of ACS.

to examine release characteristics of different growth factors but they could also demonstrate cell viability from endothelial cells encapsulated in the core [54].

Furthermore, Mistry et al., utilized the core–shell bioprinting technique to encapsulate human umbilical vein endothelial cells (HUVECs) and HepG2 cells in the core that is protected by a mechanically robust shell to demonstrate the formation of vessel-like structures and albumin secretion, respectively [56].

As mentioned earlier, cells can detect physical changes in the surrounding environment such as varying stiffness. This phenomenon and its influence on cells have been studied by many research groups. For instance, Perez et al. showed in his review how some studies demonstrated that the stiffness of core materials could directly influence the differentiation of encapsulated cells [53].

In this context, an elegant approach employed the idea of core–shell by utilizing a microfluidic device involving a double coaxial laminar flow to obtain meter-long microfibers encapsulating ECM proteins along with different types of cells with an alginate shell surrounding the core. After culturing in different core–shell devices, cells produced tubular structures after shell digestion. They could then show that certain cell types such as fibroblasts, endothelial cells, and myocytes could adhere to and proliferate when cultured on stiffer substrates and not on softer ones [57].

5.3.5 Additional Technical Equipment

Other innovative approaches that prevent structure deliquescence were also described in the literature including dispensing into hydrophobic fluorocarbon and printing in gelatin microparticle bath [37]. In addition, there are a number of solidification/gelation techniques in the form of devices coupled to the dispensing system of 3D printers such as an aerosol humidifier that continuously sprays $CaCl_2$ to the extruded strands of the alginate-based biomaterials (Figure 5.8a) [58]. For thermoresponsive materials, a cooling substrate, which is a deposition platform, can be used to assist in rapid initial solidification of gelatin Figure 5.8b [21]. Ionically cross-linked hydrogels can be printed in $CaCl_2$ or $BaCl_2$ cross-linking baths as shown in Figure 5.8c [59]. On the other hand, other equipment could also include a heating system, which allows controlling the temperature of the cartridge. This might be useful for applications involving 3D printing of gelatin-based scaffolds to allow for suitable extrusion [30].

5.3.6 Piezoelectric Pipetting Technology

Given the advancing progress and rising requirements of biofabrication, the combination of several alternative manufacturing concepts within one procedure or within one instrument offers tremendous new possibilities for scaffold production. One of those concepts is provided by piezoelectric pipetting units, a modified concept of inkjet droplet printing.

The technical background of inkjet printing was developed already some decades ago as a noncontact printing procedure based on thermally induced extrusion of liquid droplets. In general, for this fabrication method, a pipetting system is capable of processing low volumes of liquid samples, based on the

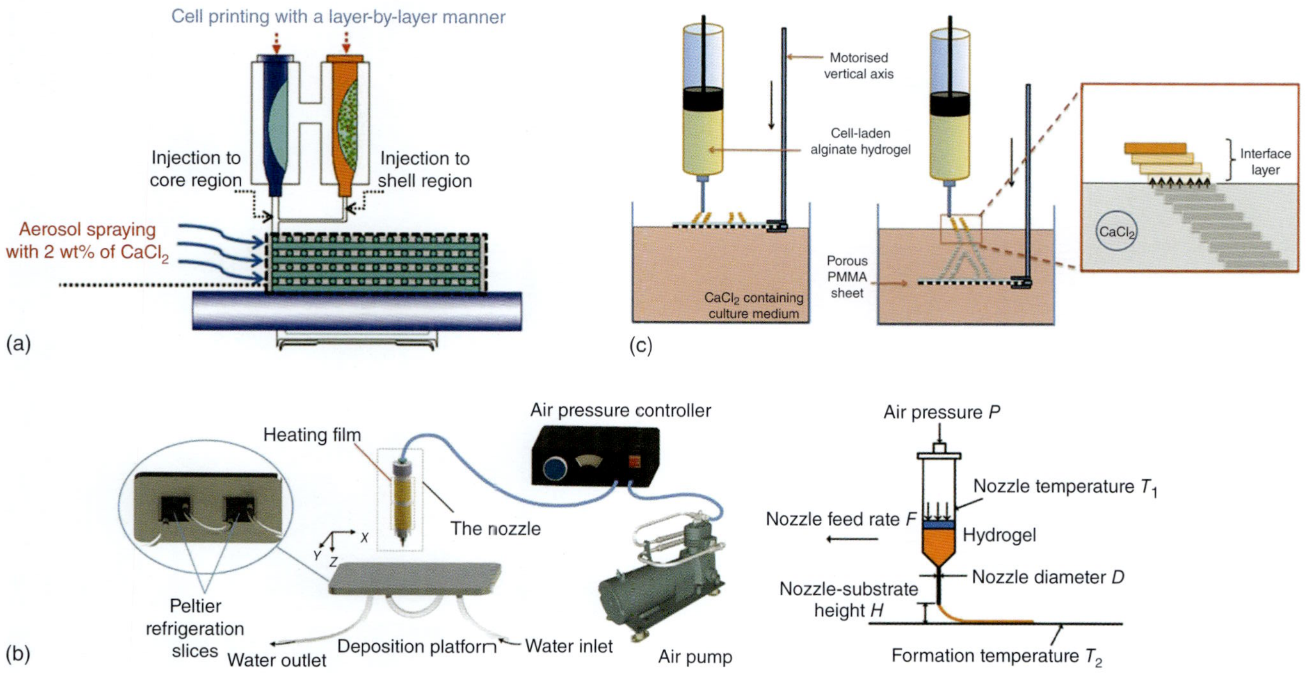

Figure 5.8 Technical equipment used during 3D printing process for solidification/gelation processes. (a) Scaffold fabrication by core–shell nozzle with an aerosol spraying device of $CaCl_2$ solution [58]. (b) Diagram of a 3D printed system with a heating film on the nozzle and a cooling deposition platform to fix the extruded bioink [21]. (c) Deposition of partially cross-linked a ginate hydrogel being submerged into $CaCl_2$ bath to create better support for upcoming layers [59]. Source: https://creativecommons.org/licenses/by/3.0/.

aspiration from microplates with dimensions designed according to the SBS (Society for Biomolecular Screening) standardization (96-well plates in the case of GeSiM BioScaffold printer). Deposition of droplets on 3D printed scaffolds could be problematic because of potential deviations of the exact Z-positions of the uppermost layers for different printing runs. Here, a noncontact and drop-on-demand approach such as inkjet printing offers benefits.

Using the GeSiM instrument, drops of volumes from 60 picoliters (pl) up to 50 nanoliters (nl) (using solenoid valve dispensers instead of piezoelectricity) can be applied at a speed of about $4\,\mathrm{m\,s^{-1}}$. The piezoelectric dispenser approaches the uppermost scaffold structure but keeps a distance of about 1 mm.

The piezoelectric inkjet technology is based on thin capillaries surrounded or covered by piezoelectric ceramic actuators. By applying electrical pulses (80–120 V), the actuator/capillary stack bends and induces an acoustic wave into the capillary structure (similar to a guitar string). The pulse duration is typically in the range of 10–50 ms. Each pulse delivers exactly one droplet. The maximum dispensing frequency depends both on the type of the dispenser and the sample properties. Frequencies of up to 1000 Hz (drops per second) are feasible, which allow coverage of a volume range up to a few microliters using those dispensers.

Another benefit of a pipetting system is the built-in rinsing bath. After inkjet printing, the remaining sample liquid will be washed off from the dispenser automatically. This system allows a subsequent and automated application of different samples to the 3D object on the printer tray (Figure 5.9).

5.3.7 Usage of Piezoelectric Inkjet Technology with Bioscaffolds

The piezoelectric or inkjet technology offers great potential for the spatially defined distribution and fine pipetting of cell suspensions as well as biologically active factors because of its highly biocompatible mechanism of placing liquid and low-viscosity polymer hydrogel droplets in a low-volume picoliter to

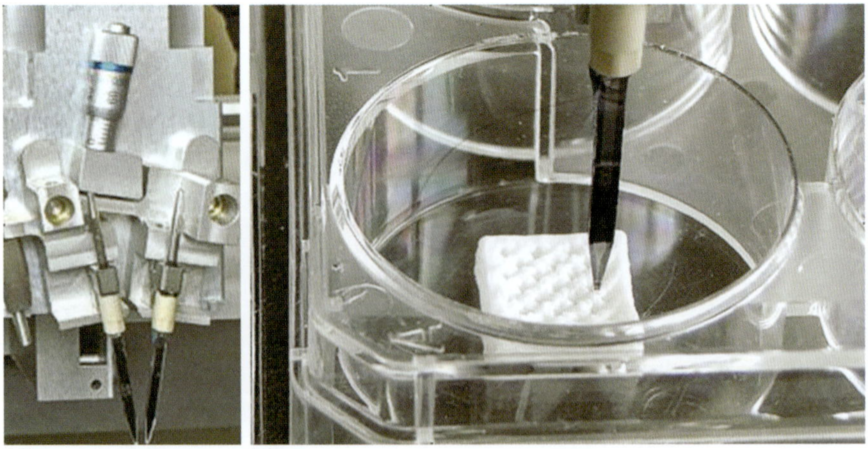

(a) (b)

Figure 5.9 Piezoelectric pipet at GeSiM BS3.2 instrument: Twin-tip pipet for mixing drops of two liquids (a); single-tip pipet (b).

nanoliter range in an ordered and defined structure. Those mechanisms enable printing of single ink droplets in a defined size controlled by the applied voltage, or larger volumes consisting of several droplets in one defined spot.

At the rising era of biofabrication, commercially available inkjet printers with usual ink cartridges were used and adjusted for a biological application. Early approaches suggested the use of the techniques to produce scaffolds and to include native ECM components or other biologically active compounds. This lead to the creation of patterns of proteins and biologically active factors [60] intended to support cell adhesion and growth, or guide cell migration via protein structures and factor gradients. The printing and micropatterning of sensitive DNA molecules on glass slides were also proven to be possible [61], enabling the formation of DNA microarrays.

Although the majority of those approaches applied thermal inkjet printing, which heats up the bioink and uses the resulting vaporizing effect inside the solution to create bubbles, early studies proved the biocompatibility of the process with mammalian cells: In hydrogels based on agar and collagen as bioinks, Xu et al. already in 2004 presented the printing and survival of cells from an ovarian hamster cell line [62] after initial studies on inkjet printing of bovine endothelial cells and smooth muscle cells into Matrigel™ and collagen gel [63]. Later, an early approach of vasculature printing was introduced, which processes endothelial cells in a thrombin/$CaCl_2$ bioink, along with a fibrin hydrogel forming microchannels that were lined with the added endothelial cells on a fibrinogen substrate [64].

Since this process of thermal droplet formation includes high temperature, it is not compatible with the printing of viable cells. Therefore, piezoelectric approaches to generate droplets gain more and more interest. The delivery of human fibroblasts via inkjet printing based on piezoelectric pipetting was presented by Saunders and Derby [65]. However, in cell solutions, agglomerates and sediments were formed during the process. To prevent this, higher viscosity of cell suspension was required as presented by Hoch et al., who modified gelatin with acryl and acetyl groups enabling photo-crosslinking after printing, proving biocompatibility in combination with porcine articular chondrocytes [66]. Fibroblasts were printed via a piezoelectric actuator in sodium alginate [67] while alginate viscosity and stability for inkjet printing have been triggered by addition of poly(vinyl alcohol) (PVA) (20%) and hyaluronic acid (3%) and printing into $CaCl_2$ solution [68]. A fine cell distribution can be achieved by lowering cell density to single cells per droplet, as suggested by Yamaguchi et al. in a saline buffer solution [69]. Another freeform approach presented the manufacturing of vessel-like structures from a fibroblast cell line processed in a sodium alginate gel with the support of a calcium chloride bath [70].

To overcome one of the limitations of such a low-volumetric technique, which is the usage in clinically relevant size, the technique was combined with a bioscaffolding process to functionalize the surface of a scaffold 3D plotted from CPC with well-defined volumes of cells or biomolecules [71] in a gradient pattern. Another study presenting piezoelectric pipetting without a bioplotted scaffold was introduced by Detsch et al. [72] who used a bioink composed of fibrinogen and thrombin with encapsulated ST-2 cells, a bone-marrow-derived stromal cell line.

All those examples reveal a great potential of the piezoelectric printing to achieve a high accuracy of cells and factors inside small volumes. However, constructs in real volumetric dimensions have not been realized by this method yet because the processing of paste-like materials that would result in a real stable 3D structure is not possible. This represents one of the challenges of current inkjet bioprinting research, improving the speed, volume, construct stability, and cell density of the created structures.

5.4 Applications of Bioscaffolder and Bioprinting Systems

5.4.1 Individualized Implants and Tissue Constructs

Because of the rapid development inside the field, highly innovative examples of specific applications that are currently realized by introduced bioscaffolding or bioprinting strategies can be highlighted:

The key goal of the entire field of TE and biofabrication is the generation of patient-individualized implants resolving the shortcomings of autologous tissue replacement or shortage of organ donors. By an automated CAD or CAM (computer-aided manufacturing) process, solid implants and also multimaterial macroporous scaffolds can be generated. A modern plotter software environment is able to generate a layered design from original STL for one material or 3MF files for multiphasic and multimaterial constructs. The idea is to use clinical imaging data (such as CT or MRI) to generate defined constructs of physiological and anatomically relevant structures [73], and particularly constructs filling up clinical size defects and replacing lost tissue parts.

To create and stabilize the overhanging structures until cross-linking, sacrificial materials need to be used, which can be easily dissolved by a medium or temperature change afterward, such as pluronic F127 or low-concentrated methylcellulose. One example that was recently realized in this way using the latest generation of GeSiM BioScaffold printer was the printing of a 3D model of a human scaphoid bone extracted from the CT data of a hand. The files were generated using publicly available open source software. The data were extracted, virtualized, segmented, and fabricated as a bone-mimicking construct from extrudable CPC paste without macropores (Figure 5.10) [74], which is able to form HAP nanocrystals after setting of the cement and is already used in the clinics as a self-hardening defect-filling material.

Another geometrical example that is often presented is the generation of a cartilage-replacing model resembling a human ear. For patients suffering from burn wounds, or children born with ear deformities because of a developmental defect, an individually designed prosthetic ear construct could be an attractive

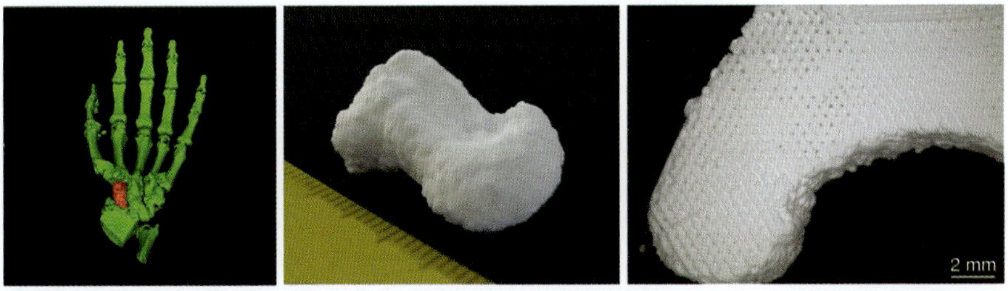

(a) (b) (c)

Figure 5.10 CT data extraction and translation to a STL file (a) to design and additively fabricate a human scaphoid bone from pasty CPC in actual physiological dimensions (b) after using open source software. Fabrication via layer-by-layer strand deposition, bar = 2 mm (c) [74].

option. Furthermore, bioprinting approaches might allow the inclusion of chondrocytes in a stable hydrogel to improve integrity and functionality. For another cell-printing approach of freeform 3D constructs from an alginate-based bioink, Hinton et al. presented the generation of branched blood vessel structures, as well as a structure resembling an embryonic chick heart created from a 3D CAD model, and downscaled the 3D model of a human brain after extraction of MRI data [75]. However, the mimicking function in complex volumetric organs still remains a tremendous challenge. Considering the possibility of including viable cells, this will certainly take more effort of optimization and combination of different techniques to ensure viability, functionality, and therefore resemblance of native tissues or organs (see Section 5.4.3).

5.4.2 Green Bioprinting

Because mammalian cells have been intensively studied for TE applications and bioprinting, it was of interest to investigate the 3D plotting of cells from other species, too. This could be of great impact for biotechnological applications in future [76]. A study demonstrated the successful bioprinting of microalgae embedded in an alginate/methylcellulose-based hydrogel blend. The team could prove the viability and survival of the algae through oxygen release experiments measuring photosynthetic activity. With the advancements of bioscaffolders equipped with a three-channel system, multichannel plotting was employed to produce a coculture system with human cells and microalgae printed simultaneously within one scaffold. This system can offer new therapeutic strategies utilizing oxygen delivery as well as secondary metabolites to the surrounding human cells [77].

Green bioprinting also extended to bioprinting of not only algae but also plant cells in another study encapsulating basil cells in a hydrogel mixture of alginate, agarose, and methylcellulose, fabricating 3D constructs by extrusion-based plotting. This model of plant-cell-laden hydrogel matrices could serve, for example, as platforms in industrial biotechnology to study cell responses to environmental influences [78].

5.4.3 Challenges for Clinical Applications of Bioprinted Scaffolds in Tissue and Organ Engineering

Despite the promising development inside the field coming up with new strategies and concepts frequently, and the enthusiastic public expectations, the actual translation of bioprinting ideas to *in vivo* studies or even application in clinical settings remains a great challenge. Completing the step from designed and manufactured (patient specific) metal implants and nonporous structures into the direction of functional tissues – cell-laden, preseeded, or *in situ*-seeded – and their clinical usage is more complex than it had been for 3D printed or laser-sintered individual implants from, e.g., titanium. The few actually promising *in vivo* studies of bioscaffolding or bioprinting had been largely performed on bone, skin, and cartilage tissue [79].

PCL is the most commonly used thermoplastic polymer implemented in 3D printing as a scaffold material or a stabilizing factor for cell-laden hydrogels. It is approved for clinical application and therefore holds a great potential to treat patients with malformations or defects, particularly in bone tissue. However, it does not really resemble the structure and function of native tissue. Besides PCL, not many materials applicable for bioscaffolding hold a permission to be applied in clinical setting. Here, the use of the approved CPCs as described in Section 5.4.1 could be favored for the fabrication of bone tissue. Other concepts combined PCL scaffolds with additional β-tricalcium phosphate [80]. However, those *in vivo* studies did not include cells inside their constructs. One pitfall of many bioprinting approaches still remains the aspect that most studies only include immortalized cell lines, derived from cancer patients or from nonhuman origin and, therefore, are less sensitive to processing and 3D cultivation conditions. To move closer to translation, primary cells should be used, potentially even from autologous source. One autologous approach on the automated generation of human skin was presented by Cubo et al. in 2016 [81]. They used fibrinogen from human plasma as basis for their bioink in that they encapsulated fibroblasts, later seeded by layers of keratinocytes. Another type of tissue for which research showed promising approaches is adipose tissue for breast reconstruction [82]. The study proved the applicability of poly(D,L)lactide scaffolds with a patient-specific customized design, after being seeded with human umbilical cord perivascular cells, in rat experiments.

Much effort is taken also on the research for providing vasculature inside the volumetric tissue constructs to prevent cells from a lack of oxygen and nutrients during long-term incubation. Scientists came up with approaches combining different concepts of biofabrication and TE [49, 83] using the interaction of endothelial cells with fibroblasts, smooth muscle cells, or stem cells to create perfusable vessels inside 3D printed constructs. Another critical aspect for tissue fabrication is presented by the high cell numbers seeded or encapsulated that are required to enable the complex functions and viability of a large construct. Expansion of largely stem cells in this context is suggested inside bioreactors over longer time intervals [84]. On the materials side, future approaches need to provide the ability of being upscaled to GMP processes fulfilling all regulatory requirements on the one hand and biological and mechanical properties on the other hand.

For complex internal organs such as liver tissues, science faced with an even greater challenge of combining different cellular systems with several large- and microscaled vessels and fluidic pathways. For some complex tissues, a combination of organoid culture and biofabrication processes was suggested for future research [85]. The fine network of renal arteries could be a potential target of core–shell bioprinting techniques when able to produce structures in very fine dimensions with a high resolution. Besides, this includes one of the most intensively studied aspects, the multicellular composition of tissue/organs, essential for the function and integration/regeneration in native environment, migration, and remodeling of cell environment. However, the requirement of the plotter being able to steer different cartridges with varying materials is not the only critical aspect. Furthermore, the coculture conditions need to be adjusted and optimized to maintain all included cell types. Even in 2D environment, this is a great task to fulfill.

A closer application of bioprinted constructs that was more feasible would be the use of 3D structures as volumetric models in basic research [86], providing a basis for drug testing or spatially defined co-culture experiments, which also marks an important milestone for biomedical research.

5.4.4 4D Printing

4D printing has recently emerged with the concept of combining a fourth dimension to a 3D bioprinting technology, which is "time". Time in this context would refer to the evolution of the 3D bioprinted construct over time after having finished the printing process. With this new factor in consideration, the shape and functionality of the printed constructs can be influenced by external stimuli such as swelling, temperature, magnetic field, or post-printing self-assembly. Two major approaches are being employed in 4D printing. First, the utilization of responsive materials in 4D printing, also known as "smart materials" that have the ability to change shape in response to the external stimuli [87]. Second is the maturation and formation of functional tissue from engineered constructs after printing through self-organization or matrix deposition in cell-laden hydrogels.

Examples on this approach include a study in which 4D printing was used to design active origami where shape memory polymer fibers were used as composites for printing in an elastomeric matrix, which enabled the formation of origami folding patterns [87].

Another popular example is poly(*N*-isopropylacrylamide) (PNIPAAm), which has been widely known for its applications in drug delivery and tissue regeneration. One of its biomedical applications in 4D printing was described by a research group, which designed a partially biodegradable thermoresponsive self-folding capsule. This was achieved by using photolithography to print a bilayered construct made of thermoresponsive PNIPAAm and water-insoluble PCL. The swelling and collapse of PNIPAAm in response to temperature caused the starlike capsule to self-fold and unfold, reversibly encapsulating/releasing yeast cells (Figure 5.11) [88].

A quite recent approach employed the utilization of shape memory polyurethane to produce 4D scaffolds in a layer-by-layer approach. The

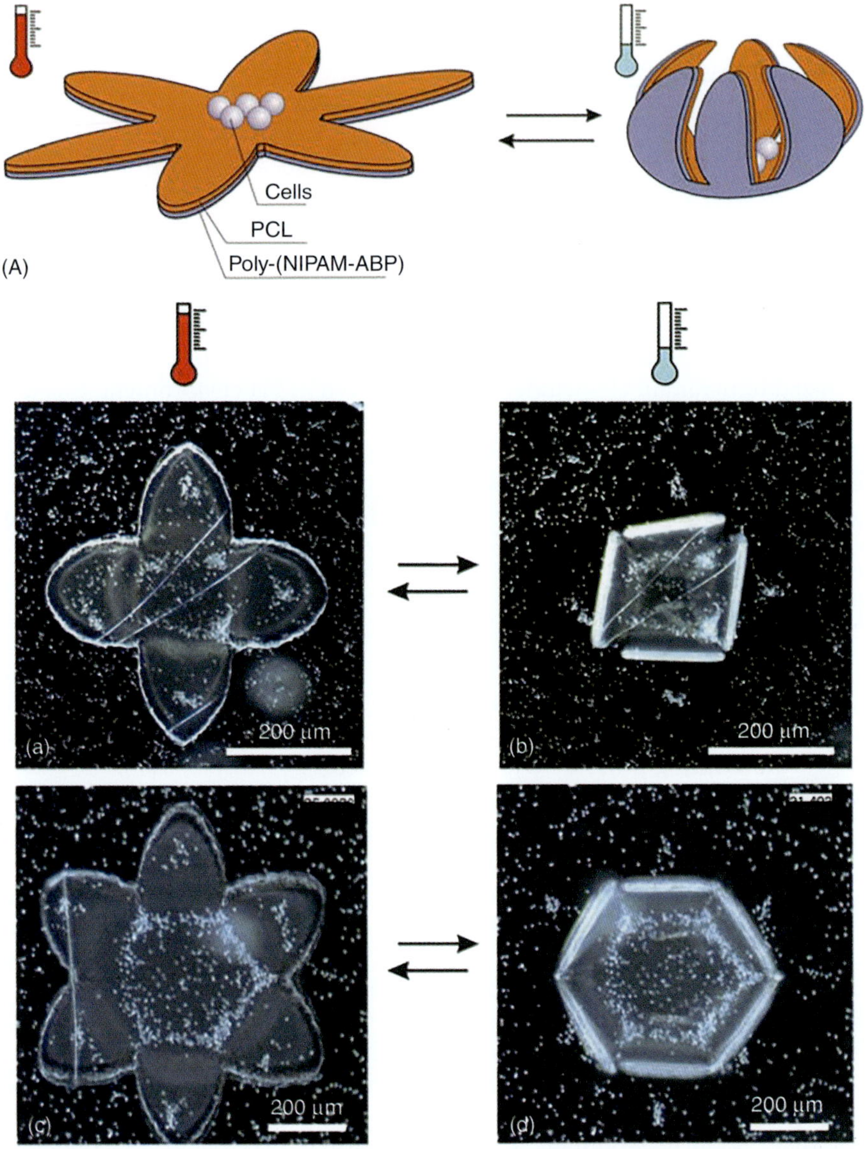

Figure 5.11 Illustration of thermoresponsive self-folding capsule. (A) A scheme of a star-shaped polymer bilayer containing cells that fold upon swelling at a lower temperature. (B) The self-folding capsules (dark field optical microscopy). Capsules containing cells are folded because of swelling at low temperature (b,d). Upon heating, unfolding of capsule and release of cells take place (a,c). Source: Stoychev et al. 2011 [88]. Reproduced with permission of Royal Society of Chemistry.

materials showed good recovery of the original shape of the scaffold in response to thermomechanical stimuli. In this study, they could also show that the shape recovery influenced the morphology of seeded cells, making these constructs important in clinical applications in the future [89].

5.5 Conclusion

The technology of bioscaffolding offers a great potential to the field of TE because of its ability of creating defined scaffold constructs from various extrudable biomaterials. Multimaterial approaches pave the way to the fabrication of cell-delivering structures for tissue interfaces and for the study of cell behavior in different environments.

Implementing bioprinting approaches even extends the possibilities of extrusion-based 3D printing technologies in biomedical research. Controlling spatial distribution of different cell types, defining cell densities, and maintaining cell viability and phenotype over longer time intervals are a great achievement. However, some limitations still exist. The challenge for current and future research is to overcome the current pitfalls: Promising research is performed on biocompatible stabilization of volumetric tissue structures, individualization of constructs, as well as on vascularization of artificial tissues, already enabling to mimic tissue structures in long-term 3D *in vitro* models. Nevertheless, a combination of several disciplines, international cooperation between engineers, computer specialists, physicians, and biomedical scientists of different fields, as well as collaboration with companies distributing and further developing innovative technologies, will still be the necessary basis for future research success in the direction of artificial tissue and organ structures.

References

1 De Deyne, P.G. and Kladakis, S.M. (2005). Bioscaffolds in tissue engineering: a rationale for use in the reconstruction of musculoskeletal soft tissues. *Clinics in Podiatric Medicine and Surgery* 22 (4): 521–532. https://doi.org/10.1016/j.cpm.2005.07.006.

2 Malda, J., Visser, J., Melchels, F.P. et al. (2013). 25th anniversary article: engineering hydrogels for biofabrication. *Advanced Materials* 25 (36): 5011–5028. https://doi.org/10.1002/adma.201302042.

3 Langer, R. and Vacanti, J.P. (1993). Tissue engineering. *Science* 260 (May): 920–926. https://doi.org/10.1126/science.8493529.

4 Do, A., Khorsand, B., Geary, S.M., and Salem, A.K. (2015). 3D printing of scaffolds for tissue regeneration applications. *Advanced Healthcare Materials* 4 (12): 1742–1762. https://doi.org/10.1002/adhm.201500168.3D.

5 Groll, J., Boland, T., Blunk, T. et al. (2016). Biofabrication: reappraising the definition of an evolving field. *Biofabrication* 8 (1): https://doi.org/10.1088/1758-5090/8/1/013001.

6 Gurtner, G.C., Callaghan, M.J., and Longaker, M.T. (2007). Progress and potential for regenerative medicine. *Annual Review of Medicine* 58 (1): 299–312. https://doi.org/10.1146/annurev.med.58.082405.095329.

7 Chen, F.M. and Liu, X. (2016). Advancing biomaterials of human origin for tissue engineering. *Progress in Polymer Science* 53: 86–168. https://doi.org/10.1016/j.progpolymsci.2015.02.004.

8 Singh, D., Singh, D., and Han, S.S. (2016). 3D printing of scaffold for cells delivery: advances in skin tissue engineering. *Polymer* 8 (1): 1–17. https://doi.org/10.3390/polym8010019.

9 Kang, H.W., Lee, S.J., Ko, I.K. et al. (2016). A 3D bioprinting system to produce human-scale tissue constructs with structural integrity. *Nature Biotechnology* 34 (3): 312–319. https://doi.org/10.1038/nbt.3413.

10 Yang, S., Leong, K.-F., Du, Z., and Chua, C.-K. (2001). The design of scaffolds for use in tissue engineering. Part I. Traditional factors. *Tissue Engineering* 7 (6): 679–689. https://doi.org/10.1089/107632701753337645.

11 Chia, H.N. and Wu, B.M. (2015). Recent advances in 3D printing of biomaterials. *Journal of Biological Engineering* 9 (1): 1–14. https://doi.org/10.1186/s13036-015-0001-4.

12 Sydney Gladman, A., Matsumoto, E.A., Nuzzo, R.G. et al. (2016). Biomimetic 4D printing. *Nature Materials* 15 (4): 413–418. https://doi.org/10.1038/nmat4544.

13 Loh, Q.L. and Choong, C. (2013). Three-dimensional scaffolds for tissue engineering applications: role of porosity and pore size. *Tissue Engineering Part B: Reviews* 19 (6): 485–502. https://doi.org/10.1089/ten.teb.2012.0437.

14 Leong, K.F., Cheah, C.M., and Chua, C.K. (2003). Solid freeform fabrication of three-dimensional scaffolds for engineering replacement tissues and organs. *Biomaterials* 24 (13): 2363–2378. https://doi.org/10.1016/S0142-9612(03)00030-9.

15 Laronda, M.M., Rutz, A.L., Xiao, S. et al. (2017). A bioprosthetic ovary created using 3D printed microporous scaffolds restores ovarian function in sterilized mice. *Nature Communications* 8 (May): 1–10. https://doi.org/10.1038/ncomms15261.

16 Chan, B.P. and Leong, K.W. (2008). Scaffolding in tissue engineering: general approaches and tissue-specific considerations. *European Spine Journal* 17: suppl. 4). https://doi.org/10.1007/s00586-008-0745-3.

17 O'Brien, F.J. (2011). Biomaterials & scaffolds for tissue engineering. *Materials Today* 14 (3): 88–95. https://doi.org/10.1016/S1369-7021(11)70058-X.

18 Gariboldi, M.I. and Best, S.M. (2015). Effect of ceramic scaffold architectural parameters on biological response. *Frontiers in Bioengineering and Biotechnology* 3 (October): 1–11. https://doi.org/10.3389/fbioe.2015.00151.

19 Murphy, S.V. and Atala, A. (2014). 3D bioprinting of tissues and organs. *Nature Biotechnology* 32 (8): 773–785. https://doi.org/10.1038/nbt.2958.

20 Chung, J.H.Y., Naficy, S., Yue, Z. et al. (2013). Bio-ink properties and printability for extrusion printing living cells. *Biomaterials Science* 1 (7): 763. https://doi.org/10.1039/c3bm00012e.

21 He, Y., Yang, F., Zhao, H. et al. (2016). Research on the printability of hydrogels in 3D bioprinting. *Scientific Reports* 6: 1–13. https://doi.org/10.1038/srep29977.
22 Nicodemus, G.D. and Bryant, S.J. (2008). Cell encapsulation in biodegradable hydrogels for tissue engineering applications. *Tissue Engineering Part B: Reviews* 14 (2): 149–165. https://doi.org/10.1089/ten.teb.2007.0332.
23 Baumann, B., Jungst, T., Stichler, S. et al. (2017). Control of nanoparticle release kinetics from 3D printed hydrogel scaffolds. *Angewandte Chemie – International Edition* 56 (16): 4623–4628. https://doi.org/10.1002/anie.201700153.
24 Axpe, E. and Oyen, M.L. (2016). Applications of alginate-based bioinks in 3D bioprinting. *International Journal of Molecular Sciences* 17.
25 Niklason, L.E. and Langer, R.S. (1997). Advances in tissue engineering of blood vessels and other tissues. *Transplant Immunology* 5 (4): 303–306. https://doi.org/10.1016/S0966-3274(97)80013-5.
26 Lode, A., Meyer, M., Brüggemeier, S. et al. (2016). Additive manufacturing of collagen scaffolds by three-dimensional plotting of highly viscous dispersions. *Biofabrication* 8 (1): https://doi.org/10.1088/1758-5090/8/1/015015.
27 Ahlfeld, T., Akkineni, A.R., Förster, Y. et al. (2017). Design and fabrication of complex scaffolds for bone defect healing: combined 3D plotting of a calcium phosphate cement and a growth factor-loaded hydrogel. *Annals of Biomedical Engineering* 45 (1): 224–236. https://doi.org/10.1007/s10439-016-1685-4.
28 Tu, X., Wang, L., Wei, J. et al. (2016). 3D printed PEGDA microstructures for gelatin scaffold integration and neuron differentiation. *Microelectronic Engineering* 158: 30–34. https://doi.org/10.1016/j.mee.2016.03.007.
29 Shipley, R.J., Jones, G.W., Dyson, R.J. et al. (2009). Design criteria for a printed tissue engineering construct: a mathematical homogenization approach. *Journal of Theoretical Biology* 259 (3): 489–502. https://doi.org/10.1016/j.jtbi.2009.03.037.
30 Luo, Y., Lode, A., Akkineni, A.R., and Gelinsky, M. (2015). Concentrated gelatin/alginate composites for fabrication of predesigned scaffolds with a favorable cell response by 3D plotting. *RSC Advances* 5 (54): 43480–43488. https://doi.org/10.1039/C5RA04308E.
31 Moroni, L., Boland, T., Burdick, J.A. et al. (2017). Biofabrication: a guide to technology and terminology. *Trends in Biotechnology* https://doi.org/10.1016/j.tibtech.2017.10.015.
32 Kilian, D., Ahlfeld, T., Akkineni, A.R. et al. (2017). Three-dimensional bioprinting of volumetric tissues and organs. *MRS Bulletin* 42 (8): 585–592.
33 Markstedt, K., Mantas, A., Tournier, I. et al. (2015). 3D bioprinting human chondrocytes with nanocellulose-alginate bioink for cartilage tissue engineering applications. *Biomacromolecules* 16 (5): 1489–1496.
34 Schütz, K., Placht, A.M., Paul, B. et al. (2017). Three-dimensional plotting of a cell-laden alginate/methylcellulose blend: towards biofabrication of tissue engineering constructs with clinically relevant dimensions. *Journal of Tissue Engineering and Regenerative Medicine* 11 (5): 1574–1587. https://doi.org/10.1002/term.2058.

35 Müller, M., Becher, J., Schnabelrauch, M., and Zenobi-Wong, M. (2015). Nanostructured pluronic hydrogels as bioinks for 3D bioprinting. *Biofabrication* 7 (3): 35006. https://doi.org/10.1088/1758-5090/7/3/035006.

36 Jia, J., Richards, D.J., Pollard, S. et al. (2014). Engineering alginate as bioink for bioprinting. *Acta Biomaterialia* 10 (10): 4323–4331. https://doi.org/10.1016/j.actbio.2014.06.034.

37 Duarte Campos, D.F., Blaeser, A., Weber, M. et al. (2013). Three-dimensional printing of stem cell-laden hydrogels submerged in a hydrophobic high-density fluid. *Biofabrication* 5 (1): https://doi.org/10.1088/1758-5082/5/1/015003.

38 Wu, Z., Su, X., Xu, Y. et al. (2016). Bioprinting three-dimensional cell-laden tissue constructs with controllable degradation. *Scientific Reports* 6 (1): 24474. https://doi.org/10.1038/srep24474.

39 Köpf, M., Campos, D.F.D., Blaeser, A. et al. (2016). A tailored three-dimensionally printable agarose – collagen blend allows encapsulation, spreading, and attachment of human umbilical artery smooth muscle cells. *Biofabrication* 8 (2): 1–15.

40 Yeo, M.G. and Kim, G.H. (2017). A cell-printing approach for obtaining hASC-laden scaffolds by using a collagen/polyphenol bioink. *Biofabrication* 9: https://doi.org/https://doi.org/10.1088/1758-5090/aa6997.

41 Kim, B.S., Kim, H., Gao, G. et al. (2017). Decellularized extracellular matrix: a step towards the next generation source for bioink manufacturing. *Biofabrication* 9 (3): https://doi.org/10.1088/1758-5090/aa7e98.

42 Ahlfeld, T., Cidonio, G., Kilian, D. et al. (2017). Development of a clay based bioink for 3D cell printing for skeletal application. *Biofabrication* 9 (3): https://doi.org/10.1088/1758-5090/aa7e96.

43 Stichler, S., Jungst, T., Schamel, M. et al. (2017). Thiol–ene clickable poly(glycidol) hydrogels for biofabrication. *Annals of Biomedical Engineering* 45 (1): 273–285. https://doi.org/10.1007/s10439-016-1633-3.

44 Kesti, M., Eberhardt, C., Pagliccia, G. et al. (2015). Bioprinting complex cartilaginous structures with clinically compliant biomaterials. *Advanced Functional Materials* 25 (48): 7406–7417. https://doi.org/10.1002/adfm.201503423.

45 Freeman, F.E. and Kelly, D.J. (2017). Tuning alginate bioink stiffness and composition for controlled growth factor delivery and to spatially direct MSC Fate within bioprinted tissues. *Scientific Reports* 7 (1): 1–12. https://doi.org/10.1038/s41598-017-17286-1.

46 Di Luca, A., Klein-Gunnewiek, M., Vancso, J.G. et al. (2017). Covalent binding of bone morphogenetic protein-2 and transforming growth factor-β3 to 3D plotted scaffolds for osteochondral tissue regeneration. *Biotechnology Journal* 12 (12): 1–9. https://doi.org/10.1002/biot.201700072.

47 Schuurman, W., Khristov, V., Pot, M.W. et al. (2011). Bioprinting of hybrid tissue constructs with tailorable mechanical properties. *Biofabrication* 3 (2): https://doi.org/10.1088/1758-5082/3/2/021001.

48 Zehnder, T., Freund, T., Demir, M. et al. (2016). Fabrication of cell-loaded two-phase 3D constructs for tissue engineering. *Materials* 9 (11): https://doi.org/10.3390/ma9110887.

49 Kolesky, D.B., Homan, K.A., Skylar-Scott, M.A., and Lewis, J.A. (2016). Three-dimensional bioprinting of thick vascularized tissues. *Proceedings of the National Academy of Sciences* 113 (12): 3179–3184. https://doi.org/10.1073/pnas.1521342113.

50 Detsch, R., Sarker, B., Zehnder, T. et al. (2014). Additive manufacturing of cell-loaded alginate enriched with alkaline phosphatase for bone tissue engineering application. *BioNanoMaterials* 15 (3–4): 79–87. https://doi.org/10.1515/bnm-2014-0007.

51 Criscenti, G., Longoni, A., Di Luca, A. et al. (2016). Triphasic scaffolds for the regeneration of the bone ligament interface. *Biofabrication* 8 (1): 015009.

52 Brown, T.D., Dalton, P.D., and Hutmacher, D.W. (2011). Direct writing by way of melt electrospinning. *Advanced Materials* 23 (47): 5651–5657. https://doi.org/10.1002/adma.201103482.

53 Perez, R.A. and Kim, H.W. (2015). Core–shell designed scaffolds for drug delivery and tissue engineering. *Acta Biomaterialia* 21: 2–19. https://doi.org/10.1016/j.actbio.2015.03.013.

54 Akkineni, A.R., Ahlfeld, T., Lode, A., and Gelinsky, M. (2016). A versatile method for combining different biopolymers in a core/shell fashion by 3D plotting to achieve mechanically robust constructs. *Biofabrication* 8 (4): 1–15. https://doi.org/10.1088/1758-5090/8/4/045001.

55 Luo, Y., Lode, A., Wu, C. et al. (2015). Alginate/nanohydroxyapatite scaffolds with designed core/shell structures fabricated by 3D plotting and in situ mineralization for bone tissue engineering. *ACS Applied Materials and Interfaces* 7 (12): 6541–6549. https://doi.org/10.1021/am508469h.

56 Mistry, P., Aied, A., Alexander, M. et al. (2017). Bioprinting using mechanically robust core–shell cell-laden hydrogel strands. *Macromolecular Bioscience* 17 (6): 1–8. https://doi.org/10.1002/mabi.201600472.

57 Onoe, H., Okitsu, T., Itou, A. et al. (2013). Metre-long cell laden microfibres exhibit tissue morphologies and functions. *Nature Materials* 12 (6): 584–590. https://doi.org/10.1038/nmat3606.

58 Ahn, S., Lee, H., and Kim, G. (2013). Functional cell-laden alginate scaffolds consisting of core/shell struts for tissue regeneration. *Carbohydrate Polymers* 98 (1): 936–942. https://doi.org/10.1016/j.carbpol.2013.07.008.

59 Tabriz, G., Ayala, H., and Nicholas, R. (2015). Three-dimensional bioprinting of complex cell laden alginate hydrogel structures. *Biofabrication* https://doi.org/10.1088/1758-5090/7/4/045012.

60 Roth, E.A., Xu, T., Das, M. et al. (2004). Inkjet printing for high-throughput cell patterning. *Biomaterials* 25 (17): 3707–3715. https://doi.org/10.1016/j.biomaterials.2003.10.052.

61 Okamoto, T., Suzuki, T., and Yamamoto, N. (2000). Microarray fabrication with covalent attachment of DNA using Bubble Jet technology. *Nature Biotechnology* 18 (4): 438–441. https://doi.org/10.1038/74507.

62 Xu, T., Jin, J., Gregory, C. et al. (2005). Inkjet printing of viable mammalian cells. *Biomaterials* 26 (1): 93–99. https://doi.org/10.1016/j.biomaterials.2004.04.011.

63 Wilson, W.C. and Boland, T. (2003). Cell and organ printing 1: protein and cell printers. *The Anatomical Record* 272A (2): 491–496. https://doi.org/10.1002/ar.a.10057.

64 Cui, X. and Boland, T. (2009). Human microvasculature fabrication using thermal inkjet printing technology. *Biomaterials* 30 (31): 6221–6227. https://doi.org/10.1016/j.biomaterials.2009.07.056.

65 Saunders, R.E. and Derby, B. (2014). Inkjet printing biomaterials for tissue engineering: bioprinting. *International Materials Reviews* 59 (8): 430–448. https://doi.org/10.1179/1743280414Y.0000000040.

66 Hoch, E., Hirth, T., Tovar, G.E.M., and Borchers, K. (2013). Chemical tailoring of gelatin to adjust its chemical and physical properties for functional bioprinting. *Journal of Materials Chemistry B* 1 (41): 5675. https://doi.org/10.1039/c3tb20745e.

67 Xu, C., Chai, W., Huang, Y., and Markwald, R.R. (2012). Scaffold-free inkjet printing of three-dimensional zigzag cellular tubes. *Biotechnology and Bioengineering* 109 (12): 3152–3160. https://doi.org/10.1002/bit.24591.

68 Arai, K., Iwanaga, S., Toda, H. et al. (2011). Three-dimensional inkjet biofabrication based on designed images. *Biofabrication* 3 (3): https://doi.org/10.1088/1758-5082/3/3/034113.

69 Yamaguchi, S., Ueno, A., Akiyama, Y., and Morishima, K. (2012). Cell patterning through inkjet printing of one cell per droplet. *Biofabrication* 4 (4): https://doi.org/10.1088/1758-5082/4/4/045005.

70 Christensen, K., Xu, C., Chai, W. et al. (2015). Freeform inkjet printing of cellular structures with bifurcations. *Biotechnology and Bioengineering* 112 (5): 1047–1055. https://doi.org/10.1002/bit.25501.

71 Giron, S., Lode, A., and Gelinsky, M. (2016). In situ functionalization of scaffolds during extrusion-based 3D plotting using a piezoelectric nanoliter pipette. *Journal of 3D Printing in Medicine* 1 (1).

72 Detsch, R., Blob, S., Zehnder, T., and Boccaccini, A.R. (2016). Evaluation of cell inkjet printing technique for biofabrication. *BioNanoMaterials* 17 (3–4): 185–191. https://doi.org/10.1515/bnm-2016-0007.

73 Hutmacher, D.W., Sittinger, M., and Risbud, M.V. (2004). Scaffold-based tissue engineering: rationale for computer-aided design and solid free-form fabrication systems. *Trends in Biotechnology* 22 (7): 354–362. https://doi.org/10.1016/j.tibtech.2004.05.005.

74 Heller, M., Bauer, H.-K., Goetze, E. et al. (2016). Materials and scaffolds in medical 3D printing and bioprinting in the context of bone regeneration. *International Journal of Computerized Dentistry* 19 (4): 301–321. Retrieved from http://www.ncbi.nlm.nih.gov/pubmed/28008428.

75 Hinton, T.J., Jallerat, Q., Palchesko, R.N. et al. (2015). Three-dimensional printing of complex biological structures by freeform reversible embedding of suspended hydrogels. *Science Advances* 1 (9): e1500758. https://doi.org/10.1126/sciadv.1500758.

76 Krujatz, F., Lode, A., Seidel, J. et al. (2017). Additive Biotech—chances, challenges, and recent applications of additive manufacturing technologies in biotechnology. *New Biotechnology* 39 (September): 222–231.

77 Lode, A., Krujatz, F., Brüggemeier, S. et al. (2015). Green bioprinting: fabrication of photosynthetic algae-laden hydrogel scaffolds for biotechnological and medical applications. *Engineering in Life Sciences* 15 (2): 177–183. https://doi.org/10.1002/elsc.201400205.
78 Seidel, J., Ahlfeld, T., Adolph, M. et al. (2017). Green bioprinting: extrusion-based fabrication of plant cell-laden biopolymer hydrogel scaffolds. *Biofabrication* 9 (4): https://doi.org/10.1088/1758-5090/aa8854.
79 Hong, N., Yang, G.-H., Lee, J., and Kim, G. (2017). 3D bioprinting and its in vivo applications. *Journal of Biomedical Materials Research Part B: Applied Biomaterials* 444–459. https://doi.org/10.1002/jbm.b.33826.
80 Shim, J.H., Won, J.Y., Park, J.H. et al. (2017). Effects of 3D-printed polycaprolactone/β-tricalcium phosphate membranes on guided bone regeneration. *International Journal of Molecular Sciences* 18 (5): https://doi.org/10.3390/ijms18050899.
81 Cubo, N., Garcia, M., Cañizo, J.F. et al. (2016). 3D bioprinting of functional human skin: production and in vivo analysis. *Biofabrication* 9 (1): 15006. https://doi.org/10.1088/1758-5090/9/1/015006.
82 Chhaya, M.P., Melchels, F.P.W., Holzapfel, B.M. et al. (2015). Sustained regeneration of high-volume adipose tissue for breast reconstruction using computer aided design and biomanufacturing. *Biomaterials* 52 (1): 551–560. https://doi.org/10.1016/j.biomaterials.2015.01.025.
83 Jia, W., Gungor-ozkerim, P.S., Zhang, Y.S. et al. (2016). Direct 3D bioprinting of perfusable vascular constructs using a blend bioink. *Biomaterials* 106: 58–68.
84 Costa, P.F., Vaquette, C., Baldwin, J. et al. (2014). Biofabrication of customized bone grafts by combination of additive manufacturing and bioreactor knowhow. *Biofabrication* 6 (3): 35006. https://doi.org/10.1088/1758-5082/6/3/035006.
85 Schneeberger, K., Spee, B., Costa, P. et al. (2017). Converging biofabrication and organoid technologies: the next frontier in hepatic and intestinal tissue engineering? *Biofabrication* 9 (1): https://doi.org/10.1088/1758-5090/aa6121.
86 Edmondson, R., Broglie, J.J., Adcock, A.F., and Yang, L. (2014). Three-dimensional cell culture systems and their applications in drug discovery and cell-based biosensors. *Assay and Drug Development Technologies* 12 (4): 207–218. https://doi.org/10.1089/adt.2014.573.
87 Gao, B., Yang, Q., Zhao, X. et al. (2016). 4D bioprinting for biomedical applications. *Trends in Biotechnology* 34 (9): 746–756. https://doi.org/10.1016/j.tibtech.2016.03.004.
88 Stoychev, G., Puretskiy, N., and Ionov, L. (2011). Self-folding all-polymer thermoresponsive microcapsules. *Soft Matter* 7 (7): 3277. https://doi.org/10.1039/c1sm05109a.
89 Hendrikson, W.J., Rouwkema, J., Clementi, F. et al. (2017). Towards 4D printed scaffolds for tissue engineering: exploiting 3D shape memory polymers to deliver time-controlled stimulus on cultured cells. *Biofabrication* 9 (3): https://doi.org/10.1088/1758-5090/aa8114.

6

Potential of 3D Printing in Pharmaceutical Drug Delivery and Manufacturing

Maren K. Preis

Åbo Akademi University, Pharmaceutical Sciences Laboratory, Turku, Finland

6.1 Introduction

This chapter focuses on the aspects of drug delivery and manufacturing by means of 3D printing technology. With new instrumentations and the evolving mindset in the field of health care that more individualized solutions are needed for patients and the treatment of different diseases, the pharmaceutical world is currently on the edge, finding a balance between established manufacturing technologies for the conventional large-scale production and the, moreover, new approach to produce medicinal products in smaller batches, on demand, personalized, and tailored for specific diseases.

The first part of this chapter will give a short overview of pharmaceutical drug delivery and the areas where 3D printing is already playing a role or will do in the future. The next part covers conventional manufacturing of drug delivery systems and where innovative approaches such as printing technologies can step in. Advances in the applications for drug delivery by means of 3D printing are discussed in the following section. The next parts cover instrumentations to enable 3D printing of pharmaceutical products. One major aspect of 3D printing of pharmaceutical products is the question where manufacturing takes place and different scenarios are displayed.

6.2 Pharmaceutical Drug Delivery

The mode of application used for a certain drug substance depends on a number of aspects such as physicochemical properties, local or systemic use, dose, indication or disease to be treated, patient, or age of the patient [1]. With growing awareness that the patient conditions are not the same, meaning that a one-size-fits-all approach is outdated, the demand for more personalized medicine increases.

The most known and convenient application of medicine is the oral route. Oral dosage forms are well known and used all over the world. Tablets and their variations, for example, orally disintegrating or effervescent ones, capsules,

3D and 4D Printing in Biomedical Applications: Process Engineering and Additive Manufacturing,
First Edition. Edited by Mohammed Maniruzzaman.

suspensions, or solutions, but also lozenges or newer forms such as buccal films, are widely accepted and established [2–5]. In most cases, drugs are available in different oral dosage forms to serve patient preferences, but also in terms of drug release properties (see also Chapter 8). Immediate and sustained release products are reasonable for the treatment of some diseases to ensure adequate drug effects and to adjust the dosing intervals.

The present interest in personalizing drug delivery and bringing therapy closer to the patient, and thereby improving treatment adherence and compliance, is one of the main challenges in today's pharmaceutical care. The one-size-fits-all approach is obsolete. Health care professionals and the pharmaceutical industry are working toward more personalized solutions, which come with the exploration of new manufacturing technologies [6]. To serve the individual needs of different patient groups, flexible manufacturing techniques with options to produce small batch sizes and the opportunity to produce on-demand and potentially on-site are required [1]. Within this context, printing technologies have been explored over the past years to fill this gap in tailored drug delivery.

By using a flexible manufacturing process, where therapeutic aspects such as dose strength, drug combinations, and release properties and personal preferences such as size, shape, or color can be considered, pharmaceutical care can be tailored according to patient needs (Figure 6.1). By using a highly flexible process like printing, it is possible to adjust the doses within the drug delivery system and to produce different dose strengths.

The combination of printing technologies and pharmaceutical active ingredients has already been explored in several studies. The first printed oral tablet has been approved in 2015 [7], and research groups worldwide work on investigating materials and applications in example for drug-eluting implants [8] and other medical devices [9].

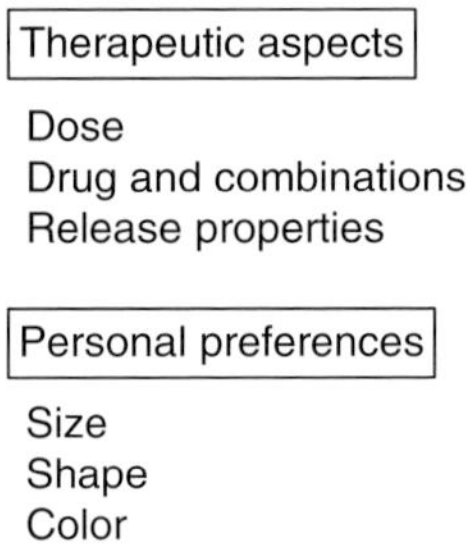

Figure 6.1 Therapeutic aspects and personal preferences by the patient, which can be considered in an individual printed drug delivery system.

The next parts in this chapter will focus on the comparison of conventional and innovative manufacturing and other aspects to consider when using printing for pharmaceutical drug delivery.

6.3 Conventional Manufacturing vs 3D Printing

Most conventional manufacturing technologies have the advantage that they are already established in pharmaceutical industry, enabling large-scale production and cost-effectiveness. One of the most asked questions regarding the production of pharmaceutical products by means of 3D printing is if this technology will replace the conventional processes. Table 6.1 gives an overview on features and limitations of conventional vs 3D printing. Printing will most likely not replace established productions of general products such as nonprescription

Table 6.1 Features and limitations of conventional manufacturing vs 3D printing.

	Conventional	**3D printing**
Features and advantages	Large scale Established Known products	Small to medium scale Customization possible Suitable for orphan drugs
Limitations and disadvantages	No individualization: Fixed doses, sized dosage forms Orphan drugs neglect	Available instrumentation Higher flexibility = higher risks New control systems, directives

drugs where large amounts of drug and preparations are required in standard strengths. The advantage of printed dosage forms lies in the aspect of customization and enabling the production of niche products or orphan drug products where the conventional large-scale production is not cost-effective and therefore neglected by most pharmaceutical manufactures [10]. Areas where printing has the potential to replace conventional manufacturing are, for example, pediatric drug delivery, where different doses and drug combinations are required, which are not available from the general suppliers, and also geriatric drug delivery where personal preferences and physical limitations play an important role in terms of therapy adherence and success [4].

To date, the used printing technologies cannot keep up with high-throughput conventional processes such as rotary tablet presses. However, the twenty-first century can be characterized as an era of technological advances. It may only be a matter of time until 3D printers go from small-scale and individual production to large-scale outputs.

The production steps of a conventionally pressed tablet and a 3D-printed tablet in a simplified depiction are shown in Figure 6.2. Once processes are established and printers are implemented, for example, in a continuous manufacturing setup, printable drug delivery systems could become part of the regular production. The instrumentation is further discussed in Section 6.5.

Figure 6.2 Example for a conventional and printing process.

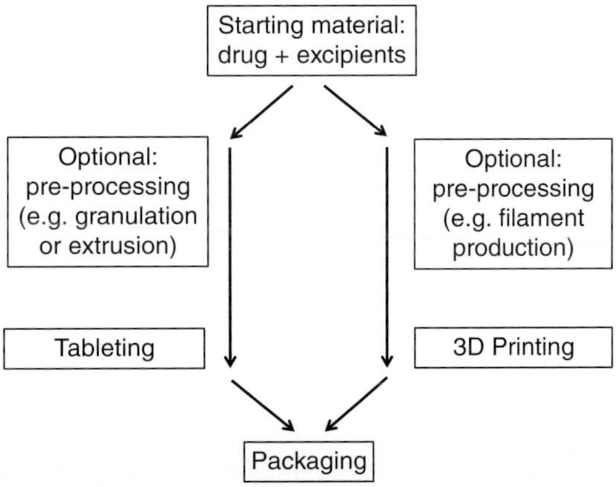

For time being, the introduction of printable drug delivery systems will most likely be applied to niches and specific indications where no suitable products are available on the market. For a product to be profitable for a pharmaceutical company when using a new technology over a conventional process, it must generate a significant advantage compared to available treatment options (as shown by ZipDose® technology, Chapter 2). A flexible manufacturing process, once it is established, either in an industrial setup or on-site in a hospital pharmacy, would be an innovative approach that brings pharmaceutical care closer to the patient.

6.4 Advanced Applications for Improved Drug Delivery

With a technology, which gives us the opportunity to create any desired shape or size of a drug delivery system by means of computer-aided design softwares, the development of tailored and patient-centered products is only the beginning.

Complex therapies and dose regimen can be facilitated by creating customized multipills for individual patients [11]. Small-sized products for children can be designed to comply with anatomic limitations. Conventional oral tablets or capsules are often too large for younger children who face swallowing issues or simply refuse the medication. Similar issues may appear for elderly patients. Swallowing dysfunctions and dysphagia are common issues for these patient populations [12]. The color and shape of a drug delivery system may appear as neglectable; however, these attributes can influence a patient's adherence to the given treatment.

As mentioned earlier, oral drug delivery is the most convenient and frequent route of administration. Other advanced applications of 3D printing include implants (Chapters 13 and 19) and other medical devices. Individual products cannot only be designed according to the patient's anatomy but they can furthermore be either imprinted with an appropriate drug or the drug is incorporated in the printable matrix during the 3D printing process.

In general, any advanced application where a tailored drug delivery system is needed can be realized using one or the combination of printing technologies by either imprinting the drug on a given device or dosage form or by incorporating the active ingredient into the printable matrix, such as a polymer filament, which is consequently used in an additive manufacturing process to print the 3D dosage form.

6.5 Instrumentations

The technical components of 3D printing in pharmaceutical drug delivery are discussed in this section. Before being able to print tailored products, suitable instruments are required. At the outset of a manufacturing process stands the design of the product and if a standard design can be considered or if it has to be patient-matched. A computer-aided design software is needed in any case to prepare the 3D file for the subsequent printing process. For patient-matched products, a 3D scanner may be necessary to assess the anatomy of the patient

Table 6.2 Exemplary instrumentation for pharmaceutical 3D printing using additive manufacturing (fused deposition modeling).

	Used for	Priority
Computer	Computer-aided design	High
(3D scanner)	Implants, custom devices	Optional, application dependent
Extrusion	Feedstock material production	Optional, depending on printing system
3D printer	Production	High
Process analytical technology (PAT)	Quality control	High
Packaging	Storage, delivery	Medium, depending on product stability and time until usage

(applicable for individual implants or devices). Softwares used within the process have to be validated and checked regarding file format converting and that the digital design ends up converting into the appropriately dimensioned physical product when printed. Furthermore, the printing system and feedstock materials need to be compatible with each other. Within these aspects, questions may arise if feedstock materials from different companies can be used with the 3D printer, and if the quality of the product remains. This point brings us to one major aspect of instrumentation: the quality control systems. Suitable in-process or post-process control tools appear appropriate to ensure the quality of printed products, in particular, when small batches of varying sizes and dose strengths are produced.

Hardware (computer, printers and printheads, quality control tools, etc.) and software (e.g. digital design) have to work appropriately, and material safety and quality have to be ensured during the printing process. Table 6.2 gives an exemplary overview of the instrumentation. The last aspect of packaging is of medium priority because it highly depends on the site of production. If the product is produced on-site in close proximity to the patient (e.g. hospital), other packaging requirements may apply compared to an industrial production and thereby longer storage (options for the location of 3D manufacturing is discussed in Section 6.6 of this chapter).

6.6 Location of 3D Printing Manufacturing

6.6.1 Pharmaceutical Industry

Pharmaceutical manufacturing takes place in industrial setups for most cases. Once 3D printers are implemented in the manufacturing process, it is possible to produce different batch sizes of 3D printed pharmaceutical products. With 3D printing as the manufacturing method, pharmaceutical companies get the opportunity to produce innovative products, small batch sizes, and to also serve

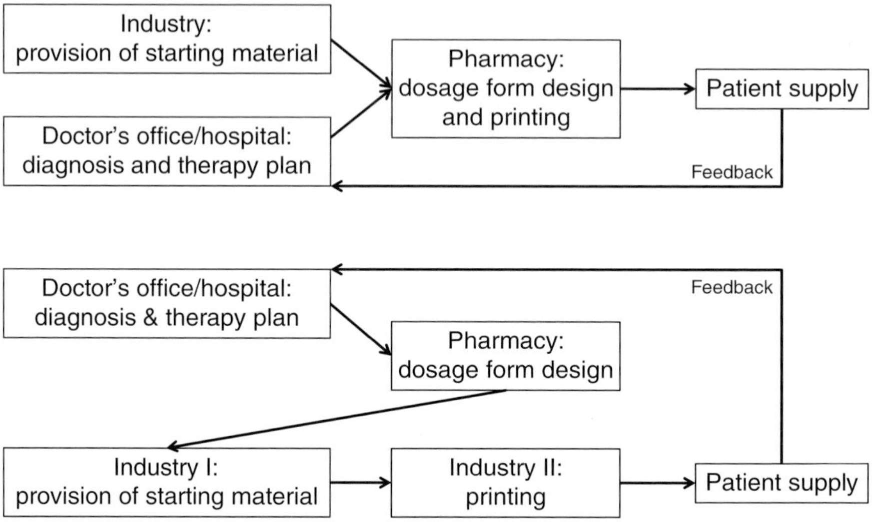

Figure 6.3 Two scenarios for 3D printing on demand.'

niche markets, which may have been neglected because of financial aspects or lack of flexibility of conventional manufacturing lines. The described scenarios in Figure 6.3 also enable new and smaller pharmaceutical manufactures, for example, with local production sites close to a point of care to provide the 3D printed products.

6.6.2 At the Point of Care

Printing at the point of care has the major advantage that the patient can be supplied with his tailored medication at the doctor's office or when hospitalized. This scenario furthermore gives the health care professionals the opportunity to react immediately to patient feedback or development (Figure 6.3). New dose strengths can be provided almost immediately after the latest checkup with the doctor. This procedure brings pharmaceutical drug delivery and manufacturing close to the patient and enables immediate therapy adjustments. Small batches, even daily printed medication, can be provided to the patient, which, in addition, avoids the disposal of unused medicine and thereby saves costs.

6.6.3 Print-at-Home

The approach to give patients the opportunity to print their own medications at home sounds futuristic and is only a vision thus far. Household-proof printers with security systems, operated and controlled for instance by the responsible doctor in his office, gives a sense of convenience that 3D printing could offer to patients and caregivers. At this point, the midterm goals are to establish 3D printing as an alternative to convenient manufacturing processes in pharmaceutical industry and other professional institutions such as hospitals and pharmacies before a print-at-home approach can be considered safe and appropriate.

6.7 Regulatory Aspects

With a new manufacturing technology for pharmaceutical products, questions regarding regulatory and safety aspects arise. Prominent authorities such as the US Food and Drug Administration are following the recent trends for printed pharmaceuticals and provide initial guidelines for 3D manufacturing. The regulatory aspects of 3D printing in additive manufacturing are discussed in Chapters 19 and 20.

6.8 Summary

Three-dimensional printing has a great potential in pharmaceutical drug delivery and production. Advantages such as flexibility and highly customized products, which can be produced closer to the patient and serve individual preferences, appear to revolutionize the pharmaceutical market in the future. The development of technologies and their applications will bring pharmaceutical drug delivery and manufacturing closer to the patient.

References

1 Sandler, N. and Preis, M. (2016). Printed drug-delivery systems for improved patient treatment. *Trends in Pharamcological Sciences* 37 (12): 1070–1080.
2 Preis, M., Woertz, C., Kleinebudde, P., and Breitkreutz, J. (2013). Oromucosal film preparations: classification and characterization methods. *Expert Opinion on Drug Delivery* 10 (9): 1303–1317.
3 Breitkreutz, J. and Boos, J. (2011). Drug delivery and formulations. In: *Handbook of Experimental Pharmacology*, 91–107. Springer.
4 Breitkreutz, J. and Boos, J. (2007). Paediatric and geriatric drug delivery. *Expert Opinion on Drug Delivery* 4 (1): 37–45.
5 Krampe, R., Visser, J.C., Frijlink, H.W. et al. (2016). Oromucosal film preparations: points to consider for patient centricity and manufacturing processes. *Expert Opinion on Drug Delivery* 13 (4): 493–506.
6 Khaled, S.A., Burley, J.C., Alexander, M.R. et al. (2015). 3D printing of tablets containing multiple drugs with defined release profiles. *International Journal of Pharmaceutics* 494: 643–650.
7 Aprecia Pharmaceuticals (2016). ZipDose Technology.
8 Water, J.J., Bohr, A., Boetker, J. et al. (2015). Three-dimensional printing of drug-eluting implants: preparation of an antimicrobial polylactide feedstock material. *Journal of Pharmaceutical Sciences* 104 (3): 1099–1107.
9 Genina, N., Holländer, J., Jukarainen, H. et al. (2016). Ethylene vinyl acetate (EVA) as a new drug carrier for 3D printed medical drug delivery devices. *European Journal of Pharmaceutical Sciences* 90: 53–63.
10 Preis, M. and Öblom, H. (2017). 3D-printed drugs for children – are we ready yet? *AAPS PharmSciTech* 18 (2): 303–308.

11 Khaled, S.A., Burley, J.C., Alexander, M.R. et al. (2015). 3D printing of five-in-one dose combination polypill with defined immediate and sustained release profiles. *Journal of Controlled Release* 217: 308–314.
12 Stegemann, S., Gosch, M., and Breitkreutz, J. (2012). Swallowing dysfunction and dysphagia is an unrecognized challenge for oral drug therapy. *International Journal of Pharmaceutics* 430 (1–2): 197–206.

7

Emerging 3D Printing Technologies to Develop Novel Pharmaceutical Formulations

Christos I. Gioumouxouzis[1], *Georgios K. Eleftheriadis*[1], *and Dimitrios G. Fatouros*[1]

Aristotle University of Thessaloniki, Department of Pharmacy, Division of Pharmaceutical Technology, 54124, Thessaloniki, Greece

7.1 Introduction

Additive manufacturing (AM) or three-dimensional (3D) printing has recently been introduced in the pharmaceutical field as an innovative manufacturing process that makes the rapid production of pharmaceutical formulations feasible with complex geometries. A novel technique allows the creation of dosage forms that formerly required costly multistage procedures, as a single-step process. Moreover, it has paved the way to personalized medicine, as it enables pharmacists to create *in situ* pharmaceutical formulations incorporating the exact amount of active pharmaceutical ingredients (APIs) that best suit each patient's individual needs. In 2015, 3D printing of medicinal products has crossed the borders of research laboratories, making its way into the market, as the first pharmaceutical dosage form (Spritam®) has been approved by the FDA [1]. In this chapter, we will present a wide variety of attempts to create 3D printed pharmaceutical formulations, categorized by the specific AM technique utilized for each one, i.e. fused deposition modeling (FDM), stereolithography (SLA), pressure-assisted microsyringes (PAM), selective laser sintering (SLS), powder bed printing (PBP), and inkjet printing (IP).

7.2 FDM 3D Printing

FDM 3D printers are the most widespread category of additive manufacturing devices. Their operating principle is the successive deposition of molten polymeric strands via one (or more) moving heated nozzle. Nozzles are fed with polymeric filament of standard diameter, and material deposition takes place on a (usually heated) bed, layer by layer. FDM printers can incorporate

[1] These authors equally contributed.

3D and 4D Printing in Biomedical Applications: Process Engineering and Additive Manufacturing, First Edition. Edited by Mohammed Maniruzzaman.

more than one printhead, each one loaded with a different filament, to achieve multimaterial printing.

The success of FDM 3D printers is based on (i) their very low cost (starting from only a few hundred dollars), (ii) their printing speed in comparison to SLA, SLS, and PBP (as they do not require extra time to spread powder or liquid resin before printing each layer), and (iii) their versatility, as a wide variety of polymers can be used as a feeding material.

Their main drawbacks are (i) the restriction of using only thermoresistant materials, as printing temperatures exceed 160 °C in most cases, resulting in the degradation of thermolabile substances, and (ii) the need of production of the feeding filament (usually by hot melt extrusion – HME), making FDM 3D printing a two-step process.

FDM 3D printing was employed in pharmaceutical research for the first time in 2014 [2]. Since then, a variety of formulations created with this technique have been investigated [3–11], including dosage forms with elaborate shapes that exhibit specific release characteristics [12–17], personalized medical devices [18–22], oral films [23, 24], and 3D printed matrices used as carriers or substrates for other drug-loaded release devices [25–28]. A summary of these investigations is presented in Table 7.1.

A variety of thermoplastic polymers were used in these studies as the main excipient–drug carrier. In most studies, APIs were incorporated into water-soluble polyvinyl alcohol (PVA) [2–5, 14, 15, 17, 23, 24], whereas recent studies investigated the printability of drug-loaded polymethacrylates (Eudragit® RL [6, 27], RS [6], E [7, 12], L [8]), polyvinylpyrrolidone (PVP) [9, 13], poly (e-caprolactone) (PCL) [21, 22, 27], hydroxypropyl cellulose (HPC) [6, 8, 16], ethyl cellulose (EC) [8], hydroxypropyl methylcellulose (HPMC) [8, 10, 19], poly(e-caprolactone)-polyvinyl acetate-polyethylene glycol (PCL-PVAc-PEG) copolymer (Soluplus®) [8], hypromellose acetate succinate (HPMCAS) [10, 11], poly(L-lactic acid) (PLA) [18, 19], polyvinylpyrrolidone-vinyl acetate copolymer (Kollidon® VA64) [10], polyvinyl alcohol-polyethylene glycol graft copolymer (Kollicoat® IR) [10], ethylene vinyl acetate (EVA) [20], and polyethylene oxide (PEO) [24].

Some of these polymers were also printed as scaffolds to create 3D matrices that enclosed APIs (insoluble PLA [26], PVA [26]), capsular devices (HPC [25], HPMCAS [28], HPMC [28], and Kollicoat IR [28]) or barriers (PLA [17] and Eudragit L [13]) in complex multilayer formulations created with dual printheads.

In early attempts, researchers tried to incorporate APIs to commercial PVA filaments by immersing them into their ethanolic or methanolic solutions. However, this method has drawbacks because the filament drug content achieved is very low (max. 1.9%) and additionally employs (generally unwanted) organic solvents in manufacturing procedure [2, 4].

Consequently, alternative methods to produce drug-loaded filaments for 3D printer feedstock and HME were utilized. As such with this approach, mixtures of API and the appropriate excipients (mainly thermoplastic polymers) are travelling through a heated barrel via a rotating screw. At the exit of the barrel, a nozzle of desired geometry and diameter shapes the extrudate into rods or sheets. After exiting the nozzle, the extrudates cool and solidify (often with the intervention of air flow or cooling water baths). Thermomixing of the blend's components inside

Table 7.1 Summary of research papers utilizing FDM 3D printing for the production of pharmaceutical formulations.

No. (References)	Title	Polymer	API	Plasticizer	Printer	T_{pr} (°C)	Infill V_{pr} (mm s^{-1})	T_{bed} (°C)	Infill (%)	Layer height (mm)	Shells	HME	Die T_{ext} (°C)	V_{ext} (rpm)	Die (mm)	Torque (Nm)
1 [2]	3D printing of modified release aminosalicylate (4-ASA and 5-ASA) tablets	PVA (com.)	4-ASA, 5-ASA	—	Makerbot Replicator® 2	210	90	N/M	10/50/90	0.2	2	—	—	—	—	—
2 [3]	Fabrication of controlled release budesonide tablets via desktop (FDM) 3D printing	PVA (com.)	Budesonide	—	Makerbot Replicator® 2X	190	90	N/M	100	0.2	2	Noztek Pro®	170	15	1.75	N/M
3 [4]	Fabrication of extended release patient-tailored prednisolone tablets via fused deposition modeling (FDM) 3D printing	PVA (com.)	Prednisolone	—	Makerbot Replicator® 2X	250	90	20	100	0.2	N/M	—	—	—	—	—

(*Continued*)

Table 7.1 (Continued)

No. (References)	Title	Polymer	API	Plasticizer	Printer	T_{pr} (°C)	Infill V_{pr} (mm s^{-1})	T_{bed} (°C)	Infill (%)	Layer height (mm)	Shells	HME	Die T_{ext} (°C)	V_{ext} (rpm)	Die (mm)	Torque (Nm)
4 [5]	Fused filament 3D printing of drug products: microstructure analysis and drug release characteristics of PVA-based caplets	PVA (com.)	Paracetamol	—	Makerbot Replicator® 2X	200	90	N/M	100	0.2	2	Noztek Pro®	180	15	1.75	N/M
5 [6]	A flexible dose dispenser for immediate and extended release 3D printed tablets	Eudragit® RL/RS/E, HPC	Theophylline	TEC/ triacetin	Makerbot Replicator® 2X	Eudragit® RL:170 Eudragit® RS:150 Eudragit® E: 140 Eudragit® RL/RS (1 : 1): 150 HPC:160	90	Eudragit® RL:90 Eudragit® RS:60 Eudragit® E: 60 Eudragit® RL/RS(1 : 1): 90 HPC:60	100	0.2	N/M	HAAKE MiniCTW®	Eudragit® RL:120 Eudragit® RS:110 Eudragit® E: 110 Eudragit® RL/RS(1 : 1): 120 HPC:110	N/M	1.5	0.6
6 [7]	Adaptation of pharmaceutical excipients to FDM 3D printing for the fabrication of patient-tailored immediate release tablets	Eudragit® E	Theophylline, 5-ASA, captopril, prednisolone	TEC	Makerbot Replicator® 2X	135	90	60	100	0.2	N/M	HAAKE MiniCTW®	90	N/M	N/M	0.8

7 [8]	Coupling 3D printing with hot melt extrusion to produce controlled-release tablets	HPC, HPMC, EC, Soluplus, Eudragit L100	Acetaminophen	N/M	Prusa i3	200	50	50	100	0.1	4	Thermo Fisher (unknown model)	HPMC:180 Others: 140-160	50	2	N/M
8 [9]	A lower temperature FDM 3D printing for the manufacture of patient-specific immediate release tablets	PVP	Theophylline, dipyridamole	TEC	Makerbot Replicator® 2X	110	90	60	100	0.2	N/M	HAAKE MiniCTW®	90	N/M	N/M	0.4
9 [10]	Formulation of 3D printed tablet for rapid drug release by fused deposition modeling (FDM): screening polymers for drug release, drug–polymer miscibility and printability	Kollidon® VA64, Kollicoat® IR, Affinsiol™15 cP, HPMCAS	Haloperidol	—	Makerbot Replicator® 2	210	45	N/M	60/100	0.1	1	Thermo Scientific Process 11	170	200	1.55	N/M

(*Continued*)

Table 7.1 (Continued)

No. (References)	Title	Polymer	API	Plasticizer	Printer	T_{pr} (°C)	Infill V_{pr} (mm s^{-1})	T_{bed} (°C)	Infill (%)	Layer height (mm)	Shells	HME	Die T_{ext} (°C)	V_{ext} (rpm)	Die (mm)	Torque (Nm)
10 [11]	Development of modified release 3D printed tablets (printlets) with pharmaceutical excipients using additive manufacturing	HPMCAS LG/MG/HG	Paracetamol	Methylparaben	Makerbot Replicator 2X	180–190	90	N/M	20/100	0.1	2	Noztek Pro®	80–110	15	1.75	N/M
11 [12]	Channeled tablets: an innovative approach to accelerating drug release from 3D printed tablets	Eudragit® E	HCTZ	TEC	Makerbot Replicator® 2X	135	90	60	100	0.2	N/M	HAAKE MiniCTW®	90	N/M	N/M	0.8
12 [13]	Fabricating a shell–core delayed release tablet using dual-FDM 3D printing for patient-centered therapy	PVP, Eudragit® L100-55	Theophylline	TEC	N/M	PVP:110 Eudragit L100-55 : 185	12	40	100	0.2	N/M	HAAKE MiniCTW®	PVP:90 Eudragit® L100-55 : 125	N/M	PVP: 1.25 Eudragit® L100-55 : 1	0.4

13 [14]	Effect of geometry on drug release from 3D printed tablets	PVA (com.)	Paracetamol	—	Makerbot Replicator® 2X	180	90	N/M	100	0.2	2	Filabot Original®	180	35	1.75	N/M
14 [15]	3D printing of medicines: engineering novel oral devices with unique design and drug release characteristics	PVA (com.)	Paracetamol, caffeine	—	Makerbot Replicator® 2X	200	90	N/M	100	0.2	2	Noztek Pro®	180	15	1.75	N/M
15 [16]	Fused deposition modeling (FDM) 3D printed tablets for intragastric floating delivery of domperidone	HPC	Domperidone	—	Makerbot Replicator® 2X	210	90	30	0	0.2	2	HAAKE MiniCTW®	145–150	20–25	N/M	0.1–0.2

(*Continued*)

Table 7.1 (Continued)

No. (References)	Title	Polymer	API	Plasticizer	Printer	T_{pr} (°C)	Infill V_{pr} (mm s^{-1})	T_{bed} (°C)	Infill (%)	Layer height (mm)	Shells	HME	Die T_{ext} (°C)	V_{ext} (rpm)	Die (mm)	Torque (Nm)
16 [17]	3D printed oral solid dosage forms containing hydrochlorothiazide for controlled drug delivery	PVA	HCTZ	Mannitol	Makerbot Replicator® 2X	PLA:220 PVA:200	90	65	100	0.3	2	Filabot Original®	170	35	1.75	N/M
17 [18]	Toward fabrication of 3D printed medical devices to prevent biofilm formation	PLA	Nitrofurantoin	—	UP Plus	200	N/M	N/M	N/M	N/M	N/M	Axon Ab® Plastics	140–180	21	1.5	N/M
18 [19]	Modifying release characteristics from 3D printed drug-eluting products	PLA(com), HPMC	Nitrofurantoin	—	MakerBot Replicator® 2	190–200	N/M	N/M	N/M	N/M	N/M	DSM Xplore®	180	30	1.5	N/M

19 [20]	Ethylene vinyl acetate (EVA) as a new drug carrier for 3D printed medical drug delivery devices	EVA, PCL(com)	Indo-methacin	—	MakerBot Replicator® 2	155	10	N/M	100	0.1	3	HAAKE MiniCTW®	100–120	10	1.5–2.5 N/M
20 [21]	Three-dimensional printed PCL-based implantable prototypes of medical devices for controlled drug delivery	PLA(com)	Indo-methacin	—	MakerBot Replicator 2	100	45	N/M	10	0.1	3	HAAKE MiniCTW®	100	10	1.5 N/M
21 [22]	Patient-specific 3D scanned and 3D printed antimicrobial polycaprolactone wound dressings	PCL	Silver nitrate, copper sulfate pentahydrate, zinc oxide	—	MakerBot Replicator® 2X	170	50	N/M	100	0.1	2	Filabot	60–80	35	1.75 N/M

(*Continued*)

Table 7.1 (Continued)

No. (References)	Title	Polymer	API	Plasticizer	Printer	T_{pr} (°C)	Infill V_{pr} (mm s^{-1})	T_{bed} (°C)	Infill (%)	Layer height (mm)	Shells	HME	Die T_{ext} (°C)	V_{ext} (rpm)	Die (mm)	Torque (Nm)
22 [23]	3D printed orodispersible films with aripiprazole	PVA	Aripiprazole	—	Zmorph 2.0S	185–190	5–10	N/M	40	0.15	2	Noztek Pro®	172	N/M	N/M	N/M
23 [24]	The application of 3D printing in the formulation of multilayered fast dissolving oral films	PEO, PVA	Ibuprofen, paracetamol	—	Wanhao Duplicator® 4	PEO:165 PVA:190	PEO:70 PVA:90	RT	40/100	0.1	2	Noztek Pro®	PEO:60 PVA:130	N/M	1.6	N/M
24 [25]	3D printing by fused deposition modeling (FDM) of a swellable/erodible capsular device for oral pulsatile release of drugs	HPC	Acetaminophen	PEG 1500	Makerbot Replicator® 2	180	90	N/M	100	0.25	2	Haake™ MiniLab II	150–165	50–60	1.75	N/M

25 [26]	Analysis of 3D prints by X-ray computed microtomography and terahertz pulsed imaging	PLA/PVA (com.)	Carbamazepine, saquinavir/halofantrine SNEDDS	—	Makerbot Replicator® 2	230	N/M	N/M	100	0.3	N/M	—	—	—	—	—
26 [27]	3D printed tablets loaded with polymeric nanocapsules: an innovative approach to produce customized drug delivery systems	Eudragit® RL 100, PCL	Deflazacort (nanocapsules)	TEC	Makerbot Replicator® 2	PCL:95 Eudragit® RL PO:170	90	N/M	50/100	0.2	2	®Noztek Pro	PCL:65 Eudragit® RL PO:110	N/M	PCL:2 Eudragit® RL PO:1.5	N/M
27 [28]	3D printed multicompartment capsular devices for two-pulse oral drug delivery	HPMC, HPMCAS, Kollicoat® IR	Acetaminophen, blue and yellow dye-containing formulations	PEG 400, PEG 8000, glycerol	Makerbot Replicator® 2	HPMC: 200, HPMCAS:200, Kollicoat® IR:180	N/M	25	100	N/M	N/M	Haake™ MiniLab II	HPMC:160, HPMCAS:180, Kollicoat® IR:160	HPMC: 70, HPMCAS: 100, Kollicoat® IR:100	1.80	HPMC: 70, HPMCAS: 100, Kollicoat® IR:80

the heated barrel results in relatively homogeneous extrudates and usually the APIs are present as amorphous dispersions inside the polymer matrix. Incorporation of two or more screws (twin-screw and multiscrew extruders respectively) in the barrel results in better mixing of the components, but substantially increases the cost of the device [29].

The versatility of HME has enabled it as a standard procedure for the creation of filaments for FDM 3D printers. That fact derives from the abundancy and diversity of polymers that can be used as carriers, the relative simplicity and speed of the method, and the potentiality to make HME a continuous procedure. These advantages overshadow the main drawback of HME, which is the limitation of using only thermostable APIs and excipients, as temperatures inside the heated barrel usually exceed 120 °C.

Two types of HMEs have been employed in the creation of filaments loaded with APIs for 3D printing: relatively simple single-screw extruders with fixed rotating screw speed [3, 5, 11, 14, 15, 17, 18, 22–24, 27] and more complex corotating twin-screw extruders with adjustable rotating speed, often operating at nitrogen environment [6–10, 12, 13, 16, 19–21, 25, 28]. Both extruder types produced acceptable filaments with drug loading up to 50% [6].

The standard diameter of the feeding filament is usually 1.75 mm and deviation from that size cannot exceed 0.05 mm because thicker filaments are too large to enter the 3D printer nozzle and thinner filaments result in poor printing quality, as the amount of the provided material is not sufficient to print layers with the desired height. Production of filament with constant diameter is tricky, as most polymers have a tendency to swell upon exiting the HME nozzle, a phenomenon called die swell. That problem can be overcome either by utilizing custom-made nozzles with reduced diameter (1–1.6 mm) [6, 10, 13, 18–21, 24, 27] or by using substances with lubricating properties such as magnesium or calcium stearate that have the ability to reduce die swell of extrudates [11]. Additionally, pulling/calibrating devices equipped with fans have been employed to ensure the production of filament suitable for 3D printing [25, 28].

HME usually results in the production of amorphous solid dispersions of the APIs inside the polymer matrix [7, 8, 10, 11, 13–17, 23, 24], although it has been reported that in some cases, the APIs can exist in a crystalline or semicrystalline state inside the extrudates (especially when API concentration is relatively high or low temperatures are used) [3, 6, 7, 9, 10, 12, 13, 15, 19–21]. Creation of amorphous dispersions facilitates dissolution of drugs, demonstrating another HME advantage.

Nevertheless, HME and FDM are procedures that employ elevated temperatures and stressing conditions as such thermogravimetric analysis (TGA) should be utilized to ensure that thermolabile substances do not degrade during manufacturing process. Moreover, differential scanning calorimetry (DSC) of the polymers used should be conducted before HME to determine glass transition temperature (T_g) of the polymers and the necessity of using plasticizers if T_g is too high.

Early published papers focused on evaluating the macroscopic properties of the dosage forms produced with 3D printing. These investigations showed that PVA-based 3D printed printlets exhibit acceptable mass and dimension deviations, very high crushing strength (>330 N), and zero friability, irresectable of the infill used [2]. Moreover, these studies demonstrated that the infill percentage modulates the dissolution profile and a faster drug release can be obtained lowering the infill percentage of the tablets. Nevertheless, it was shown that only materials and APIs that degrade at high temperatures can be fabricated with these technique, as FDM 3D printers utilize high temperatures to melt the printing polymer inside the heated nozzle (thermolabile 4-aminosalicylic acid (4-ASA) used as API degraded during 3D printing procedure) [2].

A subsequent study where FDM coupled with HME highlighted the potentiality of creating dosage forms via FDM with novel geometries previously impossible or very difficult to achieve (cube, pyramid, cylinder, sphere, and torus). In the same work, dissolution profiles of the PVA-based printlets showed that drug (paracetamol) release kinetics were not dependent on the surface area of the differently shaped dosage forms, but rather on surface area to volume ratio [14].

Additionally, PVA-based budesonide caplets produced via FDM and coated with enteric polymer Eudragit L demonstrated superior *in vitro* dissolution characteristics (rapid and complete drug release, attributed to erosion-mediated dissolution process), in comparison to marketed products [3].

The first attempt to exploit dual-printhead printer's ability of creating objects consisting of more than one layers and incorporating more than one API was conducted by Goyanes et al. [15]. Although the carrier polymer was PVA for both printheads, a different API (paracetamol and caffeine respectively) was incorporated in each printhead's filament. Two different dosage forms were created: (i) multilayer oral devices with alternate (1 mm thick) layers of paracetamol and caffeine and (ii) caplet-shaped devices (DuoCaplets®) with a core incorporating one API and a shell containing the other (and vice versa). In multilayer devices, drug release was similar for both APIs and completed after 360 minutes, whereas in DuoCaplets, a substantial delay was observed in the initiation of dissolution of the APIs enclosed in the core of the formulation. This delay can exceed 120 minutes, proving that the use of this method can protect APIs sensitive to low gastric pH [15].

Mercury intrusion porosimetry was utilized to evaluate the porosity of the aforementioned caffeine and paracetamol PVA filaments and the 3D printed caplets manufactured from them as well. The filaments exhibited very low porosity, with micropore volumes <0.008 $cm^3\ g^{-1}$. On the contrary, caplets presented more than 10-fold higher porosity (micropore volume 0.07–0.1 $cm^3\ g^{-1}$). The increased porosity values were attributed to 3D manufacturing process (deposition of cylindrical rods that connect with their adjacent rods at certain spots, leaving long internal macropores inside the structure), rather than any change in the polymer structure itself [5].

Initial attempts of producing 3D printed formulations focused on the ability to calculate with relative accuracy the mass of the 3D printed formulation (and therefore the mass of the API used) [4]. This was attained by expressing the volume (and subsequently the mass) of the dosage forms as a function of only one dimension (keeping other dimensions constant) and establishing a linear equation between these two parameters, by printing and weighting dosage forms with different values of the variable dimension. Moreover, in the same paper, important modifications were introduced, including the covering of the 3D printer building plate with Scotch Blue® painter's tape (to improve adhesion of the dosage forms on the printing surface, as Kapton tape provided did not ensure proper adhesion) and the lubrication of the filament with glycerol before printing, in order to reduce friction and prevent nozzle clogging [4].

Previous studies utilized a marketed filament, reinforced with unknown plasticizers and other additives, either as a preformulated filament or as a HME manufactured filament, deriving from pelletized marketed filament and APIs. The first attempt to create a fully custom-made filament from raw materials via HME was reported by Alhnan and coworkers [6] printing caplet-shaped tablets using filament produced from a mixture of different types of polymethacrylate polymers (Eudragit) or HPC as polymeric carriers, theophylline as API, and triethyl citrate (TEC) or triacetin as plasticizers. The result was the creation of immediate and prolonged release tablets that conveyed theophylline to the dissolution medium at different rates (100% release in one hour for Eudragit E and two hours for HPC caplets, 85% for Eudragit RL caplets, 60% for Eudragit RL caplets, and 20% for Eudragit RL caplets). Moreover, it should be mentioned that printing temperatures were significantly lower (140–170 °C) than in previous attempts that employed PVA because the polymethacrylates used have a tendency to degrade above 170 °C. This was made possible by utilizing an API (theophylline) with a high melting temperature (273 °C) that facilitated the printing procedure [6].

In a subsequent study by Goyanes et al. [11], three grades (LG, MG, and HG) of enteric polymer HPMCAS were tested, aiming to be utilized as carriers for prolonged drug release. Dosage forms exhibited distinct dissolution patterns depending on the grade of the polymer used (LG, MG, and HG grades have pH thresholds of 5.5, 6.0, and 6.5, respectively), highlighting the possibility of tuning drug dissolution according to the desired release site in the gastrointestinal (GI) tract.

Filaments must exhibit high stiffness and low brittleness in order to be suitable for 3D printing because of the fact that they have to withstand forces from the feeding gear. Taking that into consideration, a series of marketed polymers (Benecel™ HPMC E5, Klucel™ HPC EF and LF, Aqualon™ EC N14, Soluplus, and Eudragit L100) were screened (separately and as binary mixtures of different ratios) and further evaluated their suitability as drug carriers for the production of filaments for 3D printing [8]. The mechanical properties of the produced filaments (that incorporated 30% API [acetaminophen]) were assessed via a three-point bending flexural test. The results have shown that the filaments were printable if they exhibited a breaking stress $> 2941\ \mathrm{g\,mm^{-2}}$ and a breaking distance > 1 mm. The presence of 5% superdisintegrant Kollidon CL-F significantly improved the smoothness of the surface of the filaments, reducing the friction during feeding of the 3D printer and facilitating the printing procedure [8].

Another study investigated the printing suitability of filaments consisting of Kollidon VA64, Kollicoat IR, Affinisol™15 cP, and HPMCAS either individually or as binary blends (Kollidon VA64 + Affinisol15 cP, 1 : 1; Kollidon VA64 + HPMCAS, 1 : 1) [10]. Subsequently, polymeric systems exhibited a relatively rapid dissolution rate of their incorporated APIs (haloperidol was used as the model drug). A 1 : 1 mixture of Kollidon VA64 and Affinisol15 cP was identified and further evaluated as the optimum polymer carrier for immediate 3D printing pharmaceutical applications [10].

The compatibility of FDM for the creation of immediate release (IR) polymethacrylate Eudragit E PO was investigated in combination with various fillers, as the polymer alone is unsuitable for 3D printing [7]. This unsuitability is related to the branched nature of the methacrylic polymer that exhibits a lower (T_g) value compared to linear copolymers, resulting in 3D prints with a collapsed structure. An alternative to overcome this obstacle is the addition of a filler with a higher melting than the operating temperature of FDM 3D printer. Talc and tricalcium phosphate (TCP) proved to be suitable fillers, whereas 50% loading of TCP appeared to produce the optimum filaments. The presence of TCP enabled the production of 3D printed caplets of four model drugs (5-ASA, captopril, prednisolone, and theophylline), printed at relatively low temperature (135 °C), more than 85% of its drug content within 30 minutes [7].

Okwuosa et al. [9] assessed the formed filaments comprising PVP, talc, and API (dipyridamole or theophylline) suitable for printing at temperatures as low as 110 °C. This study reveals the potential of broadening the spectrum of APIs suitable for FDM 3D printing, by demonstrating the feasibility of printing relatively thermolabile APIs [9].

Manufacturing versatility of FDM 3D printers allowed the formulation of elaborate dosage forms that exhibit unique properties in terms of drug release, deriving from their sophisticated shape design [12, 13, 16, 17].

Shell–core delayed release tablets for enteric release were created enrolling a dual-extrusion FDM MakerBot Replicator® 2X 3D printer to form a core comprising plasticized PVP and talc and an external layer of plasticized Eudragit L100-55, a polymethacrylate that dissolves at pH higher than 5.5 (Figure 7.1a–c) [13]. APIs (budesonide, diclofenac, or theophylline) incorporated into the PVP/talc matrix were released after dissolution of Eudragit L100-55 shell at 120 minutes, avoiding exposure to gastric environment. Lubrication was necessary to prevent nozzle blocking caused by high friction between PVP base filament and the walls of printhead nozzles. In the produced dosage, the protective shell created by FDM was significantly thicker (<0.52 mm) than the shell used in conventionally coated tablets (40–100 μm), to ensure proper protection of the enclosed API from gastric low pH.

Floating intragastric tablets were produced as a one-stage procedure via FDM, simply by printing hollow cylinder-shaped tablets, using filament comprising unplasticized HPC and domperidone [16]. Hollow structures resulted from lowering infill values via 3D printer software. Optimized tablets with two shells and 0% infill had a relative lower density of 0.77gcm^{-3} and would be able to float for more than 10 hours *in vitro* and 8 hours *in vivo*, whereas tablets with higher infill values and density >0.9 g cm^{-3} exhibited limited buoyancy. Increased oral bioavailability and steady plasma concentration of the API in comparison to the marketed product revealed the potential of effectively manufacturing floating dosage forms via FDM 3D printing [16].

In an attempt to increase the dissolution rate of dosage forms, channeled caplets were formulated via FDM printing (Figure 7.2a–c). These caplets exhibited accelerated release of the incorporated drug (hydrochlorothiazide),

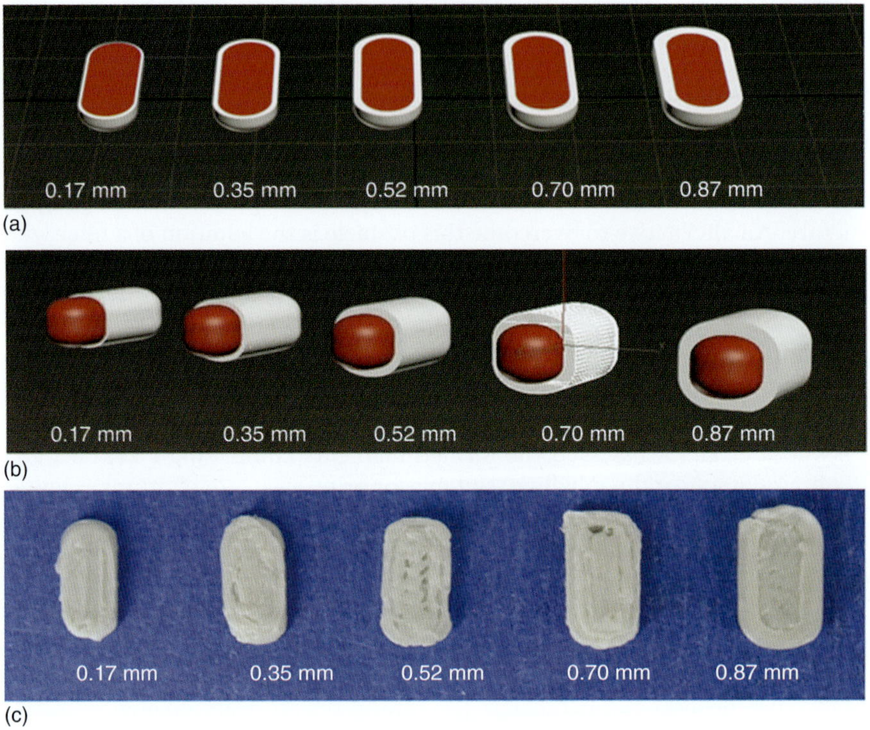

Figure 7.1 (a) and (b) Rendered images (Autodesk® 3DS Max) of shell–core designs with increasing shell thicknesses (0.17, 0.35, 0.52, 0.70, and 0.87 mm). (c) Images of 30% completed shell–core designs with theophylline core and increasing Eudragit L100-55 shell thickness. Source: Okwuosa et al. 2017 [13]. Reproduced with permission of Springer Nature.

depending on the size of the perforating channels. Specifically, formulations with channel width ≥ 0.6 mm met the USP criteria of immediate release products and also shorter multiple channels (8.6 mm) were more efficient at accelerating drug release than longer channels (18.2 mm) despite having comparable surface area/mass ratio [12].

A ring-shaped three-compartment dosage form was formulated via dual-FDM 3D printing, aiming to achieve zero-order release of the incorporated API [17]. The upper and lower compartment consisted of insoluble PLA, whereas the inner compartment was printed using water-soluble PVA, plasticized with mannitol and contained hydrochlorothiazide as the model drug (Figure 7.3a–d).

Simultaneous erosion of both external and internal lateral sides of the drug-loaded PVA layer resulted in sustaining a relative steady total release surface of the API during dissolution process, resulting in hydrochlorothiazide zero-order release. Time-lapsed X-ray microfocus computed tomography (4D-CT), visualized volumetric changes, and morphological changes of the formulations during the dissolution procedure, confirming the bidirectional homogeneous reduction of PVA layer, resulted in shape-related release profile of this particular dosage form [17].

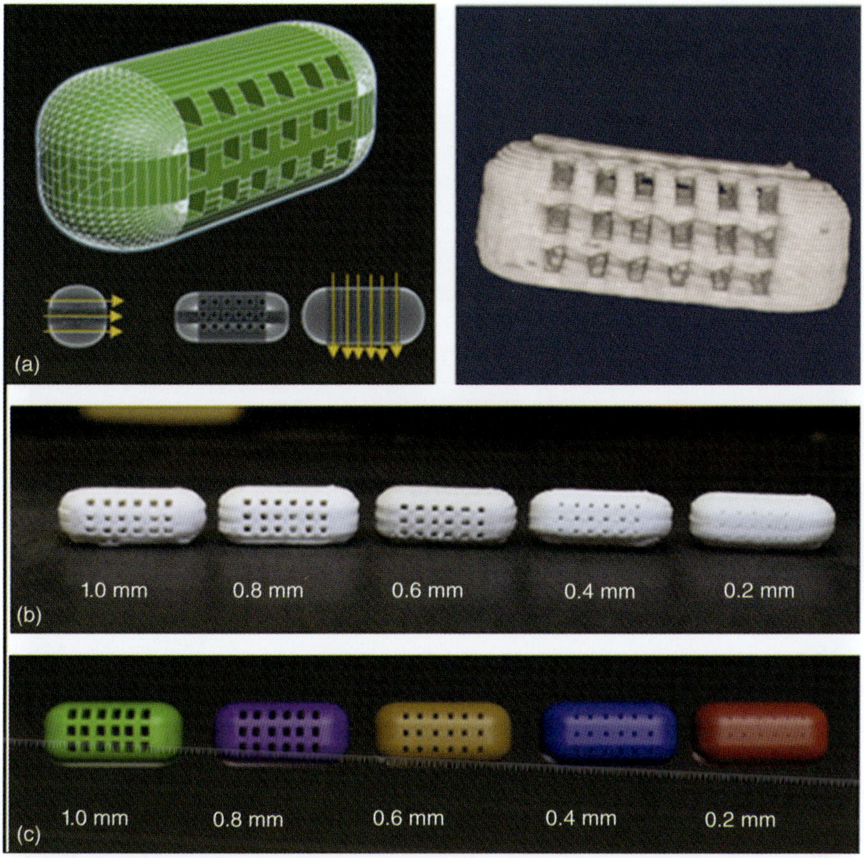

Figure 7.2 (a) Schematic illustration concept of perforating square-sectioned channels. The perforating channels were at right angle with the long axis with 3D max rendered images of frontal, side, and top view of channeled tablet designs. (b) Rendered images of tablet designs with decreasing channel size with 18-short channels. (c) Photographs of tablets with decreasing channel size of 18-short channels. Source: Sadia et al. 2018 [12]. Reproduced with permission of Elsevier.

Computed microtomography (XμCT) along with terahertz pulsed imaging (TPI) was also employed to characterize 3D printed structures used as carriers of liquid self-nanoemulsifying drug delivery system (SNEDDS) containing saquinavir and halofantrine as model drugs [26]. Cylindrical shape 3D printed structures were constructed using PVA, which consisted of two compartments one containing another. Printing was stopped in the middle of the process, to fill the compartments with the liquids and the procedure was restarted to close the samples. Moreover, cylindrical structures with only one compartment loaded with carbamazepine were printed using PVA and PLA to assess the structural differences between prints created with the two polymers. Analysis of the printed formulations with the aforementioned techniques revealed a clear difference in porosity between used polymers (5.5% for PVA and 0.2% for PLA prints), as also to the morphology of the pores inside the polymer structure (mainly long

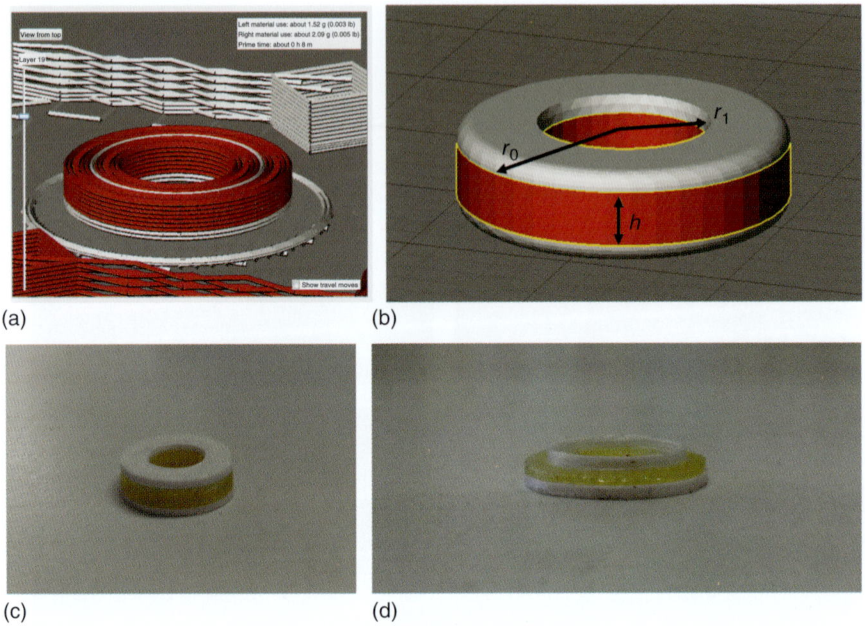

Figure 7.3 (a) Makerware®.stl model of 3D PVA-PLA three-compartment hollow cylinder formulation with inner binding PLA ring. (b) Design depicting characteristic dimensions (h, r_0, r_1) of 3D PVA-PLA formulations, (c) 3D printed three-compartment PVA-PLA printed dosage form. (d) Half-constructed 3D printed three-compartment PVA-PLA dosage form with inner binding PLA ring. Source: Gioumouxouzis et al. 2017 [17]. Reproduced with permission of Elsevier.

macro-pores for PLA, interconnected smaller pores for PVA). Additionally, the evaluation of printing accuracy indicated that compartmentalized PVA structures were $7.5 \pm 0.75\%$ larger than designed. *In vitro* dissolution tests showed that saquinavir SNEDDS, contained in the outer shell, started releasing almost simultaneously with the beginning of dissolution, whereas halofantrine SNEDDS incorporated inside the inner shell was released after 240 minutes. The produced 3D printed structures exhibited completely different dissolution behavior, in comparison to conventionally manufactured pharmaceutical dosage forms (such as compressed tablets), as their creation was based on the solidification of molten polymers instead of compression [26].

Another attempt to create 3D printed carriers using FDM led to manufacturing of oral HPC capsular devices for pulsatile release [25]. The parts comprising a capsule were separately printed, filled with acetaminophen, and assembled as conventional capsule with 600 μm wall thickness. 3D printed capsules exhibited similar dissolution profiles in comparison to capsules prepared via injection molding, indicating the advantageous nature of 3D printing for pharmaceutical applications (as it does not require manufacturing of different molds each time an altered dimension is required) [25].

Moving one step further, the same researchers designed their capsular devices as multicompartmental, aiming to incorporate APIs that could be released at different sites of the GI tract. For that reason, a third part (joint) was added to the prementioned capsular device separating the two compartments. Moreover, each compartment was manufactured by a different polymer, i.e. soluble (HPMC), gastroresistant (HPMCAS), and swellable/erodible (Kollicoat IR), which allowed immediate, enteric, and pulsatile release, respectively. Additionally, compartments were designed with 600 or 1300 μm wall thickness. A combination of compartments with different wall thicknesses and/or comprising different polymers allowed manufacturing of formulations that could deliver multiple APIs at different GI parts at variable rates [28].

In a recent study, the possibility of using 3D printed cylindrical formulations as substrates for nanocapsules loaded with deflazacort was investigated [27]. Printlets comprising from a swellable polymer (Eudragit RL 100) or a nonswellable polymer [poly(e-caprolactone)-PCL] were submerged into a nanocapsule suspension for 24 hours and a final drug loading ranging from 0.0624% to 0.620% (depending on the polymer and the excipients used) was achieved. This attempt demonstrated the feasibility of coupling FDM 3D printing with other innovative fields such as nanomedicine [27].

The capability of creating dosage forms with elaborate shapes raised questions about the acceptance of these novel shapes among patients. For that reason, a questionnaire-based survey, in which patients were asked to evaluate a variety of differently shaped 3D printlets, was conducted [30]. The options included disk, torus, sphere, tilted diamond, capsule, pentagon, heart, diamond, triangle, and cube HPC-based formulations. Conventional capsule- and disk-shaped printlets presented the best pre- and post-swallowing acceptability, along with torus-shaped formulations, a novel shaped dosage form that can only be manufactured via 3D printing methods [30].

FDM 3D printing technique has been utilized to develop dosage forms for improved patient compliance. Toward this direction, orodispersible films (ODFs) were formulated to deliver poorly soluble drug aripiprazole (BCS Class II) [23]. The polymeric films were created using PVA (a well-known water-soluble and film-forming polymer) and compared with *placebo* and solvent-casted films. Authors resolved dose-uniformity issues of the printed films by wetting the PVA flakes with ethanol and mixing it with the API before extrusion. The drug-loaded printed films resulted in acceptable dose, mass, and thickness uniformity (0.55 ± 0.08 mg cm^{-2}, 91.12 ± 4.40 mg, and 196.6 ± 12.9 μm, respectively), presenting lower data variations compared to their casted congeners. The disintegration time of the dosage forms was 27.5 and 43 seconds for the printed placebo and drug-loaded ODFs and 38 seconds for the casted films, whereas mechanical testing showed comparable Young modulus values between the printed and casted films. Comparatively, the printed films presented a higher dissolution rate, having released over 95% of the API after 15 minutes, whereas in the case of casted films, only 75% of the API was released over the same period, depicting the capability of FDM to print ODFs containing poorly soluble drugs with improved dissolution properties [23].

Similarly, fast dissolving oral films (FDFs) were printed as single-layered fast dissolving oral films (SLFDFs) or multilayered fast dissolving oral films (MLFDFs), with distinct taste-masking layers and two different textures (plain or mesh) [24]. The polymers used in the development of the dosage forms were either PEO or PVA, and ibuprofen or paracetamol were chosen as model drugs. Additionally, starch, sodium starch glycolate, and croscarmellose were incorporated in the formulation to facilitate film disintegration, whereas sodium lauryl sulfate was used to improve the drug release profile. SLFDFs presented a thickness range of 197–374 μm, with variation less than 4%, whereas the weight uniformity was greater than 97%. Mesh PVA films (either single- or multilayered) exhibited short disintegration time (42–48 seconds), whereas twofold and threefold disintegration times were observed for plain textures of either PVA or PEO. Drug release was performed at phosphate buffer (pH 7.4 for ibuprofen or 5.8 for paracetamol) and the results revealed a release rate related to the design properties. Thus, mesh designs presented faster release profiles (80% of the drug released over a period of 10 minutes), whereas plain films with additional taste-masking layers resulted in delayed release profiles [24].

FDM 3D printing was also employed in the manufacture of functional medical devices to prevent biofilm formation [18]. Researchers developed PLA disks containing poorly soluble drug nitrofurantoin, an antimicrobial agent, used in the treatment of urinary tract infections. Additional drug-free PLA samples were externally treated with the drug solution. The slow degradation rate of PLA affected the release profiles, showing an initial burst effect (2% released in 6 hours, slope 0.326) and a slower release rate for the following 186 hours (1% released, slope 0.0026). *Staphylococcus aureus* was used to study the antibiofilm properties of the devices. In drug-loaded PLA samples, an 85% inhibition of biofilm formation was evidenced, compared to 24.6% for the drug-treated PLA samples [18].

In a subsequent work, 3D printed PLA–HPMC devices were fabricated to fine-tune the release of nitrofurantoin [19]. The examined formulations contained an increasing content of the water-soluble component HPMC (up to 40% of total solids). The release of API was studied in phosphate buffer saline, pH 7.4, over a period of 24 days and the release profiles were recorded. SEM imaging and image analysis of the different formulations indicated similar morphological characteristics and porosity. However, an improved release rate was observed for the 40% HPMC containing formulation, linked to the higher content of the water-soluble excipient, compared to the 20% HPMC sample, thus highlighting the successful tailoring of drug release characteristics [19].

Genina et al. [20] investigated the printability of different grades of vinyl acetate in the copolymer, to develop T-shaped intrauterine systems (IUS) and subcutaneous rods. Filaments intended for medical devices, containing 5% or 15% of the model drug indomethacin, were assessed and optimized for their melting index and flexural modulus. Following the manufacturing process, printing temperatures varied in the range 145–215 °C, directly connected to the viscosity of the polymer melt for different EVA grades. Solid-state characterization performed on drug-loaded filaments and printed prototypes revealed the presence of the drug in crystalline state in the former and both crystalline and amorphous (or dissolved) states in the latter. Furthermore, the presence of amorphous or dissolved drug in the IUS and subcutaneous rods prototypes enhanced the release rate, presenting a burst effect during the first days, followed by diffusion of the drug from the polymer matrix [20].

Controlled release implantable PCL devices were developed by Hollander et al. [21], highlighting the applicability of FDM 3D printing process on fabricating drug-loaded T-shaped prototypes of IUS. The quality of the printed prototypes was optimized, based on filament characteristics, polymer's melt viscosity, and settings of the printing device. All filaments and printed formulations presented similar behavior, with an initial burst release, followed by a slower release rate over time.

In another work, Muwaffak et al. [22] incorporated antimicrobial metal ions, (zinc, copper, and silver) into a PCL filament to manufacture wound dressings. 3D templates were formatted by 3D scanning a nose and an ear to highlight the potential of coupling 3D scanning and FDM 3D printing on developing customizable wound dressings to fit the needs of an individual patient. The antibacterial activity was tested against a common skin-infective bacterium, *S. aureus*, presenting the strongest bactericidal potential of silver lasting c. 16 hours.

7.3 Pressure-Assisted Microsyringe

PAM is a 3D printing method that enables the layer-by-layer formation of 3D objects via extrusion of a viscous (semisolid or paste) polymeric ink on a building platform. Extrusion of the ink is accomplished by mechanically generating direct pressure or pumping pressurized air in a syringe containing the printing ink, thus forcing the viscous material out of the syringe.

The first attempt enrolling the PAM 3D printing technology for pharmaceutical dosage forms was reported by Rattanakit et al. 2012 [31]. The ink, comprising a PVA-API paste, was either extruded on a PLGA casted film or enclosed between the printed PLGA layers. After printing, the former formulation was rolled as a scroll, whereas the latter remained in a sandwich configuration. Both formulations exhibited prolonged drug release profiles depending on their geometric characteristics, i.e. up to 150 days for the scroll configuration and >100 days for the sandwich formulation. In the latter, an increase of PLGA layers evidenced the reduction of the release rate, thus expanding the period of drug release.

Another study highlighted the capability of the PAM method to fabricate formulations fulfilling the USP regulations [32]. Bilayer tablets were prepared, comprising an IR layer containing API, HPMC 2910 (binder), and microcrystalline cellulose/sodium starch glycolate (disintegrants) and a sustained release (SR) layer composed of API, HPMC 2208 (binder/matrix), and poly(acrylic acid)-(PAA) (hydrophilic matrix).

All formulations satisfied the USP standards, by means of weight variation, hardness, friability, and thickness. Moreover, the dissolution studies presented SR over a period of 12 hours, as required to be comparable to the release profile of the commercial tablet.

In a subsequent study from the same group [33], PAM printed multi-API formulations with defined release profiles were created. The design of tablets included three distinct compartments, enabling osmotic release of captopril and diffusion-associated SR of nifedipine and glipizide. The printing ink was based on HPMC as a hydrophilic matrix and each compartment comprised different

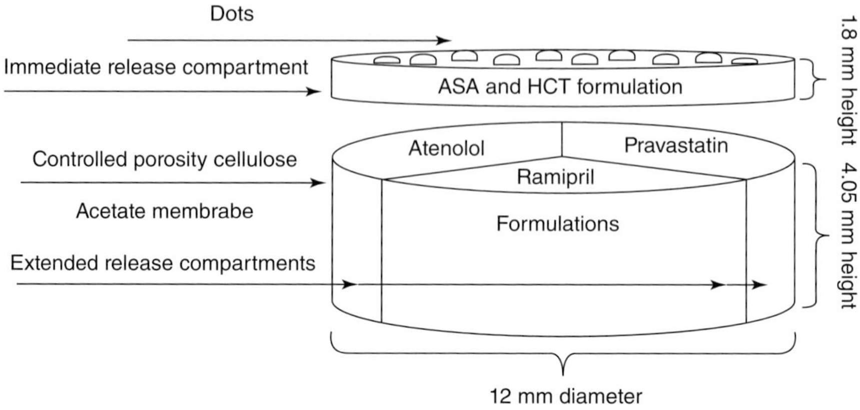

Figure 7.4 Illustration of the polypill structural design, incorporating the aspirin and hydrochlorothiazide in the immediate release compartment, and atenolol, pravastatin, and ramipril in the sustained release compartments. Source: Khaled et al. 2015 [34]. Reproduced with permission of Elsevier.

excipients to provide the desired release characteristics. The same researchers reported the fabrication of a polypill tablet, containing five compartmentalized APIs [34]. The design included an immediate release compartment for aspirin and hydrochlorothiazide, whereas atenolol, pravastatin, and ramipril were infused in three distinct compartments, in an attempt to provide SR of the drugs (Figure 7.4).

Carriers of cyclosporin A have also been developed via PAM to assist the effective transplantation of xenogeneic cells [35]. To enable loading of a hydrophobic drug into a hydrophilic carrier, the investigators prepared a mixture of alginate hydrogel and drug-loaded PLGA microspheres and the carrier was printed with the addition of an external PCL–PLGA framework. *In vitro* and *in vivo* assessment of the carriers exhibited suppression of the immune response, thus providing feasibility of xenogeneic cell transplantation in the absence of rejection response.

Bone regeneration and drug delivery were the subject of further studies reported on PAM 3D printing [36, 37]. A macro/mesoporous scaffold was manufactured, incorporating methyl-functionalized mesoporous silica nanoparticles and carboxyl-functionalized mesoporous bioactive glasses, loaded with rifampin and isoniazid, respectively [36]. The mesoporous materials were bounded with poly(3-hydroxybutyrate-*co*-3-hydroxyhexanoate) to fabricate the drug-loaded scaffolds. The printed scaffolds presented good regenerative ability and mechanical properties and prolonged release profiles compared to commercial calcium phosphate scaffolds. In the direction of bone regeneration and drug delivery, vancomycin-loaded hydroxyapatite/gelatin scaffolds were printed via PAM [37]. Fabricated scaffolds presented good mechanical properties and first-order release profile. Moreover, the release of the drug inhibited the growth of *S. aureus* bacteria. In a recent study [38], anticancer agent 5-fluorouracil was incorporated in a PLGA–PCL patch, manufactured by PAM. The patch exhibited prolonged release over four weeks, thus effectively suppressing the growth of subcutaneous pancreatic cancer xenografts in mice.

7.4 SLA 3D Printing

Another 3D printing method is SLA, in which a laser beam solidifies a liquid resin consisting of monomers by polymerizing them layer by layer, according to the desired pattern, in order to create a solid object. The advantages of SLA are the excellent printing resolution (20 μm), limited only by the width of the focused laser and the avoidance of using high temperatures, that can degrade APIs and excipients [39]. The unique characteristic of this method is that APIs and/or water are entrapped in the polymer matrix simply by the photopolymerization of the monomers [40]. The main drawback of SLA is the relatively limited number of available photo-crosslinkable substances (that include polyethylene glycol diacrylate, PEGDA; poly-2-hydroxyethyl methacrylate, pHEMA; polyethylene glycol dimethacrylate, PEGDMA; and polypropylene fumarate-diethyl fumarate, PPF-DEF)[39]. From the aforementioned materials, only PEGDA has been employed in the creation of pharmaceutical formulations using 3D printing, either for the creation of torus-shaped dosage forms loaded with paracetamol or 4-ASA [39] or ibuprofen hydrogels (using water entrapment in the polymer matrix) [40].

Release of the API from the polymer matrix can be modified by adding polyethylene glycol (PEG 300) into the formulation [39]. Because PEG is not photo-crosslinkable, its increased presence in the polymer matrix will result in the formation of a less dense structure, from which the APIs can escape more easily (and thus increase its dissolution rate). Also, PEG acts as a plasticizer decreasing brittleness of the dosage forms formulated as hydrogels with SLA [40]. The major advantage of creating preswollen hydrogels is the increase of the rate of drug release from the matrices because the wetting stage necessary to facilitate diffusion in the GI tract is omitted, as these formulations are prehydrated. In such formulations, water content plays a crucial role in mass uniformity (achieved only with relatively low water content (below 20%)) because of the fact that these printers are optimized to work with resins of high viscosity (the addition of excessive amount of water lowers the viscosity values, thereby deteriorating printing) [40].

The SLA technique seems to produce amorphous drug dispersions and also does not require preprocessing procedures (such as HME in FDM 3D printing) that can lead to the reduction of API contained in the formulation [39]. One of the major concerns regarding the use of SLA 3D printing for biomedical applications is the toxicity of the substances used to trigger photopolymerization (photoinitiators, PI) by converting monomers to free radicals (i.e. diphenyl(2,4,6-trimethylbenzoyl)phosphine oxide, DPPO). This problem was recently solved by using pharmacologically nontoxic riboflavin and triethanolamine (as PI and co-photoinitiator, respectively) [40].

7.5 Powder Bed 3D Printing

Powder bed 3D printing is a method of rapid prototyping by selectively dispensing liquid binders on powder materials. In brief, a powder substrate is deposited

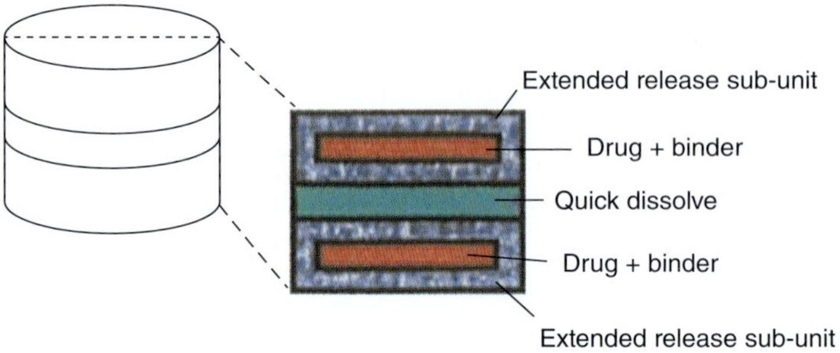

Figure 7.5 Illustration of the layers forming the breakaway tablet. Source: Rowe et al. 2000 [43]. Reproduced with permission of Elsevier.

on the printing platform, whereas a dispenser deposits a binder solution on predefined areas. The binder droplets dissolve the powder components, following recrystallization and a new powder layer is then deposited on the platform. By sequentially repeating these steps, 2D powder layers are formed and bounded to each other, whereas the unbound powder is removed to obtain the 3D printed object.

The first report on 3D printed drug delivery devices manufactured with this technique highlighted the ability to fine-tune the release profiles of dyes (incorporated in the binder) from different formulations [41]. This was achieved by designing a nonresorbable PCL exterior barrier (top and bottom), whereas the middle layers were constructed from PEO powder (serving as diffusion promoter). The same group using the same technology investigated the control of dose precision and the aftereffects of variations in process parameters on the release properties of chlorpheniramine maleate from oral dosage forms [42].

Rowe et al. [43] reported effective modifications of the release properties of complex oral solid dosage forms following different release patterns (Figure 7.5). These characteristics were attributed to the geometry of the formulations, as well as the use of different types of excipients.

Biodegradable subcutaneous implants, containing ethinylestradiol, were fabricated by PBP and evaluated by *in vitro* release studies, as well as *in vivo* release kinetics in ovariectomized rabbits [44]. Three implant designs were investigated, containing (i) a single channel of API in a PCL matrix (I), (ii) a homogeneous distribution of API in the polymer matrix (II), and (iii) a concentration gradient of API in a PCL–PLGA matrix (III). In all cases, *in vitro* release profiles were continuous, but nonlinear, over a period of 13 weeks for implant I or 7 weeks for implants II and III. Different behavior was also noticed in *in vivo* studies, with API release from implants II and III reaching a peak level in four days, followed by a gradual decline for six weeks. In contrast, implant I showed a lower drug release, however, maintaining a constant profile for 13 weeks. Additionally, a good agreement was succeeded between *in vitro*/*in vivo* studies, highlighting the potential of predicting plasma concentrations by *in vitro* release data.

An experimental design approach was utilized on fabricated orally dispersed captopril formulations for sublingual delivery [45]. The independent variables were the type and the level of powder (maltitol, maltodextrin, and polyvinylpyrrolidone), as well as the binder saturation level, and their evaluation included second-level interactions. Several physical properties (output parameters) were measured, such as dose accuracy, content uniformity, hardness, density, friability, dispersion time, and moisture absorption. The researchers identified the set of independent variables that mostly affect the manufacturing process and presented a mathematical model predicting data of the dependent parameters, thus highlighting the strategical design potential of orally dispersing formulations with this technique.

Yu et al. developed oral fast disintegrating tablets, with specific porosity [46, 47]. The design of the tablet had compact top and bottom section, whereas the core of the middle section comprised loose powder, to promote fast disintegration. The formulations exhibited good pharmacotechnical and mechanical properties, and the dissolution studies evidenced the fast release of the drug (c. 98%, in the initial two minutes).

Controlling the release of highly soluble APIs was the objective of a core–shell matrix containing pseudoephedrine hydrochloride [48]. The powder mixture was composed of different HPMC and polyvinyl acetate–polyvinylpyrrolidone ratios, to modify the release profiles. The binder liquid was an aqueous API or an ethanolic-triethyl citrate solution, to achieve an immediate release core or a limiting release shell, respectively. The formulations presented near zero-order release profiles and successfully correlated with additional *in vivo* studies.

Zero-order release profiles were also obtained from hydrophilic tablets with material gradient, containing HPMC, ethyl cellulose, Eudragit RS-100, stearic acid, and sodium lauryl sulfate as release retardants of acetaminophen, which was dispersed in the powder bed [49].

The same group developed multilayered doughnut-shaped tablets, including the API in both powder mixture and binder solution [50]. The powder mixture comprised HPMC and ethyl cellulose, as release retardants. The combined features of special geometry, multiple layers, and concentration variations of the drug in the formulation, resulted in desired drug release profiles and adjustable dosage.

PLA implants were developed by a PBP process in an attempt to generate complex release profiles of levofloxacin, by optimizing the composition of the binder solution (acetone–ethanol mixture) [51]. The researchers successfully printed predefined structures that resulted in distinct pulsatile and steady-state release profiles.

In a similar way, the capability of printing complex patient-tailored geometric structures was investigated in the case of API-loaded bioceramics and biocomposites [52]. In brief, vancomycin hydrochloride, ofloxacin, and tetracycline hydrochloride were post-printing loaded on microporous dicalcium phosphate anhydrous, dicalcium phosphate dihydrate, or hydroxyapatite scaffolds. Controllable 3D morphologies with on-demand release profiles were successfully fabricated, presenting a great potential toward personalized implants via 3D printing technologies.

PLLA implants containing isoniazid were fabricated as columnar-shaped tablets, doughnut-shaped tablets, and multilayer doughnut-shaped tablets with top and bottom barrier layers [53]. All prints were biocompatible and presented a burst release, whereas the release of the API from the multilayered implant was stabilized after the sixth day, as a result of the formulations' special geometric characteristics.

In a recent study, low-temperature 3D printed bioceramic implants, incorporating two antibiotics, were developed for the treatment of osteomyelitis [54]. Vancomycin was spread in the powder substrate (tricalcium phosphate), whereas rifampin was either spread in the powder bed or infused as a solution. *In vivo* studies in a mouse model compared the fabricated implants with a poly(methyl methacrylate) bone cement. The 3D printed implants exhibited enhanced bactericidal activity, although post-printing coating of the implants with PLGA improved efficient treatment of implant-associated bone infections.

Focusing on the treatment of chronic osteomyelitis, a multidrug implant was manufactured as a concentric cylinder, comprising poly-D,L-lactic acid (PDLLA) [55]. The polymeric powder was bounded layer by layer through deposition of levofloxacin and tobramycin binder solutions. The developed formulations presented a sustained and programmed release profile *in vitro*. Moreover, the effect of implants on chronic osteomyelitis was investigated with contribution of an animal model. The researchers observed an enhanced pharmacodynamic action sufficient to treat chronic osteomyelitis, similar to a tumor chemotherapy treatment.

7.6 SLS 3D Printing

SLS 3D printing is a technique similar to PBP (powder layers solidified one after another), but instead of using liquid binders, solidification of the desired powder regions is achieved by using a laser beam. After the solidification of all powder layers in the desired shape, the 3D printed object is retrieved from the bottom of the powder bed.

High precision is one of the advantages of that technique (as "printhead" is a laser beam). Moreover, SLS 3D printing is solvent free and faster than PBP (as PBP prints require a drying period of 48 hours after their formulation to expel liquid binder via evaporation). In comparison to FDM 3D printing, SLS has the advantage of a one-stage process (whereas FDM requires previous filament production). On the other hand, high-energy lasers are required to sinter the powder (usually comprising grinded plastic or ceramic materials), limiting possible applications to drug manufacturing, because of possible degradation of labile APIs.

A recent study has demonstrated the feasibility of manufacturing pharmaceutical formulations using SLS, by incorporating paracetamol into immediate (Kollicoat IR) and intestinal (Eudragit L100-55) release carriers [56]. Kollicoat IR is graft copolymer consisting of 75% polyvinyl alcohol and 25% polyethylene glycol with an average MW of 45 000 Da, whereas Eudragit L100-55 is a methacrylic acid/ethyl acrylate (1,1) copolymer dissolving at pH 5.5 and higher. A small quantity (3% w/w) of a commercial energy absorption enhancer (Candurin® Gold Sheen) was used to facilitate 3D printing, as the aforementioned copolymers do not absorb the blue laser light (445 nm) used by the 3D printing device to achieve powder sintering.

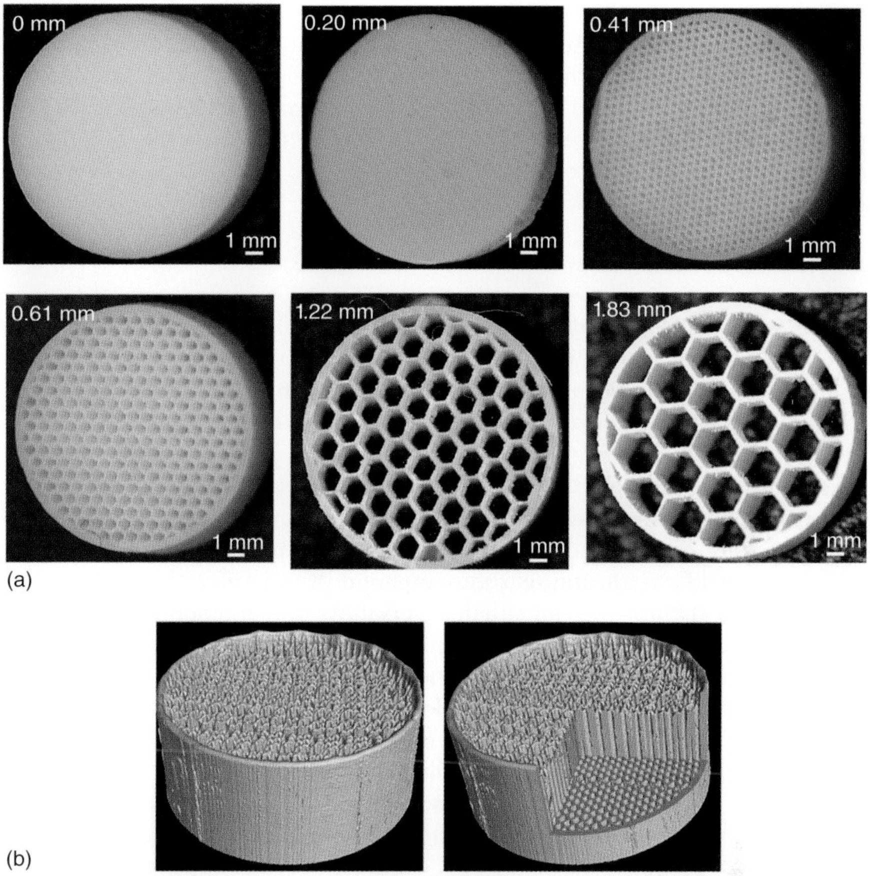

Figure 7.6 Images of inkjet printed tablets with honeycomb patterns. (a) Photographs illustrating various honeycomb cell sizes between 0.20 and 1.83 mm. (b) μCT images of the honeycomb configuration [58]. Source: https://creativecommons.org/licenses/by/3.0/.

7.7 3D Inkjet Printing

3D inkjet printing is an alternative drop-on-demand approach on fabricating 3D structures. Besides the available inkjet printing techniques, piezoelectric systems have been recently incorporated in the manufacture of pharmaceutical dosage forms.

Paclitaxel-loaded PLGA inks were prepared and printed onto glass slides in various patterns, i.e. circles, grids, honeycombs, and rings [57]. The developed microparticles presented homogeneous morphological characteristics. The release studies were conducted in phosphate buffered saline pH 7.4, evidencing a biphasic profile of burst release during the first day, attributed to diffusion of the API and slow release for up to six days, because of degradation of the polymer.

In a recent study, a solvent-free inkjet printing method was developed to produce drug-loaded tablets, comprising beeswax and fenofibrate in a honeycomb pattern [56]. To achieve this, a hot melt chamber was coupled to the piezoelectric printer. The tablets showed homogeneous distribution of the drug within beeswax. Moreover, the ability to modify the release characteristics of the formulation was investigated by controlling the honeycomb cell size (Figure 7.6), thus altering the available surface area for interaction with the dissolution medium [58].

7.8 Conclusions

Additive manufacturing technologies are receiving growing attention for the creation of medical applications. Introduction of 3D printing is expected to revolutionize the field of drug production, personalizing medicinal treatment at an industrial, hospital, and even community pharmacy level. At their infancy, 3D printing technologies are very promising for patients requiring individual and tailored doses depending on their age, weight, and possible defective hepatic or renal functionality. Coupling of 3D printing with pharmacogenomics can lead to the administration of exact drug doses, meeting each patient's individual needs and helping avoid overdosing or subtherapeutic drug levels. Moreover, on an industrial level, 3D printing can enhance versatility of manufacturing products, as changing drug loading of a formulation will only require changing 3D printer's configuration and not employing additional molds, devices, and procedures (allowing companies to offer a much wider range of drug dosages for each API). Additionally, AM makes the production of dosage forms feasible with elaborate geometries and complex material combinations, resulting in formulations with unique release characteristics. It must also be mentioned that AM is a one-stage process, a feature of paramount importance, as it reduces time and cost expenses associated with complex multistage procedures, necessary to obtain modern drug delivery systems. In summary, AM is expected to direct drug therapy toward more anthropocentric medication practices and lead to a new era of pharmaceutical technology.

References

1 ApreciaPhamaceuticals (2015). FDA approves the first 3D printed drug product. *Aprecia Pharm. Co.* 44 (5054433): 2–4.

2 Goyanes, A., Buanz, A.B.M., Hatton, G.B. et al. (2015). 3D printing of modified-release aminosalicylate (4-ASA and 5-ASA) tablets. *European Journal of Pharmaceutics and Biopharmaceutics* 89: 157–162.

3 Goyanes, A., Chang, H., Sedough, D. et al. (2015). Fabrication of controlled-release budesonide tablets via desktop (FDM) 3D printing. *International Journal of Pharmaceutics* 496 (2): 414–420.

4 Skowyra, J., Pietrzak, K., and Alhnan, M.A. (2015). Fabrication of extended-release patient-tailored prednisolone tablets via fused deposition modelling (FDM) 3D printing. *European Journal of Pharmaceutical Sciences* 68: 11–17.

5 Goyanes, A., Kobayashi, M., Martínez-pacheco, R. et al. (2016). Fused-filament 3D printing of drug products: microstructure analysis and drug release characteristics of PVA-based caplets. *International Journal of Pharmaceutics* 514 (1): 290–295.

6 Pietrzak, K., Isreb, A., and Alhnan, M.A. (2015). A flexible-dose dispenser for immediate and extended release 3D printed tablets. *European Journal of Pharmaceutics and Biopharmaceutics* 96: 380–387.

7 Sadia, M., Sośnicka, A., Arafat, B. et al. (2016). Adaptation of pharmaceutical excipients to FDM 3D printing for the fabrication of patient-tailored immediate release tablets. *International Journal of Pharmaceutics* 513 (1–2): 659–668.

8 Zhang, J., Feng, X., Patil, H. et al. (2017). Coupling 3D printing with hot-melt extrusion to produce controlled-release tablets. *International Journal of Pharmaceutics* 519 (1–2): 186–197.

9 Okwuosa, T.C., Stefaniak, D., Arafat, B. et al. (2016). A lower temperature FDM 3D printing for the manufacture of patient-specific immediate release tablets. *Pharmaceutical Research* 33 (11): 2704–2712.

10 Solanki, N., Tahsin, M., Shah, A., and Serajuddin, A.T.M. (2018). Formulation of 3D printed tablet for rapid drug release by fused deposition modeling (FDM): screening polymers for drug release, drug-polymer miscibility and printability. *Journal of Pharmaceutical Sciences* 107 (1): 390–401.

11 Goyanes, A., Fina, F., Martorana, A. et al. (2017). Development of modified release 3D printed tablets (printlets) with pharmaceutical excipients using additive manufacturing. *International Journal of Pharmaceutics* 527 (1–2): 21–30.

12 Sadia, M., Arafat, B., Ahmed, W. et al. (2018). Channelled tablets: an innovative approach to accelerating drug release from 3D printed tablets. *Journal of Controlled Release* 239: 355–363.

13 Okwuosa, T.C., Pereira, B.C., Arafat, B. et al. (2017). Fabricating a shell-core delayed release tablet using Ddal FDM 3D printing for patient-centred therapy. *Pharmaceutical Research* 34 (2): 427–437.

14 Goyanes, A., Robles Martinez, P., Buanz, A. et al. (2015). Effect of geometry on drug release from 3D printed tablets. *International Journal of Pharmaceutics* 494 (2): 657–663.

15 Goyanes, A., Wang, J., Buanz, A. et al. (2015). 3D printing of medicines: engineering novel oral devices with unique design and drug release characteristics. *Molecular Pharmaceutics* 12 (11): 4077–4084.

16 Chai, X., Chai, H., Wang, X. et al. (2017). Fused deposition modeling (FDM) 3D printed tablets for intragastric floating delivery of domperidone. *Scientific Reports* 7 (1): 2829.

17 Gioumouxouzis, C.I., Katsamenis, O.L., Bouropoulos, N., and Fatouros, D.G. (2017). 3D printed oral solid dosage forms containing hydrochlorothiazide for controlled drug delivery. *Journal of Drug Delivery Science and Technology* 40: 164–171.

18 Sandler, N., Salmela, I., Fallarero, A. et al. (2014). Towards fabrication of 3D printed medical devices to prevent biofilm formation. *International Journal of Pharmaceutics* 459 (1–2): 62–64.

19 Boetker, J., Water, J.J., Aho, J. et al. (2016). Modifying release characteristics from 3D printed drug-eluting products. *European Journal of Pharmaceutical Sciences* 90: 47–52.
20 Genina, N., Hollander, J., Jukarainen, H. et al. (2016). Ethylene vinyl acetate (EVA) as a new drug carrier for 3D printed medical drug delivery devices. *European Journal of Pharmaceutical Sciences* 90: 53–63.
21 Holländer, J., Genina, N., Jukarainen, H. et al. (2016). Three-dimensional printed PCL-based implantable prototypes of medical devices for controlled drug delivery. *Journal of Pharmaceutical Sciences* 105 (9): 2665–2676.
22 Muwaffak, Z., Goyanes, A., Clark, V. et al. (2017). Patient-specific 3D scanned and 3D printed antimicrobial polycaprolactone wound dressings. *International Journal of Pharmaceutics* 527 (1–2): 161–170.
23 Jamróz, W., Kurek, M., Łyszczarz, E. et al. (2017). 3D printed orodispersible films with aripiprazole. *International Journal of Pharmaceutics* 533 (2): 413–420.
24 Ehtezazi, T., Algellay, M., Islam, Y. et al. (2017). The application of 3D printing in the formulation of multilayered fast dissolving oral films. *Journal of Pharmaceutical Sciences.*
25 Melocchi, A., Parietti, F., Loreti, G. et al. (2015). 3D printing by fused deposition modeling (FDM) of a swellable/erodible capsular device for oral pulsatile release of drugs. *Journal of Drug Delivery Science and Technology* 30: 360–367.
26 Markl, D., Zeitler, J.A., Rasch, C. et al. (2017). Analysis of 3D prints by X-ray computed microtomography and terahertz pulsed imaging. *Pharmaceutical Research* 34 (5): 1037–1052.
27 Beck, R.C.R., Chaves, P.S., Goyanes, A. et al. (2017). 3D printed tablets loaded with polymeric nanocapsules: an innovative approach to produce customized drug delivery systems. *International Journal of Pharmaceutics* 528 (1–2): 268–279.
28 Maroni, A., Melocchi, A., Parietti, F. et al. (2017). 3D printed multi-compartment capsular devices for two-pulse oral drug delivery. *Journal of Controlled Release* 268: 10–18.
29 Douroumis, D. (2012). *Hot-Melt Extrusion: Pharmaceutical Applications.* Chichester, UK: Wiley.
30 Goyanes, A., Scarpa, M., Kamlow, M. et al. (2017). Patient acceptability of 3D printed medicines. *International Journal of Pharmaceutics* 530 (1–2): 71–78.
31 Rattanakit, P., Moulton, S.E., Santiago, K.S. et al. (2012). Extrusion printed polymer structures: a facile and versatile approach to tailored drug delivery platforms. *International Journal of Pharmaceutics* 422 (1–2): 254–263.
32 Khaled, S.A., Burley, J.C., Alexander, M.R., and Roberts, C.J. (2014). Desktop 3D printing of controlled release pharmaceutical bilayer tablets. *International Journal of Pharmaceutics* 461 (1–2): 105–111.
33 Khaled, S.A., Burley, J.C., Alexander, M.R. et al. (2015). 3D printing of tablets containing multiple drugs with defined release profiles. *International Journal of Pharmaceutics* 494 (2): 643–650.

34 Khaled, S.A., Burley, J.C., Alexander, M.R. et al. (2015). 3D printing of five-in-one dose combination polypill with defined immediate and sustained release profiles. *Journal of Controlled Release* 217: 308–314.

35 Song, T.-H., Jang, J., Choi, Y.-J. et al. (2015). 3D-printed drug/cell carrier enabling effective release of cyclosporin a for xenogeneic cell-based therapy. *Cell Transplantation* 24 (12): 2513–2525.

36 Zhu, M., Li, K., Zhu, Y. et al. (2015). 3D-printed hierarchical scaffold for localized isoniazid/rifampin drug delivery and osteoarticular tuberculosis therapy. *Acta Biomaterialia* 16: 145–155.

37 Martínez-Vázquez, F.J., Cabañas, M.V., Paris, J.L. et al. (2015). Fabrication of novel Si-doped hydroxyapatite/gelatine scaffolds by rapid prototyping for drug delivery and bone regeneration. *Acta Biomaterialia* 15: 200–209.

38 Yi, H.-G., Choi, Y.-J., Kang, K.S. et al. (2016). A 3D-printed local drug delivery patch for pancreatic cancer growth suppression. *Journal of Controlled Release* 238: 231–241.

39 Wang, J., Goyanes, A., Gaisford, S., and Basit, A.W. (2016). Stereolithographic (SLA) 3D printing of oral modified-release dosage forms. *International Journal of Pharmaceutics* 503 (1–2): 207–212.

40 Martinez, P.R., Goyanes, A., Basit, A.W., and Gaisford, S. (2017). Fabrication of drug-loaded hydrogels with stereolithographic 3D printing. *International Journal of Pharmaceutics* 532 (1): 313–317.

41 Wu, B.M., Borland, S.W., Giordano, R.A. et al. (1996). Solid free-form fabrication of drug delivery devices. *Journal of Controlled Release* 40 (1–2): 77–87.

42 Katstra, W., Palazzolo, R., Rowe, C. et al. (2000). Oral dosage forms fabricated by three dimensional printing™. *Journal of Controlled Release* 66 (1): 1–9.

43 Rowe, C., Katstra, W., Palazzolo, R. et al. (2000). Multimechanism oral dosage forms fabricated by three dimensional printing™ *Journal of Controlled Release* 66 (1): 11–17.

44 Lin, S., Chao, P.-Y., Chien, Y.W. et al. (2001). In vitro and in vivo evaluations of biodegradable implants for hormone replacement therapy: Effect of system design and PK-PD relationship. *AAPS PharmSciTech* 2 (3): 55–65.

45 Lee, K.-J., Kang, A., Delfino, J.J. et al. (2003). Evaluation of critical formulation factors in the development of a rapidly dispersing captopril oral dosage form. *Drug Development and Industrial Pharmacy* 29 (9): 967–979.

46 Yu, D.-G., Branford-White, C., Yang, Y.-C. et al. (2009). A novel fast disintegrating tablet fabricated by three-dimensional printing. *Drug Development and Industrial Pharmacy* 35 (12): 1530–1536.

47 Yu, D.-G., Shen, X.-X., Branford-White, C. et al. (2009). Novel oral fast-disintegrating drug delivery devices with predefined inner structure fabricated by three-dimensional printing. *Journal of Pharmacy and Pharmacology* 61 (3): 323–329.

48 Wang, C.-C., Tejwani (Motwani), M.R., Roach, W.J. et al. (2006). Development of near zero-order release dosage forms using three-dimensional printing (3-DPTM) technology. *Drug Development and Industrial Pharmacy* 32 (3): 367–376.

49 Yu, D.G., Yang, X.L., Huang, W.D. et al. (2007). Tablets with material gradients fabricated by three-dimensional printing. *Journal of Pharmaceutical Sciences* 96 (9): 2446–2456.

50 Yu, D.-G., Branford-White, C., Ma, Z.-H. et al. (2009). Novel drug delivery devices for providing linear release profiles fabricated by 3DP. *International Journal of Pharmaceutics* 370 (1–2): 160–166.

51 Huang, W., Zheng, Q., Sun, W. et al. (2007). Levofloxacin implants with predefined microstructure fabricated by three-dimensional printing technique. *International Journal of Pharmaceutics* 339 (1–2): 33–38.

52 Gbureck, U., Vorndran, E., Müller, F.A., and Barralet, J.E. (2007). Low temperature direct 3D printed bioceramics and biocomposites as drug release matrices. *Journal of Controlled Release* 122 (2): 173–180.

53 Wu, G., Wu, W., Zheng, Q. et al. (2014). Experimental study of PLLA/INH slow release implant fabricated by three dimensional printing technique and drug release characteristics in vitro. *Biomedical Engineering Online* 13 (1): 97.

54 Inzana, J., Trombetta, R., Schwarz, E. et al. (2015). 3D printed bioceramics for dual antibiotic delivery to treat implant-associated bone infection. *European Cells and Materials* 30: 232–247.

55 Wu, W., Ye, C., Zheng, Q. et al. (2016). A therapeutic delivery system for chronic osteomyelitis via a multi-drug implant based on three-dimensional printing technology. *Journal of Biomaterials Applications* 31 (2): 250–260.

56 Fina, F., Goyanes, A., Gaisford, S., and Basit, A.W. (2017). Selective laser sintering (SLS) 3D printing of medicines. *International Journal of Pharmaceutics* 529 (1–2): 285–293.

57 Lee, B.K., Yun, Y.H., Choi, J.S. et al. (2012). Fabrication of drug-loaded polymer microparticles with arbitrary geometries using a piezoelectric inkjet printing system. *International Journal of Pharmaceutics* 427 (2): 305–310.

58 Kyobula, M., Adedeji, A., Alexander, M.R. et al. (2017). 3D inkjet printing of tablets exploiting bespoke complex geometries for controlled and tuneable drug release. *Journal of Controlled Release* 261: 207–215.

8
Modulating Drug Release from 3D Printed Pharmaceutical Products

Julian Quodbach

Heinrich Heine University Düsseldorf, Institute of Pharmaceutics and Biopharmaceutics, Building 26.22.00.38, Universitätsstr. 1, 40225, Düsseldorf, Germany

8.1 Introduction

Pharmaceutical dosage forms are complex devices to deliver active pharmaceutical ingredients (APIs) to patients. Their purpose is to ensure API concentrations above the minimal effective yet below the lowest toxic blood plasma concentration in the human body for an effective treatment. Therefore, a properly designed dosage form increases patient safety and decreases the severity of adverse effects. Many dosage forms govern the plasma concentration levels by controlling the release rate of the API. If a drug is readily absorbed, a sustained API release over a prolonged period can assure that the plasma concentrations do not reach harmful levels. The same is true if the elimination of the API is slow, which can lead to an unsafe API accumulation when API absorption continues. Nifedipine is a common example for a drug showing a fast absorption with potentially severe adverse effects if the release rate from the dosage form is not controlled [1]. The development of sustained release formulations of nifedipine increased patient safety and paved the way to the success of the drug substance. It has also been demonstrated that the introduction of modified release dosage forms does not only increase the safety but also plays a big role in improving patient compliance [2]. Frequently, the development of dosage forms showing a suitable release profile is key to a successful therapy. This understanding led to the development of several approaches to modify the drug release. In Figure 8.1 a nonexhaustive overview of these techniques is shown.

Traditionally, modified release techniques are mostly applied to tablets, but they are also used for inserts (e.g. Nuva Ring®) and implants (e.g. Implanon®). The application of these techniques requires in-depth knowledge of the processes as well as specialized equipment. For example, drum coaters or fluid bed processors are used to process hundreds of kilograms of tablets or excipient mixtures at the same time. They are inherently not suited for processing small batches. With the emergence of increasingly cheap genomic and metabolomic analyses [3], a better understanding of the therapeutic mechanisms of drug substances, and the recognition of patient individuality, the call for personalized

3D and 4D Printing in Biomedical Applications: Process Engineering and Additive Manufacturing,
First Edition. Edited by Mohammed Maniruzzaman.

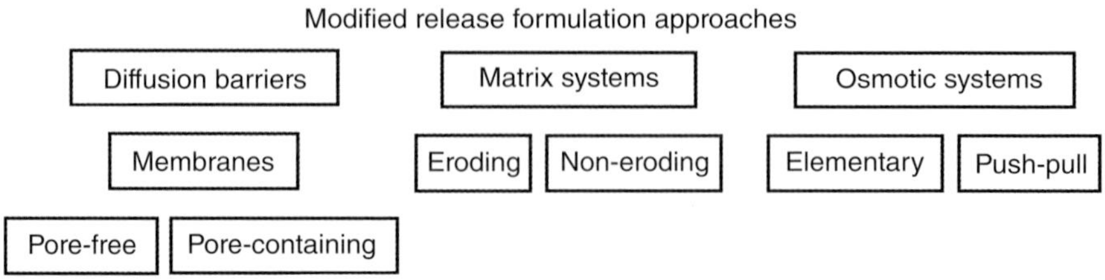

Figure 8.1 Common modified release approaches for solid oral dosage forms.

and individualized medicines is getting louder. In 2015, Schork described that the 10 highest grossing drugs in the USA failed to help 3–24 people for every person where they improved the condition [4]. The main reason was found to be the heterogeneity of patients, who cannot always be treated with the same medications or same APIs for a condition. Thorough screenings of patients are necessary to determine which drug improves the outcome most, what the required dose is, and over what period of time it should be delivered. Therefore, new technologies are necessary to enable the manufacturing of patient-specific dosage forms in small batches. Besides just containing the correct dose of selected API(s), these techniques need to be capable of modifying the release profile to at least a similar degree as traditional release modification techniques.

3D printing holds the potential to deliver on those requirements. The next chapter gives an introduction in 3D printing techniques with (potential) pharmaceutical applications and discusses strategies to modify drug release from 3D printed (3DP) dosage forms.

8.2 Pharmaceutically Used 3D Printing Processes and Techniques

8.2.1 Process Flow of 3D Printing Processes

The process flow of 3D printing processes is similar for all printing techniques. First, the desired object is designed in a computer-aided design (CAD) program. Even though virtually every shape can be designed, the final dosage form should be created with rational consideration. A safe and unproblematic administration should have highest priority. After completion of the design, the object is then saved, usually as a stl-file (stl is the short form for stereolithography). During the saving process, the object surface is triangulated and transformed to a 3D surface representation of the original object. This means that all further information of the original object (e.g. color or other material attributes) except the surface is discarded. As the surface is only represented by differently sized triangles, the settings for the triangulation have to be chosen carefully to not lose relevant surface details. The stl-file is imported to the printer software, and the necessary print settings, such as the print temperature, layer height, scaling, *infill* pattern and density, and print speed are selected. The infill describes the pattern, which is used to fill the printed outlines of an object. Normally, printed objects are not printed solidly, but rather as hollow shells with a more or less complete infill to ensure mechanical resilience and safe handling. The shape of this pattern and the degree of filling can be selected. The last step before the actual printing process is the *slicing* of the object. 3D printing processes do not grow the desired object from the center, but by successively printing thin layers on top of each other until the object is finished. In the slicing process, the object is divided into these layers and the control code for the printer kinematics is generated, the so-called G-code. The G-code is subsequently transferred to the printer and the printing process starts. Although these steps are the same for most printing techniques, the actual printing processes differ greatly. For pharmaceutical relevant techniques, the individual layers can be produced via inkjet-based systems, via extrusion and by the use of electromagnetic radiation. As a basic understanding of the technological differences of these techniques is necessary to understand their potential, they will be introduced in the following sections. For a more thorough introduction into 3D printing techniques, the study of one of the numerous reviews is recommended, e.g. [5–7].

8.2.2 Inkjet-Based Printing Technologies

Inkjet printing is the process of jetting small droplets of ink on a substrate. Piezoelectric or thermal inkjet printer heads are equipped with an ink reservoir and eject the ink by the use of either a piezoelectric or a thermal process. Piezoelectric printer heads are more frequently used in pharmaceutics as they allow the use of a wide range of inks and solvents compared to thermal printer heads, which allow only a limited range because of excessive heat generation during the printing process. Inkjet printer heads are used for drop-on-drop (DoD) and drop-on-powder (DoP) 3D printing. For DoD deposition, only the first layer of ink is ejected on a predefined substrate. The following layers are deposited on top of the formerly printed layers and the object is built further with every additional pass of the printer head. The ink may be an API solution or suspension based on water or organic solvents. During the DoP process, an ink consisting of a binder solution is deposited on an even powder bed. After evaporation of the solvent, the powder is fused together. Instead of a binder solution, a solvent for (parts of) the powder formulation could be potentially used to dissolve low amounts of excipients or APIs that form solid bridges upon solvent evaporation. After the binder has solidified, a new powder layer is created on top of the powder bed and the printing process starts again. To speed up the drying process, the powder bed can be continuously heated. After the printing process, the excessive powder between the solidified parts has to be removed. This printing process is also referred to as powder bed printing.

8.2.3 Extrusion-Based Printing Techniques

The common characteristic of extrusion-based printing is the forced movement of plasticized material through a nozzle and the deposition of the extruded material on a printing bed. Two techniques are based on extrusion principles: fused filament fabrication (FFF; also called Fused Deposition Modeling™, FDM) and the application of pressure-assisted microsyringes (PAM). FFF requires polymer filaments that are commonly fed into the printer head with a gear drive. Below the gear drive, the filament is plasticized in a heating element. The continuous feeding process creates excess pressure within the heating element and causes the extrusion of the plasticized material through the attached nozzle. The mass is then deposited on a temperature-controlled printer bed according to the G-code instructions. This technique requires the preparation of drug-containing filaments before the actual printing step, commonly by hot melt extrusion (HME). The amount of extruded filament depends on the diameter of the filament and the nozzle as well as the speed of the feeding system into the printer head. Nozzle diameters range from 0.1 to 2.0 mm, with 0.35 and 0.4 mm being the most frequently used.

Whereas the manufacturing of filaments narrows the selection of pharmaceutical grade substances because of their mechanical properties, most thermoplastic materials, semisolids, and even liquids can be printed without extensive preprocesses via PAM. In PAM, material is extruded through needle-shaped nozzles from syringes and cartridges by the application of pressure. If a thermoplastic material is used, it is stored in cartridges and the temperature is adjusted to achieve a suitable melt viscosity. If the material is semisolid or liquid, the cartridges can also be cooled to set the viscosity for highest stability and print quality. Once the selected temperature is reached, a pressure is created and the material is extruded through the nozzle on a printer bed. Once the material is deposited on the printer bed, it needs to solidify by cooling, drying, or a chemical cross-linking process similar to the stereolithography process (see Section 8.2.4). For PAM, the amount of extruded mass depends on the viscosity of the extruded mass and the pressure exerted on the material.

8.2.4 Laser-Based Techniques

Stereolithography (SLA) and selective laser sintering (SLS) utilize lasers to solidify materials to 3D objects. SLA is the oldest 3D printing technique with the first patent issued in 1986 [8] and the first commercially available 3D printer in 1987, the SLA-1 by 3D systems. In SLA, the laser is placed above or below a basin filled with a photopolymerizable polymer solution. A UV laser generates free radicals by irradiation of a photoinitiator in the polymer solution [9]. The free radicals cause the chemical cross-linking of the polymers and thereby a solidification of the prior dissolved polymer. The laser traces the outline and infill of the required object and solidifies the first layer in contact with the print bed. Depending on the position of the laser, the print bed is then moved down (laser source above the basin) or up (laser source below the basin), creating a new layer of non-crosslinked solution on top or below the prior layer. The laser traces the outline and infill of the next layer, and the steps are repeated until the object is finished. Because of the use of highly reactive molecules and polymers and potential residuals thereof, it is questionable whether this technique could get permission by the appropriate authoritative bodies for the printing of dosage forms.

SLS is very similar to DoP printing. The main difference is that for SLS the powder layers are not solidified by a printed ink, but by sintering from a heat impulse generated from a laser source. Therefore, an even powder bed is created and a laser traces the shape of the current layer on the powder. The temperature increase because of the laser irradiation causes the powder particles to sinter together and form a solid layer structure. After a new powder layer has been created on top of the solidified one, the process starts again. After the object is finished, excessive powder has to be removed to access the solid object. To decrease the necessary energy input of the laser for the sinter process, the powder bed can be heated constantly.

8.3 Modifying the Drug Release Profile from 3D Printed Dosage Forms

8.3.1 Approaches to Modify the Drug Release

3D printing offers several approaches to modify the release profile from dosage forms. The most basic approach is the modification of the formulation itself. By changing the quantity of or replacing and introducing other excipients, the release profile can be altered. In this case, *most basic* shall not imply that formulation modifications are easier to implement than other strategies, on the contrary. Choosing the most suitable excipients to achieve the required result is the main responsibility of formulation scientists and is applicable to most manufacturing techniques. However, for 3D printing, the formulation is not only restricted by traditional requirements, e.g. powder flowability, particle size, or incompatibilities, but also special requirements of the printing process. The unconventional advantage of 3D printing over established manufacturing techniques is the control over the geometry of the printed dosage form. The surface area and therefore the drug release rate can be specifically adjusted during the design process without altering the API dose. This approach allows the modification of the drug release from one feedstock. If the utilized printer offers the opportunity to employ two or more different feedstock materials, the combination of shape and material adjustment allows precise control over the direction of diffusion and thereby also the release kinetic.

8.3.2 Modifying the Drug Release by Formulation Variation

8.3.2.1 Fused Filament Fabrication

Most studies that investigate the formulation influence in 3D printed dosage forms rely on FFF. The necessary filaments for FFF can be prepared in various ways. Premanufactured filaments can be soaked in an API solution, which results in low drug loadings. Alternatively, filaments might be coated with a thin layer of API containing excipients. The most versatile manufacturing process for filaments, however, is HME. HME is also frequently used in the manufacturing of more traditional dosage forms. A preblend or individual excipients are dosed in the extruder, where they are conveyed and mixed by screw elements. The heated barrel, in which the screws are mounted, provides additional energy to plasticize the material. After the material is thoroughly mixed, it is forced through the extruder die and the resulting filament is collected. The filament is then used as the feedstock for the FFF process. The formulation can be methodically varied to achieve the desired quality attributes of the filaments. A typical HME blend contains at least one or more API, one or more thermoplastic material, and a plasticizer to decrease the glass transition temperature T_g. A large number of studies have been conducted to assess drug stability and drug release in extruded masses. However, for FFF 3D printing, one additional quality attribute has to be met – the mechanical properties of the filament to ensure constant conveying in the printer head and reliable printing.

Water et al. investigated the applicability of non-pharma grade polylactic acid (PLA) as the feedstock material for drug-eluting implants [10]. The authors varied the amount of API (nitrofurantoin) in a blend of PLA with or without 5% hydroxyapatite and investigated the changes in the dissolution profile of FFF-printed disks. As displayed in Figure 8.2a, an initial burst release within the first three hours of dissolution is apparent in all cases. After the burst phase, the drug dissolution rate mainly depends on the API concentration. The higher the API concentration, the higher the amount of dissolved API after 45 days. The overall low amount of dissolved API of <10% is attributed to the inaccessibility of API entrapped in the insoluble PLA matrix. The addition of 5% HA corroborates this hypothesis as it increases the surface roughness and thereby the accessible surface area, which is especially apparent in the case of 30% API in the printed object. The *in vitro* effect of the dissolution variations on the planktonic growth of *Staphylococcus aureus* bacteria is shown in Figure 8.2b. The release profiles match the growth inhibition of *S. aureus*, where disks with nitrofurantoin concentrations >10% lead to a prolonged growth inhibition. Similar results regarding the impact of API concentration are reported from Holländer et al. for poly(ε-carpolactone) (PCL) filaments containing γ-indomethacin [11]. Increasing the amount of indomethacin from 5% to 30% led to overall higher dissolution rates. Therefore, changing the amount of incorporated API can be

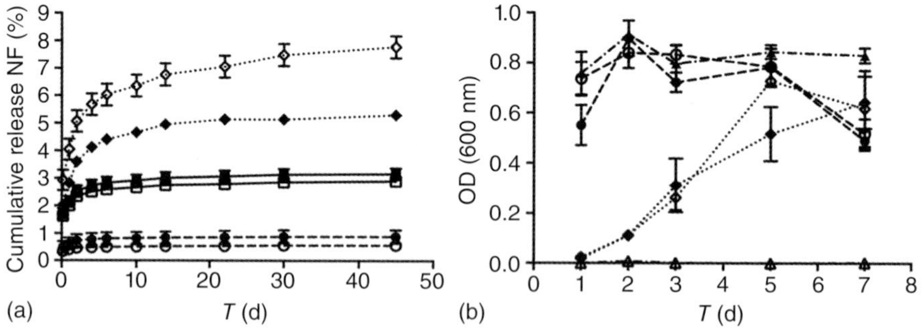

Figure 8.2 The release profiles of printed disks with a diameter of 10 mm and a height of 2 mm over 45 days (a) and planktonic growth inhibition over seven days induced by printed disks (b). Samples with 30%, 20%, and 10% nitrofurantoin are shown as solid diamonds, squares, and circles. Samples containing the same amount of nitrofurantoin and an additional 5% of hydroxyapatite are shown as the corresponding open symbols. Phosphate-buffered saline pH 7.4 at 37 °C; $n = 3$, mean ± SD. Source: Water et al. 2015 [10]. Reprint with permission of Elsevier.

a viable strategy to modify the dissolution rate. However, depending on the selected API concentration in the filament, the size of the printed dosage form would need to be adapted to ensure the correct dose.

If the API concentration is to remain constant in the filaments but the release rate shall be modified, the composition of the excipients has to be modified. This approach was studied by several working groups, e.g. Kempin et al. [12], Zhang et al. [13], and Alhijjaj et al. [14]. Kempin et al. assessed different polymers and drug loads for FFF of implants. The authors used 5% (w/w) of quinine as model drug in combination with Eudragit® RL, PCL, poly(L-lactide) (PLLA), and ethyl cellulose (EC). Extruded filaments and printed hollow cylinders are shown in Figure 8.3. The filaments produced from Eudragit RS and EC appear clear in contrast to the filaments of PCL and PLLA, which the authors explain with higher polymer crystallinity. Because of a lower polymer viscosity upon printing, the hollow cylinders of PCL and PLLA also display a widening of the lowest layers. This is due to a spreading of the first layers on the heated printing bed because of the weight of successively printed layers. The fluorescence microscopic images indicate a uniform drug distribution throughout the cylinders. The extent of drug dissolution and the dissolution rate greatly depend on the applied polymer (Figure 8.4). 3.7 ± 0.2% of API were released from Eudragit RS-based cylinders within 78 days, 76.4 ± 1.8% within 58 days from PCL cylinders, 53.2 ± 3.7% from PLLA cylinders in 177 days, and 4.5 ± 0.8% from EC implants in 100 days. Similar to the reports of Holländer et al. [11], a burst release is detected for API from the surface of a PCL matrix, followed by a sustained release released due to a slow drug diffusion process through the polymeric matrix. In contrast to the works of Water et al. [10], no burst release was observed for PLLA cylinders. This

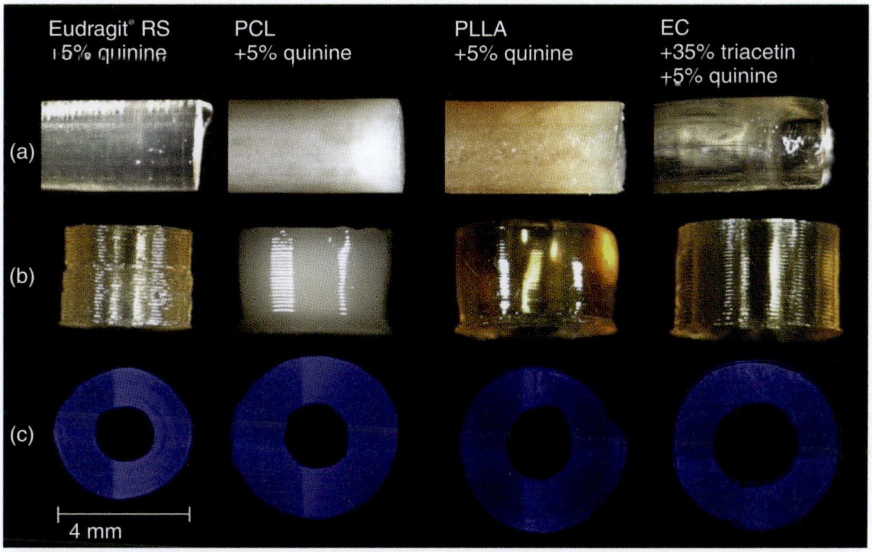

Figure 8.3 Extrudates (a), FFF printed hollow cylinders in side view (b), and fluorescence microscopic images of printed cylinders from the top (c). Source: Kempin et al. 2017 [12]. Reprint with permission of Elsevier.

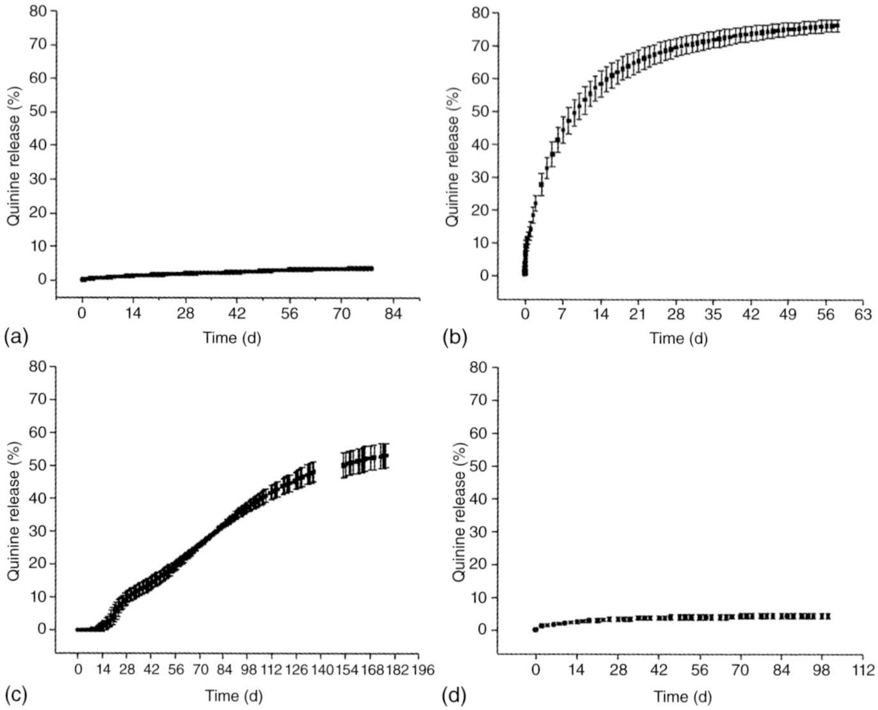

Figure 8.4 Dissolution curves of hollow cylinders consisting of 5% quinine and Eudragit RS (a), PCL (b), PLLA (c), and EC with additional 35% of triacetin (d). Phosphate-buffered saline pH 7.4 at 37 °C, sink conditions; $n = 3$, mean ± SD. Source: Kempin et al. 2017 [12]. Reprint with permission of Elsevier.

is explained by using the more crystalline pharma grade PLLA, which generally shows a slower drug release, in contrast to the more amorphous poly(D,L-lactide) in the study of Water et al. The results shown in Figure 8.4c display an almost constant drug release after a lag time of two weeks. In all cases, the API was molecularly dispersed in the matrices, indicating that the variations between the formulations are mainly because of the polymer properties.

Zhang et al. presented a three-staged formulation optimization strategy for 3D printed dosage forms containing 30% acetaminophen (=paracetamol) [13]. In the first stage, 30% API was extruded with pure polymers and the obtained filaments were mechanically tested with a three-point bend test (Table 8.1). Depending on the results of the three-point bend test, binary polymer mixtures were extruded in the second stage to achieve more suitable mechanical properties. The last formulation adjustments were performed in the third stage according to results from *in vitro* dissolution studies and the three-point bend test. For the dissolution studies, three different samples of each formulation were produced. Tablets were printed via FFF and compressed to similar dimensions and weight from milled filaments as well as physical mixtures (PM) of the excipients. 3D printed tablets III-1 (for the composition, see Table 8.1) released 87% of API within 10 hours, III-2 63% and III-3 72% (Figure 8.5).

Table 8.1 Three-stage formulation strategy according to Zhang et al. [13].

Formulation		Drug (w/v)	Polymer (w/w)	Disintegrator (w/w)
1		0	100% PLA	—
	I-1	30%	70% HPC LF	—
	I-2	30%	70% HPC EF	—
Stage 1	I-3	30%	70% HPMC E5	—
	I-4	30%	70% EC N14	—
	I-5	30%	70% Soluplus	—
	I-6	30%	70% Eudragit L100	—
	II-1	30%	35% HPMC E5 + 35% EC N14	—
	II-2	30%	35% HPMC E5 + 35% HPC EF	—
	II-3	30%	35% HPMC E5 + 35% HPC LF	—
Stage 2	II-4	30%	35% HPMC E5 + 35% Soluplus	—
	II-5	30%	35% HPMC E5 + 35% Eudragit L100	—
	II-6	30%	35% EC N14 + 35% Soluplus	—
	II-7	30%	35% HPC LF + 35% EC N14	—
	III-1	30%	45.5% HPMC E5 + 19.5% EC N14	5%
	III-2	30%	45.5% HPMC E5 + 19.5% HPC EF	5%
Stage 3	III-3	30%	45.5% HPMC E5 + 19.5% HPC LF	5%
	III-4	30%	50% HPMC E5 + 15% Soluplus	5%
	III-5	30%	50% HPMC E5 + 15% Eudragit L100	5%
	III-6	30%	50% EC N14 + 15% Eudragit L100	5%

HPC, hydroxypropyl cellulose; HPMC, hydroxypropyl methylcellulose.
Source: Zhang et al. 2016 [13]. Reprint with permission of Elsevier.

Drug release from printed tablets based on EC (III-6) was slow, releasing only 8.9% within 24 hours. Acetaminophen formed a partial solid dispersion with evenly distributed particles throughout the EC matrix. In combination with the nonexistent water solubility of the polymer, this leads to the observed slow drug release. Tablets from milled extrudates (EXT) and PMs released API faster than the 3D printed tablets. This is mostly because of a quick disintegration of the compressed tablets, whereas the 3D printed tablets barely disintegrated. By changing the formulation, a wide range of dissolution rates could be realized. Alhijjaj et al. also investigated the polymer blends as a means to influence dissolution behavior [14]. The authors extruded 10% of felodipine with Eudragit E, Soluplus®, and PVA in combination with Tween 80, polyethylene glycol (PEG) 4000, and polyethylene oxide 100000. Dissolution profiles (not shown) of printed disks showed an increase in the dissolution rate for all formulations over six hours in pH 1.2 HCl and pH 6.8 phosphate buffer saline compared to crystalline felodipine. The most pronounced dissolution rate improvement was shown by Eudragit E (85% dissolved in 60 minutes in pH 1.2), followed by Soluplus (10% dissolved in 60 minutes). PVA led only to a slight increase in the dissolution rate, independent of medium pH. The faster dissolution performance of Eudragit E

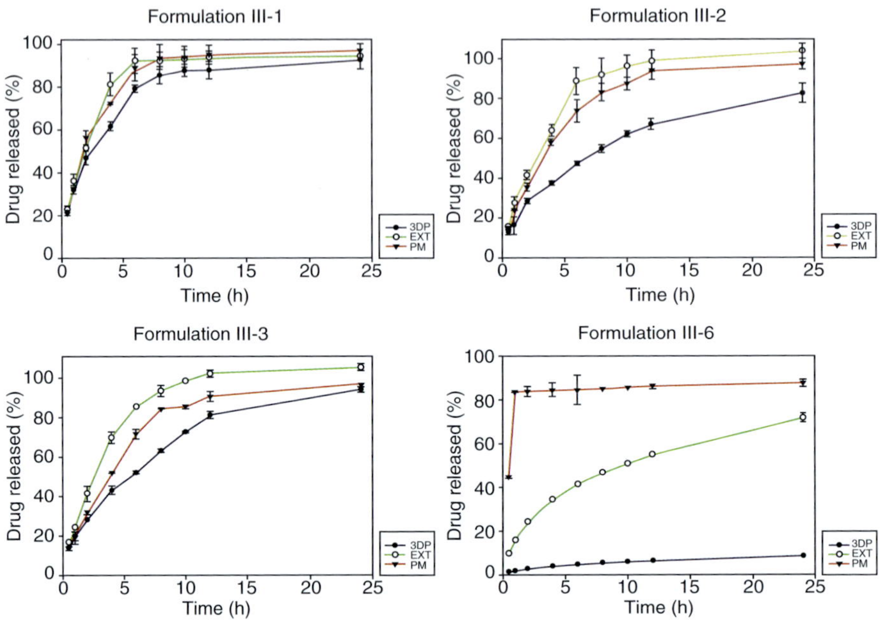

Figure 8.5 Dissolution profiles of 3D printed tablets (3DP), tablets compressed from milled extrudates (EXT), and physical mixtures (PM). The composition is shown in Table 8.1. Simulated intestinal fluid without pancreatin, pH 6.8 at 37 °C; $n = 3$, mean ± SD. Source: Zhang et al. 2016 [13]. Reprint with permission of Elsevier.

is credited to a lower solubility of Soluplus and a larger crystalline fraction of PEG/PEO in the printed Soluplus-based objects.

8.3.2.2 Other Printing Techniques

Less research has been conducted with printing techniques aside FFF. Higher costs of purchase for equipment and less experience with these techniques might be contributing factors to the focus on FFF. However, some working groups investigated the applicability of other techniques. Khaled et al. employed PAM to print tablets from hydroxypropyl methylcellulose (HPMC) pastes containing 82–90% guaifenesin as API and 6–14% HPMC (w/w). Fickian diffusion from the HPMC matrices was found to be the main mechanism for drug release. The release rates decreased according to increasing HPMC content, as the diffusional release process is slowed down by a denser polymer matrix (not shown). SLS was used by Fina et al. to 3D print tablets with 5%, 20%, and 35% of paracetamol [15]. Kollicoat® IR and Eudragit L100-55 were used as thermoplastic materials. As the polymers do not absorb laser light at the applied wavelength of 445 nm, 3% of Candurin® gold sheen, a mineral pigment, was added to all formulations. The pigment colorant absorbed enough energy of the laser light to trigger sintering of the polymers. It has to be noted that the API did not degrade during the brief irradiation period with the laser and that the friability for all formulations was below 1%. Figure 8.6 displays the dissolution profiles of tablets produced with Kollicoat IR (for the composition, refer to the caption). The low-dose

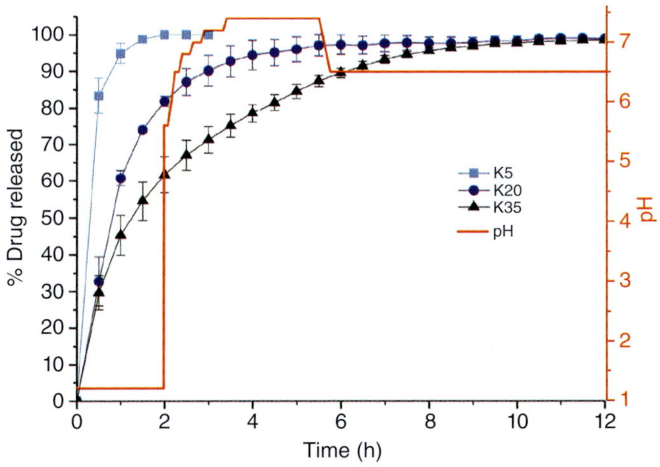

Figure 8.6 Dissolution profiles of tablets prepared by SLS. Tablets contained 5%, 20%, and 35% of paracetamol, 3% of Candurin gold sheen, and Kollicoat IR. In the legend, K stands for Kollicoat IR and the number for the API percentage. pH was adapted according to the red line at 37 °C. $n = 3$, mean ± SD. Source: Fina et al. 2017 [15]. Reprint with permission of Elsevier.

formulation released 80% of API in 30 minutes, the medium-dose formulation in two hours, and the high-dose formulation in five hours. Higher drug contents led to more complete sintering and less porous tablets, which results in a slower drug release independent of the medium pH. Although at least parts of the API remained in a crystalline state in Kollicoat IR prints, the brief sintering process was sufficient to dissolve paracetamol in Eudragit L100-55 and form a solid solution. Dissolution differences of Eudragit formulations were apparent but not as pronounced as for Kollicoat IR (not shown). Wang et al. investigated the dissolution variability of paracetamol and 4-aminosalicylic acid tablets produced with SLA [16]. The photopolymer solution contained varying amounts of PEG diacrylate and PEG 300 and 1% of diphenyl(2,4,6-trimethylbenzoyl) phosphine oxide as the photoinitiator. Printed, torus-shaped tablets are shown in Figure 8.7. The higher the PEG diacrylate concentration in the photopolymer solution, the lower was the dissolution rate (not shown). It is hypothesized that lower concentrations of the reactive polymer PEG diacrylate are accompanied with lower degrees of cross-linking of the polymer, creating a less dense matrix for polymer diffusion.

8.3.3 Manipulating the Dosage Form Geometry as a Means to Modify API Release

The dissolution rate from dosage forms does not only depend on the formulation properties but also on the geometry of the dosage form, i.e. the surface area accessible to the dissolution medium. Some of the fundamental theories describing dissolution behavior were published more than a hundred years ago. A well-known example is the Noyes–Whitney equation, which was modified in

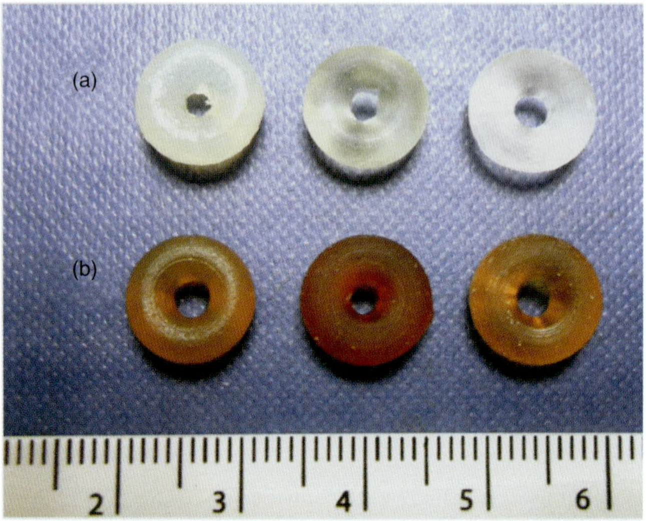

Figure 8.7 SLA printed tablets containing approximately 5% API with decreasing amounts of PEG diacrylate and increasing amounts of PEG 300 (left to right) with paracetamol (a) and 4-aminosalicylic acid (b). Source: Wang et al. 2016 [16]. Reprint with permission of Elsevier.

1903 by Nernst and Brunner to include the surface area as a proportionality constant (Eq. (8.1)) [17, 18].

$$\frac{\mathrm{d}c}{\mathrm{d}t} = \frac{A * D * (c_\mathrm{s} - c_\mathrm{t})}{\delta * V} \tag{8.1}$$

where dc/dt represents the change in concentration over time $t_x - t_{x-1}$, A the surface area accessible to dissolution, D the diffusion coefficient of the substance in the solvent, δ the thickness of the diffusive layer, V the volume of the solution, c_s the saturation concentration, and c_t the concentration in the solvent at time t. Equation (8.1) describes a first-order dissolution kinetic, which can be applied to many immediate release and unmodified dosage forms. The Higuchi model, which describes drug release via Fickian diffusion from matrices, also incorporates the area [19, 20]. Based on the Higuchi model, Lapidus and Lordi developed the equation further to introduce the surface-to-volume ratio rather than the surface area [21]. Other models, such as the Hixson–Crowell model, further reflect the change in surface area during dissolution processes [22]. Using 3D printing to control the surface area or specific surface area (SSA), defined as surface per unit volume or mass, is therefore an obvious approach. Nonetheless, only few studies were conducted into this specific field of research until now.

8.3.3.1 Fused Filament Fabrication

Goyanes et al. incorporated approximately 4% (w/w) paracetamol in non-pharma grade PLA and 3D printed different geometries from the obtained filament [23]. The selected shapes were cube, pyramid, cylinder, sphere, and torus (Figure 8.8). The sizes were adjusted to create three different test groups. In the first group, the surface area was kept constant, in the second group the surface area-to-volume

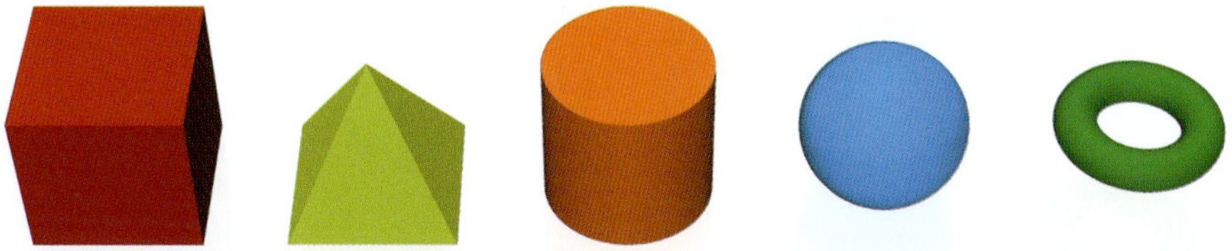

Figure 8.8 CAD designs of printed objects.

ratio, and in the third group the weight. Intragroup variances of the SSA were below 8%. Dissolution measurements for the first group, in which the surface area was kept constant, are shown in Figure 8.9a. At first glance, the results might be surprising, as similar dissolution profiles could be expected from different geometries if the surface area is kept constant. However, the drug release differs greatly with the geometry of the printed object, the time to 90% release (t_{90}) ranging from 2 hours for the pyramid to almost 10 hours for cylinder and sphere. Because the surface area is kept constant, the volume of the objects had to be adjusted. Therefore, the surface-to-volume ratio, or SSA, is changing. The order of highest to slowest dissolution rate matches the surface-to-volume ratios of the printed objects. The pyramid has the highest surface-to-volume ratio with a value of 1.169 $mm^2\ mm^{-3}$, whereas the cylinder and sphere have the lowest surface-to-volume ratios with values 0.854 and 0.634 $mm^2\ mm^{-3}$, respectively. When the authors printed objects with similar surface-to-volume ratios, most release profiles showed a higher resemblance with t_{90} values ranging from two to four hours. These findings are in agreement with the results from Reynolds et al., who demonstrated almost identical dissolution profiles from differently sized tablets with varying API doses if the surface-to-volume ratio was kept constant [24]. Reynolds et al. explicitly state that the surface-to-volume relationship is only a valid parameter if API diffusion is the fundamental release mechanism from the dosage form. This might explain why the results from Goyanes et al. (Figure 8.9b) display scattering. The authors mention that the main drug release mechanism in their study was not diffusion but erosion. Pietrzak et al. printed caplets of the same shape but different dimensions based on Eudragit RL [25]. Their dissolution results from the insoluble Eudragit matrix further support that the SSA has a significant influence. With increasing print size and API dose and therefore decreasing surface-to-volume ratio, the dissolution rate decreased. The smallest caplets with the largest surface-to-volume ratio released 50% of the API within 4 hours, the largest caplets with the lowest surface-to-volume ratio required almost 11 hours for the same relative API release.

8.3.3.2 Drop-on-Drop Printing

Kyobula et al. manufactured tablets via DoD 3D printing [26]. A PiXDRO LP50 inkjet printer was fitted with a hot melt chamber and white beeswax loaded with 5% fenofibrate was used as the ink. The hot melt chamber and printer head had to be heated to 90 °C to facilitate the printing process due to the high melt viscosity of the beeswax. Round tablets with a diameter of 10 mm were

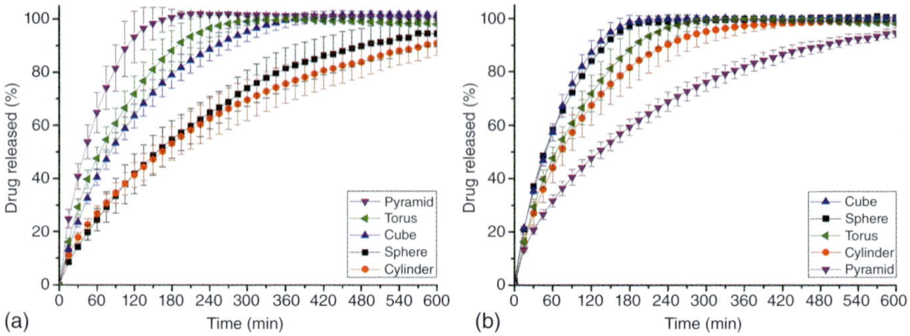

Figure 8.9 Dissolution profiles of printed geometries with constant surface area (a) and constant surface-to-volume ratio (b). Phosphate buffer pH 6.8, 37 °C, sink conditions. $n = 3$, mean ± SD. Source: Goyanes et al. 2015 [23]. Modified and reprinted with permission of Elsevier.

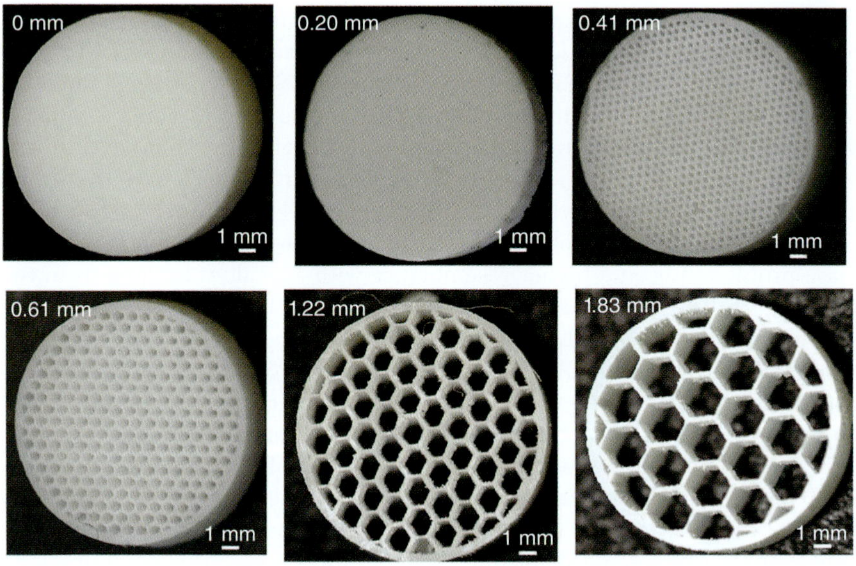

Figure 8.10 DoD 3D printed tablets with different infill patterns. The diameter of the honeycomb cells is indicated in the top left corner of every image. Source: Used from [26] under the Creative Commons Attribution License (CC BY 4).

printed with a honeycomb infill pattern. The weight of the tablets was kept constant and, therefore, the tablet height varied between 9.73 and 1.98 mm depending on the diameter of the honeycomb cells, which ranged from 1.83 mm for the lowest infill level to a solid tablet (Figure 8.10). The surface-to-volume ratio was determined via X-ray microtomography and increased from 5.25 for tablets with a cell diameter of 1.83 mm to 7.07 for tablets with a cell diameter of 0.406 mm. The ratio decreased strongly for tablets with the lowest cell diameter of 0.203 mm to 5.80, which is attributed to imprecise printing at this resolution. The solid tablet had a surface-to-volume ratio of 1.41. According to the prior

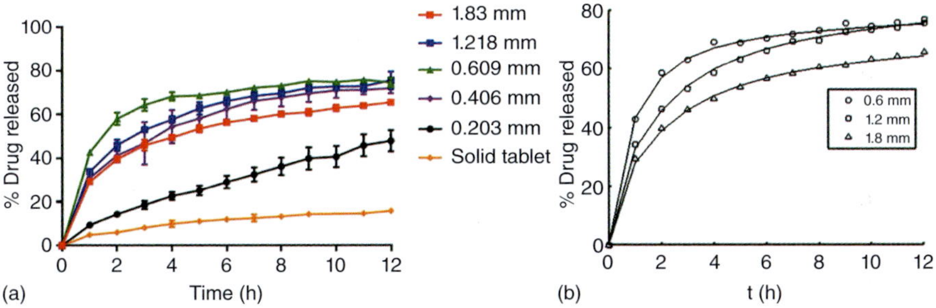

Figure 8.11 Drug dissolution profiles of printed tablets according to their cell diameter (a) and fit of the developed numerical model (b). Source: Modified from [26] under the Creative Commons Attribution License (CC BY 4).

considerations that the surface-to-volume ratio should be the determining factor in Fickian diffusion-controlled drug release, which applies for wax-based dosage forms, the drug release should increase with decreasing cell diameter of the honeycomb structure. The results of the dissolution studies are shown in Figure 8.11a and deviate from the expectations. The dissolution increases with the surface-to-volume ratio until a maximum is reached at a cell diameter of 0.609 mm. Below this cell diameter, the dissolution rate decreases slightly for a cell diameter of 0.406 mm and strongly for a cell diameter of 0.203 mm. The authors determined the wetting times of the hexagonal pores and found that wetting times increase distinctly when the pore size is <0.5 mm. Together with a decreased medium flow through small pores, this effect is superimposed on the dissolution curves. The result is a decreased dissolution rate even though the surface-to-volume ratio increases. Given that Fickian diffusion is the dissolution rate-controlling factor for cell diameters of 0.6–1.8 mm, the authors used a diffusion-based model to fit the dissolution profiles with good agreement (Figure 8.11b). Using the fit parameters, predictions for different cell diameters and cell wall diameters were calculated, but not experimentally validated. Nonetheless, precise control over the surface-to-volume ratio allows the adjustment of dissolution profiles over a wide range.

8.3.4 Dissolution Control via Directed Diffusion and Compartmentalization

As discussed before, modification of the surface-to-volume ratio is a capable means to modify the release rate from printed dosage forms. Combining API-free compartments with API-loaded compartments in a single dosage form allows the targeted modification of diffusional area and direction of diffusion. As a result, it should be possible to manipulate the dissolution rate and the dissolution kinetic.

8.3.4.1 Drop-on-Powder Printing

The impact of formulation variations and compartmentalization on the release profile was already investigated in the first published studies on pharmaceutical

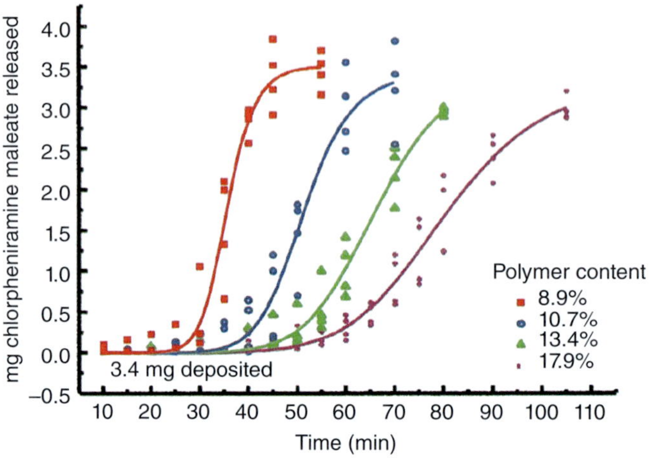

Figure 8.12 API release from DoP printed dosage forms with increasing amounts of polymer content. 0.1 N HCl for 1 h followed by a medium change to pH 7.5 phosphate buffer at 37 °C. Source: Katstra et al. 2000 [27]. Reprint with permission of Elsevier.

3D printing. DoP printing was used by Katstra et al. to manufacture microcrystalline cellulose (MCC)-based tablets [27]. The authors used solutions of Eudragit E-100 in ethanol (20% w/w) and Eudragit RL PO in acetone (20% w/w) as inks and printed increasing amounts on the MCC powder, resulting in binder concentrations ranging from 8.9% to 17.9% by volume. Chloramphenicol maleate (3.4 mg) was printed from a 30% water solution (w/w) in a circular region in the center of each dosage form. The API was printed in a second pass of the printer head on regions that already contained binder. Results of the dissolution curves of dosage forms containing Eudragit E-100 are shown in Figure 8.12. As the drug substance was placed in the center of the dosage form, tablets exhibited a lag time because of the erosion of the drug-free shell. As the erosion rate decreased with increasing polymer content, the lag time increased, whereas the peak release rate decreased in consequence. Tablets printed with a Eudragit RL PO solution did not erode but swelled during the dissolution process (not shown). Even though the dissolution was prolonged three- to fivefold compared to the tablets printed on with Eudragit E-100, the results were similar. Increasing the amount of binder from 9% to 17% led to an increased lag time from 30 to 60 minutes and to an increased overall dissolution time from 3 to 10 hours. The same research group also printed tablets with immediate + extended release [28]. The formulation was the same as in [27] with MCC as the main solid and Eudragit E-100 and Eudragit RL PO solutions as the ink. Both polymer solutions were printed in different compartments in one dosage form (Figure 8.13a). When API (chlorpheniramine) was included only in the fast degrading region based on Eudragit E-100, the API was released completely within 45 minutes (Figure 8.13b, blue). Similarly, when the API was included only in the slowly eroding Eudragit RL PO part, dissolution took place over the whole measurement time (Figure 8.13b, red). By adding API into both, the immediate and extended release regions, a biphasic dissolution profile is resulted (Figure 8.13b, purple). The authors also used Eudragit L-100,

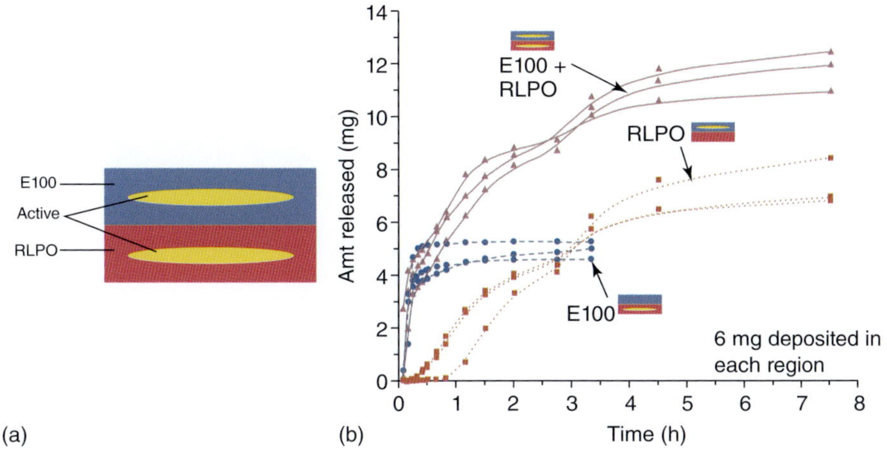

Figure 8.13 Compartmentalized tablet design (a) and dissolution curves of powder-printed tablets (b). 0.1 N HCl for 1 h followed by a medium change to pH 7.5 phosphate buffer at 37 °C. Source: Rowe et al. 2000 [28]. Reprint with permission of Elsevier.

an anionic methacrylic ester copolymer, which is soluble above pH 6, to print enteric pulsatile release dosage forms. The tablet was printed completely with Eudragit L-100 and two API containing regions, one at the surface and one at the center of the tablet. After the release of the API on the surface within one hour, the second dissolution pulse only started after a total of six hours. The use of inks with tailor-made properties for one print job makes this technique highly versatile in the manufacturing of specialized dosage forms.

Yu et al. also used DoP printing to realize zero-order release kinetics of acetaminophen [29]. In order to achieve linear release profiles, the authors printed drug delivery devices (DDDs) in the shape of hollow cylinders (Figure 8.14a). Top and bottom of the DDDs were printed with an EC containing ink, thereby forming barrier layers, which prohibit diffusion and dissolution in these directions. Drug release was therefore constricted to the inner and peripheral annular regions of the dosage form. To prevent a burst release from drug particles on the surface of the peripheral annular region, the amount of EC was increased in a layer of 200 µm. The powder mixtures contained acetaminophen, HPMC, EC, polyvinylpyrrolidone (PVP) K30, and colloidal silicon dioxide (60 : 20 : 10 : 9.5 : 0.5 w/w). The EC containing ink solution consisted of 2% EC in an ethanol/water mixture, the API containing ink of 10% API also in a water/ethanol mixture. After printing, the dosage forms were dried overnight in vacuum at 35 °C. DDDs with constant overall diameter but different radii of the aperture (0.5–2.0 mm), with constant radius of the aperture but different overall diameter (4–5.5 mm), and varying heights (4–6 mm) were printed and evaluated. In all but one cases, the friability (determined with 4.5–8.7 g of tablets and 100 revolutions in four minutes) was below 1% with tensile strengths ranging from 0.51 to 0.86 MPa. The drug release from DDDs of different height was identical and is therefore not shown. The dissolution results of DDDs with varying aperture diameters are shown in Figure 8.14b.

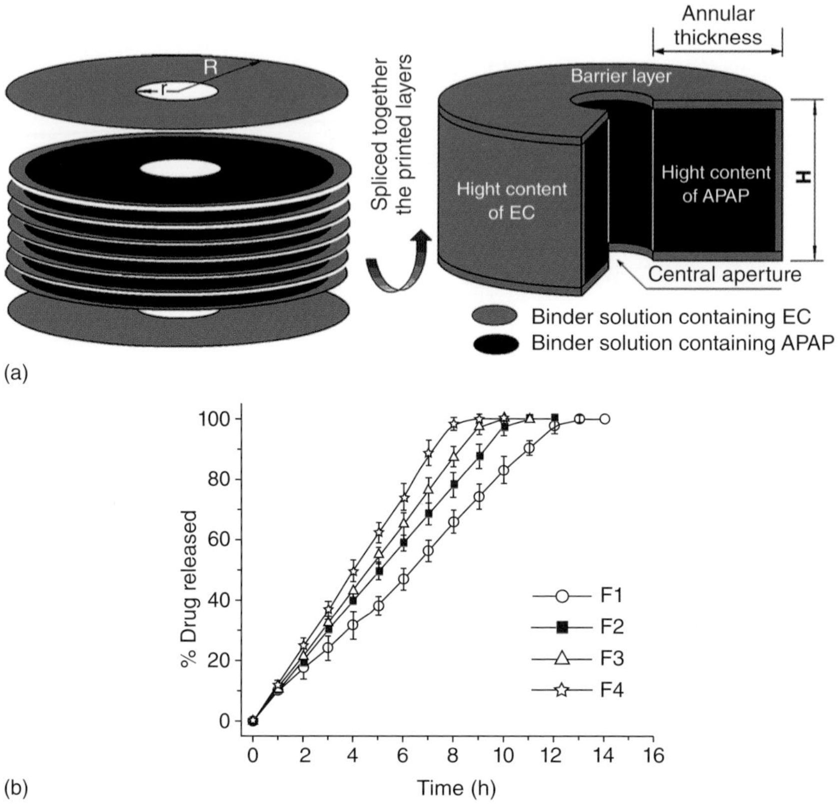

Figure 8.14 DDD design (a) and dissolution profiles of DDDs with increasing aperture diameters from F1 = 0.5 mm to F4 = 2.0 mm (b). Phosphate buffer pH 6.8; $n = 6$, mean ± SD. Source: Yu et al. 2009 [29]. Modified and reprinted with permission of Elsevier.

The authors describe that the annular thickness, i.e. the distance from aperture to outer perimeter, seems to be the determining influence on the dissolution rate. As shown in Figure 8.14b, the relative dissolution rate increases with increasing aperture radius. Because the overall diameter remains the same, the dissolution rate increases with decreasing annular thickness. The same impact and relative change of dissolution rate was found for the last set of experiments, where the aperture was set and the overall diameter varied (not shown). The annular thicknesses ranged from 3.0 to 4.5 mm in both experiments, and the times of 30%, 50%, and 80% drug release were very similar to those discussed before (not shown). Controlling the annular thickness of hollow cylinder-shaped DDDs seems to allow a release kinetic independent dose adjustment, which could be of great benefit for patients.

8.3.4.2 Fused Filament Fabrication

A similar dosage form design was used in a more recent study with FFF [30]. In this case, the bottom and top layers were printed with regular PLA filament and the API containing middle layer was printed from a filament consisting of

PVA, mannitol, and hydrochlorothiazide. During FFF, the polymeric material is molten and the layers are fused together simply by depositing a polymer wire from the nozzle on the layer below. This can lead to poor layer adhesion, especially between chemically diverse materials, and delamination. To prevent delamination, the authors introduced a 400 μm thick PLA ring in the center of the polymeric annular region. Dissolution profiles of printed dosage forms with drug containing layers of 250 and 300 μm thickness were compared (not shown) and displayed a zero-order API release. The approach in this study was to have a fixed API dose of 25 mg and by varying the DDD thickness, the outer and aperture diameters had to be adapted. In consequence, the annular thicknesses were not constant but changed from 3.37 to 3.69 mm, which might explain observed variations of the dissolution curves. Thus, the approach to print hollow cylinders to ascertain a prolonged drug release with zero-order kinetic can be realized also via FFF.

By printing dosage forms based on a core–shell design (Figure 8.15a), delayed release can also be achieved. In Figure 8.15b, the dissolution data of printed tablets containing acetaminophen (paracetamol) in the outer layer and caffeine in the core are shown [31]. In this case, non-pharma grade PVA filament was cut into small pieces and extruded with the APIs. The release rates changed with the percentage of included drug with higher drug loadings leading to a quicker API release and a shorter lag time for the release from the inner core. The decreased lag time is likely caused by a faster erosion and dissolution of the shell at the higher API loading, exposing the core quicker to the dissolution medium. Similar results were observed when the shell was printed with caffeine and the core with paracetamol. Higher API loadings also decreased the lag time of the release from the core. Another study investigated the influence of the thickness of the printed shell in a gastric resistant dosage form [32]. The outer shell represented the gastric resistant barrier of the formulation and consisted of 50% Eudragit L100-55, 33% talc, and 17% of triethyl citrate as plasticizer. The core contained APIs in a soluble PVP matrix. The thickness of the shell varied from 0.17 to 0.87 mm (Figure 8.16a,b). The dissolution results shown in

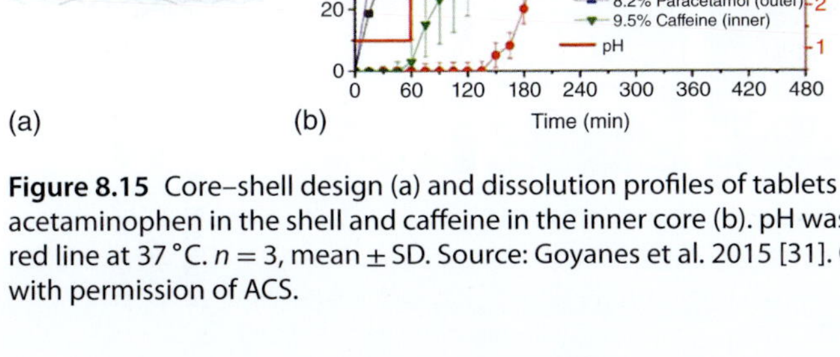

Figure 8.15 Core–shell design (a) and dissolution profiles of tablets containing acetaminophen in the shell and caffeine in the inner core (b). pH was adapted according to the red line at 37 °C. $n = 3$, mean ± SD. Source: Goyanes et al. 2015 [31]. Copyright (2015). Adapted with permission of ACS.

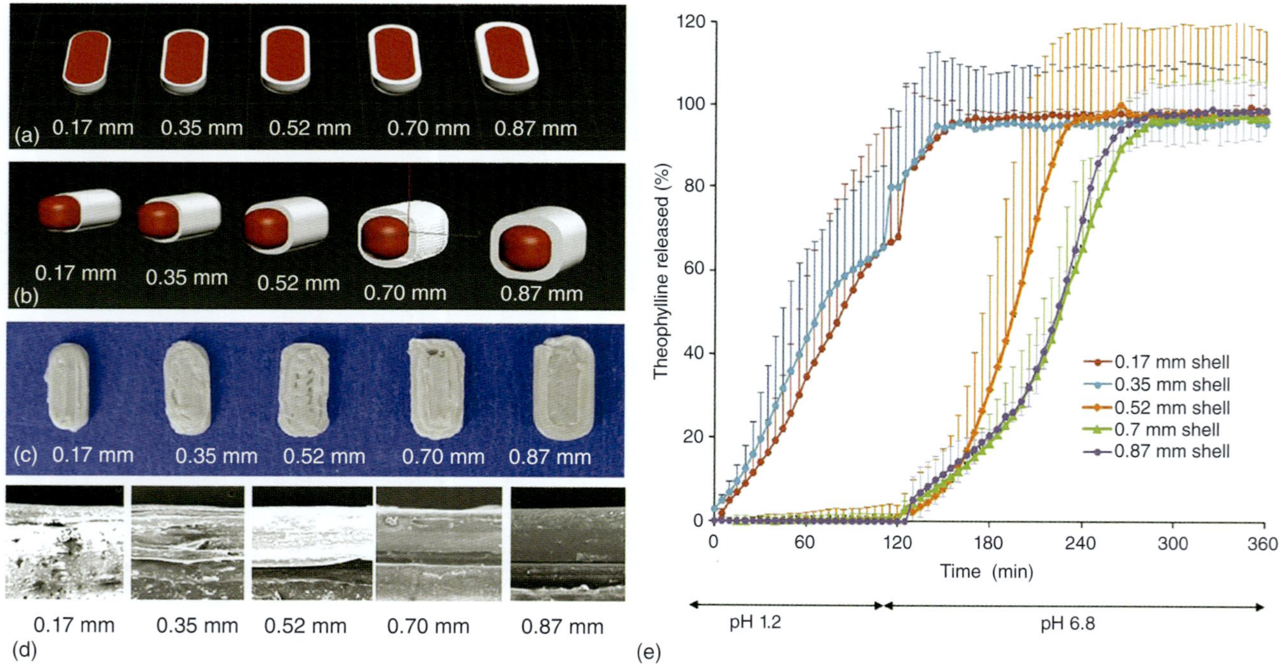

Figure 8.16 Renders of shell–core design with increasing shell thickness (a,b), images of tablets after printing 30% of the dosage forms (c), SEM images of tablets surfaces (d), and release profiles of theophylline (e). $n = 3$, mean ± SD; 2 h at pH 1.2 followed by 4 h at pH 6.8 at 37 °C. Source: Reprint from [32]. With permission of Springer.

Figure 8.16e demonstrate that with shell thicknesses of 0.17 and 0.35 mm, no gastric resistance could be accomplished, whereas a complete gastric resistant shell is formed for thicknesses >0.52 mm. In contrast to conventional coating techniques, where a sufficient functional coating thickness is >40 μm, the shell formed by FFF has to be almost an order of magnitude thicker. If the shell was printed without holes, a gastric resistance would be expected even for the lowest wall thickness. That this is not the case is likely because of the limitations of the printing technique and the experimental setup in this study. Holes and other irregularities in the shell can be due to the comparably low resolution of the FFF process. It is possible that incomplete seams were formed where the printing of a new shell layer starts and ends. The dissolution medium might then access the core at the seams and cause a premature API release. Another reason might be a small nozzle diameter. Even though the nozzle diameter is not stated in the article, a wall thickness in multiples of approximately 0.17 mm suggests the use of a nozzle with a comparable diameter. To ensure a constant polymer flow, a high pressure and very homogeneous distribution of the talc in the filament is necessary. Nests formed by talc might lead to transient nozzle blockage and in consequence introduce holes in the shells. Yet, if the shell is printed with sufficient thickness, gastric resistant dosage forms can be printed.

8.3.4.3 Printing with Pressure-Assisted Microsyringes

The formulation development of semisolids for PAM-based printing is somewhat less demanding compared to FFF. Gels and pastes over a comparably wide range of properties can be printed and formed to dosage forms. Khaled et al. combined an immediate release compartment with three sustained release compartments to realize a five-API-in-one dosage form [33]. Two APIs (acetylsalicylic acid and hydrochlorothiazide) were printed in the immediate release compartment, and atenolol, pravastatin, and ramipril in the sustained release compartments, respectively (Figure 8.17a). These APIs represent a frequently used combination for the treatment of cardiovascular diseases. To achieve an immediate drug release, both APIs (34.48% w/w) were mixed with 55.18% of sodium starch glycolate, a hydrophilic disintegrant, and 10.34% of PVP K30 as the binder. To achieve sustained release properties for the remaining three APIs, a hydrophobic shell was printed containing cellulose acetate (22.64% w/w), D-mannitol (62.26%) as the filler, and PEG 6000 (15.10%) as the plasticizer. HPMC- and lactose-based mixtures with atenolol, pravastatin, and ramipril were printed from separate cartridges. Water was added to the mixtures containing APIs, whereas a mixture of acetone and dimethyl sulfoxide was necessary to ensure printability of the sustained release shell. More than 75% of API was released from the immediate release compartment within 30 minutes (Figure 8.17b) and more than 65% of API from the sustained release compartments in 12 hours. The rapid release of acetylsalicylic acid and hydrochlorothiazide is attributed to the high amount of disintegrant, which absorbs high amounts of water, and the subsequent swelling leads to the disintegration of this part of the dosage form. The sustained release is the result from a gel formation of the HPMC in combination with the API diffusion from this gel through the cellulose acetate membrane. This system allows further modifications of the release profiles depending on the cellulose acetate

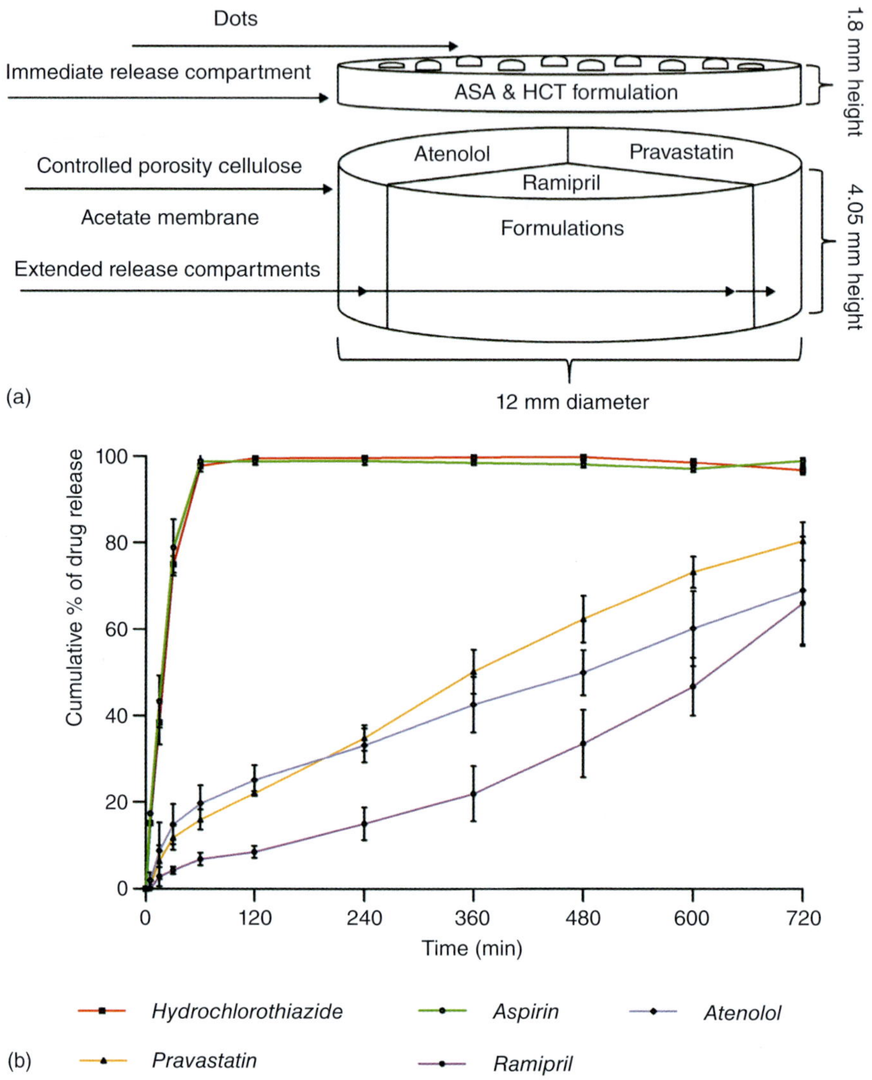

Figure 8.17 Design of five-in-one dosage form with modified release properties (a) and the dissolution profiles (b). Phosphate buffer pH 6.8 at 37 °C. $n = 3$, mean ± SD. Source: Khaled et al. 2015 [33]. Modified and reprinted with permission of Elsevier.

shell thickness and variations of excipient concentrations. Such modifications might be beneficial and necessary if other APIs are necessary.

8.4 Conclusion

Modifications of the dissolution behavior have been used for several decades to improve treatments and patient adherence. The traditional manufacturing techniques, however, are not able to meet the needs of individualized medicine. 3D printing, on the other hand, is inherently capable to produce small batches of dosage forms and allows adjusting API doses with specifically modified API release behavior. Three viable strategies to control drug dissolution from printed dosage forms were introduced. If 3D printing is adopted by hospitals and pharmacies, not all strategies are equally likely to be adopted. From a standpoint of manufacturing safety and complexity, it is doubtful that multiple intermediate products of the same API, such as filaments or pastes, will be prepared or kept on stock. Rather, only one feedstock or intermediate will be used per API. The use of formulation variations as presented in Section 8.3.2 is thus improbable. Changing the CAD design file to manipulate the surface-to-volume ratio or control the direction and surface area available for diffusion is more likely and offers a wider range of possibilities. Whatever approaches will be adapted, 3D printing is a powerful tool to modify dissolution characteristics of individualized medicines.

References

1 Sorkin, E.M., Clissold, S.P., and Brogden, R.N. (1985). Nifedipine a review of its pharmacodynamic and pharmacokinetic properties, and therapeutic efficacy, in ischaemic heart disease, hypertension and related cardiovascular disorders. *Drugs* 30 (3): 182–274.

2 Dwyer, M.S., Levy, R.A., and Menander, K.B. (1986). Compliance through the use of modern dosage forms. *Journal of Pharmacy Technology* 2 (4): 166–170.

3 Institute, N.H.G.R. (2016). The cost of sequencing a human genome. https://www.genome.gov/27565109/the-cost-of-sequencing-a-human-genome/ (accessed 06 November 2017).

4 Schork, N. (2015). Personalized medicine: time for one-person trials. *Nature* 520 (7549): 609–611.

5 Goole, J. and Amighi, K. (2016). 3D printing in pharmaceutics: a new tool for designing customized drug delivery systems. *International Journal of Pharmaceutics* 499 (1–2): 376–394.

6 Zema, L., Melocchi, A., Maroni, A. et al. (2017). Three-dimensional printing of medicinal products and the challenge of personalized therapy. *Journal of Pharmaceutical Sciences* 106 (7): 1697–1705.

7 Norman, J., Madurawe, R., Moore, C.M.V. et al. (2016). A new chapter in pharmaceutical manufacturing: 3D-printed drug products. *Advanced Drug Delivery Reviews* 108: 39–50.

8 Hull, C.W. (1986). Apparatus for production of three-dimensional objects by stereolithography. US Patent 4, 575,330 and issued 11 March 1986.

9 Gittard, S. and Narayan, R. (2010). Laser direct writing of micro- and nano-scale medical devices. *Expert Review of Medical Devices* 7 (3): 343–356.

10 Water, J.J., Water, J.J., Bohr, A. et al. (2015). Three-dimensional printing of drug-eluting implants: preparation of an antimicrobial polylactide feedstock material. *Journal of Pharmaceutical Sciences* 104 (3): 1099–1107.

11 Holländer, J., Genina, N., Jukarainen, H. et al. (2016). Three-dimensional printed PCL-based implantable prototypes of medical devices for controlled drug delivery. *Journal of Pharmaceutical Sciences* 105 (9): 2665–2676.
12 Kempin, W., Franz, C., Koster, L.-C. et al. (2017). Assessment of different polymers and drug loads for fused deposition modeling of drug loaded implants. *European Journal of Pharmaceutics and Biopharmaceutics* 115: 84–93.
13 Zhang, J., Feng, X., Patil, H. et al. (2016). Coupling 3D printing with hot-melt extrusion to produce controlled-release tablets. *International Journal of Pharmaceutics* 519 (1–2): 186–197.
14 Alhijjaj, M., Belton, P., and Qi, S. (2016). An investigation into the use of polymer blends to improve the printability of and regulate drug release from pharmaceutical solid dispersions prepared via fused deposition modeling (FDM) 3D printing. *European Journal of Pharmaceutics and Biopharmaceutics* 108: 111–125.
15 Fina, F., Goyanes, A., Gaisford, S. et al. (2017). Selective laser sintering (SLS) 3D printing of medicines. *International Journal of Pharmaceutics* 529 (1–2): 285–293.
16 Wang, J., Goyanes, A., Gaisford, S. et al. (2016). Stereolithographic (SLA) 3D printing of oral modified-release dosage forms. *International Journal of Pharmaceutics* 503 (1–2): 207–212.
17 Nernst, W. (1903). Theorie der Reaktionsgeschwindigkeit in heterogenen Systemen. *Zeitschrift für Physikalische Chemie* 47: 52–55.
18 Brunner, E. (1903). Reaktionsgeschwindigkeit in heterogenen Systemen. *Zeitschrift für Physikalische Chemie* 47: 56–102.
19 Higuchi, T. (1963). Mechanism of sustained-action medication. Theoretical analysis of rate of release of solid drugs dispersed in solid matrices. *Journal of Pharmaceutical Sciences* 52 (12): 1145–1149.
20 Higuchi, T. (1961). Rate of release of medicaments from ointment bases containing drugs in suspension. *Journal of Pharmaceutical Sciences* 50 (10): 874–875.
21 Lapidus, H. and Lordi, N.G. (1968). Drug Release from Compressed Hydrophilic Matrices. *Journal of Pharmaceutical Sciences* 57 (8): 1292–1301.
22 Hixson, A.W. and Crowell, J.H. (1931). Dependence of reaction velocity upon surface and agitation. *Industrial and Engineering Chemistry* 23 (8): 923–931.
23 Goyanes, A., Robles Martinez, P., Buanz, A. et al. (2015). Effect of geometry on drug release from 3D printed tablets. *International Journal of Pharmaceutics* 494 (2): 657–663.
24 Reynolds, T.D., Mitchell, S.A., and Balwinski, K.M. (2002). Investigation of the effect of tablet surface area/volume on drug release from hydroxypropylmethylcellulose controlled-release matrix tablets. *Drug Development and Industrial Pharmacy* 28 (4): 457–466.
25 Pietrzak, K., Isreb, A., and Alhnan, M. (2015). A flexible-dose dispenser for immediate and extended release 3D printed tablets. *European Journal of Pharmaceutics and Biopharmaceutics* 96: 380–387.

26 Kyobula, M., Adedeji, A., Alexander, M. et al. (2017). 3D inkjet printing of tablets exploiting bespoke complex geometries for controlled and tuneable drug release. *Journal of Controlled Release* 261: 207–215.
27 Katstra, W.E., Palazzolo, R.D., Rowe, C.W. et al. (2000). Oral dosage forms fabricated by Three Dimensional Printing™. *Journal of Controlled Release* 66 (1): 1–9.
28 Rowe, C.W., Katstra, W.E., Palazzolo, R.D. et al. (2000). Multimechanism oral dosage forms fabricated by three dimensional printing™. *Journal of Controlled Release* 66 (1): 11–17.
29 Yu, D.G., Yu, D.-G., Branford-White, C. et al. (2009). Novel drug delivery devices for providing linear release profiles fabricated by 3DP. *International Journal of Pharmaceutics* 370 (1–2): 160–166.
30 Gioumouxouzis, C., Katsamenis, O., Bouropoulos, N. et al. (2017). 3D printed oral solid dosage forms containing hydrochlorothiazide for controlled drug delivery. *Journal of Drug Delivery Science and Technology* 40: 164–171.
31 Goyanes, A., Wang, J., Buanz, A. et al. (2015). 3D printing of medicines: engineering novel oral devices with unique design and drug release characteristics. *Molecular Pharmaceutics* 12 (11): 4077–4084.
32 Okwuosa, T., Pereira, B., Arafat, B. et al. (2017). Fabricating a shell-core delayed release tablet using dual FDM 3D printing for patient-centered therapy. *Pharmaceutical Research* 34 (2): 427–437.
33 Khaled, S., Burley, J., Alexander, M. et al. (2015). 3D printing of tablets containing multiple drugs with defined release profiles. *International Journal of Pharmaceutics* 494 (2): 643–650.

9

Novel Excipients and Materials Used in FDM 3D Printing of Pharmaceutical Dosage Forms

Ming Lu

Sun Yat-sen University, School of Pharmaceutical Sciences, Department of Pharmaceutics, No. 132, Waihuan Dong Road, Guangzhou, 510006, China

9.1 Introduction

In this chapter, we highlight the polymeric materials and excipients used in pharmaceutical fused deposition modeling (FDM) 3D printing.

FDM 3D printing is one of the most extensively used 3D printing techniques. It was introduced during the 1980s as a branch of the additive manufacturing technology. Recently, this technology has been employed to manufacture pharmaceutical dosage forms [1–6], enabling to fabricate patient-tailored tablets with high accuracy, incorporate different drugs in one dose with different release behaviors, and produce release-tailored medications.

The nature of FDM 3D printing is a melting–extrusion–solidification process. As shown in Figure 9.1, the filament is fed into the heater block by feeding gears, softened, and then melted. The molten filament is extruded from the nozzle to deposit on the platform or on the previous layer, followed by solidification to form a designed structure. The nozzle moves in the $X-Y$ plane to create the first layer on the surface of the platform or on the raft [7, 8]. Successive layers are printed by moving the platform or the nozzle in the Z-plane with the distance equivalent to the layer thickness [9], which was determined by the diameter, the extruding speed, and the travelling speed of the nozzle.

Extrusion is the key step of the printing process. However, there is no effective extrusion element in a commercially available FDM 3D printer [9]. Instead, the filament, driven by the feeding gears, acts as a piston to push the molten filament through the nozzle. Therefore, column strength and melt viscosity of the filament were considered as two critical properties for successful printing. Firstly, the filament should be not only tough enough to be rolled up on the spool and to go through the feeding gears but also stiff enough to push the melts through the nozzle (Figure 9.2b). If the filament is very brittle, it will be difficult to be collected using the spool and will break when goes through the feeding gears (Figure 9.2a). On the contrary, a very soft filament will bend or buckle above the heating element while feeding (Figure 9.2c). Secondly, the melt viscosity should be low enough because it determines the pressure drop in the nozzle and thus the

3D and 4D Printing in Biomedical Applications: Process Engineering and Additive Manufacturing, First Edition. Edited by Mohammed Maniruzzaman.

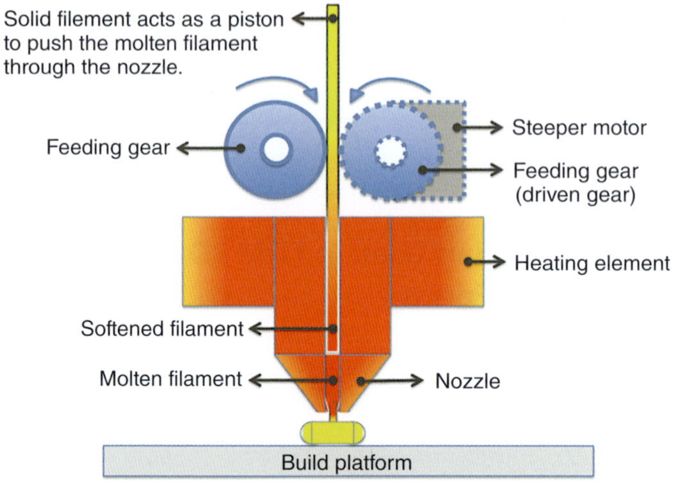

Figure 9.1 Schematic illustration of a FDM 3D printer.

Figure 9.2 Filament properties during 3D printing process. (a) Filaments very brittle; (b) filaments good for printing; and (c) filaments very soft. Source: Zhang et al. 2017 [12]. Reprinted with permission of Elsevier.

force that is needed to push the melts through the nozzle. However, if the melt viscosity is very low, the material might be extruded as droplets rather than as a continuous rod.

The solidification rate of the molten polymer was another important factor for successful printing. The printed strips should rapidly solidify to avoid the collapse of the structure [10]. However, very fast solidification will lead to the better-defined interfaces and thus poor adhesion between the strips and layers [9].

The surface roughness of the filament also has a significant influence on the printing process. Moderate roughness facilitates the continuous printing [11] because proper friction helps the feeding gears to push the filament forward. However, overly rough surface makes printing difficult because of very high friction [12]. The surface roughness of the filament can be adjusted by adding a non-melt drug [13] and a filler [11] for roughening or blending with flexible materials [12] for smoothening.

Accurate control of the filament diameter at a correct and uniform size is very important for continuous printing, fine structure, and precise dose [14–18]. It is because the 3D printer had deposited the calculated volume of the filament based on the default diameter [14] (usually 1.75 or 2.85 mm for a commercially available FDM 3D printer). The diameter of the extruded filament was mainly determined by the size of the die, the pulling speed, and the swelling ability of the polymer. The selection of the correct size of the die was based on the swelling ability of the polymer material, whereas the degree of swelling can be adjusted by controlling the parameters, such as the length-to-diameter ratio of the die, the screw speed, and the extrusion temperature [14]. Therefore, all factors need to be considered together to achieve the correct size of the filament. A pulling/calibrating device might be helpful to precisely control the filament diameter [1, 8, 19].

Based on the above-mentioned principles for FDM 3D printing, this technology has two main limitations when applied in the pharmaceutical field.

The first limitation is the potential thermal degradation of the drug. Although many pharmaceutical scientists successfully decreased the printing temperature from 200–230 °C (the recommended temperature for traditional poly (L-lactic acid) (PLA) and poly(vinyl alcohol) (PVA) filament by the manufacturer) to 100–170 °C (Table 9.1), thermal degradation of the drug during printing process was still observed in many research studies [10, 20].

The second limitation is the poor thermoplasticity and mechanical properties of the FDA-approved pharmaceutical grade polymers. The commonly available filaments for FDM 3D printing were made of acrylonitrile butadiene styrene (ABS), PLA, poly(ε-caprolactone) (PCL), and PVA. These polymers have good thermoplasticity, stiffness, and flexibility. However, ABS has toxicity, whereas PLA and PCL are not the popular polymeric matrices for pharmaceutical dosage forms. PVA was first employed as the matrix to print the oral dosage form because of its high water solubility. Incorporation of a drug into the PVA filament was initially achieved by soaking the filament in a high-concentration drug solution, followed by drying. However, the drug load is very low and hard to control. Alternatively, hot melt extrusion (HME) was employed to prepare the drug-loaded filaments, enabling high drug loading. Since 2015, many research studies have focused on the application of pharmaceutical grade polymers in FDM 3D printing. By now, most of the pharmaceutical polymers were successfully printed into various dosage forms with immediate or controlled release behavior (listed in Table 9.1), including cellulosic polymers (Hydroxypropyl cellulose [HPC], hydroxypropyl methylcellulose [HPMC], hypromellose acetate succinate [HPMCAS], and ethyl cellulose [EC]), polyvinyl polymers (PVA, ethylene vinyl acetate [EVA], and polyvinylpyrolidone [PVP], Soluplus®), and polymethacrylate-based polymers (Eudragit® RL, Eudragit® RS, Eudragit® L10-55, and Eudragit® EPO).

In this chapter, the latest developments in polymers and excipients used in FDM 3D printing of pharmaceutical dosage forms were reviewed to illustrate the areas of research advancing pharmaceutical 3D printing. The key physicochemical properties of pharmaceutical polymers reviewed in this chapter are listed in Table 9.2.

Table 9.1 Cases of FDM 3D printing based on various polymers.

Polymer	API[a)]	Drug loading	Other excipients	Dosage form	$T_{extrusion}$[b)] (°C)	$T_{printing}$[c)] (°C)	T_{plate}[d)] (°C)	Die (mm)	References
Biodegradable polymers									
PLA	Nitrofurantoin	5%	None	Disk	180	200	RT[e)]	NA[f)]	[22]
	Nitrofurantoin	10%, 20%, 30%	5% Hydroxyapatite	Disk	180	NA	RT	1.5	[14]
	Nitrofurantoin	10%, 20%	5% Hydroxyapatite	Disk	180	NA	RT	NA	[48]
	Nitrofurantoin	5%	20% and 40% HPMC	Disk	180	190–200	RT	NA	[23]
	Salicylic acid	0.63%	None	Nose mask	190	230	RT	NA	[10]
PCL	Salicylic acid	1.34%	None	Nose mask	60	170	RT	NA	[10]
	Indomethacin	5%, 15%, 30%	None	Intrauterine system	100	100 165	RT	1.5	[26]
Polyvinyl polymers									
PVA	Fluorescein	0.29%	None	Tablet	No	190–220	RT	NA	[28]
	4-Aminosalicylic acid	0.06%	None	Tablet	No	190–220	RT	NA	[20]
	5-Aminosalicylic acid	0.25%							
	Prednisolone	1.89–1.97%	None	Tablet	No	250	20	NA	[29]

	Budesonide	4.14%	None	Caplet	170	190	RT	1.75	[30]
	Acetaminophen	3.95%	None	Various geometries	180	180	RT	1.75	[31]
	Acetaminophen	4.3% and 8.2%	None	Multilayer device	180	200	RT	1.75	[32]
	Caffeine	4.7% and 9.5%		DuoTablet					[33]
	None	—	5% Glycerol	Disk	190	225	RT	1.80	[19]
	Felodipine	8.6%	22.5% Tween 80	Disk	130	150	RT	1.75	[9]
	Glipizide	2.2% and 4.8%	None	DuoTablet	180	195	RT	1.75	[34]
EVA	Indomethacin	5% and 15%	None	T-shaped rod	105–120	150–215	RT	1.5–2.0	[7]
PVP	Theophylline	10%	12.5% TEC[g)]	Caplet	90–100	110	RT	NA	[37]
	Dipyridamole		27.5% Talc						
	Theophylline	10%	12.5% TEC 27.5% Talc/TBP[h)]	Caplet (as the core)	90–100	110	40	1.25	[47]
	Budesonide	2.3%	12.5% TEC 35.2% Talc						
	Diclofenac sodium	20%	17.5% TEC 17% Talc						

(Continued)

Table 9.1 (Continued)

Polymer	API[a)]	Drug loading	Other excipients	Dosage form	$T_{extrusion}$[b)] (°C)	$T_{printing}$[c)] (°C)	T_{plate}[d)] (°C)	Die (mm)	References
Soluplus®	Felodipine	9.6%	15% Tween 80 10% PEG 4000 15% PEO	Disk	120	150	RT	1.75	[9]
	None	—	10% PEG 400	Disk	80	200	RT	1.80	[19]
Cellulosic polymers									
HPC	Theophylline	50%	4% Triacetin	Caplet	110–125	160	60	1.5	[13]
	None	—	0–10% PEG 1500	Capsular device	150–165	180–210	RT	1.75	[8]
	None	—	None	Disk	165	180	RT	1.80	[19]
HPMC	None	—	5% PEG 400	Disk	160	200	RT	1.80	[19]
	Acetaminophen	10%	20% Soluplus®	Tablet	160	200	50	2	[43]
HPMCAS	None	—	5%PEG 8000	Disk	180	200	RT	1.80	[19]
	Acetaminophen	4.9–5.1%	5% Magnesium stearate + 15% methylparaben	Tablet	80–110	180–190	RT	1.75	[45]
		46.0–49.0%	5% Magnesium stearate + 5% methylparaben						
	Haloperidol	10% and 20%	PVP VA64 (1 : 1)	Tablet	150–170	210	RT	1.55	[49]

Polymethacrylate-based polymers									
Eudragit® RL100	Theophylline	50%	5% TEC	Caplet	120[i]	170	90	1.5	[13]
	None	—	15% TEC	Disk	120	160	RT	1.80	[19]
Eudragit® RS	Theophylline	50%	7.5% TEC	Caplet	110[i]	150	60	1.5	[13]
Eudragit® L100-55	None	—	20% TEC	Disk	160	160	RT	1.80	[19]
	None	—	16.67% TEC 33.33% Talc	Caplet (as the shell)	125–135	185	40	1	[47]
Eudragit® EPO	Theophylline	50%	3.5% TEC	Caplet	110[i]	140	60	1.5	[13]
	Felodipine	9.5%	10% Tween 80 15% PEG4000 15% PEO	Disk	100	150	RT	1.75	[9]
	5-Aminosalicylic acid Theophylline Prednisolone Captopril	12.5%	3.25% TEC 37.5% TCP[j]	Caplet	90	135	60	NA	[11]
	Hydrochlorothiazide	12.5%	3.25% TEC 37.5% TCP	Caplet with channels	90	135	60	NA	[50]

a) API – active pharmaceutical ingredient.
b) $T_{extrusion}$ – extrusion temperature.
c) $T_{printing}$ – printing temperature.
d) T_{plate} – plate temperature.
e) RT – room temperature.
f) NA – not available.
g) TEC – triethyl citrate.
h) TBP – tribasic sodium phosphate.
i) the formulation was first mixed at 130 °C for five minutes and then extruded at the noted temperature.
j) TCP – tricalcium phosphate.

Table 9.2 Physicochemical properties of pharmaceutical polymers in FDM 3D printing.

Polymer	M_w [a] ($g\,mol^{-1}$)	C[b]/A[c]	T_g [d] (°C)	T_m [e] (°C)	T_d [f] (°C)	Water solubility	Bio degradable
Biodegradable polymers							
PLA	—	C	55–65	150–190	>300	Insoluble	Yes
PCL	80 000	C	−60[2]	60[2]	385[25]	Insoluble	Yes
Polyvinyl polymers							
PVA	7 000–18 000	C	55–75	190–230	>300	Insoluble	Yes
EVA	NA	C	−22–38	60–102[7]	>300[51]	Insoluble	No
PVP K30	50 000	A	149[35]	—	175[35]	Soluble	No
PVPVA64	45 000	A	101[35]	—	230[35]	Soluble	No
Soluplus®	118 000	A	70[35]	—	250[35]	Soluble	No
Cellulosic polymers							
HPC LF	95 000	A	111[41]	—	227[41]	Soluble	No
HPMC E5	10 000	A	173[52]	—	228[52]	Soluble	No
HPMCAS MF	18 000	A	122[44]	—	267[44]	pH > 6.0	No
EC 7p	NA	C	128[41]	169[41]	205[41]	Insoluble	No
Polymethacrylate-based polymers							
Eudragit® L100-55	320 000	A	96 ± 5[46]	—	159[46]	pH > 5.5	No
Eudragit® EPO	47 000	A	45 ± 5[46]	—	220[46]	pH < 5.0	No
Eudragit® RL100	32 000	A	63 ± 5[46]	—	140[46]	Insoluble	No
Eudragit® RS100	32 000	A	58 ± 5[46]	—	110–120[46]	Insoluble	No

a) M_w – weight-average molecular weight.
b) C – crystalline.
c) A – amorphous.
d) T_g – glass transition temperature.
e) T_m – melting temperature.
f) T_d – degradation temperature.

9.2 Biodegradable Polyester

9.2.1 Polylactic Acid (PLA)

PLA and ABS are the most widely used materials in a 3D printing technique. Considering the biodegradability, biocompatibility, and regular approval, PLA was first considered as the polymer material to apply a 3D printing technique in the pharmaceutical field.

The chemical structure is shown in Figure 9.3. Given the stereospecific characteristics, PLA can be divided into poly-D-lactic acid (PDLA) (from D-LA), poly-(L-lactic) acid (PLLA) (form L-LA), and poly-D,L-lactic acid (PDLLA). PDLA and PLLA are semicrystalline with a melting point around 170 °C [21], whereas PDLLA is amorphous with a glass transition temperature around 55 °C [21]. The PLA material used in 3D printing was generally PLLA. The PLA filament could be extruded around 180 °C and printed between 190 and 230 °C [10, 22, 23]. It is worth noting that the crystallization of PLLA is very slow to print. Therefore, if the crystalline product is desired, a crystallization acceleration strategy (either a faster crystallization-grade PLA or introducing a pretreatment to speed it up) should be considered. For example, Ingeo™ 3D850 and 3D870, the special PLA grades for 3D printing manufactured by NatureWorks company, have faster crystallization than other grades (these two resins are not to be used within the bodies of humans or animals). Moreover, post-annealing is necessary to promote the crystallization.

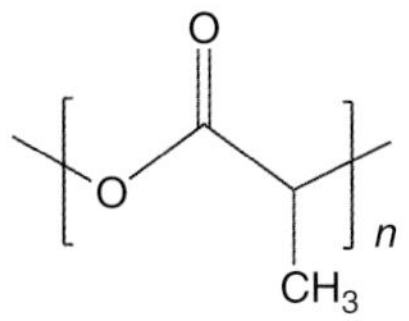

Figure 9.3 Chemical structure of PLA.

PLA can be hydrolyzed by an acid- or a base-catalyzed reaction in the biological system. Lactic acid is produced upon degradation, which can be transformed into CO_2 and H_2O by the Krebs cycle (the citric acid cycle). Therefore, this material is biodegradable, but the degradation rate is very slow. A complete degradation might take several years.

Sandler et al. [22] first reported the pharmaceutical application of FDM 3D printing technique using PLA as a sustained release matrix. The authors extruded 5% nitrofurantoin-loaded PLA filament at 180 °C and then printed disks with a diameter of 2.4 cm and a thickness of 3 mm at 200 °C. The bacterial attachment experiment results indicated that the 3D printed nitrofurantoin–PLA disks much more effectively inhibited the formation of biofilm (85% inhibition for 220 minutes) than the blank PLA disks treated with the nitrofurantoin solution (24.6% inhibition for 220 minutes). The slow release of nitrofurantoin was suggested to be beneficial in this inhibition effect. However, the drug release is very slow (only 2% released during the first 6 hours and an additional 1% for the next 186 hours). In a follow-up study [23], 20% and 40% HPMC were introduced into the 5% nitrofurantoin–PLA formulation, resulting in ~30% and 45–50% accumulated release after 24 days, respectively. Increasing drug loading can also improve the accumulated release of nitrofurantoin from the PLA disks [14].

Goyanes et al. [10] employed PLA and PCL as a matrix to print personalized shape patches loaded with salicylic acid for the treatment of acne. The drug-loaded polymer was prepared by dissolving the polymer and salicylic acid in an organic solvent and then casting into the film, followed by cutting into small pieces before extrusion. The filaments, extruded at 190 °C for PLA-based formulation and 60 °C for PCL-based formulation, were printed into circular patches. The cumulative drug released from the printed PLA–salicylic acid patch (16 and 22 μg cm^{-2} at 15 and 60 minutes, respectively) was lower than that released from the PCL–salicylic acid patch (40 and 66 μg cm^{-2} at 15 and 60 minutes, respectively). The PLA–salicylic acid nose mask was successfully printed at 230 °C with enough flexibility and clearly defined shape, but the PCL–salicylic acid filament failed because of the slow solidification of layer when printing at 170 °C and the nozzle blockage at lower printing temperature. Considering the slow drug release and the thermal degradation of drug at high printing temperature, stereolithography printing was suggested to be a better method for fabricating salicylic acid-loaded nose mask in comparison of FDM 3D printing.

Generally, PLA is an ideal polymer material for the fabrication of medical devices with sustained drug release.

9.2.2 Poly(ε-caprolactone) (PCL)

PCL is a biodegradable polymer approved by the FDA for therapeutic application in surgery, sustained drug delivery, and tissue engineering. The chemical structure is shown in Figure 9.4. PCL can be degraded by hydrolysis in aqueous media. The degradation products are metabolized by organisms [24] and do not result in an acidic environment, which is different from PLA, therefore allowing for use *in vivo*. PCL is semicrystalline, with the melting point of 60 °C [2]. Because of the very low T_g of −60 °C [2], PCL stays in rubber state at room temperature. The degradation temperature of PCL was reported as 385 °C [25], and thus, it has a very wide temperature window for processing. In comparison of PLA, PCL is more hydrophobic and thus degrades more slowly, resulting in slower drug release.

Goyanes et al. [10] tried to use PCL to print salicylic acid-loaded nose mask but failed because PCL solidified very slowly when printed at 170 °C (lowering the print temperature resulted in the blockage of the nozzle). Holländer et al. [26] successfully fabricated T-shaped PCL prototypes of intrauterine system loaded with 5%, 15%, and 30% indomethacin at 100 °C, much lower than that in their previous work (170 °C) [10]. The diameter of the filament was optimized by adjusting the screw speed, die diameter, and fill level of the extruder. The 1.5 mm die was suggested as the best choice to produce the indomethacin–PCL filament with the desired 1.75 ± 0.05 mm diameter considering the melt viscosity (~10^3 Pa s at 100 °C) and the swelling ability of PCL. The layer height was

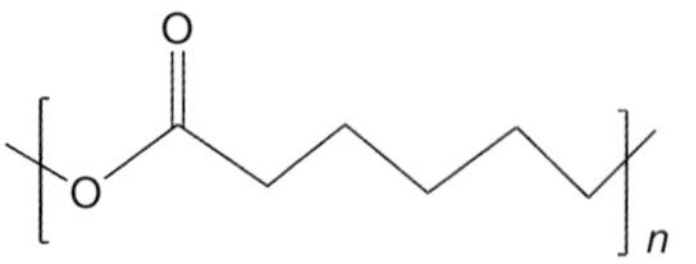

Figure 9.4 Chemical structure of PCL.

0.1 mm and the fill percentage was 10%. The authors mentioned that the PCL filament loaded with 30% indomethacin can also be printed at 165 °C (beyond the melting point of the raw indomethacin, 160 °C), but the product has poorer quality than that printed at 100 °C because of the poor adhesion of the polymer melt between the printed layers. However, the physical state of indomethacin was not well controlled. Indomethacin partially remained in the crystalline state in the printed samples with different crystallinity. The amorphous drug partially recrystallized during the storage. The actual drug contents were much lower than the theory values for all printed samples. The drug was released extendedly for 30 days from the printed samples with the predominant mechanism of diffusion. This study [26] indicated that PCL is a good polymeric matrix to fabricate pharmaceutical implant by using FDM 3D printing.

9.3 Polyvinyl Polymer

9.3.1 Polyvinyl Alcohol (PVA)

PVA is manufactured by the polymerization of vinyl acetate (VA) followed by partial or full hydrolysis (shown in Figure 9.5). The degree of hydrolysis is determined by the stop time of the saponification reaction. The water solubility of PVA depends on the degree of polymerization and the degree of hydrolysis. The highest solubility can be achieved with the hydrolysis degree between 87% and 89% [27]. The printability of PVA is excellent with a good thermoplasticity, a low melt viscosity, and a suitable solidification rate. Therefore, this polymer is extensively used in FDM 3D printing as a supporting material, which helps to fabricate complicated structures and can then be easily removed by dissolving.

Goyanes et al. [20, 28] have, for the first time, explored PVA as a carrier in FDM 3D printing pharmaceuticals. In the earliest stage, the model drug was incorporated into the commercially available PVA filament by incubating the filament in a concentrated drug solution overnight. However, the actual drug loading was very low (0.29 ± 0.01% for fluorescein, 0.063 ± 0.001% for 5-aminosalicylic acid, and 0.236 ± 0.004% for 4-aminosalicylic acid). Drug-loaded tablets were successfully printed at 210–220 °C. The main release mechanisms were suggested to be erosion for fluorescein-loaded tablets and the combination of erosion and diffusion for 5-aminosalicylic acid and 4-aminosalicylic acid formulations. The infill percentage can significantly modulate the drug release, a lower infill percentage facilitating faster release. This solvent immersion method for drug incorporation was also employed by Skowyra et al. [29] with the actual drug loading of 1.86–1.97%. The desired drug contents in the final printed products were achieved by adjusting the dimension of the printed tablets. The solvent immersion method is beneficial for the heat-sensitive drug because it proceeds at room temperature. However, the incorporated drug content was very low and

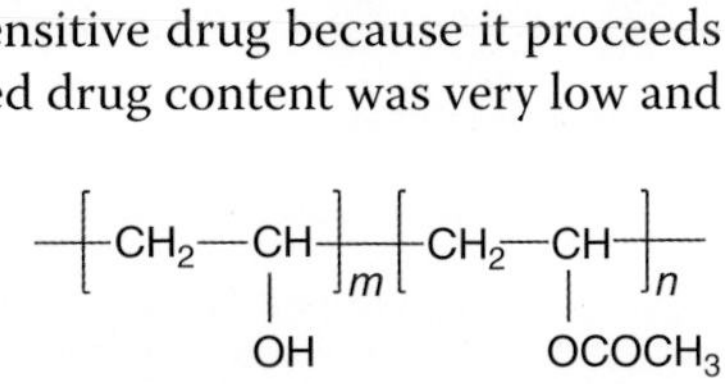

Figure 9.5 Chemical structure of PVA.

could not be quantitatively controlled. In addition, it requires a large amount of organic solvent and a follow-up drying process.

Therefore, HME was explored to fabricate the drug-loaded PVA filament by extruding PVA with budesonide [30]. The commercial PVA filament was cut into small pieces and milled into fine powder (<1000 μm). The mixtures of PVA and budesonide were extruded into the filaments at 170 °C and then printed into caplets at 190 °C, followed by coating with Eudragit L100 for enteric release using a fluidized bed. This study, for the first time, prepared the drug-loaded PVA filaments using HME, by which the drug content of the filament can be accurately controlled and a high drug loading can be achieved. However, the significant drug loss was observed after extrusion, whereas the drug load in the printed caplet was very close to that in the filament, indicating that the drug loss mainly occurred in the extrusion process. A similar phenomenon was observed in the follow-up research of this group [31–33]. The authors attributed drug loss to the adherence of the powdered drug to the walls of the container and the barrel of the extruder. This result was also confirmed by other research groups [26, 34], where a single screw extruder was used with a poorer mixing ability in comparison of a twin-screw extruder.

PVA was further employed to fabricate two kinds of capsule-shaped solid devices with different internal structures (as shown in Figure 9.6) [32]. One is a multiplayer device with each layer containing drugs, which are different to the drug present in the adjacent layers. Another is a two-compartment device comprising a caplet embedded within a larger caplet (DuoCaplet). Each compartment contained a different drug. Dissolution test in biorelevant bicarbonate media revealed an erosion-mediated release process. In multilayer devices, the drug release rate was similar for both drugs but faster at higher drug loading. In DuoCaplet, the drug incorporated in the external layer was first released and then the drug in the internal layer was liberated. The bilayer hardness and layer

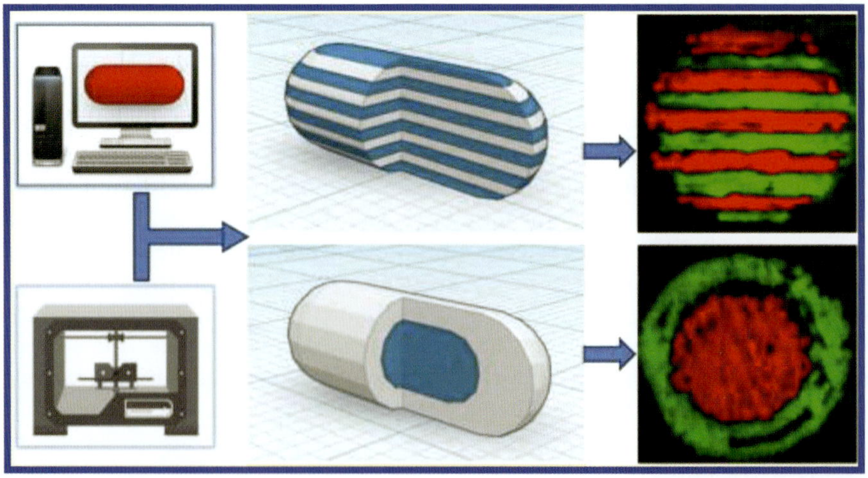

Figure 9.6 3D representation of the printed solid dosage forms and the corresponding Raman mapping: (middle-up) sectioned multilayer device and (middle-down) sectioned DuoCaplet (caplet in caplet). Source: Goyanes et al. 2015 [32]. Reprinted with permission of ACS.

bonding were sufficient enough. This work illustrated the possibility to fabricate the formulations incorporating different drugs with different doses and release behaviors.

Li et al. [34] also fabricated a glipizide–PVA DuoTablet with different drug loadings in the internal (4.76%) and external (2.18%) layers. The filament was extruded at 180 °C and then printed at 200 °C with the dimension of 8.25 ± 0.07 mm (diameter) $\times 2.48 \pm 0.06$ mm (height) for the inner layer and 10.5 ± 0.01 mm (diameter) $\times 3.95 \pm 0.12$ mm (height) for the outer layer. The external layer exhibited a fast dissolving of drug (90% at two hours), whereas the internal layer showed a lag time of 85 minutes before drug release. The DuoTablet exhibited a controlled release for five hours.

PVA, employed in the aforementioned research, is a commercially available filament, which has been formulated into a flexible filament with other excipients. Actually, pure PVA is very brittle to be printed. Alhijjaj et al. [9] improved the printability of pure PVA (33–38% partially hydrolyzed, number-average molecular weight of 18 000–25 000 g mol^{-1}) by adding 22.5% Tween 80 as the plasticizer. The 10% felodipine-loaded PVA filament was extruded at 130 °C and then successfully printed into disks at 150 °C. Thus, it could conclude that neat PVA is 3D printable after formulating.

9.3.2 Ethylene Vinyl Acetate (EVA)

EVA, a copolymer of ethylene and VA (shown in Figure 9.7), is one of the most extensively used nondegradable polymers in pharmaceutical implant because of its excellent flexibility and good biocompatability. The properties highly depend on the VA content (usually 9–40% [18]) and molecular weight. At the same molecular weight, the higher VA content favors increased flexibility, higher water solubility, and decreased crystallinity [18]. EVA has excellent extrudability. Two commercial EVA-based contraceptive implants, Nuvaring® and Implanon®, were manufactured by HME.

Recently, Genina et al. [7] explored EVA as a controlled release matrix for FDM 3D printing, with indomethacin as a model drug. Ten grades of EVA, with different VA contents and melting indexes (MIs), were evaluated for the extrudability and printability. MI is usually employed to rapidly evaluate the melt viscosities of the polymer in plastic industry. High VA content and low molecular weight favor big MI (low melt viscosity) and low stiffness. The swelling abilities of EVA are quiet different for various grades. EVA with lower MI (higher melt viscosity) has bigger swelling ability and thus needs smaller size die to keep the diameter of the final filament at 1.75 mm. Considering the proper stiffness of the filament and the suitable melt viscosity, EVA with the VA content of 16% and MI of 28 g/10 min was selected as the matrix material. Filaments, loaded with 5% and 15% indomethacin, were individually extruded at 110 °C and then printed into prototypes at 165 °C. Because EVA did not adhere to the build plate and

Figure 9.7 Chemical structure of EVA.

the polyimide tape, low density polyethylene (LDPE) film was used as the raft and removed by cutting after printing. Indomethacin partly remained amorphous state in the printed prototypes, which might lead to the issue of physical stability. The prototypes containing 5% indomethacin released faster than that those containing 15% indomethacin. Drug diffusion was suggested to be the predominant release mechanism. This work illustrated that EVA is a 3D printable polymeric material for controlled release, but the grade should be carefully selected.

9.3.3 Polyvinylpyrrolidone (PVP)

PVP was a synthetic polymer from free radical polymerization of vinylpyrrolidone. The chemical structure is shown in Figure 9.8. With the increasing molecular weight, the PVP homopolymers exhibited increased T_g and melt viscosity [35]. PVP K30 is the most extensively used grade of the PVP polymer in solid dispersion for immediate release. It has excellent capacity to inhibit the recrystallization of drug from amorphous solid dispersion [16] and supersaturated solution [36]. However, the process window of PVP is very narrow. The T_g and T_d of Kollidon® K30 ($M_w = 50\,000\,g\,mol^{-1}$) were reported as 149 and 175 °C, respectively [35]. Actually, pure PVP K30 was not extrudable unless a proper plasticizer was introduced [15]. Considering the poor extrudability and thermoplasticity, the application of PVP in FDM 3D printing is a big challenge.

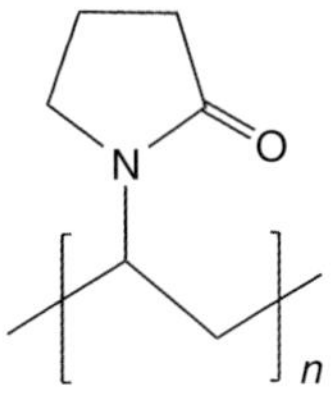

Figure 9.8 Chemical structure of PVP.

Okwuosa et al. [37] explored the possibility of formulating PVP ($M_w = 40\,000\,g\,mol^{-1}$) with other excipients to print immediate release tablets at a low temperature. Theophylline and dipyridamole were used as model drugs with the drug loading of 10%. A 12.5% of triethyl citrate (TEC) was added as the plasticizer. The drug-loaded filament was successfully extruded but failed to fabricate a structure via FDM 3D printing because of the poor flow from the nozzle and the collapsion of the structure. Talc, a nontoxic filler, was introduced to allow the formation of continuous flow and rapid solidification of the structure. The T_g of the formulation of PVP/TEC/Talc 55.6 : 13.9 : 30.6 was 89.78 °C, much lower than that of pure PVP (169.92 °C). The drug-loaded formulation was PVP/TEC/Talc/API = 50 : 12.5 : 27.5 : 10. PVP-based tablets with immediate release properties were successfully printed at a very low printing temperature of 110 °C. A percentage of theophylline (T_m = 270 °C) remained in the crystalline form, whereas dipyridamole (T_m = 165 °C) remained amorphous state in the printed tablets. Dissolution test indicated that more than 85% of theophylline and dipyridamole was released within the first 30 minutes. Further study demonstrated that the drug-loaded PVP solid dispersion was successfully printed at 110 °C as the core of a delayed release tablet with a shell–core structure (see details in Section 9.5.2).

With the facilitation of a plasticizer and filler, PVP was revealed as a candidate polymer carrier for fabricating the immediate release dosage forms by FDM 3D printing.

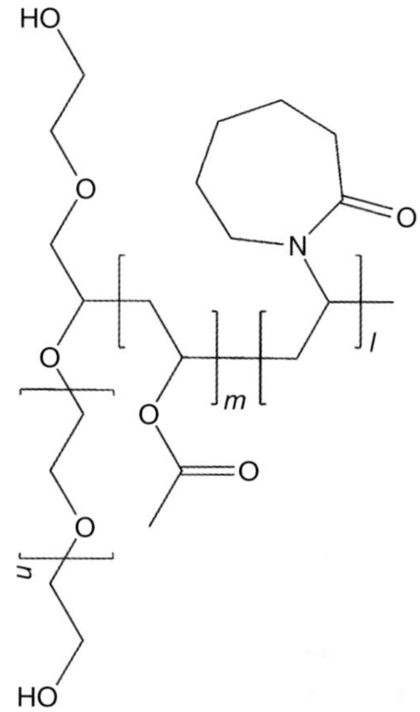

Figure 9.9 Chemical structure of Soluplus®.

9.3.4 Soluplus

Soluplus is a polyvinyl caprolactam–polyvinyl acetate–polyethylene glycol graft copolymer (weight ratio: 57/30/13) with PEG 6000 as the backbone (shown in Figure 9.9) [35]. This polymer was extensively investigated as a solid dispersion matrix with solubilization effect because of its amphiphilic structure [17]. The T_g of Soluplus was relatively low (70 °C) [35], attributed to the self-plasticization of the PEG segment. Moreover, it keeps thermally stable up to 250 °C [35] and thus has excellent extrudability with wide temperature window for processing. However, Soluplus filament was not FDM printable even loading with 10% drug [9].

Alhijjaj et al. [9] blended Soluplus with polyethylene oxide (PEO) (M_w = 100 000 g mol^{-1}, 15%), PEG4000 (10%), and Tween 80 (15%). The filament, composed of the polymer blends and 10% felodipine, was successfully printed into a designed disk because of the improved thermoplasticity. However, only 28% felodipine was released after six hours. The limiting step of the dissolution was suggested as the disintegration process, during which the disk first unraveled into strips. This "peeling effect" was attributed to well-defined interfaces between strips and layers, resulting from the low melt viscosity and the rapid solidification.

9.4 Cellulosic Polymers

Cellulose is a natural polysaccharide composed of linearly linked units of D-glucose unit. There are plenty of intra- and intermolecular hydrogen bonds between cellulose chains, resulting in high crystallinity (40–90%) [38] in native cellulose and poor water solubility. Therefore, various cellulose derivatives were synthesized by partially or fully reacting the hydroxyl groups of glucose units with reagents. The substituents broke part of hydrogen bonds, making the material amorphous. The water solubility and the physicochemical properties of these cellulose ethers depended on the chemical structure of the substituent groups and the degree of substitution.

9.4.1 Hydroxypropyl Cellulose (HPC)

HPC is a cellulose derivative, in which three reactive hydroxyls were fully or partially substituted with hydroxyl propoxy groups through an ether linkage on each monomer unit (shown in Figure 9.10). The remaining hydroxyl groups in glucose unit and the aliphatic hydroxyl groups in the side chains are accessible for the hydration and the formation of hydrogen bonds with water. This makes HPC soluble in cold water. The hydrogen bonds between polymer chains are weakened, which makes crystallization more difficult and thus keeps HPC in amorphous state. The substituent groups impart this material with high flexibility. Therefore, HPC can be hot melt extruded into opaque and flexible films [39] with bioadhesive property [40]. The T_g of HPC (Klucel®LF, $M_w = 95\,000\ g\ mol^{-1}$) was reported as 111 °C [38, 41]. Considering its high degradation temperature of 227 °C [38, 41], a wide extrudable range (melt viscosity of 1000–10 000 Pa s) of 170–200 °C was predicted [38, 41].

Melocchi et al. [8] successfully printed hollow capsular devices for oral pulsatile release using HPC as the polymeric material. The authors extruded the HPC filaments at 150–165 °C with a Φ1.75 mm die and a lab-made pulling/calibrating device. The printing temperature was set as 180 °C to achieve an acceptable balance between the melt flowing and the thermal stability of HPC. A PLA raft was first printed as a raft to connect the building platform with the HPC device during the printing process and then cut away. The printed bodies and caps can be assembled to tight devices. The release test exhibited a lag phase before rapid release of acetaminophen, the model drug. Zhang et al. [12] failed to print the HPC LF and HPC EF filaments loaded with 30% acetaminophen because the filaments were very soft and flexible to be fed into the printer. The authors did not describe the physical state of acetaminophen in the HPC filaments. Thus, it is difficult to evaluate the influence of acetaminophen on the printability of HPC.

As most pharmaceutical grade polymers are very brittle and HPC is very flexible, blending polymers with HPC might be an effective method to enhance the flexibility of the filament.

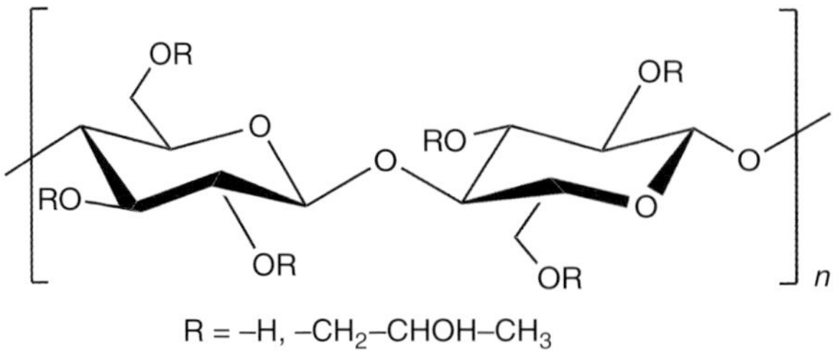

Figure 9.10 Chemical structure of HPC.

Figure 9.11 Chemical structure of HPMC.

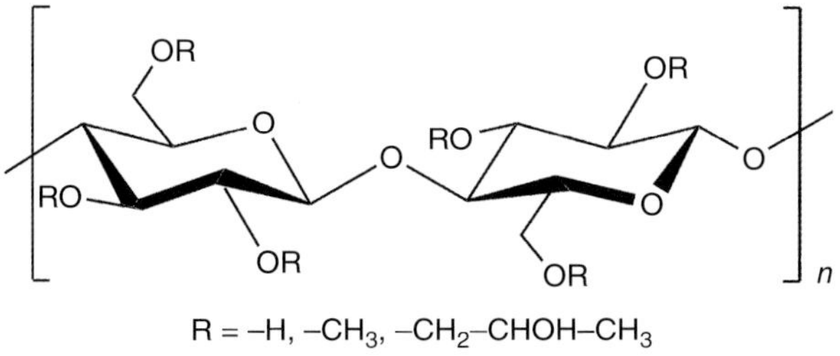

9.4.2 Hydroxypropyl Methylcellulose (HPMC)

HPMC is one of the dominant polymers in amorphous solid dispersion. There are many commercially available solid dispersion products based on the HPMC matrix, including Sporanox®, Prograf®, Intelence®, Zortress®, Onmel®, and Afinitor® [42]. HPMC is available in a variety of grades with different molecular weights and substitution degrees of hydroxyl and methyl groups (shown in Figure 9.11). HPMC E5 is the most commonly used grade in solid dispersions. It has a very high T_g of 170–180 °C [15] and a poor thermoplasticity, but an excellent capacity to inhibit nucleation of amorphous drug. Because the reactive hydroxyl groups in glucose unit were substituted with methyl and hydroxypropyl groups, HPMC is much more brittle than HPC, but still tougher than other pharmaceutical polymers, such as EC, Soluplus, and Eudragit L100 [12].

Zhang et al. [12] investigated the printability and mechanical properties of six pharmaceutical grade polymers (HPMC, EC, Soluplus, HPC LF, HPC EF, and Eudragit L100). A three-point bend test method, a commonly used method in plastic industry, was employed for the first time to evaluate the stiffness and toughness of pharmaceutical polymers, which are considered as the most important mechanical properties of the printable filament besides the melt viscosity. The filament should be stiff enough to be fed into a 3D printer and then push the molten filament to the printer nozzle. The stiffness can be quantitatively evaluated by measuring the "load" needed to achieve a certain "deformation". Brittleness is another important parameter for evaluating filaments because brittle filament could not be rolled up on the spool after extrusion and will break when passing the gear [12]. Brittleness can be evaluated by measuring the deformation of material before it fractures. After applied with force, the brittle structure breaks without significant deformation, whereas tough structure breaks after large deformation. In this work, breaking stress (representing stiffness) and breaking distance (representing brittleness) were used to evaluate the printability of the filament. PLA was used as the reference, which has a very high stiffness and a good toughness. All formulations were loaded with 30% acetaminophen as the model drug. The EC filaments displayed good stiffness but high brittleness, being easily broken by the feeding gears, whereas the HPC filaments were very soft to be fed into the printer. The HPMC filament exhibited high toughness and stiffness, but rough surfaces and high melt viscosity made the printing difficult [12]. Thus, HPMC was individually blended with 19.5% EC and 19.5% HPC, respectively, whereas EC was blended with 15% Eudragit L100. The addition of 5% superdisintegrant Kollidon CL-F smoothened the surface of the filaments and thus reduced the friction while feeding, making the filament printable. The authors concluded that a filament must simultaneously be adequately stiff (breaking stress >2941 g mm^{-2}) and tough enough (breaking distance >1 mm) with smooth surface and low melt viscosity to allow for successful printing.

In the continuation of this work [43], HPMC was employed as the matrix, blending with Soluplus at the weight ratio of 7 : 2, to print controlled release tablets with various structures and different release profiles. The model drug was acetaminophen with the melting point of 169–170 °C, but the drug load decreased to 10%. The filaments were extruded at 160 °C, forming an amorphous solid dispersion. The printing temperature was 200 °C. The dissolution test indicated that a lower infill percentage led to more pores in the tablet and thus accelerated the drug release, while tight shell structure effectively controlled the release rate. All tablets achieved ~100% drug release within 24 hours. The tablet, printed at 20% infill without the shell, completed the release within three to four hours. This work revealed HPMC as a potential pharmaceutical polymer to print immediate release dosage form.

9.4.3 Hydroxypropyl Methylcellulose Acetate Succinate (HPMCAS)

Hydroxypropyl methylcellulose acetate succinate (HPMCAS) has a cellulose backbone with a substitution of hydroxypropoxy, methoxy, acetyl, and succinyl groups (shown in Figure 9.12). It was originally developed and marketed in 1986 by Shin-Etsu Chemical Co. Ltd. (Japan) as an enteric polymer. Six grades are commercially available. The F (fine) and G (granular) grades differ only in their particle sizes, whereas the L, M, and H grades are chemically different and vary in their pH solubility (shown in Table 9.3). The L, M, and H grades dissolve at pH ≥ 5.5, 6.0, and 6.5, respectively. The pH-dependent solubility can be attributed to the ratio of succinyl and acetyl groups. Thus, the release of drug in gastrointestinal (GI) tract can be controlled by selecting a suitable grade of HPMCAS.

As an amorphous polymer, HPMCAS has a medium T_g of 119–122 °C [44] and remains stable up to 190 °C [15]. The degradation temperature of HPMCAS

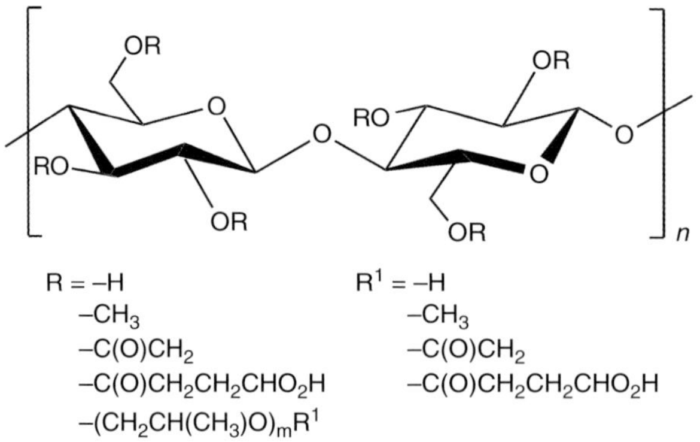

Figure 9.12 Chemical structure of HPMCAS.

Table 9.3 Chemical structure, glass transition temperature, and thermal degradation temperature of AquaSolve™ HPMCAS.

Grade	Acetyl content (%)	Succinyl content (%)	Methoxy content (%)	Hydroxypropoxy content (%)	T_g (°C)	T_d [a)] (°C)
L	5–9	14–18	20–24	5–9	119	258
M	7–11	10–14	21–25	5–9	120	267
H	10–14	4–8	22–26	6–10	122	276

a) In this table, T_d was the temperature, at which 5% weight loss was measured.
Source: Adapted from Reference [44].

was also reported as 257–276 °C [44], at which 5% weight loss was measured excluding moisture loss. HPMCAS has an excellent extrudability benefiting from the wide temperature window between T_g and T_d, but a poor printability because of the high brittleness of the filament.

Goyanes et al. [45] successfully printed tablets using three different HPMCASs (LG, MG, and HG) as the matrix. Acetaminophen was used as the model drug with drug loads of 5% and 50%, respectively. The addition of 5% magnesium stearate as the lubricant facilitated the successful extrusion for all formulations at a low temperature range (80–110 °C). Methylparaben, as the plasticizer, facilitated to improve the flexibility and resistance of the HPMCAS filament. Consequently, the tablets were printed at 180–190 °C with two different infill percentages (20% and 100%). The formulations were tested at pH 1 simulating gastric environment for 2 hours, and then pH 5.6–7.4 representing the small intestinal environment for 3.5 hours, followed by a pH at 6.5 representing the colonic environment. Dissolution test revealed faster drug release from the tablet prepared with polymers with a lower pH threshold (HPMCAS LG > HPMC AS MG > HPMCAS HG). All the formulations released less than 10% drug in the acidic environment. Tablets loaded with 50% acetaminophen released the drug faster in comparison of 5% formulations at the buffer stage. All the formulations printed with 20% infill dissolved faster than the tablets with 100% infill. X-ray microcomputed tomography analysis revealed that the tablets with 20% infill contained many cavities in the core. Drug diffusion and polymer dissolution (surface erosion) were suggested to be the main mechanism of drug release from the printed tablets. These results indicated that HPMCAS is a polymer candidate to fabricated enteric tablets by FDM 3D printing technology.

9.5 Polymethacrylate-Based Polymers

Polymethacrylates are a group of pH-sensitive polymers. The solubility of these polymers highly depends on the chemical structure and the pH value, which makes them suitable for different purposes, from gastric or intestinal soluble formulations to insoluble but swellable delivery systems. Four Eudragit

Table 9.4 Molecular weight, T_g measured at 20 °C min^{-1} heating rate, T_{max} for functional groups at a residence time of 4.5 minutes, and extrusion temperature of Eudragit® polymers.

	Molecular weight (g mol^{-1})	$T_g \pm 5$ (°C)	T_{max} for functional groups (°C)	$T_{extrusion}$ (°C)
Eudragit® L 100-55	320 000	96	159	141–158
Eudragit® E 100	47 000	45	220	120–205
Eudragit® RL 100	32 000	63	140	120–140
Eudragit® RS 100	32 000	58	110–120	120–140

Source: Nollenberger and Albers 2012 [46]. Reprinted with permission of John Wiley & Sons.

polymers have been reported to be polymer carriers or gastric-resistant shell in FDM 3D printed dosage forms, i.e. Eudragit E (soluble at pH < 5.0), Eudragit L100-55 (soluble at pH > 5.5), Eudragit RL, and Eudragit RS (both insoluble). The functional groups of the four polymers were quite different, resulting in different water solubility and physical properties. The molecular weight, T_g, thermal damaging temperature of functional groups, and extrusion temperature are summarized [46] in Table 9.4.

9.5.1 Eudragit RL/RS

Eudragit RL and Eudragit RS are insoluble but permeable in gastrointestinal tract. Their chemical structures are shown in Figure 9.13. The permeability of Eudragit RL is higher than Eudragit RS, resulting from more quaternary amino groups in Eudragit RL chains, which was more hydrophilic and promote the interaction of polymer with water. Therefore, Eudragit RL and RS can be used to formulate extended release dosage form.

Both the polymers have very low T_gs (63 °C for Eudragit RL and 58 °C for Eudragit RS) [46]. However, considering the poor thermal stability, the temperature window for processing is very narrow (120–140 °C) [46]. Considering that the printing temperature was usually 40–50 °C higher than the extrusion temperature, both Eudragit RL and Eudragit RS could not be printed without a plasticizer.

Pietrzak et al. [13] found that the addition of drug with a high melting temperature, acting as filler, can promote the constant flow and rapid solidification.

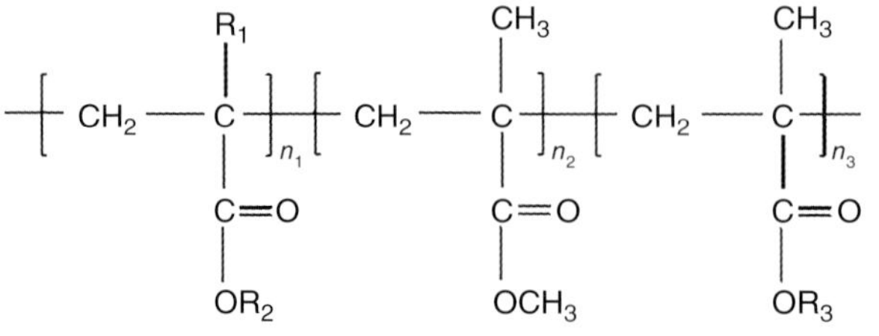

Figure 9.13 Chemical structure of Eudragit® RL/RS ($R_1 = H$, $R_2 = C_2H_5$, $R_3 = C_2H_5N(CH_3)_3Cl^-$; RL: n_1:n_2:n_3 = 1 : 2 : 0.2; RS: n_1:n_2:n_3 = 1 : 2 : 0.1).

Theophylline was selected as a model drug because of its high melt temperature of 273 °C and a good water solubility. The extended release caplet was successfully printed between 150 and 170 °C by formulating Eudragit RL and Eudragit RS with 50% theophylline and 5–7.5% TEC. The platform was kept at 60–90 °C. Dissolution test indicated that the theophylline release from Eudragit RL caplets extended for 16 hours, faster than that from Eudragit RS caplets. Blending Eudragit RL with Eudragit RS or changing the caplet size can modify the drug release behavior. Diffusion was considered as the main release mechanism. The results indicated that Eudragit RL and Eudragit RS can act as the sustained release matrix fabricated by FDM 3D printing.

9.5.2 Eudragit L100-55

Eudragit L100-55 is an enteric soluble polymer because the carboxylate group can be negatively charged at pH > 5.5. Its chemical structure is shown in Figure 9.14. Therefore, it has been extensively used as the film coating material for enteric-coated tablet. The T_g and T_d of Eudragit L100-55 were reported to be 96 and 159 °C (read from the figure in Ref. [46]), respectively. High T_g, poor thermal stability, combined with high melt viscosity and slow solidification, imparts Eudragit L100-55 very poor extrudability and printability. Therefore, a plasticizer must be added for continuous printing.

Okwuosa et al. [47] employed Eudragit L100-55 and PVP to print gastric-resistant tablet with shell–core structure. Blank Eudragit L100-55 shell acted as shell material for gastric protection, whereas PVP was the core matrix incorporated individually with three model drugs at different drug loadings (10% for theophylline, 2.3% for budesonide, and 20% for diclofenac sodium). TEC (plasticizer, 16.67–35.2%) and talc (lubricant, 17–33.33%) were incorporated to decrease the melt viscosity and accelerate the solidification. Consequently, Eudragit L100 filament and drug-loaded PVP filament were smoothly extruded at 125 °C with Φ1 mm die and 90 °C with Φ1.25 mm die, respectively. The author mentioned that frequent blocking of the PVP filament was encountered in dual FDM 3D printing as the material was often adhered to the inner wall of the nozzle head when the first nozzle was printing. The addition of oleic acid, a lubricant liquid with high boiling point, can facilitate the smooth alternations between nozzles by physically preventing the sticking of the filament to the internal wall of the nozzle. The platform was kept at 40 °C. Finally, the shell–core structure was successfully printed at 185 °C for shell and 110 °C for core, respectively. The pH change dissolution test indicated that a shell thickness over 0.52 mm is necessary to achieve sufficient protection in the

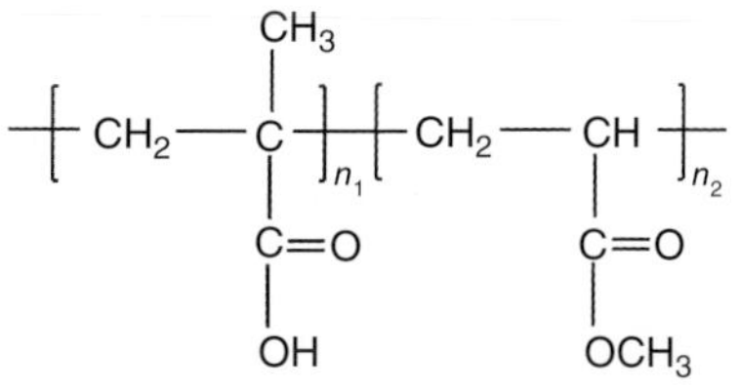

Figure 9.14 Chemical structure of Eudragit® L100-55 (n_1:n_2 = 1 : 1).

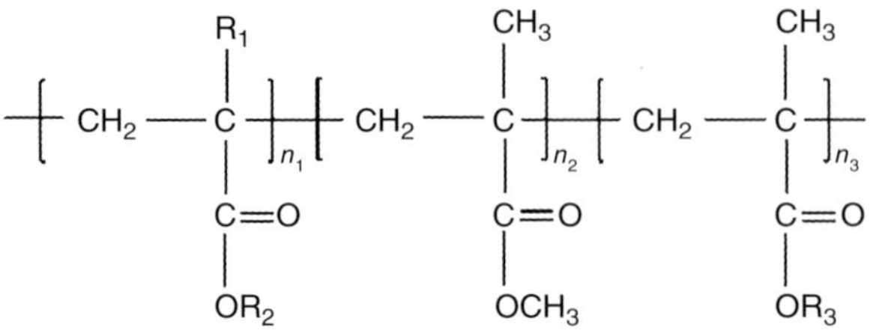

Figure 9.15 Chemical structure of Eudragit® EPO ($R_1 = CH_3$, $R_2 = C_4H_9$, $R_3 = C_2H_5N(CH_3)_2$, n_1:n_2:n_3 = 1 : 1 : 2)

acid medium. This work illustrated the potential of FDM 3D printing technique to fabricate gastric-resistant tablets by printing Eudragit L100 as the shell.

9.5.3 Eudragit E 100

Eudragit E is a gastric soluble polymer because the dimethylamino group can be positively charged at pH < 5.0. The chemical structure is shown in Figure 9.15. Compared with Eudragit RS/RL and Eudragit L100-55, Eudragit E has the lowest T_g, best thermal stability and widest process temperature window, indicating a best processability. However, the FDM 3D printing Eudragit E filament resulted in a collapsed structure with poor features and deformable build, which might be attributed to the branch structure of polymer chains [11].

Pietrzak et al. [13] printed immediately release caplets at 140 °C by blending Eudragit E with 3.5% TEC and 50% theophylline. The platform was kept at 60 °C. The caplet exhibited complete release within 25 minutes. Because the printing temperature is much lower than T_m of theophylline, the drug acted as a filler to facilitate the continuous printing. In the continuation of this work, five pharmaceutical fillers were investigated to improve the printability of Eudragit E, including microcrystalline cellulose, spray dried and directly compressible lactose, talc, and tricalcium phosphate (TCP) [11]. Both talc and TCP were effective. The latter was selected as the model filler for further investigation. The addition of 50% TCP made the smooth surface of the filament become rough, which facilitated continuous printing. TEC as the plasticizer was incorporated to improve the flexibility of the filament. Insufficient plasticizer resulted in brittle filament, whereas very much plasticizer led to an excessively flexible filament and thus a poor pressure in the nozzle. The formulation was optimized as Eudragit EPO:TEC:TCP:drug = 46.75 : 3.25 : 37.5 : 12.5. Four model drugs were selected to test the release behavior from the gastric soluble matrix, including 5-ASA(T_m = 283 °C), theophylline (T_m = 273 °C), prednisolone (T_m = 230 °C), and captopril (T_m = 103 °C). The drug-loaded filaments were extruded at 90 °C and then printed into caplets at 135 °C with a platform at 60 °C. 5-ASA, theophylline, and prednisolone existed in crystalline state in the caplets with the drug content higher than 93%, whereas captopril was amorphous in the caplets with a lower content of 88%. High-performance liquid chromatography results indicated that HME process is the major degradation step rather than the FDM

3D printing step. In this work, the authors mixed the materials in extruder at 100 °C for five minutes before extrusion at 90 °C. The drugs were exposed to high temperature much longer in the extruder than in the hot nozzle of the 3D printer (fraction of a second). All caplets disintegrated within 12 minutes and the majority of the drug was released within 30 minutes.

Alhijjaj et al. [9] blended Eudragit E with PEG4000 (15%), PEO (M_w = 100 000, 15%), and Tween 80 (10%). The polymer blends were printed into a designed disk. The dissolution test indicated that a rapid and complete release was achieved in pH 1.2 HCl (84.3% within 30 minutes), whereas a delayed but 100% release of felodipine was obtained in pH 6.8 buffer. The rapid disintegration and dissolution in pH 1.2, together with the bulk erosion and complete disintegration of disk in pH 6.8, were observed from the images captured during the dissolution process.

These studies indicated that Eudragit E could be used in FDM 3D printing as an immediate release matrix former.

9.6 Conclusion

Most pharmaceutical grade polymers were successfully FDM 3D printed by formulating with various excipients, such as plasticizer and non-melting filler. This progress opens new avenues for pharmaceutical application of FDM 3D printing technology. Various dosage forms have been fabricated using this technique, such as tablets/caplets with various release profiles (immediate, sustained, and enteric release), polypills with the structure of multiple layers, and Duo Tablet/caplet, together with implant.

However, pharmaceutical FDM 3D printing is still in its infancy. Firstly, thermal degradation of drug was observed in printing process [10, 20], limiting the pharmaceutical application of this technology. More work need to be done on this big issue. Secondly, the physical state of API in printed dosage form is difficult to control [7, 26, 28, 37] and thus should be given full attention because it might have significant influence on the dissolution behavior, the physical stability, and thus the bioavailability of the drug. Thirdly, thermoplasticity and mechanical properties of pharmaceutical polymers need to be improved by adding plasticizer/filler or blending with other thermoplastic polymers, which might influence the release behavior of the drug. The balance between printability and drug release might reduce the design flexibility for the formulation optimization. Therefore, the influence of the excipients on drug release and its mechanism should be investigated in detail. In addition, the melt viscosity of the molten filament (suitable for printing) and the mechanical properties of the solid filament (suitable for feeding) need to be quantitatively evaluated to instruct the formulation optimization.

Although there is a long way to go toward the extensive application for pharmaceutical FDM 3D printing, the bright prospect can be predicted, especially for personalized products, complex products, and immediate consumption.

References

1 Maroni, A., Melocchi, A., Parietti, F. et al. (2017). 3D printed multi-compartment capsular devices for two-pulse oral drug delivery. *Journal of Controlled Release* 268: 10–18.

2 Long, J., Gholizadeh, H., Lu, J. et al. (2017). Application of fused deposition modelling (FDM) method of 3D printing in drug delivery. *Current Pharmaceutical Design* 23 (3): 433–439.

3 Norman, J., Madurawe, R.D., Moore, C.M.V. et al. (2017). A new chapter in pharmaceutical manufacturing: 3D-printed drug products. *Advanced Drug Delivery Reviews* 108: 39–50.

4 Goyanes, A., Fernández-Ferreiro, A., Majeed, A. et al. (2018). PET/CT imaging of 3D printed devices in the gastrointestinal tract of rodents. *International Journal of Pharmaceutics* 536 (1): 158–164.

5 Li, Q., Guan, X., Cui, M. et al. (2018). Preparation and investigation of novel gastro-floating tablets with 3D extrusion-based printing. *International Journal of Pharmaceutics* 535 (1–2): 325–332.

6 Yang, Y., Wang, H., Li, H. et al. (2018). 3D printed tablets with internal scaffold structure using ethyl cellulose to achieve sustained ibuprofen release. *European Journal of Pharmaceutical Sciences* 115: 11–18.

7 Genina, N., Hollander, J., Jukarainen, H. et al. (2016). Ethylene vinyl acetate (EVA) as a new drug carrier for 3D printed medical drug delivery devices. *European Journal of Pharmaceutical Sciences* 90: 53–63.

8 Melocchi, A., Parietti, F., Loreti, G. et al. (2015). 3D printing by fused deposition modeling (FDM) of a swellable/erodible capsular device for oral pulsatile release of drugs. *Journal of Drug Delivery Science and Technology* 30 (Part B): 360–367.

9 Alhijjaj, M., Belton, P., and Qi, S. (2016). An investigation into the use of polymer blends to improve the printability of and regulate drug release from pharmaceutical solid dispersions prepared via fused deposition modeling (FDM) 3D printing. *European Journal of Pharmaceutics and Biopharmaceutics* 108: 111–125.

10 Goyanes, A., Det-Amornrat, U., Wang, J. et al. (2016). 3D scanning and 3D printing as innovative technologies for fabricating personalized topical drug delivery systems. *Journal of Controlled Release* 234: 41–48.

11 Sadia, M., Sosnicka, A., Arafat, B. et al. (2016). Adaptation of pharmaceutical excipients to FDM 3D printing for the fabrication of patient-tailored immediate release tablets. *International Journal of Pharmaceutics* 513 (1-2): 659–668.

12 Zhang, J., Feng, X., Patil, H. et al. (2017). Coupling 3D printing with hot-melt extrusion to produce controlled-release tablets. *International Journal of Pharmaceutics* 519 (1–2): 186–197.

13 Pietrzak, K., Isreb, A., and Alhnan, M.A. (2015). A flexible-dose dispenser for immediate and extended release 3D printed tablets. *European Journal of Pharmaceutics and Biopharmaceutics* 96: 380–387.

14 Water, J.J., Bohr, A., Boetker, J. et al. (2015). Three-dimensional printing of drug-eluting implants: preparation of an antimicrobial polylactide feedstock material. *Journal of Pharmaceutical Sciences* 104 (3): 1099–1107.
15 DiNunzio, J.C., Zhang, F., Martin, C., and McGinity, J.W. (2012). Melt extrusion. In: *Formulating Poorly Water Soluble Drugs*, 1e (ed. R.O. Williams, A.B. Watts and D.A. Miller), 311–362. New York: Springer.
16 Wang, B., Wang, D., Zhao, S. et al. (2017). Evaluate the ability of PVP to inhibit crystallization of amorphous solid dispersions by density functional theory and experimental verify. *European Journal of Pharmaceutical Sciences* 96: 45–52.
17 Kojo, Y., Matsunaga, S., Suzuki, H. et al. (2017). Improved oral absorption profile of itraconazole in hypochlorhydria by self-micellizing solid dispersion approach. *European Journal of Pharmaceutical Sciences* 97: 55–61.
18 Yewale, C., Tandel, H., and Misra, A. (2014). Application of polymers in transdermal drug delivery. In: *Applications of Polymers in Drug Delivery* (ed. A. Misra and A. Shahiwala). Shawbury, UK: Smithers Rapra Technology Ltd.
19 Melocchi, A., Parietti, F., Maroni, A. et al. (2016). Hot-melt extruded filaments based on pharmaceutical grade polymers for 3D printing by fused deposition modeling. *International Journal of Pharmaceutics* 509 (1–2): 255–263.
20 Goyanes, A., Buanz, A.B.M., Hatton, G.B. et al. (2015). 3D printing of modified-release aminosalicylate (4-ASA and 5-ASA) tablets. *European Journal of Pharmaceutics and Biopharmaceutics* 89: 157–162.
21 Williams, C.K. and Hillmyer, M.A. (2008). Polymers from renewable resources: a perspective for a special issue of polymer reviews. *Polymer Reviews* 48 (1): 1–10.
22 Sandler, N., Salmela, I., Fallarero, A. et al. (2014). Towards fabrication of 3D printed medical devices to prevent biofilm formation. *International Journal of Pharmaceutics* 459 (1–2): 62–64.
23 Boetker, J., Water, J.J., Aho, J. et al. (2016). Modifying release characteristics from 3D printed drug-eluting products. *European Journal of Pharmaceutical Sciences* 90: 47–52.
24 Chen, D.R., Chen, H.L., Bei, J.Z., and Wang, S.G. (2000). Morphology and biodegradation of microspheres of polyester-polyether block copolymer based on polycaprolactone/polylactide/poly(ethylene oxide). *Polymer International* 49 (3): 269–276.
25 Sivalingam, G., Karthik, R., and Madras, G. (2004). Blends of poly(epsilon-caprolactone) and poly(vinyl acetate): mechanical properties and thermal degradation. *Polymer Degradation and Stability* 84 (2): 345–351.
26 Holländer, J., Genina, N., Jukarainen, H. et al. (2016). Three-dimensional printed PCL-based implantable prototypes of medical devices for controlled drug delivery. *Journal of Pharmaceutical Sciences* 105 (9): 2665–2676.
27 Park, J.C., Ito, T., Kim, K.O. et al. (2010). Electrospun poly(vinyl alcohol) nanofibers: effects of degree of hydrolysis and enhanced water stability. *Polymer Journal* 42 (3): 273–276.

28 Goyanes, A., Buanz, A.B.M., Basit, A.W., and Gaisford, S. (2014). Fused-filament 3D printing (3DP) for fabrication of tablets. *International Journal of Pharmaceutics* 476 (1–2): 88–92.
29 Skowyra, J., Pietrzak, K., and Alhnan, M.A. (2015). Fabrication of extended-release patient-tailored prednisolone tablets via fused deposition modelling (FDM) 3D printing. *European Journal of Pharmaceutical Sciences* 68: 11–17.
30 Goyanes, A., Chang, H., Sedough, D. et al. (2015). Fabrication of controlled-release budesonide tablets via desktop (FDM) 3D printing. *International Journal of Pharmaceutis* 496 (2): 414–420.
31 Goyanes, A., Martinez, P.R., Buanz, A. et al. (2015). Effect of geometry on drug release from 3D printed tablets. *International Journal of Pharmaceutics* 494 (2): 657–663.
32 Goyanes, A., Wang, J., Buanz, A. et al. (2015). 3D printing of medicines: engineering novel oral devices with unique design and drug release characteristics. *Molecular Pharmaceutics* 12 (11): 4077–4084.
33 Goyanes, A., Kobayashi, M., Martinez-Pacheco, R. et al. (2016). Fused-filament 3D printing of drug products: Microstructure analysis and drug release characteristics of PVA-based caplets. *International Journal of Pharmaceutics* 514 (1): 290–295.
34 Li, Q., Wen, H., Ji, D. et al. (2017). Preparation and investigation of controlled-release glipizide novel oral device with three-dimensional printing. *International Journal of Pharmaceutics* 525 (1): 5–11.
35 Kolter, K., Karl, M., and Gryczke, A. (eds.) (2012). *Hot-melt Extrusion with BASF Pharma Polymers: Extrusion Compendium*. Germany: Ludwigshafen.
36 Knopp, M.M., Nguyen, J.H., Becker, C. et al. (2016). Influence of polymer molecular weight on in vitro dissolution behavior and in vivo performance of celecoxib:PVP amorphous solid dispersions. *European Journal of Pharmaceutics and Biopharmaceutics* 101: 145–151.
37 Okwuosa, T.C., Stefaniak, D., Arafat, B. et al. (2016). A lower temperature FDM 3D printing for the manufacture of patient-specific immediate release tablets. *Pharmaceutical Research* 33 (11): 2704–2712.
38 Meena, A., Parikh, T., Gupta, S.S., and Serajuddin, A.T.M. (2014). Investigation of thermal and viscoelastic properties of polymers relevant to hot melt extrusion – II: cellulosic polymers. *Journal of Excipients and Food Chemicals* 5 (1): 46–55.
39 Repka, M.A., Gerding, T.G., Repka, S.L., and McGinity, J.W. (1999). Influence of plasticizers and drugs on the physical-mechanical properties of hydroxypropylcellulose films prepared by hot melt extrusion. *Drug Development and Industrial Pharmacy* 25 (5): 625–633.
40 Repka, M.A. and McGinity, J.W. (2001). Bioadhesive properties of hydroxypropylcellulose topical films produced by hot-melt extrusion. *Journal of Controlled Release* 70 (3): 341–351.
41 LaFountaine, J.S., McGinity, J.W., and Williams, R.O. (2016). Challenges and strategies in thermal processing of amorphous solid dispersions: a review. *AAPS PharmSciTech* 17 (1): 43–55.

42 Jermain, S.V., Brough, C., and Williams, R.O. (2018). Amorphous solid dispersions and nanocrystal technologies for poorly water-soluble drug delivery – an update. *International Journal of Pharmaceutics* 535 (102): 379–392.

43 Zhang, J., Yang, W., Vo, A.Q. et al. (2017). Hydroxypropyl methylcellulose-based controlled release dosage by melt extrusion and 3D printing: Structure and drug release correlation. *Carbohydrate Polymers* 177 (1): 49–57.

44 AquaSolve™ hydroxypropylmethylcellulose acetate succinate – Physical and chemical properties handbook. http://www.ashland.com/file_source/Ashland/Industries/Pharmaceutical/Links/PC-12624.6_AquaSolve_HPMCAS_Physical_Chemical_Properties.pdf (accessed 01 January 2018).

45 Goyanes, A., Fina, F., Martorana, A. et al. (2017). Development of modified release 3D printed tablets (printlets) with pharmaceutical excipients using additive manufacturing. *International Journal of Pharmaceutics* 527 (1–2): 21–30.

46 Nollenberger, K. and Albers, J. (2012). Applications of poly(meth)acrylate polymers in melt extrusion. In: *Hot-melt Extrusion: Pharmaceutical Applications* (ed. D. Douroumis). Chichester, UK: Wiley.

47 Okwuosa, T.C., Pereira, B.C., Arafat, B. et al. (2017). Fabricating a shell–core delayed release tablet using dual FDM 3D printing for patient-centred therapy. *Pharmaceutical Research* 34 (2): 427–437.

48 Khorasani, M., Edinger, M., Raijada, D. et al. (2016). Near-infrared chemical imaging (NIR-CI) of 3D printed pharmaceuticals. *International Journal of Pharmaceutics* 515 (1–2): 324–330.

49 Solanki, N.G., Tahsin, M., Shah, A.V., and Serajuddin, A.T.M. (2018). Formulation of 3D printed tablet for rapid drug release by fused deposition modeling: screening polymers for drug release, drug-polymer miscibility and printability. *Journal of Pharmaceutical Sciences* 107 (1): 390–401.

50 Sadia, M., Arafat, B., Ahmed, W. et al. (2018). Channelled tablets: an innovative approach to accelerating drug release from 3D printed tablets. *Journal of Controlled Release* 269: 355–363.

51 Riva, A., Zanetti, M., Braglia, M. et al. (2002). Thermal degradation and rheological behaviour of EVA/montmorillonite nanocomposites. *Polymer Degradation and Stability* 77 (2): 299–304.

52 Hall, M. and Read, M. (2012). Hot-melt extrusion of ethylcellulose, hypromellose and polyethylene oxide. In: *Hot-melt Extrusion: Pharmaceutical Applications* (ed. D. Douroumis). Chichester, UK: Wiley.

10

Recent Advances of Novel Materials for 3D/4D Printing in Biomedical Applications

Jasim Ahmed

Kuwait Institute for Scientific Research, Environment & Life Sciences Research Center, PO Box 24885, Safat 13109, Kuwait

10.1 Introduction

Three-dimensional printing (3DP) is a recognized additive manufacturing (AM) or rapid prototyping (RP) technology, which allows the manufacturer to construct custom 3D objects using computer software and computer-aided design (CAD) [1]. AM is believed to improve precision along with a significant reduction of costs and time as compared to the conventional subtractive manufacturing. In anticipation of a fast-growing market with a massive global impact on major industries, the transition from research to manufacturing has emerged a new phase in which the complex design of intricately small 3D structures at the nanoscale has taken the path to a renewed approach to processing and fabrication [2]. In terms of manufacturing technology, 3DP is an AM process during which the mass change of the part is positive ($\Delta M > 0$), whereas processes with $\Delta M = 0$ and $\Delta M < 0$ are termed equivalent manufacturing and subtractive manufacturing, respectively [3]. In 3DP, 3D structures are fabricated by the controlled layering of materials that merge into an engineered object with desired dimensions encompassing three-dimensional (3D) geometries and properties [4]. AM technologies including direct ink writing (DIW), laser-based polymerization, standard lithography, or epitaxial assembly techniques can be used to create 3D structured objects [5]. The 3D printing technology is used for both prototyping and distributed manufacturing, with applications in industrial design, automotive industry, aerospace, architecture, medical industries, tissue engineering, and even food industry.

Recently, 3DP moves a step further into a new dimension to 4D printing (4DP). 4DP has the environmental, economic, and strategic implications of 3DP and provides exceptional capabilities in transforming digital information of the virtual world into physical objects of the material world [6]. In 4DP, the smart materials are combined with 3DP to capture their responses to external stimuli (e.g. heat, current, UV light, or other energy sources) that can be utilized for shape recovery, sensors, and actuators [7, 8]. These stimuli-responsive 3DP objects were dubbed "4DP objects," highlighting time as the fourth dimension [9]. Shape morphing

3D and 4D Printing in Biomedical Applications: Process Engineering and Additive Manufacturing, First Edition. Edited by Mohammed Maniruzzaman.

after 3DP is the key characteristic of 4DP. 3DP of shape memory materials has been considered as novel 4DP, and plenty of information is available in the literature on the subject.

The most promising applications of 3DP have been reported in the area of biomedical engineering including human health. AM has found extensive use as a tool to bioengineer tissue, varying in composition from bone and teeth to vascular and organ scaffolding, and successfully applied in dental implants and customized prostheses [10, 11]. In tissue engineering, the 3DP technology involves collecting correct information of tissues and organs for designing the model, conveying the information into an electrical signal to control the printer, and developing a printing process, which maintains the cell viability during the fabrication process. The 3D bioprinting has been employed to fabricate 3D cell scaffolds or medical implants for the field of regenerative medicine [12]. Additionally, 3DP has assisted advancement in individualized patient care, allowing for the development of patient-specific treatment plans via the printing of patient anatomy. It is quite convenient to study a tangible model of the patient's anatomy before surgery that could help medical professionals to prepare well than relying solely on 3D images acquired by MRI or CT scans [13]. The technology can be expected to grow faster for the development of human organ regenerations, implantable devices, otolaryngology, and precise drug development.

The real challenge of AM lies in the selection of suitable materials that can be fabricated into new hardware components. Such limitations can be avoided by following bottom-up fabrication processes, where arbitrary 3D mesoscale structures with microscale architectures and submicron precision could be generated. The fabrication processes would have a large impact on manufacturing if they can accommodate a large range of test materials (e.g. metals, ceramics, and polymers), which can further allow for rapid translation from computer model to fabricated component, and are scalable to large numbers of components or bulk material billets.

This chapter describes various materials used for AM and provides an array of information on the properties of materials used in 3D and 4D printing.

10.2 Materials for 3DP

The major part of the 3DP industry mostly depends on single material printing. It provides limited choices of available materials compatible with commercial printers, and eventually, it strictly limits the variations in the physical and chemical properties of 3D printed objects. To overcome these limitations, multimaterial printers have been developed with partial control on material composition and properties, offering layered composite materials. Materials intended for its use in 3DP has to pass a series of tests: (i) firstly, the material has the capability to pass through the extrusion to produce a plastic filament, (ii) in the subsequent extrusion capable of trace-binding during the 3D printing process, and (iii) finally it should be appropriate for the end use application as a 3D printed part/object. Some studies have been carried out on 3DP for a series of materials, including metal, ceramics, natural and synthetic polymers, wood, magnetic materials, and liquids in order to achieve the desired result. Materials are used as a powder, resin, pellets, and granules required as per the printing process. Metals have the potential for use in 3D printing of scaffolds including iron, cobalt, chromium, stainless steel, and titanium alloy because of their high mechanical strengths which have been shown to be similar to that of the bone [14]. However, limited 3D metal printing techniques are available. Furthermore, metal corrosion and aging and potential toxicity of metal ions are serious thoughts being further evaluated for long-term implantation of metal in 3DP. In 3DP, fabrication of 3D components can be transformed into the creation of multiple 2D cross sections, and even the parts with any complex shapes can be fabricated, even parts with a cavity inside [15]. By using smart and biological materials, 3DP is presently emerging from 3D to 4D and is going as far as fabrications that comprise living things. In this section, a series of materials are discussed leading to produce 3DP, which can be used for biomedical applications with their merits and demerits.

The 3D printing materials are available in various material types along with their different states. With the rapid development of 3D manufacturing technology, more advanced materials are developed, and new applications are emerging.

10.3 Rheology

Rheology measures the flow and behavior of the materials. It measures the material flow during the AM precisely whether it is Newtonian or non-Newtonian. Many printing materials display non-Newtonian, viscoelastic behaviors including shear thinning, yielding, or shear thickening. The ability and quality of 3D printing are closely dependent on material rheology. To obtain a desired and continuous printing or flow of ink, the 3D printer needs an optimization. Printability is also affected by how an extruded material exits a nozzle of a printer. An extrusion-based printing becomes smooth for a material/s with a Newtonian flow and without a yield stress. Final print quality is primarily determined by material viscoelasticity, yield stress, and relaxation time [16].

Gelation methods (e.g. physical, chemical, and photo-crosslinking) are employed to ensure the stability of bioprinted constructs. Bioinks are produced from natural and/or synthetic polymers [e.g. gelatin/collagen, hyaluronic acid (HA), and PEG]. The characterization of these bioinks is important so that it can pass through the printing head and avoid the blockages in the printer nozzle. Gelation of hydrogels can also be detected by the rheological measurement. The rheological behavior of a hydrogel precursor and the gelation kinetics are key factors for its printability.

10.4 Ceramics for 3D Printing

Ceramics comprise both metallic and nonmetallic elements and have been used as materials for 3D printed scaffolds because of their high mechanical strength and biocompatibility [17]. Hydroxyapatite (HAp), commonly found in human teeth and bones, is a ceramic itself, and therefore, making the use of HAp, or similar ceramics, attractive materials for creating scaffolds with strong mechanical properties similar to that of natural bone.

Ceramics are mostly suitable for biochemical, biomedical, and diagnostic applications because of their esthetic value, biocompatibility, and physicochemical properties [18]. Comparing with metals and polymers, the extremely high melting point of ceramic materials is one of the most critical challenges in the field of AM. Some of the important bioceramics with potential applications in biomedical devices are pure alumina (Al_2O_3), pure zirconia (ZrO_2), aluminosilicates, zirconia doped with yttria, and vitreous carbon and calcium phosphate-based composites. Among these, alumina and zirconia show their suitability in implants than in other metals because of their resistance to corrosion and the ability to withstand very high temperatures [19].

The ceramics are used as powders, and their material characteristics are the major concern in the fabrication of 3DP, in many cases. Selection of raw material is crucial because it influences the functionality and mechanical and structural properties of the fabricated part/s. The selection of raw materials depends on the machine's limitations and the specific requirements of the final parts. Some deposition of these ceramic materials requires the modification of industrial 3D printer machines. The materials are normally loaded into the 3D printer and the parts with predefined shape, pore, and strut size are printed. A list of ceramics used for 3DP is given in Table 10.1.

3DP can produce ceramic parts without any cracks or large pores by optimization of the process parameters, and their mechanical properties resemble those of conventionally fabricated ceramic parts. It is also possible to produce pore-free

Table 10.1 Raw ceramic materials with particle size and mechanical strength used in the 3D printing process.

Material	Particle size (μm)	Mechanical strength	References
$SiO_2 + CaCO_3$	25	47 MPa compressive ($CaCO_3$ 5 wt%) at 1300 °C	[20]
Leucite – glass ceramic (Vita VM-13)	18	≈120 MPa flexural	[21]
Alumina (RC172-DBM)	75–150	231 MPa flexural (undoped)	[22]
Hydroxyapatite (HAp SP19)	63–80	13.7 MPa compressive	[23]
β-Tricalcium phosphate	10–25	2.3–8.7 MPa compressive	[24, 25]
Bioactive glass ceramic	<45	16.10 ± 1.53 MPa compressive	[26]
Mg-doped wollastonite + β-tricalcium phosphate	<2	80–100 MPa compressive	[27]
Zirconia	90	763 MPa flexural	[28]
α/β-Tricalcium phosphate	550	1.73–5.52 MPa compressive (undoped); 4.31–10.32 MPa compressive (doped)	[29]

ceramic parts by incorporating colloidal processing techniques in the AM process. The major limitations in the AM of ceramics are the availability of starting materials for the production of the feedstock.

The 3D printed ceramic parts were fabricated by mixing specific pre-ceramic monomers with a UV photoinitiator (PI) in a stereolithography (SLA) printer [30]. These polymer structures were pyrolyzed to a ceramic with uniform shrinkage without any pore. Silicon oxycarbide microlattice and honeycomb cellular materials fabricated with this approach, exhibiting higher mechanical strength than ceramic foams of similar density. The green bodies (advanced ceramics) were fabricated by stereolithography-based AM technique using photosensitive resins loaded with 25–60 wt% of alumina, 3- and 8-YSZ [31]. Photogrammetry was employed to assess the accuracy of the 3D printing process, and it showed a marked deviation between the CAD model and the resulting object, particularly in the top part of the specimens, possibly because of the use of volatile solvents, which cause changes in the photoresins used. The obtained results could be implemented in dental applications.

10.5 Polymers and Biopolymers for 3D Printing

The monomer mixtures are capable of fabricating 3D printed materials, which later on changes to 4D shapes with the action of suitable stimulants. Hydroxyethyl acrylate (HEA), hydroxyethyl methacrylate (HEMA), and potassium 3-sulfopropylmethacrylate (PSPMA) are examples of monomers used for 3DP. Thermoplastic polymers including acrylonitrile butadiene styrene (ABS), polyamide (PA), poly(methyl methacrylate) (PMMA), polyether ether ketone (PEEK), poly(ε-caprolactone) (PCL), polylactic acid/polylactide (PLA), polyethylene glycol (PEG), and polycarbonate (PC) along with thermosetting polymers including epoxy resins have been commonly used in 3DP. However, epoxy resins require thermal or UV-assisted curing to complete the polymerization process, and therefore, epoxy resins are suitable for heat or UV-assisted 4DP process.

Polymers with the low melting point are extensively used in 3DP because of their low weight, low cost, and processing flexibility. The glass transition temperature, T_g, of the polymer has a tremendous influence on the development of 3D and 4D printing materials. However, 3DP polymers could have geometric complexity, but lack of mechanical strength and functionality is a challenge for their wide applications [32]. Therefore, in recent years, development of composite materials that are compatible with the available printers has attracted serious attentions. A combination of polymeric materials could achieve the desired mechanical and functional properties for 3DP. One approach is the reinforcement of particle/s into the polymer matrix to advance the properties of the blend. Particles are easy to be reinforced with polymers, either in dry powder form for selective laser sintering (SLS) or liquid form for SLA, or further to be extruded into printable filaments for fused deposition modeling (FDM) process [32]. The selected particle-reinforced polymer materials used for 3D

Table 10.2 Materials used for 3D printing of particle-reinforced polymer composites and their property improvements.

Material/s	Property improvement	References
Iron/ABS and copper/ABS	Improved storage modulus and thermal conductivity, reduced coefficient of thermal expansion	[33, 34]
$BaTiO_3$/ABS and $CaTiO_3$/polypropylene	Improved dielectric permittivity and controllable resonance frequency	[35]
$BaTiO_3$/ABS	Improved and tunable dielectric permittivity, periodic and graded structures were printed for demonstration	[36]
Thermoplastic elastomer/ABS	Reduced anisotropy of printed components	[37]
Al and Al_2O_3/Nylon-6	Reduced frictional coefficient	[38]
$Al2O_3$/UV cured resin	Improved dielectric permittivity and reduced dielectric loss tangent	[39]
Alumina/UV-sensitive resin	Controlled orientation of magnetized particle and bioinspired composite microarchitectures were printed	[40]
Alumina/polyurethane acrylate	Controlled local composition and orientation of magnetized particle	[41]
Diamond microparticle/acrylate based resin	Improve heat transfer, heat sink, and cooling system were printed for demonstration	[42]
Tungsten/PC	Improved dielectric permittivity, X-ray attenuation factor, and impact resistance	[43]
Glass bead/Nylon-11	Improved tensile modulus and compressive modulus, whereas reduced elongation at break	[44]

printing and the property improvements of the resulting composites are shown in Table 10.2. The major thrusts of particle-reinforced composites for consideration in the 3D printing are an improvement in mechanical properties, wear resistance, and dielectric permittivity.

Biopolymers get special interest in the 3DP/4DP intended for biomedical applications because of its biodegradability and biocompatibility. Among natural biopolymers, gelatin, alginate, collagen, agarose, chitosan, cellulose, and HA are important, whereas PLA, polyglycolic acid (PGA), PLGA–PCL, poly-l-lactide/ε-caprolactone-tricalcium phosphate (PCL–TCP), poly(lactic-co-glycolic acid) PLGA, PLGA–TCP, and PCL–PLGA–TCP have been reported for 3D printed biomedical applications. Several biopolymers have been tested for bioinks in 3DP processes, including gelatin, collagen, alginate, fibrin, chitosan, and agarose. Each of these biopolymers has some specific advantages and disadvantages toward 3DP, and combinations of bioinks can be used to promote their specific advantages and to limit their disadvantages. Starch-derived materials (e.g. nanocellulose) are also found as good candidates for 3DP. Many biomaterials have been used.

10.5.1 Polylactide (PLA)

PLAs, a corn-based polymer, are industrially synthesized either by condensation polymerization of the lactic acid monomers (2-hydroxypropanoic acid) or by ring-opening polymerization (ROP) of lactides (3,6-dimethyl-1,4-dioxane-2, 5-dione). There are three isomers of PLA, namely, D, L, and DL. Among these isomers, the thermal, mechanical, and biological properties vary significantly because of their crystallinity and molecular structures [45, 46]. The L-form is more active for their biological activity and is currently used as a printing material in 3DP. PLA and their copolymers have been employed for the manufacturing of medical devices including drug delivery systems, sutures, and surgical implants. The novelty of the PLA is that it becomes a Newtonian fluid above the melting temperature, and it can be molded to desirable shapes after cooling. Additionally, the glass transition temperature, T_g, of some of the PLAs is near about the body temperature of the human, so it can be easily adapted within the human body.

10.5.2 Poly(ε-caprolactone) (PCL)

PCL is a biodegradable and biocompatible semicrystalline linear aliphatic polyester with potential applications in drug delivery and tissue engineering [47, 48]. The glass transition temperature, T_g, of PCL is too low (−60 °C to −65 °C) [49], proving it to be a very strong and elastic material, and can be used successfully in clinical applications. PCL has exceptional rheological and viscoelastic properties, so it can be easily fabricated into implants and medical devices. However, the biodegradability of PCL, which is believed to be about two years, becomes the major concern for its application. Different approaches have been made to improve the mechanical and crystallization properties of PCL so that the polymer can be used in biomedical engineering successfully. Blending with nanoscale materials (e.g. layered silicates, multiwalled carbon nanotubes, and metal and alloy nanoparticles) is an effective approach to extend the applications of the polymer to produce new biodegradable and biocompatible nanocomposites with improved or even unexpected properties. In addition, copolymerization with a suitable polymer like PLA is another approach to use the polymer effectively.

10.5.3 Hyaluronic Acid

Hyaluronic acid (HA) or hyaluronan is an anionic, non-sulfated, natural glycosaminoglycans (GAG). It can be obtained either from animal sources or produced by microbial fermentation (e.g. *Bacillus subtilis*). It is a linear polysaccharide consisting of alternating disaccharide units of D-glucuronic acid and *N*-acetyl-D-glucosamine linked together by β-1,4- and β-1,3-glycosidic bonds. HA is naturally biocompatible and important in regulating many cellular behaviors and tissue functions, including cell migration, proliferation, differentiation, and angiogenesis [50]. It is a biocompatible and hydrophilic polymer that forms viscous solutions at relatively low concentrations, playing a key role in wound healing by promoting cell motility and proliferation. The polymer is incorporated in bioinks by blending with printable hydrogels, including methacrylated gelatin (GelMA) [51], and via conjugation with thermoresponsive poly(*N*-isopropylacrylamide) [52].

10.6 4D Printing

Four-dimensional (4D) printing is a novel technology with enormous capability for fabricating complex, stimuli-responsive 3D structures, providing unlimited potential for tissue and organ engineering applications. However, an interaction mechanism needs to be identified for which the printed smart structure can respond to stimulus in a proper way. In 4DP, the fabricated structure accomplishes a self-assembly process instantly after 3DP; this self-assembly is a process in which a pre-existing form dynamically moves to another structure as a consequence of external stimuli. It is very important for a 4DP to retain the configuration or function before and after the stimulation while the surrounding environments may change significantly [29]. For example, if a 3D fabricated construct has a primary conformation "A" in air and shifts to a conformation "B" in water, both conformations should be structurally stable, without external forces to maintain its structure [53]. Materials used for 4DP with potential applications to biomedical engineering and their characteristics are summarized in Table 10.3.

10.6.1 Bioprinting

Bioprinting technologies allow the automated biofabrication of cell-laden constructs through the layer-by-layer deposition of bioinks in both *in vivo* and *in vitro* [67, 68]. As a novel technique, 4D bioprinting could bring a solution to critical medical needs, such as tissue regeneration, and may find extensive applications in the biomedical field. Because bioprinting technologies are monitored by computer, there is a huge potential for the technology that can be combined with medical image systems (e.g. computed tomography and magnetic resonance imaging) and CAD/CAM tools to generate personalized constructs organized at different length scales [69].

Mostly, bioinks are made from natural and/or synthetic polymers including collagen/gelatin, HA, PEG, etc. Collagen is the most abundant protein present in the extracellular matrix (ECM) of many tissues. Collagen can produce a hydrogel at physiological conditions by triple helix formation, and therefore, it has been employed for cell encapsulation purposes because of the presence of cell-interactive RGD (arginine–glycine–aspartic acid) sequences in their backbone, which stimulate cell adhesion [70].

The physicochemical properties of a hydrogel precursor, particularly the rheological behavior, the swelling properties, the surface tension, and the gelation

Table 10.3 Materials used for 4D printing with potential of applications to biomedical engineering and their characteristics.

Material/s	Printing process	Characteristics	References
A composite of SMP fibers in an elastomeric matrix	Polyjet	T_g: elastomeric matrix: -5 °C; fibers: 35 °C Young's modulus: elastomeric matrix: 0.7 MPa at 15 °C; fibers: 3.3 MPa at 60 °C and 13.3 MPa at 15 °C.	[54]
A composite of SMP fibers in an elastomeric matrix: TangoBlack as the elastomeric matrix and Gray 60 as the SMP fiber	Polyjet	T_g: TangoBlack: −5 °C; Gray 60: 47 °C Young's modulus: TangoBlack: 0.7 MPa; Gray 60: 6 MPa	[55]
A composite of two SMP fibers for multiple transition temperatures. TangoBlack+ as the elastomeric matrix; DM 8530 (Gray 60), DM 9895 as the SMP fibers	Polyjet	T_g: TangoBlack: 2 °C; DM 8530: 57 °C and DM 9895: 38 °C Tensile strength: DM 8530: 29–38 MPa; DM 9895: 8.5–10 MPa	[56]
Inhomogeneous with a hinge section. Epoxy polymers as the hinge	Polyjet	T_g of epoxys: 32, 35, 55, 60, 65 °C	[57]
Soybean oil acrylate; Ciba Irgacure 819	Customized printer using stereolithography	T_g of soybean oil: 20 °C	[58]
A PCL-based composite: PCL, PCL triol and Caster oil	Rhinoceros 3D	T_g of PCL: −8, 9, 21, 35, 0 °C Recovery speed (o/s): 69.5, 30.0, 23.1, 3.9, 75.5	[59]
A composite of hydrogels: Alginate/PNIPAAm	3D bioplotter	Young's modulus: 10% of NiPAAm: 0.29 MPa at 20 °C; 1.4 MPa at 60 °C 15% of NiPAAm: ~0.26 MPa at 20 °C; 1.8 MPa at 60 °C; 20% of NiPAAm: 0.17 MPa at 20 °C; 1.9 MPa at 60 °C	[60]
PLA	3D polymer printer	T_g: 60–65 °C	[61]
Phenylbis(2, 4, 6-trimethylbenzoyl) phosphine oxide as a photoinitiator; methacrylate-based polymer; sudan I and rhodamine B as photoabsorber	Customized SLA 3D printer	T_g: TangoBlack: −5 °C; Gray 60: 47 °C Young's modulus Sample A: 16.5 MPa with 5% failure strain Sample B: 0.92 MPa with 99% failure strain	[62]

(*Continued*)

Table 10.3 (Continued)

Material/s	Printing process	Characteristics	References
Hydrophilic polymer: vinyl caprolactam, polyethylene, epoxy diacrylate oligomer, Irgacure 819, wetting agent	Polyjet	Young's modulus Rigid material: 2 GPa with Poisson's ratio of 0.4 Expanding material: 40 MPa in the dry state and 5 MPa when fully swollen	[63]
A composite with particles: polyurethane acrylate; fumed silica particles; Irgacure 819 as a photoinitiator	Customized magnetic jet 3D printer	Young's modulus Matrix polymer: 5 MPa Particle composite polymer: ~5, 8.5 MPa	[41]
A composite of wood fibers PLA and poly(hydroxyalkanoate): 40 wt% wood fiber; 40 wt% coconut fiber	3D printer	Young's modulus: 0° compressed composite: 4000 MPa with ~0.15% of swelling 90° compressed composite: 3500 MPa with ~0.43% of swelling 90° with high speed printing: 900 MPa with ~0.5% of swelling	[64]
A composite of wood-derived fibrils; cellulose microfibrils; hydrogels ((poly-*N*,*N*-dimethylacrylamide) matrix with PNIPAAm)	Inkjet printer	Young's modulus: Stiff filler : >100 GPa Printed composite: 40 kPa, 10% swelling strain (longitudinal) 20 kPa, 40% swelling strain (transverse)	[53]
A composite of fibers; hydrogels (alginate(Alg)/poly(acrylamide) (PAAm))	3D bioplotter	Young's modulus: hydrogels (Alg/PAAm): 26 kPa (cast); 66 kPa (printed)	[65]
PDMS; rigid part	3D SLA printer	Young's modulus Rigid part: 2700 MPa (Poisson's ratio of 0.3) PDMS : ~2 MPa	[66]

kinetics, are important factors for its printability [70]. This mainly applies for biofabrication techniques that rely on bioink dispensing. The bioprinting processing methodologies can be subdivided into three groups, namely, extrusion bioprinting (pneumatic and mechanical), orifice-free bioprinting (laser-induced forward transfer (LIFT) and printing by surface acoustic waves), and inkjet bioprinting (piezoelectric and thermal) [70]. However, the concept and mechanism of 4D bioprinting are not yet fully understood. In the following sections, recent advances in materials for use in tissue engineering and drug delivery systems are discussed.

10.6.2 Smart or Intelligent Materials

Intelligent materials have been extensively encouraged as a key technology that will improve all manners of novel products with exceptional competencies. Smart or intelligent materials are defined as those materials have the capability to convert their geometry/shape and even change their physical properties such as Young's modulus, stiffness, and resistance under the influence of external stimuli such as temperature, magnetic fields, irradiation, and light [71]. The shape memory effect (SME) is defined as the ability, which has been quasi-plastically and severely pre-deformed, to recover original shape under the correct stimulus [72, 73]. Sometimes, these materials are termed as soft active materials (SAMs). With the introduction of smart materials, the "smartness" of materials, capturing their responses to external stimuli, has been utilized for the shape recovery, sensors, and actuators [7, 8]. The most common stimuli to trigger the SME are heating/cooling (thermoresponsive), light (photoresponsive), chemicals (chemoresponsive), and mechanical loading (mechanoresponsive) [74]. Materials exhibiting different actions with the application of stimuli are shown in Table 10.4. These stimuli-responsive 3D printed objects were dubbed "4D printed objects," highlighting time as the fourth dimension [9]. Intelligent materials have been widely promoted as a key technology that will enhance all manners of novel products with unique capabilities. Smart materials can be divided into two subclasses: (i) the programmable shape memory polymers (SMPs) and (ii) the artificial hydrogels. However, they differ significantly in the mechanism of the shape change. Shape morphing after printing is the main characteristic of 4DP. Shape variations in 4D printed objects can be induced by different external stimuli, to cause expansion, shrinkage, or folding of the printed objects. SAMs can be chemically tuned to attain biocompatibility and biodegradability and widely studied for the fabrication of biomedical devices, wearable devices, artificial muscles, and other smart products. SMP fibers were specifically printed into an elastomeric matrix to generate an active composite that was used to fabricate integrated hinges to enable origami folding structures.

10.6.2.1 Thermal Stimuli-Induced Transformation

A useful activation method for 4DP is the practice of high temperatures to trigger the smart material for its shape change and the concept of self-assembling origami. Thermoresponsive materials have the ability to shrink, fold, or swell

Table 10.4 Intelligent materials type and their actions.

Terminology	Description
Shape memory	Material shape changes in response to external stimuli
Self-assembly	Permits self-folding for assembly
Self-actuating	Self-actuation in response to an external stimulus
Self-sensing	Allows programmed detection and quantification of external stimuli

Table 10.5 Glass transition temperatures of some selected polymers used in 3D/4D printing.

Polymer	T_g (°C)[a)]
Acrylonitrile butadiene styrene (ABS)	105
Polyamide 6	47
Polycarbonate (PC)	147
High-impact polystyrene	100
Polyurethane	−63
Poly(*N*-isopropylacrylamide) (NIPAM)	90–130
Poly(ethylacrylate)	−25
Poly(acrylic acid) ($M_n = 2\,000$)	70
Poly(methacrylic acid) ($M_n = 40\,000$)	140
Polyethylene glycol (PEG)	−60 to −70
Polyethylene glycol methacrylate methyl ether	40
Polylactic acid/polylactides (PLA)	60–65
Poly(ε-caprolactone) (PCL)	−60 to −70
Gelatin (Pig skin Type A)	110, 197
Gelatin (Fish skin)	60
Chitosan	50–56
2-Hydroxyethyl methacrylate (HEMA)	55
Poly(2-hydroxyethyl methacrylate) (pHEMA)	99

a) T_g depends on the number-average molecular weight and data presented are representative for the polymer; the presented data are based on the literature and author's unpublished work.

upon changes in temperature, whereas some polymers have phase transition temperatures close to physiological temperature. Tailoring of thermal stimuli-induced composite materials, which are known as printed active composites (PAC), is based on a large difference in the glass transition temperature, T_g, among the constituents. Polymers with triple-shape memory structures were made by fabricating the composite materials with well-separated T_g [55, 75]. A list of polymers with their T_g are shown in Table 10.5. Ge et al. [54] fabricated a thermoresponsive PAC by directly imprinting fibers of SMPs into an elastomeric

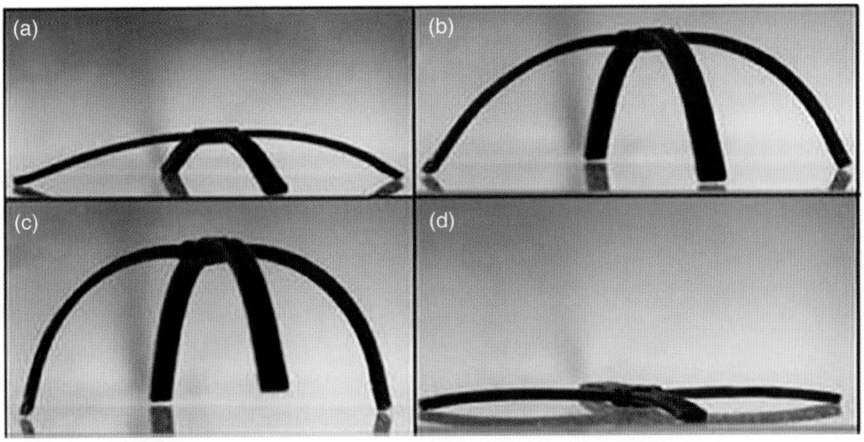

Figure 10.1 Multishape memory effect from 2D to 3D by self-bending in a smart trestle [56]. Source: https://creativecommons.org/licenses/by/4.0/.

matrix. The T_g values of the elastomeric matrix and the SMP fibers were about −5 and 35 °C, respectively. The PAC was employed in a hinge section as an active part of a 3D printed rod. Furthermore, PAC hinges were fabricated with SMP fibers consisting of the material coined "Gray 60" with the T_g of about 47 °C [55]. The deformation angle of the hinge was measured with respect to the strain and hinge's length, and the maximal deformation angle was found about −20° and was obtained for a 10-mm-long hinge. Applying thermoresponsive SMPs to attain two or more shape changes with increasing temperature, Wu et al. [56] had fabricated composite materials using more than one type of SMP fibers. The multishape memory effects for 2D to 3D transformations in various structures, such as the active trestle and active helix shape, are illustrated in Figures 10.1 and 10.2, respectively. Because SMPs have unique transition temperatures, multishape active composites should be fabricated with multimaterials. They used two fibers with the T_g of 57 and 38 °C, respectively. The matrix was prepared with TangoBlack+, which has the lowest glass transition temperature (2 °C) among the materials that were used by the researchers. Based on heat, the stimulus and these shape-shifting behaviors were achieved in the usual shape memory cycles with well-known programming and recovery steps.

3D printed layered composite structures with multiple families of SMP fibers were fabricated with a wide range of T_g to control the transformation of the structure [56]. After a one-step thermomechanical programming process, the fiber families were sequentially activated to bend when the temperature was increased (Figure 10.3). By tuning the volume fraction of the fibers, bending deformation can be controlled. Adoption of flat 2D structures that can be programmed to fold and open themselves when subjected to heat can be implied to 4DP applications.

The 3D printed PLA-based polymers showed SME and displayed a heat-shrinkable characteristic with a significant transformation of shape while heated. The concept of internal strain/stress during FDM printing of PLA has been used to induce patterned transformation(s) [76]. The internal

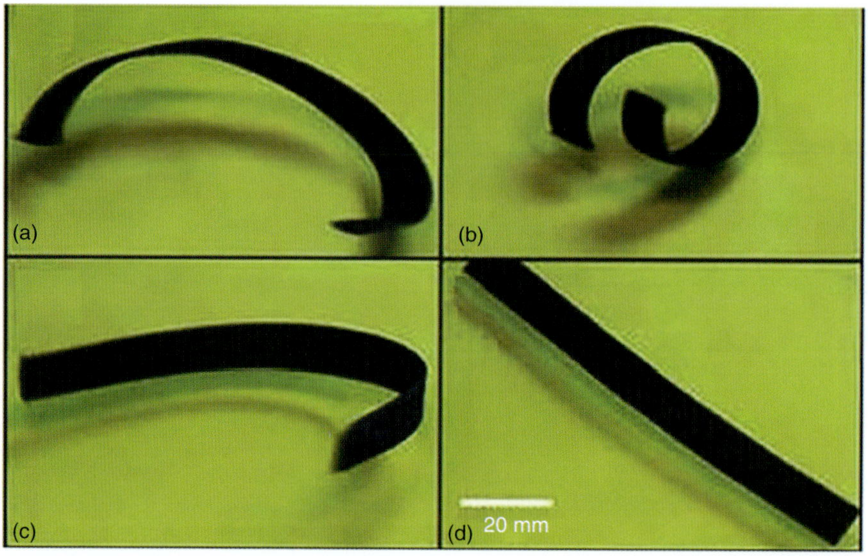

Figure 10.2 Multishape memory effect from 2D to 3D by self-bending in an active helix shape [56]. Source: https://creativecommons.org/licenses/by/4.0/.

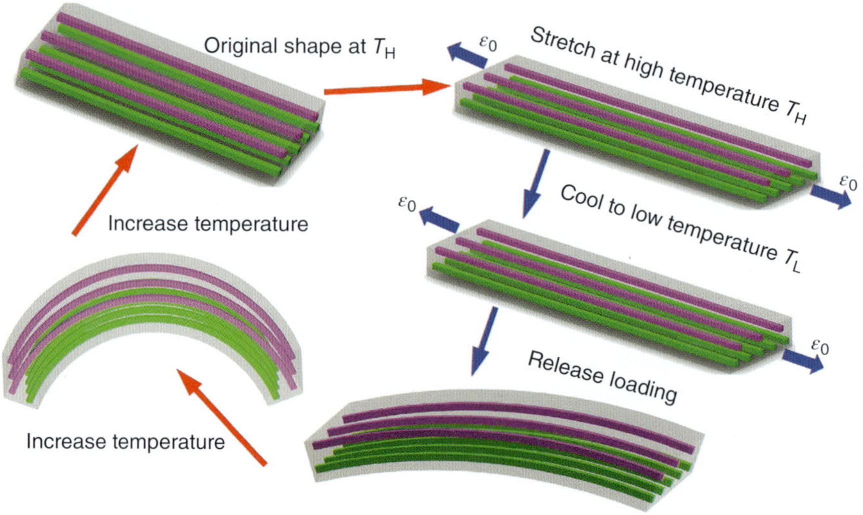

Figure 10.3 The typical programming steps for SMPs and the desired response [56]. Source: https://creativecommons.org/licenses/by/4.0/.

stress/strain can be maintained for an extended period at temperatures below the T_g, and the release of the internal stress/strain is achieved by heating the printed architecture above the T_g.

Composite porous scaffolds with the shape memory were fabricated by blending PLA and HA (PLA/15 wt% HA) for the repair of bone defects. The addition of HA improves the modulus of the composites and consequently contributes to the maintenance of the permanent shape. During compression–heating–compression cycles of the 3D printed shape memory scaffolds, the HA particles were found to resist the growth of the cracks. A 98% shape recovery of the scaffold triggered by heating is assumed to have a self-healing function by narrowing and preventing crack propagation. Triblock amphiphilic copolymers such as poly(D,L-lactic acid-co-ethylene glycol-*co*-D,L-lactic acid) (PELA) exhibit shape memory behavior during a phase separation between the hydrophilic and hydrophobic blocks [77]. Mostly, the phase having a higher T_g acts as the physical cross-links or hard segments, whereas the lower T_g acts as the soft segments. These SMPs can be deformed and fixed into temporary shapes at room temperature with rapid shape recovery (<3 s) at 50 °C. The polymer can be printed in a commercial FDM 3D printer into 3D macroporous scaffolds with shape memory properties due to its thermoplastic feature.

10.6.2.2 Hydrogel

The fundamental of tissue engineering is based on building 3D complex structures composed of cells and scaffold composition. Hydrogels are interesting as an active material in morphing structures because they undergo huge deformations because of their high degree of swelling. Hydrogel materials produce a 3D network structure, which has excellent water-holding capabilities, providing conditions similar to those of the ECM, and therefore, soft hydrogels have received increasing attention in regenerative medicine and tissue engineering, in particular, for use as scaffolds [78, 79]. Hydrogels have many advantages for use in biomedical applications because of their biocompatibility, viscoelasticity, and ease of fabrication. Hydrogels swell when solvent molecules diffuse into polymer network, and furthermore, they are used as fillers between layers in the tissue or as support to the 3D printed tissues. Recently, 4DP significantly inspires the exploration of self-healing hydrogels in the field of tissue and organ regeneration [80].

Recently, the fabrication of double-network (DN) hydrogels has been reported using different polymers and biopolymers. A stretchable DN hydrogel was fabricated by combining an ionically cross-linked κ-carrageenan network with a covalently cross-linked polyacrylamide (PAAm) network [81]. The κ-carrageenan/PAAm DN hydrogel demonstrated an excellent recoverability (84–91%) and significant self-healing capability. The 3D printability of the pre-gel solution of κ-carrageenan/AAm exhibited a liquid-like behavior ($G'' > G'$) at elevated temperature (Figure 10.4). Upon cooling, both G' and G'' increased gradually, and a sharp transition occurred at the critical temperature of 44.7 °C because of the formation of a gel (crossover point; $G' = G''$), followed by a predominating solid-like behavior ($G' > G''$). The κ-carrageenan/AAm solution showed a quick gelation rate (Figure 10.4b), which allowed the κ-carrageenan/AAm solution to form a gelled string on the substrate quickly. The viscosity increased significantly (0.7–2000 Pa s) during the sol–gel transition (Figure 10.3c), which is desirable for the fabrication of precise 3D structures. Moreover, the warm pre-gel solution of κ-carrageenan/AAm can be used as an ink for printing complex 3D structures with robust mechanical strength obtained after the UV exposure. The κ-carrageenan/PAAm DN hydrogel exhibited high

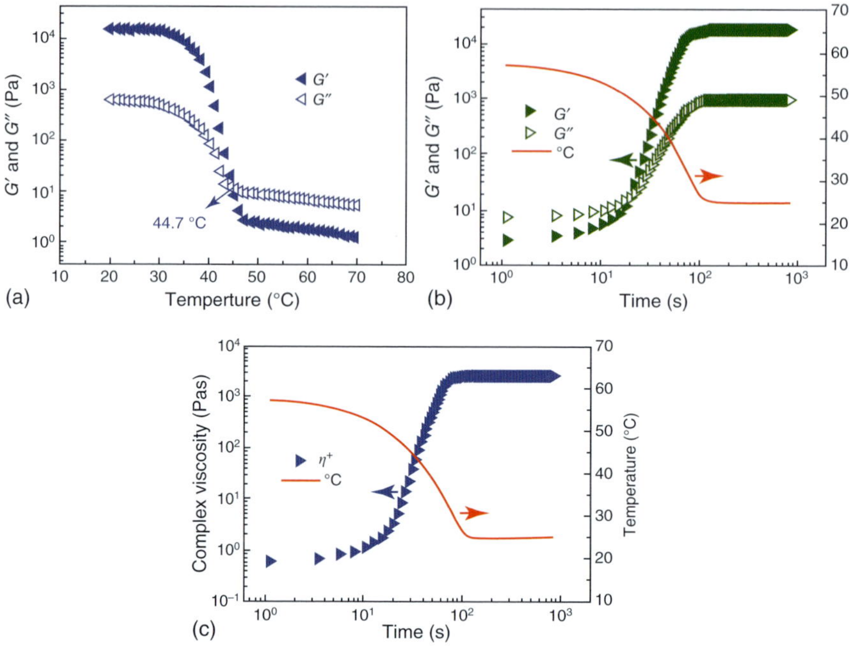

Figure 10.4 Mechanical spectra of k-carrageenan/AAm pre-gel solution at a constant frequency of 1 Hz (a) temperature ramp from 70 °C to 20 °C; (b) the elastic and the viscous modulus; and (c) complex viscosity as a function of time and temperature. Source: Liu and Li 2017 [81]. Reproduced with permission of ACS.

strain sensitivity with a gauge factor of 0.63 at the strain of 1000%, and therefore, the hydrogel can be used as sensitive strain sensors for biomedical applications.

A two-step method was employed for the fabrication of DN hydrogels at room temperature using a low-cost 3D printer [82]. In the beginning, the network precursor solution was fabricated into 3D printable via extrusion fitted with a nozzle by adding a layered silicate to make it shear-thinning. After printing and UV curing, the objects were soaked in a second network precursor solution and further UV cured to create an interpenetrating network of poly(2-acrylamido-2-methylpropanesulfonate) and PAAm. The stiffness and maximum elongation of the gel were manipulated by varying the ratio of PAAm to cross-linker to yield compression strength and elastic modulus of 61.9 and 0.44 MPa, respectively, which were greater than those reported for bovine cartilage. The maximum compressive (93.5 MPa) and tensile (1.4 MPa) strengths of the gel are double than the other reported 3D printed gels, and the gel did not deform after it is soaked in water. An X-ray computed tomography image indicated that the developed hydrogel has the potential to customize hydrogel implants based on 3D images of a patient's anatomy.

A tissue engineering scaffold was fabricated from clinically approved materials with the capability of delivering biomolecules and direct cell fate. The 3DP process combines polymer casting with supercritical fluid technology to produce 3D interpenetrating polymer network (IPN) scaffold of silicone-poly(2-hydroxyethyl methacrylate)-*co*-poly(ethylene glycol) methyl ether acrylate (pHEMA-*co*-PEGMEA) [83]. The pHEMA-*co*-PEGMEA IPN materials were used to support the growth of human mesenchymal stem cells (hMSC), resulting in high cell viability and metabolic activity over a three-week period. Additionally, the IPN scaffolds support 3D tissue formation inside the porous scaffold with well-spread cell morphology on the surface of the scaffold. As a proof of concept, sustained doxycycline (DOX) release from pHEMA-*co*-PEGMEA IPN was confirmed, and the biological activity of released drug from IPN was established using a DOX-regulated green fluorescent reporter (GFP) gene expression assay with HeLa cells. Because of its unique mechanical and drug-releasing characteristics, IPN scaffolds could be employed for directing stem cell differentiation by releasing various chemicals from its hydrogel network.

He et al. [84] developed a 3D printed hydrogel scaffold successfully by mixing sodium alginate (SA) and gelatin with a proper ratio and feed rate. The mixture was extruded on a cool substrate for the solidification of gelatin and fixing the biostructures. After printing, the structures were immersed in the calcium chloride ($CaCl_2$) solution for the cross-linking of SA and acquiring the cell-laden hydrogel structures. The cell viability after printing was compared with the casting, and the results showed that the bioprinting method almost had no extra damage to the cells.

10.7 3D and 4D Printed Bone Scaffolds with Novel Materials

Scaffolds are an essential part of bone tissue engineering. These are 3D biocompatible structures that can provide mechanical support, cellular activity, and protein production through biochemical and mechanical interactions and offer a template for cell attachment and stimulate bone tissue formation *in vivo* [85, 86]. Pore size, porosity, pore interconnectivity, and mechanical strength are important parameters that finally influence the performance of a scaffold. Biocompatibility and biodegradability are important properties for scaffold materials to possess, confirming that they are degraded into nontoxic products while leaving behind only the desirable living tissue [14]. Furthermore, the porosity of the implant influences the implant adhesion to the bone through bone ingrowth into the pores of the prosthetic, and it facilitates biodegradability of the scaffold because of the reduced material presence [17]. Ideal porosity and pore size of the 3D printed scaffold to encourage bone ingrowth are in the range of 30–70% and 500–1000 μm, respectively [87]. The choice of bone scaffold material depends largely on the purpose of the graft, as the materials used to form 3D grafts have variable resorption times.

Although a significant number of published works and patents are available in the development of 3DP, however, individual ceramic implants and scaffolds for bone tissue engineering have not been completely solved yet. The use of toxic organic solvents for obtaining these objects and their poor mechanical strength until now has restricted application of 3DP in biomedical research and clinical practice. Materials with nanostructure show excellent biological properties: high protein adsorption, improved function of stimulation of osteoclasts and osteoblasts, cells involved in remodeling of the bone tissue, and decreased function of the proliferation of competing cells, especially, fibroblasts responsible for the formation of the connective tissue [88]. Currently, the composite systems with nanostructured surface based on calcium phosphates are the most promising materials for the creation of bioactive scaffolds for the replacement of defects and directed regeneration of bone tissues [89].

Mostly, a large variety of metallic, ceramic, polymeric, and composite materials are processed through 3DP for bone scaffold; however, the selection of binder and optimization of process parameter are the means to successful part fabrication. A list of material-binder system combinations of bone scaffolds using 3DP is presented in Table 10.6. Biodegradable starch-derived binders can be used for bone replacement applications because of its mechanical strength, which is close to the trabecular bone [103]. Both structural designs and post-processing conditions can influence the mechanical properties of 3D-printed starch-based scaffolds [96]. 3D printed polyethylene (PE) scaffolds with 22.3–49.7% porosity have shown a tensile strength up to 4 MPa and no toxicity to human osteoblasts [97].

Recently, new combinations of materials are being tested for increased bioresorption and biocompatibility, including a β-TCP and bioactive glass mixture. These materials are fabricated into 3D scaffolds by selective 3DP processes or printing the powdered form of the chosen material with an organic binder to form a ceramic 3D scaffold [24]. calcium phosphate (CaP) ceramics are widely used in bone tissue engineering because of their excellent bioactivity, osteoconductivity, and similarities in composition to bone. Capillaries and vessel formation and homogeneous osteoconduction from central channels have previously been observed in 3D-printed HA blocks [104].

PEEK is a semicrystalline thermoplastic with excellent fibroblasts and osteoblasts biocompatibility, and with desired mechanical properties. PEEK is currently used in a variety of biomedical orthopedic applications. A bioactive PEEK/HAp composite was fabricated with a unique configuration in which the HAp (bioactive phase) distribution was computer controlled within a PEEK matrix [105]. This novel process results in complete interconnectivity of the HAp network within a composite material, representing a superior advantage over alternative forms of the product. The technique combines free-forming extrusion, a type of AM, and compression molding. PEEK/HAp composites with a range of HAp were produced using static pressure loading to minimize air entrapment within the PEEK matrix. In addition, the technique can also be employed to produce porous PEEK structures with controlled pore size and distribution. A direct fabrication of 3D printed porous ceramic scaffolds for bone tissue engineering was made using HAp powder with designed internal architecture [92]. Cytotoxicity tests and cell proliferation studies confirmed that the scaffold showed good cell viability as well as good proliferation behavior.

Incorporation of PEG into the PLA facilitates the fabrication of scaffold printing process as well as plasticizing effect and also introduces structural and physicochemical changes to the resultant scaffolds. Plenty of information is

Table 10.6 3D printed materials with binders for bone tissue engineering.

Material	Layer thickness	Binder	References
TCP	20 μm	Aqueous based	[90]
α/β-TCP modified with 5 wt% hydroxypropymethylcellulose	100 mm	Water	[91]
CaP mixture with Ca/P ratio of 1.7	100 mm	10% phosphoric acid	[91]
Tetracalcium phosphate (TTCP), dicalcium phosphate and TCP	100 mm	25% citric acid	[91]
HA	300 mm	Schelofix (water-soluble polymeric compound)	[92]
TTCP/β-TCP	100 mm	25 wt% of citric acid	[93]
TTCP/calcium sulfate dihydrate	100 mm	25 wt% of citric acid	[93]
HA	100 mm	No information	[94]
TCP	100 mm	No information	[94]
Biphasic calcium phosphate (BCP)	100 mm	No information	[94]
α/β-TCP (final product: dicalcium phosphate dihydrate (DCPD))	No information	20% phosphoric acid	[95]
Starch/PLLA + PCL	No information	Distilled water + blue dye	[96]
HDPE	0.175 mm	Maltodextrin + polyvinyl alcohol (PVA)	[97]
SiO_2–ZnO-doped TCP	20 mm	Aqueous based	[29]
PE/HDPE	0.175 mm	Water-based binder	[98, 99]
PLA	No information	Chloroform	[100]
HA/maltodextrin	0.175	Water-based binder	[101]
TTCP (final product: HA)	100 mm	0.5 mol l^{-1} $Ca(H_2PO_4)_2$ + 10% H_3PO_4	[24]
TCP (final product: brushite)	100 mm	0.5 mol l^{-1} $Ca(H_2PO_4)_2$ + 10% H_3PO_4	[24]
HA/A-W glass	0.1 mm	Water based	[102]

available on the thermomechanical properties of PLA/PEG blends and block copolymer systems of PLA with PEG [106, 107]. However, the effect of PEG on the final structural, surface, and mechanical properties of 3D scaffolds processed by RP has not been explored well. Serra et al. [108] developed 3D rapid prototyping scaffolds using PLA/PEG blends (5, 10, and 20% w/w of PEG) and PLA/PEG/bioactive CaP glass composites. Surface analysis and DSC revealed a rearrangement of polymer chains, and topography, wettability, and elastic modulus increase of the studied surfaces as PEG was incorporated. Moreover, addition of 10% and 20% PEG led to nonuniform 3D structures with poorer mechanical properties. *In vitro* degradation studies showed that the inclusion of PEG significantly enhanced the degradation rate of the material. Results indicated that the presence of PEG not only improves PLA processing but also leads to relevant surface and geometrical and structural changes including modulation of the degradation rate of PLA-based 3D printed scaffolds.

The cytokine profile of human monocytes/macrophages in contact with biodegradable 3D scaffolds with different surface properties, architecture, and controlled pore geometry was fabricated by 3D printing technology [109]. Fabrication processes were optimized to create four different 3D platforms based on PLA, PLA/calcium phosphate glass (G5 molar composition: $44.5P_2O_5$–$44.5Ca_2O$–$6Na_2O$–$5TiO_2$), or chitosan. Although all scaffolds supported monocyte/macrophage adhesion and stimulated cytokine production, striking differences between PLA-based and chitosan scaffolds were found, with chitosan eliciting increased secretion of tumor necrosis factor (TNF)-a, whereas PLA-based scaffolds induced higher production of interleukin (IL)-6, IL-12/23, and IL-10. Even though the material itself induced the biggest differences, the scaffold geometry also impacted on TNF-a and IL-12/23 production, with chitosan scaffolds having larger pores and wider angles, leading to a higher secretion of these pro-inflammatory cytokines. These findings supported the appropriateness of these 3D platforms to study the modulation of macrophage responses by specific parameters.

The 3D customized antibiotic-incorporated scaffold made of PCL/ poly(lactic-*co*-glycolic acid) (PLGA)/tobramycin (TM) was fabricated using a 3DP system for the treatment of osteomyelitis and regeneration of debrided defect [110]. The TM powders were incorporated into the PCL/PLGA blend melt at 110 °C without heat denaturation (decomposition temperature of TM $\approx$ 224 °C). The blend (PCL/PLGA/TM) was transferred into steel syringe in dispensing head of a 3DP system and melted at 130 °C in the steel syringe for dispensing fibers using a multihead deposition system. Both disk diffusion and broth dilution tests confirmed the bactericidal activity of the PCL/PLGA/TM scaffold, which was maintained without heat denaturation. In *in vitro* experiments, DNA contents of seeded cells in the scaffold did not decrease during seven days of culture period, indicating that 3DP fabricated scaffold is not toxic to cells. Because PCL is hydrophobic with a long degradation time (two years), whereas PLGA is relatively hydrophilic with six-month degradation time with high water uptake ability, therefore, the release could be a longer period. The scaffolds further demonstrated the anti-inflammatory ability and antibacterial efficacy. In *in vivo* efficacy, the drug-loaded scaffolds were examined in a rat femur model for the treatment of chronic osteomyelitis at a period of eight weeks after the implantation. It was observed that the tissue swelling in the vicinity of the infected lesion was remarkably alleviated within four weeks after scaffold implantation, whereas new bone formation was obvious in PCL/PLGA/TM scaffold at week 8 after the implantation. The duration was not adequate to observe a complete bone regeneration in the infected sites. The obtained results indicate that the 3D printed PCL/PLGA/TM scaffold capable of eradicating osteomyelitis and regenerating bone tissue would be a promising solution as a carrier for the delivery of antibiotics in orthopedics.

10.7.1 3DP/4DP for Drug Delivery and Bioprinting

3D printed scaffolds are used for growth factor and drug delivery to improve bone growth in scaffolds. The localized delivery of growth factors and drugs has attracted significant attention because of the potential for dose reduction, controlled release pattern, and the insignificant side effects than the conventional delivery. Three different CaPs, namely, brushite, monetite, and HAp, have been fabricated using 3DP. Vancomycin hydrochloride, ofloxacin, and tetracycline hydrochloride were loaded onto these compositions via immersion/vacuum impregnation. Drug absorption is generally depended on the definite surface area and the release followed an exponential pattern. Additionally, drug immersion in a PLA/PGA matrix delayed the release [111]. The polymer incorporation in 3D-printed scaffolds could delay drug release kinetics and shifted the reaction rate from first to zero order. In addition, vancomycin, heparin, and rhBMP-2 incorporation during printing revealed a reduction in biological activity because of the degradation of drugs during spraying through the nozzles [112]. It is concluded that 3D printed scaffolds can be used in drug delivery.

The 3D macroporous gelatin methacrylamide constructs were fabricated using the bioplotter pneumatic dispensing system [113]. Photosensitive gelatin/gelatin methacrylamide was synthesized from gelatin (type B, bloom strength of 257) and two types of photoinitiators (PIs), namely, 1-[4-(2-hydroxyethoxy)-phenyl]-2-hydroxy-2-methyl-1-propane-1-one (Irgacure 2959, I2959) and 2,20-azobis [2-methyl-*N*-(2-hydroxyethyl)propionamide] (VA-086). It was observed that the scaffolds could be designed having a 100% interconnected pore network in the gelatin concentration range of 10–20 w/v%. The fabrication of cell-laden scaffolds was applied for tissue engineering by encapsulation of the hepatocarcinoma cell line (HepG2). Control over the deposited strand dimensions can be assured because of the physical properties of gelatin methacrylamide hydrogels and machine-operating parameters. The constructs with the desired stiffness and high shape reliability can be designed. The induced shear stress, curing irradiation dose, and the applied photoinitiator are influential for the resulting cell viability. High viability, >97%, constructs displaying a maintained expression of liver-specific functions were obtained using the VA-086 photoinitiator.

Novel biocompatible supramolecular polymers have been synthesized and tested for application in a 3D inkjet printer to generate features that represent the first steps toward supramolecular polymer-based hybrid scaffolds for regenerative medicine [114]. PCL-diol with a number-average molecular weight (M_n) of 2000 Da was attached to a series of hydrogen bonding moieties yielding the supramolecular polymer with desirable solubility characteristics. A benzyl terminated polymer showed the best-printed features concerning image resolution, and subsequently, 5–7% (w/v) silica particles were loaded into the polymer formulation, which successfully deposited. It was found that the polymer was biocompatible and nontoxic, and the cell attachment was not influenced by the addition of hydrogen bonding motifs to the biocompatible PCL. This proof-of-concept study has confirmed the use of polymer hybrid materials, which may be 3D printed to form biomedical scaffolds for regenerative medicine in the 3D printing of complex structures.

The FDM 3D printing has been extensively used to fabricate immediate release pharmaceutical tablets with several model drugs. One-step 3DP was employed to process paracetamol-loaded filaments into 3D printed tablets (printlets) incorporating up to 50% drug loading with two different infill percentages (20% and 100%) [115]. Hot melt extrusion (HME) was used to produce paracetamol-loaded filaments from three different grades of the pharmaceutical excipient hypromellose acetate succinate (HPMCAS), grades LG, MG, and HG. Micro-CT analysis indicated that printlets with 20% infill had cavities in the core compared to 100% infill and that the density of the 50% drug loading printlets was higher than the equivalent formulations loaded with 5% drug. In biorelevant bicarbonate dissolution media, drug release from the printlets was dependent on the polymer composition, drug loading, and the internal structure of the formulations. All HPMCAS-based printlets showed delayed drug release properties, and in the intestinal conditions, drug release was faster from the printlets prepared with polymers with a lower pH threshold. These results confirmed that FDM 3D printing could manufacture delayed release printlets without the need for an outer enteric coating. Furthermore, the release profile could be translated into the development of personalized dose medicines.

The impact of adding a nonmelting component to the methacrylic matrix to facilitate FDM 3D printing at different ratios using various pharmaceutical fillers on the production of the compatible filament was assessed [116]. The filament was considered compatible with the gears of the FDM 3D printer's head. Among the investigated fillers in this work, directly compressible lactose, spray-dried lactose, and microcrystalline cellulose showed a level of degradation at 135 °C, whereas talc and TCP allowed the consistent flow of the filament and a successful 3D printing of the tablet. Following the two thermal processes (HME and FDM 3D printing), drug contents were greater than 89% in all cases. XRPD indicated that a fraction of 5-ASA, theophylline, and prednisolone remained crystalline, whereas captopril was in the amorphous form. This unique approach provides a low-cost production method for on-demand manufacturing of individualized dosage forms.

Stents are important devices to expand the human vessels and has been the attention of many studies from different viewpoints. To overcome the difficulties of traditional fabrication methods, a 4D printed, thermoresponsive, SMP-based, cardiovascular stent was fabricated by Ge et al. [62] (Figure 10.5). Based on a high-resolution PμSL AM system, Ge et al. [62] manufactured the high-resolution stents with various diameters, heights, number of joints, ligament diameters, and inter-ligament angles. These stents also have reversible thermoresponsive shape-shifting properties after printing.

10.7.2 Polyurethane-Based Scaffolds for Tissue Engineering

Thermoplastic polyurethane (PU) is a highly elastic linear polymer composed of soft segments (e.g. flexible polyester or polyethers) and hard segments (e.g. diisocyanates with benzyl structure). PU displays good biocompatibility and excellent mechanical properties including good abrasion resistance, high elongation, and moderate tensile and compression strength [117]. These properties facilitate PU

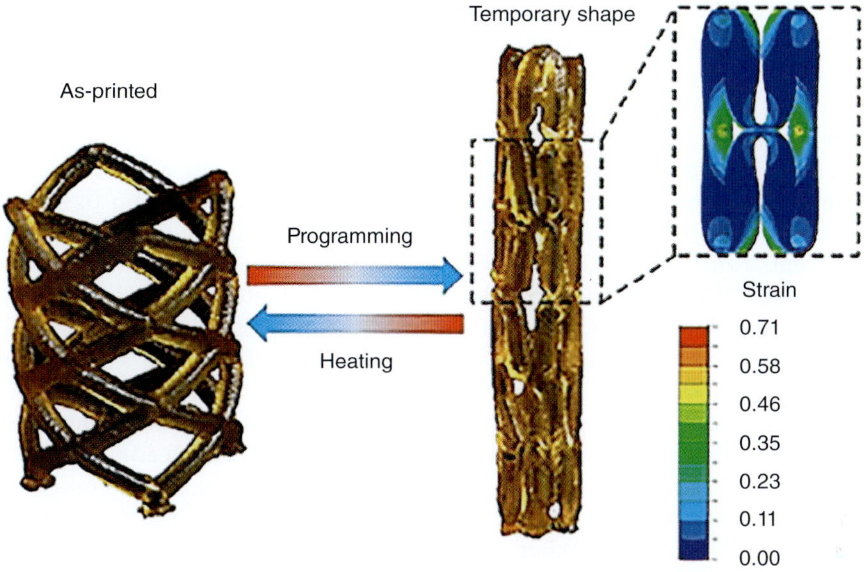

Figure 10.5 A 4D-printed, thermoresponsive stent, which is able to reversibly change its diameter and height [62]. Source: https://creativecommons.org/licenses/by/4.0/.

to be widely employed in many fields including coatings, foaming, adhesives, and tissue engineering. PU is one of the best synthetic polymers with excellent biocompatibility and is a good candidate for 3D printing materials for biomedical applications. Unfortunately, PU is mostly synthesized with the toxic organic solvents; as a result, there may be a chance of remaining solvent residue at the end product. It limits its application to combine cell printing.

Hung et al. [118] developed water-based 3D printing materials with controlled bioactivity for customized cartilage tissue engineering. The printing ink contains the water dispersion of synthetic biodegradable polyurethane (PU) elastic nanoparticles, hyaluronan, and bioactive ingredients TGFβ3 or a small-molecule drug Y27632 to replace TGFβ3. The PU elastomer as the feeding stock for 3D printing was synthesized from mixed soft segment of PCL diol and polyethylene butylene adipate diol (PEBA diol) in 2 : 3 molar ratio. The hard segment was isophorone diisocyanate (IPDI) and two-chain extenders were 2,2-bis(hydroxymethyl) propionic acid (DMPA) and ethylenediamine (EDA). The detailed synthetic processes are described elsewhere [111]. A self-developed low-temperature fused deposition manufacturing (LFDM) system (3DP) was employed with CAM/CAD technology equipped with an x–y–z motion platform, a nozzle with a heater, a cooling system, and pressure and temperature controllers. To prepare scaffolds with controlled release of chondrogenic induction factor, Y27632 or TGFβ3 was embedded in PU/HAp scaffolds. For embedding Y27632 in scaffolds, PU dispersion (30 wt%) was blended with different volumes of HA solution (2 wt%) and Y27632 solution, so the final scaffolds contained 5–25 ppm of Y27632. For embedding TGFβ in scaffolds, PU dispersion was blended with HA solution and TGFβ3 solution.

PU/HAp scaffolds of various shapes and dimensions were made from the process and even a large-size 3D printed scaffold ($3.5 \times 3.5\,cm^2$ with 1.5 cm heights) was successfully fabricated. After printing, the scaffold was resistant to water and the organic solvent. The cyclic loading on PU/HAp scaffolds as measured by a DMA exhibited an excellent elasticity ($E' = 0.33\,MPa$). These scaffolds improve the self-aggregation of mesenchymal stem cells (MSCs) and, also a timely release of the bioactive ingredients, induce the chondrogenic differentiation of MSCs and produce the matrix for cartilage repair. Rabbit knee implantation supports the potential of the novel 3DP scaffolds in cartilage regeneration. The rabbit MSC-seeded 3DP PU/HAp/Y scaffolds were implanted into the chondral defects of rabbit knees for evaluation *in vivo*. Based on the histological staining of GAG (safranin O) after one month, PU/HAp/Y scaffolds seeded with MSCs had better regeneration effect than PU/HAp seeded with MSCs or PLGA scaffolds seeded with MSCs. Authors advocated that the 3DP composite scaffolds with controlled release bioactivity may have potential in customized tissue engineering.

PU shows poor shape fixity and poor mechanical strength, and, therefore, it requires to incorporate/blend with compatible materials. A blend of PU/PLA could improve the overall mechanical strength, impact resistance, and also shape memory property. Furthermore, a loading of a trace amount of graphene oxide (GO), which is a 2D single layer of sp^2 hybridized carbon atoms into polymer matrix, improves the composite material properties tremendously. Chen et al. [119] successfully employed the FDM 3DP of TPU/PLA/GO nanocomposites and its potential application as biocompatible materials. Nanocomposites were prepared by solvent-based mixing process and extruded into filaments for FDM printing. The addition of GO has significantly improved the mechanical properties of the polymer matrix, 167% in compression modulus and 75.5% in tensile modulus. The printing orientation leads to different mechanical responses because of the weak adhesion strength between layers during 3DP. Thermal stability has also been improved, with 90 °C increase in degradation temperature as well as the new formation of better crystalline structures. Additionally, the 3D printed nanocomposites exhibit good biocompatibility with NIH3T3 cells, indicating promise as biomaterials scaffold for tissue engineering applications.

Nanocomposites can be easily printed into complex shapes with high quality. FTIR and SEM images reveal good dispersion of GO in polymer matrix. Cell culturing results reveal excellent cell viability of 3D printed scaffolds, indicating that the limited addition of GO has no obvious toxicity to cell growth, and a small amount of GO is beneficial for cell proliferation. Based on the above results, the 3D printed TPU/PLA/GO nanocomposite exhibits excellent mechanical properties, thermal stabilities, and cell viability, which allows it to be widely applied in many fields, especially as a good potential candidate in tissue engineering scaffolds.

A dual-crosslinking HA system was fabricated as a printable hydrogel ink, which exhibited both shear-thinning and self-healing behaviors via guest–host bonding, as well as covalent cross-linking for stabilization using photopolymerization [120]. The dual-crosslinking hydrogel filaments formed structures with more than 16 layers that were stable over a month with a marginal loss in mechanical properties, and the printed filament size ranged from 100 to

500 μm, depending on the printing parameters. The printed structures were functionalized to support cell adhesion.

10.8 Future and Prospects

The materials used for 3D and 4D printing intended for biomedical applications have been described in this chapter. Metallic, polymer-based composite materials are predominantly used for the fabrication of the construct. Those developing materials to be utilized for 3D printing should take into account variety, composition, strength, and finishing procedures in order to increase the versatility of the technology. Presently, the variety of materials is limited; it is either powder based or has low enough viscosities to be extruded from the printing head. Most of the manufacturers recommend proprietary materials to be used in their 3D printers. As a result, the material pool becomes limited for 3D printing especially for a newly developed printer, which immensely requires growing, the quantity and diversity of materials must increase. Understanding the rheological behavior of the new materials in the 3DP is very crucial for a smooth operation of the printer. Similarly, the tensile and thermal properties of the materials provide strength and applicability for biomedical applications. 3DP can fabricate scaffolds with defined shapes with newly developed materials, with controlled and interconnected porous structures. The most critical issue in ceramics for implementing 3DP is the mechanical strengths of porous scaffolds because an increase in the porosity decreases the strength of the scaffolds significantly. Additionally, the brittleness makes these scaffolds more difficult to handle during processing. Resorbable polymer penetration and the use of resorbable glassy materials could improve the mechanical strength and toughness of these scaffolds viewed, and the smartness of the materials are imparted by a careful selection of stimuli.

A wide difference in the glass transition temperature, T_g, of the participating materials in the composite produces the smart materials. Bioprinting is now a reality; although getting suitable materials for fabrication; it remains a challenge for real applications. Well-designed hydrogels have potentials to work as inks for bioprinting. Currently, biomedical and pharmaceutical companies are in a position to create more specific drugs, enabling the generation and transplantation of several tissues, including multilayered skin, bone, vascular grafts, tracheal splints, heart tissues, and cartilaginous structures, and changing the way that doctors and surgeons plan procedures. However, the development and translation of multifunctional smart materials for 3D and 4D bioprinting into tissue engineering and regenerative medicine should be based on studying the relevant interdisciplinary literature thoroughly and thereafter implementing the major design parameters into the print of an architecture and morphology, which allows cell migration, proliferation, and subsequently vascularized tissue formation and remodeling [121].

As current techniques are more fine-tuned and more bioink materials become available, the design of effective ECM-like scaffolds becomes progressively possible. The emphasis of 3D printing techniques in medical science to date has mostly been aimed at regenerating or replacing tissue *in vivo*; however, alternative approaches also being investigated include the printing of functional tissues *in vitro*. Demand for processes such as 3DP will increase in the coming years because of their ability to make custom medical devices that can be tailored to patient-specific and defect-specific clinical needs. Extensive process–property optimization is still needed to accomplish this goal. Finally, printing live cells or adding growth factors/drugs is another fascinating area of growth.

References

1 Kang, H.S., Lee, J.Y., Choi, S. et al. (2016). Smart manufacturing: past research, present findings, and future directions. *International Journal of Precision Engineering and Manufacturing-Green Technology* 3 (1): 111–128.

2 Khare, V., Sonkaria, S., Lee, G.-Y. et al. (2017). From 3D to 4D printing – design, material and fabrication for multi-functional multi-materials. *International Journal of Precision Engineering and Manufacturing-Green Technology* 4 (3): 291–299.

3 Lu, B., Li, D., and Tian, X. (2015). Development trends in additive manufacturing and 3D printing. *Engineering* 1 (1): 85–89.

4 Wong, K.V. and Hernandez, A. (2012). A review of additive manufacturing. *ISRN Mechanical Engineering* 10. doi: 10.5402/2012/208760.

5 (a) 3D Systems. Solutions. http://www.3dsystems.com/solutions/overview (accessed 30 November 2017); (b) Bak, D. (2003). Rapid prototyping or rapid production? 3D printing processes move industry towards the latter. *Assembly Automation* 23: 340–345.

6 Campbell, T.A., Tibbits, S., and Garrett, B. (2016). *The Next Wave: 4D Printing Programming the Material World*. Atlantic Council http://www.atlanticcouncil.org/images/publications/The_Next_Wave_4D_Printing_Programming_the_Material_World.pdf (accessed 4 January 2016.

7 Yang, W.G., Lu, H., Huang, W.M. et al. (2014). Advanced shape memory technology to reshape product design, manufacturing and recycling. *Polymers* 6 (8): 2287–2308.

8 Lee, Y., Lee, H., Hwang, T. et al. (2015). Sequential folding using light-activated polystyrene sheet. *Scientific Reports* 5: doi: https://doi.org/10.1038/srep16544.

9 Tibbits, S. (2014). 4D printing: multi-material shape change. *Architectural Design* 84 (1): 116–121.

10 Gross, B.C., Erkal, J.L., Lockwood, S.Y. et al. (2014). Evaluation of 3D printing and its potential impact on biotechnology and the chemical sciences. *Analytical Chemistry* 86 (7): 3240.

11 Matsumoto, K., Ishiduka, T., Yamada, H. et al. (2014). Clinical use of three dimensional models of the temporomandibular joint established by rapid prototyping based on cone-beam computed tomography imaging data. *Oral Radiology* 30: 98.

12 Cui, H., Nowicki, M., Fisher, J.P., and Zhang, L.G. (2017). 3D bioprinting for organ regeneration. *Advanced Healthcare Materials* 6: 1601118.

13 Rengier, F., Mehndiratta, A., von Tengg-Kobligk, H. et al. (2010). 3D printing based on imaging data: review of medical applications. *International Journal of Computer Assisted Radiology and Surgery* 5: 335–341.

14 Do, A.-V., Khorsand, B., Geary, S.M., and Salem, A.K. (2015). 3D printing of scaffolds for tissue regeneration applications. *Advanced Healthcare Materials* 4: 1742–1762.

15 Xie, Y.M. (2014). Designing orthotropic materials for negative or zero compressibility. *International Journal of Solids and Structures* 51 (23–24): 4038–4051.

16 Sweeney, M., Campbell, L.L., Hanson, J. et al. (2017). Characterizing the feasibility of processing wet granular materials to improve rheology for 3D printing. *Journal of Materials Science* 52: 13040–13053.

17 Seitz, H., Rieder, W., Irsen, S. et al. (2005). Three-dimensional printing of porous ceramic scaffolds for bone tissue engineering. *Journal of Biomedical Materials Research Part B: Applied Biomaterials* 74: 782–788.

18 Huebshch, N. and Mooney, D.J. (2010). Inspiration and application in the evolution of biomaterials. *Nature* 462: 426–432.

19 Vagkopoulou, T., Koutayas, S.O., Koidis, P., and Strub, J.R. (2009). Zirconiain dentistry: Part 1. Discovering the nature of an upcoming bioceramic. *European Journal of Esthetic Dentistry* 4 (2): 135–161.

20 Chang, C.-H., Lin, C.-Y., Liu, F.-H. et al. (2015). 3D printing bioceramic porous scaffolds with good mechanical property and cell affinity. *PLoS ONE* 10 (11): 143713.

21 Yang, L.Z., Oliveira, S.G., and Stuckera, B. (2013). Development of a 3D printing method for production of dental application. In: *Proceedings of the 24th International Solid Freeform Fabrication Symposium*, 346–353.

22 Yoo, J., Cima, M.J., Khanuja, S., and Sachs, E.M. (1993). Structural ceramic components by 3D printing. In: *Solid Freeform Fabrication Symposium*, 40–50. DTIC Document.

23 Spath, S., Drescher, P., and Seitz, H. (2015). Impact of particle size of ceramic granule blends on mechanical strength and porosity of 3D printed scaffolds. *Materials* 8 (8): 4720–4732.

24 Gbureck, U., Hölzel, T., Klammert, U. et al. (2007). Resorbable dicalcium phosphate bone substitutes prepared by 3D powder printing. *Advanced Functional Materials* 17 (18): 3940–3945.

25 Santos, C.F., Silva, A.P., Lopes, L. et al. (2012). Design and production of sintered b-tricalcium phosphate 3D scaffolds for bone tissue regeneration. *Materials Science and Engineering C* 32 (5): 1293–1298.

26 Wu, C., Luo, Y., Cuniberti, G. et al. (2011). Three-dimensional printing of hierarchical and tough mesoporous bioactive glass scaffolds with a controllable pore architecture, excellent mechanical strength and mineralization ability. *Acta Biomaterials* 7 (6): 2644–2650.

27 Shao, H., He, Y., Fu, J. et al. (2016). 3D printing magnesium-doped wollastonite/β-TCP bioceramics scaffolds with high strength and adjustable degradation. *Journal of the European Ceramic Society* 36 (6): 1495–1503.

28 Ebert, J., Özkol, E., Zeichner, A. et al. (2009). Direct inkjet printing of dental prostheses made of zirconia. *Journal of Dental Research* 88 (7): 673–676.

29 Fielding, G.A., Bandyopadhyay, A., and Bose, S. (2012). Effects of SiO_2 and ZnO doping on mechanical and biological properties of 3D printed TCP scaffolds. *Dental Materials* 28 (2): 113–122.

30 Eckel, Z.C., Zhou, C., Martin, J.H. et al. (2016). Additive manufacturing of polymer-derived ceramics. *Science* 351 (6268): 58–62.

31 Varghese, G., Moral, M., Castro-García, M. et al. (2017). Fabrication and characterisation of ceramics via low-cost DLP 3D printing. *Boletín de la Sociedad Española de Cerámica y Vidrio* 57 (1): 9–18. doi: https://doi.org/10.1016/j.bsecv.2017.09.004.

32 Wang, X., Jiang, M., Zhou, Z. et al. (2017). 3D printing of polymer matrix composites: a review and prospective. *Composites Part B: Engineering* 110: 442–458.

33 Nikzad, M., Masood, S., and Sbarski, I. (2011). Thermo-mechanical properties of a highly filled polymeric composites for fused deposition modeling. *Materials & Design* 32 (6): 3448e56.

34 Hwang, S., Reyes, E.I., K-S, M. et al. (2015). Thermo-mechanical characterization of metal/polymer composite filaments and printing parameter study for fused deposition modeling in the 3D printing process. *Journal of Electronic Materials* 44 (3): 771–777.

35 Isakov, D., Lei, Q., Castles, F. et al. (2016). 3D printed anisotropic dielectric composite with meta-material features. *Materials & Design* 93: 423–430.

36 Castles, F., Isakov, D., Lui, A. et al. (2016). Microwave dielectric characterisation of 3D-printed $BaTiO_3$/ABS polymer composites. *Scientific Reports* 6 (1): doi: https://doi.org/10.1038/srep22714.

37 Perez, A.R.T., Roberson, D.A., and Wicker, R.B. (2014). Fracture surface analysis of 3D-printed tensile specimens of novel ABS-based materials. *Journal of Failure Analysis and Prevention* 14 (3): 343–353.

38 Boparai, K., Singh, R., and Singh, H. (2015). Comparison of tribological behaviour for Nylon6–Al–Al_2O_3 and ABS parts fabricated by fused deposition modelling: this paper reports a low cost composite material that is more wear-resistant than conventional ABS. *Virtual and Physical Prototyping* 10 (2): 59–66.

39 Kurimoto, M., Yamashita, Y., Ozaki, H. et al. (2015). 3D printing of conical insulating spacer using alumina/UV-cured-resin composite. In: *2015 IEEE Conference on Electrical Insulation and Dielectric Phenomena (CEIDP)*. IEEE.

40 Martin, J.J., Fiore, B.E., and Erb, R.M.D. (2015). Designing bioinspired composite reinforcement architectures via 3D magnetic printing. *Nature Communications* 6 (1): doi: https://doi.org/10.1038/ncomms9641.

41 Kokkinis, D., Schaffner, M., and Studart, A.R. (2015). Multimaterial magnetically assisted 3D printing of composite materials. *Nature Communications* 6: (Art. No.: 8643).

42 Kalsoom, U., Peristyy, A., Nesterenko, P., and Paull, B. (2016). A 3D printable diamond polymer composite: a novel material for fabrication of low cost thermally conducting devices. *RSC Advances* 6 (44): 38140e7.

43 Shemelya, C.M., Rivera, A., Perez, A.T. et al. (2015). Mechanical, electromagnetic, and X-ray shielding characterization of a 3D printable tungsten

polycarbonate polymer matrix composite for space-based applications. *Journal of Electronic Materials* 44 (8): 2598–2607.
44 Chung, H. and Das, S. (2006). Processing and properties of glass bead particulate-filled functionally graded Nylon-11 composites produced by selective laser sintering. *Materials Science and Engineering A* 437 (2): 226–234.
45 Ahmed, J., Jhang, X.J., Song, J., and Varhney, S.K. (2009). Thermal characteristics of polylactides. *Journal of Thermal Analysis and Calorimetry* 95 (3): 957–964.
46 Ahmed, J. and Varhney, S.K. (2011). Polylactides-chemistry, properties and green packaging – a review. *International Journal of Food Properties* 14: 37–58.
47 Schlesinger, E., Ciaccio, N., and Desai, T.A. (2015). Polycaprolactone thin-film drug delivery systems: empirical and predictive models for device design. *Materials Science and Engineering C* 57: 232–239.
48 Nair, B.P., Sindhu, M., and Nair, P.D. (2016). Polycaprolactone–laponite composite scaffold releasing strontium ranelate for bone tissue engineering applications. *Colloids and Surfaces B: Biointerfaces* 143: 423–430.
49 Ahmed, J., Auras, R., Kijchavengkul, T., and Varhney, S.K. (2012). Rheological, thermal and structural behavior of poly(ε-caprolactone) and nanoclay blended films. *Journal of Food Engineering* 111: 580–589.
50 Burdick, J.A., Chung, C., Jia, X. et al. (2005). Controlled degradation and mechanical behavior of photopolymerized hyaluronic acid networks. *Biomacromolecules* 6 (1): 386–391.
51 Schuurman, W., Levett, P.A., Pot, M.W. et al. (2013). Gelatin-methacrylamide hydrogels as potential biomaterials for fabrication of tissue-engineered cartilage constructs. *Macromolecular Bioscience* 13 (5): 551–561.
52 Skardal, A., Devarasetty, M., Kang, H.W. et al. (2015). A hydrogel bioink toolkit for mimicking native tissue biochemical and mechanical properties in bioprinted tissue constructs. *Acta Biomaterials* 25: 24–34.
53 Gladman, A.S., Matsumoto, E.A., Nuzzo, R.G. et al. (2016). Biomimetic 4D printing. *Nature Materials* 15 (4): 413–418.
54 Ge, Q., Qi, H.J., and Dunn, M.L. (2013). Active materials by four-dimension printing. *Applied Physics Letters* 103 (13): 131901.
55 Ge, Q., Dunn, C.K., Qi, H.J., and Dunn, M.L. (2014). Active origami by 4D printing. *Smart Materials and Structures* 23 (9): 094007.
56 Wu, J., Yuan, C., Ding, Z. et al. (2016). Multi-shape active composites by 3D printing of digital shape memory polymers. *Scientific Reports* 6 (1): doi: https://doi.org/10.1038/srep24224.
57 Mao, Y., Yu, K., Isakov, M.S. et al. (2015). Sequential self-folding structures by 3D printed digital shape memory polymers. *Scientific Reports* 5: 13616.
58 Miao, S., Zhu, W., Castro, N.J. et al. (2016). 4D printing smart biomedical scaffolds with novel soybean oil epoxidized acrylate. *Scientific Reports* 6: 27226.
59 Miao, S., Zhu, W., Castro, N.J. et al. (2016). Four-dimensional printing hierarchy scaffolds with highly biocompatible smart polymers for tissue engineering applications. *Tissue Engineering Part C: Methods* 22 (10): 952–963.

60 Bakarich, S.E., Gorkin, R., and Spinks, G.M. (2015). 4D printing with mechanically robust, thermally actuating hydrogels. *Macromolecular Rapid Communications* 36 (12): 1211–1217.
61 Zhang, Q., Zhang, K., and Hu, G. (2016). Smart three-dimensional lightweight structure triggered from a thin composite sheet via 3D printing technique. *Scientific Reports* 6: 22431.
62 Ge, Q., Sakhaei, A.H., Lee, H. et al. (2016). Multimaterial 4D printing with tailorable shape memory polymers. *Scientific Reports* 6: 31110.
63 Raviv, D., Zhao, W., McKnelly, C. et al. (2014). Active printed materials for complex self-evolving deformations. *Scientific Reports* 4: 7422.
64 Le Duigou, A., Castro, M., Bevan, R., and Martin, N. (2016). 3D printing of wood fibre biocomposites: from mechanical to actuation functionality. *Materials & Design* 96: 106–114.
65 Naficy, S., Gately, R., Gorkin, R. et al. (2017). 4D printing of reversible shape morphing hydrogel structures. *Macromolecular Materials and Engineering* 302 (1): doi: https://doi.org/10.1002/mame.201600212.
66 Au, A.K., Bhattacharjee, N., Horowitz, L.F. et al. (2012, 2015). 3D-printed microfluidic automation. *Lab on a Chip* 15 (8): 1934–1941.
67 Skardal, A., Mack, D., Kapetanovic, E. et al. (2012). Bioprinted amniotic fluid-derived stem cells accelerate healing of large skin wounds. *Stem Cells Translational Medicine* 1: 792.
68 Xu, T., Zhao, W., Zhu, J.-M. et al. (2013). Complex heterogeneous tissue constructs containing multiple cell types prepared by inkjet printing technology. *Biomaterials* 34: 130.
69 Murphy, S.V. and Atala, A. (2014). 3D bioprinting of tissues and organs. *Nature Biotechnology* 32: 773.
70 Hölzl, K., Lin, S., Tytgat, L. et al. (2016). Bioink properties before, during and after 3D bioprinting. *Biofabrication* 8: 032002.
71 Shin, D.-G., Kim, T.-H., and Kim, D.-E. Review of 4D printing materials and their properties. *International Journal of Precision Engineering and Manufacturing-Green Technology* 4 (3): 349–357.
72 Huang, W.M., Ding, Z., Wang, C.C. et al. (2010). Shape memory materials. *Materials Today* 13: 7, 54–8, 61.
73 Otsuka, K. and Wayman, C.M. (1998). *Materials Science Forum*. Cambridge University Press.
74 Sun, L., Huang, W.M., Ding, Z. et al. (2012). Stimulus-responsive shape memory materials: a review. *Materials & Design* 33: 577–640.
75 Xie, T., Xiao, X., and Cheng, Y.T. (2009). Revealing triple-shape memory effect by polymer bilayers. *Macromolecular Rapid Communications* 30: 1823–1827.
76 Zhang, Q., Yan, D., Zhang, K., and Hu, G. (2015). Pattern transformation of heat-shrinkable polymer by three-dimensional (3D) printing technique. *Scientific Reports* 5 (1): 8936.
77 Gu, X. and Mather, P.T. (2012). Entanglement-based shape memory polyurethanes: synthesis and characterization. *Polymer* 53 (25): 5924–5934.

78 Ferris, C.J., Gilmore, K.J., Wallace, G.G., and in het Panhuis, M. (2013). Modified gellan gum hydrogels for tissue engineering applications. *Soft Matter* 9 (14): 3705–3711.
79 Hou, S., Wang, X., Park, S. et al. (2015). Rapid self-integrating, injectable hydrogel for tissue complex regeneration. *Advanced Healthcare Materials* 4 (10): 1491.
80 Taylor, D.L. and in het Panhuis, M. (2016, 2016). Self-healing hydrogels. *Advanced Materials* 28 (41): 9060–9093.
81 Liu, S. and Li, L. (2017). Ultrastretchable and self-healing double-network hydrogel for 3D printing and strain sensor. *ACS Applied Materials & Interfaces* 9: 26429–26437.
82 Yang, F., Tadepalli, V., and Benjamin, J. (2017). 3D printing of a double network hydrogel with a compression strength and elastic modulus greater than those of cartilage. *ACS Biomaterials Science & Engineering* 3: 863–869.
83 Mohanty, S., Alm, M., Hemmingsen, M. et al. (2016). 3D printed silicone–hydrogel scaffold with enhanced physicochemical properties. *Biomacromolecules* 17: 1321––1329.
84 He, Y., Yang, F.F., Zhao, H.M. et al. (2016). Research on the printability of hydrogels in 3D bioprinting. *Scientific Reports* 6: (Art. No: 29977).
85 Rezwan, K., Chen, Q.Z., Blaker, J.J., and Boccaccini, A.R. (2006). Biodegradable and bioactive porous polymer/inorganic composite scaffolds for bone tissue engineering. *Biomaterials* 27: 3413–3431.
86 Müller, A.M., Davenport, M., Verrier, S. et al. (2009). Platelet lysate as a serum substitute for 2D static and 3D perfusion culture of stromal vascular fraction cells from human adipose tissue. *Tissue Engineering Part A* 15: 869–875.
87 Cuodeau, A., Sachs, E., and Cadarise, S. (2000). Design and fabrication of cast orthopedic implants with freeform surface textures from 3-D printed ceramic shell. *Journal of Biomedical Materials Research* 53: 525–535.
88 Dorozhkin, S.V. (2011). Calcium orthophosphates: occurrence, properties, biomineralization, pathological calcification and biomimetic applications. *Biomaterials* 1–2: 121–164.
89 Dorozhkin, S.V. (2009). Nanodimensional and nanocrystalline apatites and other calcium orthophosphates in biomedical engineering, biology and medicine. *Materials* 2: 1975–2045.
90 Tarafder, S., Balla, V.K., Davies, N.M. et al. (2012). Microwave sintered 3D printed tricalcium phosphate scaffolds for bone tissue engineering. *Journal of Tissue Engineering and Regenerative Medicine* 7: 631–641.
91 Vorndran, E., Klarner, M., Klammert, U. et al. (2008, 2008). 3D powder printing of β-tricalcium phosphate ceramics using different strategies. *Advanced Engineering Materials* 10: B67–B71.
92 Leukers, B., Gulkan, H., Irsen, S.H. et al. (2005). Biocompatibility of ceramic scaffolds for bone replacement made by 3D printing. *Materialwissenschaft und Werkstofftechnik* 36 (12): 781–787.
93 Khalyfa, A., Vogt, S., Weisser, J. et al. (2007). Development of a new calcium phosphate powder-binder system for the 3D printing of patient specific

implants. *Journal of Materials Science: Materials in Medicine* 18: 909–916. doi: 10.1007/s10856-006-0073-2.
94 Detsch, R., Schaefer, S., Deisinger, U. et al. (2011). In vitro-osteoclastic activity studies on surfaces of 3D printed calcium phosphate scaffolds. *Journal of Biomaterials Applications* 26: 359–380.
95 Klammert, U., Gbureck, U., Vorndran, E. et al. (2010). 3D powder printed calcium phosphate implants for reconstruction of cranial and maxillofacial defects. *Journal of Cranio-Maxillofacial Surgery* 38: 565–570.
96 Lam, C.X.F., Mo, X.M., Teoh, S.H., and Hutmacher, D.W. (2002). Scaffold development using 3D printing with a starch-based polymer. *Materials Science and Engineering C* 20: 49–56.
97 Suwanprateeb, J., Thammarakcharoen, F., Wongsuvan, V., and Chokevivat, W. (2012). Development of porous powder printed high density polyethylene for personalized bone implants. *Journal of Porous Materials* 19 (5): 623–632.
98 Suwanprateeb, J., Suvannapruk, W., Wasoontararat, K. et al. (2011). Preparation and comparative study of a new porous polyethylene ocular implant using powder printing technology. *Journal of Bioactive and Compatable Polymers* 26: 317–331.
99 Suwanprateeb, J., Kerdsook, S., Boonsiri, T., and Pratumpong, P. (2011). Evaluation of heat treatment regimes and their influences on the properties of powder-printed high-density polyethylene bone implant. *Polymer International* 60: 758–764.
100 Giordano, R.A., Wu, B.M., Borland, S.W. et al. (1996). Mechanical properties of dense polylactic acid structures fabricated by three dimensional printing. *Journal of Biomaterials Science Polymer Edition* 8: 63–75.
101 Suwanprateeb, J. and Chumnanklang, R. (2006). Three-dimensional printing of porous polyethylene structure using water-based binders. *Journal of Biomedical Materials Research Part B: Applied Biomaterials* 78B: 138–145.
102 Suwanprateeb, J., Sanngam, R., Suvannapruk, W., and Panyathanmaporn, T. (2009). Mechanical and in vitro performance of apatite-wollastonite glass ceramic reinforced hydroxyapatite composite fabricated by 3D-printing. *Journal of Materials Science Materials in Medicine* 20: 1281–1289.
103 Salgado, A.J., Coutinho, O.P., and Reis, R.L. (2004). Novel starch-based scaffolds for bone tissue engineering: cytotoxicity, cell culture, and protein expression. *Tissue Engineering* 10: 465–474.
104 Becker, S.T., Bolte, H., Schünemann, K. et al. (2012). Endocultivation: the influence of delayed vs. simultaneous application of BMP-2 onto individually formed hydroxyapatite matrices for heterotopic bone induction. *International Journal of Oral and Maxillofacial Surgery* 41: 1153–1160.
105 Vaezi, M. and Yang, S.F. (2015). A novel bioactive PEEK/HA composite with controlled 3D interconnected HA network. *International Journal of Bioprinting* 1 (1): 66–76.
106 Ahmed, J., Varhney, S.K., Auras, R., and Hwang, S.W. (2010). Thermal and rheological properties of L-polylactide/polyethylene glycol/silicate nanocomposites films. *Journal of Food Science* 75: N97–N108.
107 Ahmed, J., Hiremath, N., and Jacob, H. (2016). Antimicrobial, rheological and thermal properties of plasticized polylactide films incorporated with

essential oils to inhibit *Staphylococcus aureus* and *Campylobacter jejuni*. *Journal of Food Science* 81: E419–E429.
108 Serra, T., Ortiz-Hernandez, M., Engel, E. et al. (2014). Relevance of PEG in PLA-based blends for tissue engineering 3D-printed scaffolds. *Materials Science and Engineering C* 38: 55–62.
109 Almeida, C.R., Serra, T., Oliveira, M.I. et al. (2014). Impact of 3-D printed PLA- and chitosan-based scaffolds on human monocyte/macrophage responses: unraveling the effect of 3-D structures on inflammation. *Acta Biomaterialia* 10: 613–622.
110 Shim, J.-H., Kim, M.-J., Park, J.Y. et al. (2015). Three-dimensional printing of antibiotics-loaded poly-ε-caprolactone/poly(lactic-*co*-glycolic acid) scaffolds for treatment of chronic osteomyelitis. *Tissue Engineering and Regenerative Medicine* 2 (5): 283–293.
111 Hsu, S.H., Hung, K.C., Lin, Y.Y. et al. (2014). Water-based synthesis and processing of novel biodegradable elastomers for medical applications. *Journal of Materials Chemistry B* 2: 5083–5092.
112 Vorndran, E., Klammert, U., Ewald, A. et al. (2010). Simultaneous immobilization of bioactives during 3D powder printing of bioceramic drug-release matrices. *Advanced Functional Materials* 20: 1585–2159.
113 Billiet, T., Gevaert, E., De Schryver, T. et al. (2014). The 3D printing of gelatin methacrylamide cell-laden tissue-engineered constructs with high cell viability. *Biomaterials* 35: 49–62.
114 Hart, L.R., Li, S., Sturgess, C. et al. (2016). 3D printing of biocompatible supramolecular polymers and their composites. *ACS Applied Materials & Interfaces* 8: 3115–3122.
115 Goyanes, A., Fina, F., Martorana, A. et al. (2017). Development of modified release 3D printed tablets (printlets) with pharmaceutical excipients using additive manufacturing. *International Journal of Pharmaceutics* 527: 21–30.
116 Muzna, S., Sosnicka, A., Arafat, B. et al. (2016). Adaptation of pharmaceutical excipients to FDM 3D printing for the fabrication of patient-tailored immediate release tablets. *International Journal of Pharmaceutics* 513: 659–668.
117 De Leon, A.C., Chen, Q., Palaganas, N.B. et al. (2016). High performance polymer nanocomposites for additive manufacturing applications. *Reactive and Functional Polymers* 103: 141–155.
118 Hung, K.-C., Tseng, C.-S., Dai, L.-G., and S-h, H. (2016). Water-based polyurethane 3D printed scaffolds with controlled release function for customized cartilage tissue engineering. *Biomaterials* 83: 156–168.
119 Chen, Q., Mangadlao, J.D., Wallat, J. et al. (2017). 3D printing biocompatible polyurethane/poly(lactic acid)/graphene oxide nanocomposites: anisotropic properties. *ACS Applied Materials & Interfaces* 9: 4015–4023.
120 Ouyang, L., Highley, C.B., Rodell, C.B. et al. (2016). 3D printing of shear-thinning hyaluronic acid hydrogels with secondary crosslinking. *ACS Biomaterials Science & Engineering* 2 (10): 1743–1751.
121 Castro, N.J., Meinert, C., Levett, P., and Hutmacher, D.W. (2017). Current developments in multifunctional smart materials for 3D/4D bioprinting. *Current Opinion in Biomedical Engineering* 2: 67–75.

11

Personalized Polypills Produced by Fused Deposition Modeling 3D Printing

Sheng Qi, Jehad Nasereddin, and Fahad Alqahtani

University of East Anglia, School of Pharmacy, Earlham Road, Norwich, NR4 7TJ, UK

11.1 Introduction

Oral medications such as tablets and capsules have been well recognized as the most convenient way of administering drugs, offering the best patient adherence because of their noninvasive and convenient nature. A "one-size-fits-all" approach has been the general principle behind the design of most modern oral medicines. Fixed dose medications can be rapidly mass-produced at low cost using processes such as tableting and capsule filling. Such manufacturing technologies are well understood and with established regulatory pathways [1]. However, the drawbacks of this approach become apparent when patients need frequent dosing modifications in order to manage the side effects and when patients with chronic conditions are overloaded with taking multiple oral medications multiple times a day. The limited numbers of oral fixed dose combination (FDC) products that exist today on the market have demonstrated the significant advantages for improving the treatment adherence rates [2]. The lack of flexibility in dosing and the combination of drugs available still make them "one-size-fits-all." With the aging global population, there is an increase in the proportion of population with multiple disease conditions. The large pill burden these patients face lowers their adherence to the treatment. Poor adherence leads to poor treatment outcomes and high readmission rate and is associated with social and economic pressures on individuals, their families, and the health care provider.

Personalized medicine in the literature commonly refers to using patient's genetic information to enable therapeutic decisions tailored to an individual patient. Recently, the term has also been applied to produce a personalized pill containing all polypharmacy medications in a single unit dose in a physical form that can be easily administered to patients who require the combination of drugs and their doses to be suitable for each individual patient's disease types and progression. However, it should be stressed that such personalized complex products have their unique clinical value in treating the patient groups who have severe pill burden issues and are not to replace the existing highly cost-effective normal oral medicines. Because of the highly individualized

3D and 4D Printing in Biomedical Applications: Process Engineering and Additive Manufacturing,
First Edition. Edited by Mohammed Maniruzzaman.

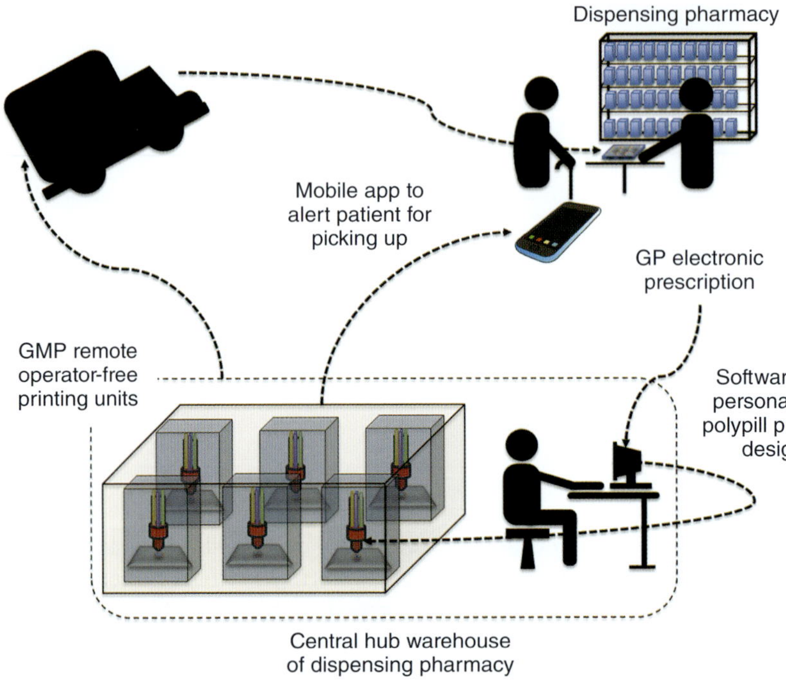

Figure 11.1 Illustration of an example of personalized polypills produced and dispensed by FDM 3D printing at the point of care.

nature of such products, their fabrication at the point of care, for example, in hospitals and pharmacies, would be more practical than the current model of long chain of manufacturing and distribution, as illustrated in Figure 11.1. In this chapter, such personalized pills containing more than one drug are referred as polypills, which is a term exclusively used in the treatment of cardiovascular diseases (CVD) in the literature, but the usage is expanded here to cover a range of pills that combine many medicines to provide a single solid dosage form that allows the patients to self-administer easily. 3D printing is a range of fabrication technologies that involve layer-by-layer building of materials into a three-dimensional geometry [3]. Recently, a number of excellent review articles on 3D printed drug formulations have been published [1, 4–10]. There are two methods that stand out because of their low cost and the modular configuration of the printers, inkjet (IJ) printing and fused deposition modeling (FDM) printing. Between the two, a wide range of film-based and bulky solid dosage forms can be produced at the point of care to suit an individual patient's clinical needs and physical abilities to take solid medicines, e.g. pediatric and elderly patients with swallowing difficulties are more likely to find oral films acceptable in comparison to bulky tablets. This new way of manufacturing and dispensing the final dosage form at the point of care will also create new business models and opportunities for medical and retail industries [9]. As IJ printing has been intensively reviewed elsewhere [11, 12], this chapter will focus solely on FDM 3D printing. An increased number of research publications has demonstrated the potential of FDM 3D printing as an enabling technology to produce personalized polypills. In this chapter, the current state of development of using FDM 3D printing to produce personalized polypills is reviewed, and the challenges and barriers in the translation of this technology from the laboratory bench to the patient's bedside of such polypills are discussed.

11.2 Polypharmacy and Polypills

11.2.1 Clinical Evidence and Current State of the Art

Polypharmacy is used in the medical literature to describe the coadministration of multiple medications to patients who may have multiple comorbidities [13]. Patient groups with chronic and life-limiting conditions such as CVD, HIV, cancer, and neurological conditions may require polypharmacy. Many of these medications are administered orally and lead to patients having high daily pill burden. Such regimens often result poor adherence to the treatment. Therefore, the concept to formulate multiple pharmaceutical active ingredients into a single pill has been proposed with the overall aim to improve patient adherence and therefore clinical outcomes. Currently, this formulation approach has been translated into commercially available products as FDC. Communicable diseases such as HIV, tuberculosis, and malaria have already established high uptake of such FDC products. For noncommunicable diseases such as CVD, there are a wide range of FDC products approved worldwide, which are often referred to as polypills primarily for the prevention of CVD [14].

Wald and Law first reported the effectiveness of polypills containing six different drugs for reducing CVD risks in 2003 [15]. Since then, the World Health Organization (WHO) has recommended polypills with a combination of several medications commonly used to treat heart diseases and high blood pressure as the best secondary prevention treatment for reducing the premature deaths for the population <70 years old by 25% [14]. The usual combinations include antiplatelet (i.e. aspirin), antihypertensive (i.e. beta-blockers and angiotensin-converting enzyme (ACE) inhibitors and lipid-lowering agents (i.e. statins) at a fixed dose. Despite it is unclear and still under investigation that whether the polypills are cost-effective and have good long-term effectiveness in reducing cardiovascular events, it is conclusively from all existing clinical trial data that polypills approach can significantly increase patient adherence in all patients [16]. Patients in a primary care setting and patients with poor baseline adherence are the main groups who could benefit most from taking polypills [16]. However, most of the marketed FDC products only contain two drugs. The polypills in the true sense of multiple drug formulations (with more than two active drugs) are not easy to manufacture. There have been a few efforts aimed at developing dosage forms that allow dosing flexibility and personalization; such techniques include formulation of drugs into liquid dosage forms (in which the volume corresponding to the desired dose is measured and administered to the patient), multiparticulate solid dosage forms such as granules, pellets, or mini-tablets, in which a single unit (i.e. a single granule) contains a fraction

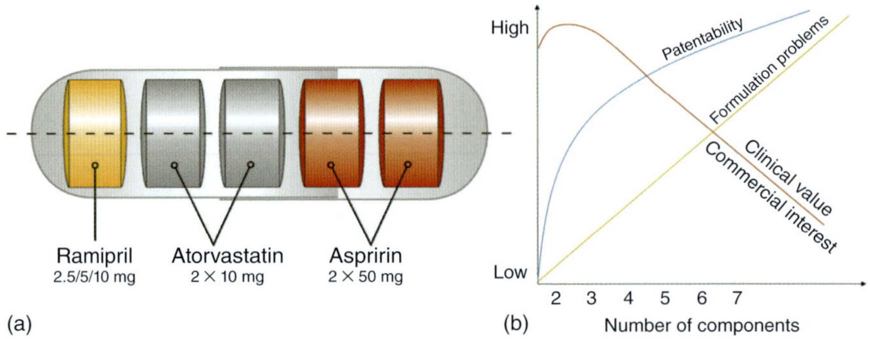

Figure 11.2 (a) Illustration of the Fuster-CNIC-Ferrer CV Polypill (Trinomia®, Sincronium®, and Iltria®) technology and (b) challenges faced by the development of polypills containing multiple drugs. Source: Tamargo et al. 2015 [18]. Reproduced with permission of Elsevier.

of the dose, with the desired dose being achieved by counting or weighing the number of units required to achieve the prescribed dose and recombine them into a single unit using a capsule, and scored tablets (which can be broken or cut to achieve fractions of the larger dose contained within an entire tablet) [17].

The most significant challenge is to minimize the chemical instability/incompatibility between the different drugs and excipients. To tackle this challenge, a Spanish pharmaceutical company, Ferrer, developed and licensed their CVD polypills (Trinomia®, Sincronium®, and Iltria®) in several counties using a patented technology platform. As illustrated in Figure 11.2a, this technology collectively put a combination of different active ingredients in a single capsule with separated compartments [18]. These compartments within each capsule can avoid the physicochemical incompatibilities between the active pharmaceutical ingredients (APIs) and preserve the biopharmaceutical and pharmacokinetic properties of each drug. However, the drawback of this technology being used for personalized medicine dispensing is that the capsule filling based on individual patient medication needs at the point of care is time-consuming and labor-intensive. In addition, the quality control of the capsules as an additional material complicates the process and increases the risks of introducing error and tampering contamination.

11.2.2 Future Personalization

Expanding the CVD polypill approach to the treatment and management of other chronic diseases could benefit patients through improvement in adherence. However, the development of FDC for different diseases all facing a common issue is associated with the fine balance of the market potential (with consideration of developmental costs) and the clinical value as described by Gulietta and Guerrero (Figure 11.2b) [18, 19]. Future personalized medicine should not only consider the individualized drug combination but also consider the individualized dose for each drug. The true sense of individualized dosing is customized in accordance with a patient's medical information including body weight, age, history, comorbidities, and easy dosing adjustability in response to therapeutic drug-monitoring results. However, the development of such personalized medical products will face significant practical issues including possible chemical incompatibility between the drugs and drug excipients during manufacture and on aging; *in vivo* undesired drug–drug interactions; accurate, precise, and specific assays for the mixed drug products; development of pharmacopeial specifications; and quality control (QC) standards [18].

In terms of the fabrication method, none of the existing polypill manufacturing technology can be easily standardized to be used at the point care. Automated, portable, and modular designs of the manufacturing equipment would be much more suitable for producing personalized drug and dose combinations as a single unit dose at the point of care. One group of techniques that fits with these criteria is 3D printing, which has generated considerable interest in the past decade [20]. 3D printing was pioneered in the late twentieth century as a rapid prototyping, high-throughput fabrication method for engineering applications. However, the past few years have seen a rise in the number of affordable "desktop" 3D printers aimed at the general public. The technique becoming less specialized for niche applications and more suitable (both in terms of cost and complexity) for hobbyists has led to a greater level of exploration of its usefulness for applications within a number of different industries [21]. The technique has since found applications in the health care industry, being used fabricate medical devices, artificial limbs, pharmaceutical dosage forms, and even live human tissues [20, 22–24]. Using 3D printing to produce personalized polypills at the point of care is conceptually different from the traditional FDC polypills. It allows the production of personalized combination of medications for each patient at the point of care – not a mass production of FDC products. It also allows the printing of different drugs in different parts of the polypills using a one-step process.

In addition, a number of researchers have suggested that 3D printing could be particularly useful for providing dosing accuracy in pediatrics and allowing personalized dosing for narrow therapeutic index drugs [8, 12, 22, 25, 26]. Table 11.1 summarizes the views expressed in the literature on the potential advantages of 3D printing for personalized medicine production [1, 12]. Moreover, 3D printed products have the capability to avoid problems with traditional tablet splitting and to reduce dose variation or miscalculation [8]. To further improve the adherence, 3D printing can produce the suitable size, shape, and appearance of the solid dosage form for each individual patient. For example, to support children's adherence, it has been suggested that printed animal shaped or figurines could potentially have benefits [1]. However, it should be recognized that significant technical and regulatory challenges are faced by 3D printing as a new group of manufacturing techniques. Currently, there is a great interest in modernizing conventional pharmaceutical manufacturing and regulation. 3D printing could be a critical factor facilitating this change and may motivate the pharmaceutical industry to develop new smart products that make production of more personalized dosage forms feasible and alleviate manufacturing and inventory waste [7, 8].

Table 11.1 Potential advantages of 3D printing in drug manufacturing.

Advantage	3D printing capability	Application in drug delivery	Medical and economic benefits
Increased product complexity	Printing unmoldable or difficult-to-make shapes	• Highly porous products that orally disintegrate • Toroidal products with zero-order release • Tablet shapes for pediatrics	• Improved patients' adherence and compliance • Enhanced effectiveness • Reduced side effects • New therapy based on combination products • Reduced operation cost • Precise dosing of low drug dose • Rapid manufacturing • Decreased production steps • Continuous manufacturing
	Computationally the design of object	• Excipient distribution for modifying drug release • Complex drug combination	
Personalization	Printing an infinite variety of shapes and drug loading	• Personalized dosing for potent drugs • Personalized dosing for pediatrics and geriatrics • Customized shapes for children	• Improved pediatrics and geriatrics compliance • Reduced side effect of potent drugs • Appropriate drug dosing and release according to patient's medical information • Decreased medical burden for elderly • Avoid splitting conventional tablets and errors in dose calculation
	Varying composition within simple and modular design	• "Polypills" that contained multiple drugs in single dose unit • Hollow pills with adjustable infill percent to control release rate	
On-demand manufacturing	Rapid printing from digital design (no intermediate machinery)	• Printing in emergency settings • Printing immediately • Reduced barriers to experimentation during drug product development	• Reduced time to market for new drugs • New drugs produced and brought to market
	Printing at the point of care	Short shelf life medications	

Source: Reproduced with permission from [1, 12].

11.3 FDM 3D Printing of Pharmaceutical Solid Dosage Forms

In the pharmaceutical industry, 3D printing has generated a lot of research interest as it is seemingly holding the key to solve the challenge the industry is facing in terms of developing personalized medicines. Because of the sequence of events through which 3D printing fabricates an object, it is possible to manufacture dosage forms consisting of multiple drugs with flexible doses that are adapted to the needs of a particular patient by embedding different drugs in different layers of the same unit dosage form [27]. The prospects offered by 3D printing were further boosted in 2015 when the United States Food and Drug Administration (FDA) approved for commercialization the first 3D printed pharmaceutical product, Spritam®, manufactured by a pharmaceutical company Aprecia Pharmaceuticals [28]. Although Spritam does not utilize the full capabilities made possible by 3D printing in terms of personalization of the dosage form (it is manufactured as a single, 1000 mg dose of a single drug, levitracetam), its approval indicates general acceptance of marketing dosage forms manufactured by 3D printing. Here, we exclusively discuss the potential of FDM 3D printing, also known as fused filament fabrication (FFF), which is a variant of 3D printing that utilizes thermoplastic polymers for constructing the three-dimensional object as a future method for producing personalized oral polypills. FDM printing used for drug-containing samples has been continuously evolving over the past 10 years in terms of drug-loading methodology and the complexity of the formulations [29]. However, it is also clear that the major challenges for taking this new manufacturing technique forward are from the aspects of printable materials, process and quality control to pharmaceutical standards, and the establishment of relevant regulations for ruling this new manufacturing method.

11.3.1 Basic Principle of FDM 3D Printing

In 1989, Scott Crump patented FDM, which is a process using a movable headed dispensing head to build three-dimensional solid materials in a layer-by-layer manner [30]. It is an extrusion-based, economic, simply accessible, and versatile 3D printing technique that can fabricate a solid dosage form [7]. In a typical FDM printing run, a thermoplastic polymer (which had previously been shaped into a thin strand called a filament) is fed into the printer by the action of two counter-rotating rollers. As illustrated in Figure 11.3a, the feeding rollers carry the filament to a heating zone where the filament is heated to a temperature beyond its melting/softening point to form a flowable melt, which is then deposited as a layer on the build plate through a nozzle. The printhead moves within the x and y-axes, whereas the build plate "platform" can move on the z-axis, which can be temperature controlled at below the printing temperature for allowing the printed materials to solidify between each layer [33]. The mechanical assembly consisting of the feeding rollers, heating zone, and the nozzle are collectively referred to as the printing head. After the melt exits the printing head, it is deposited onto the build plate to form a layer; this

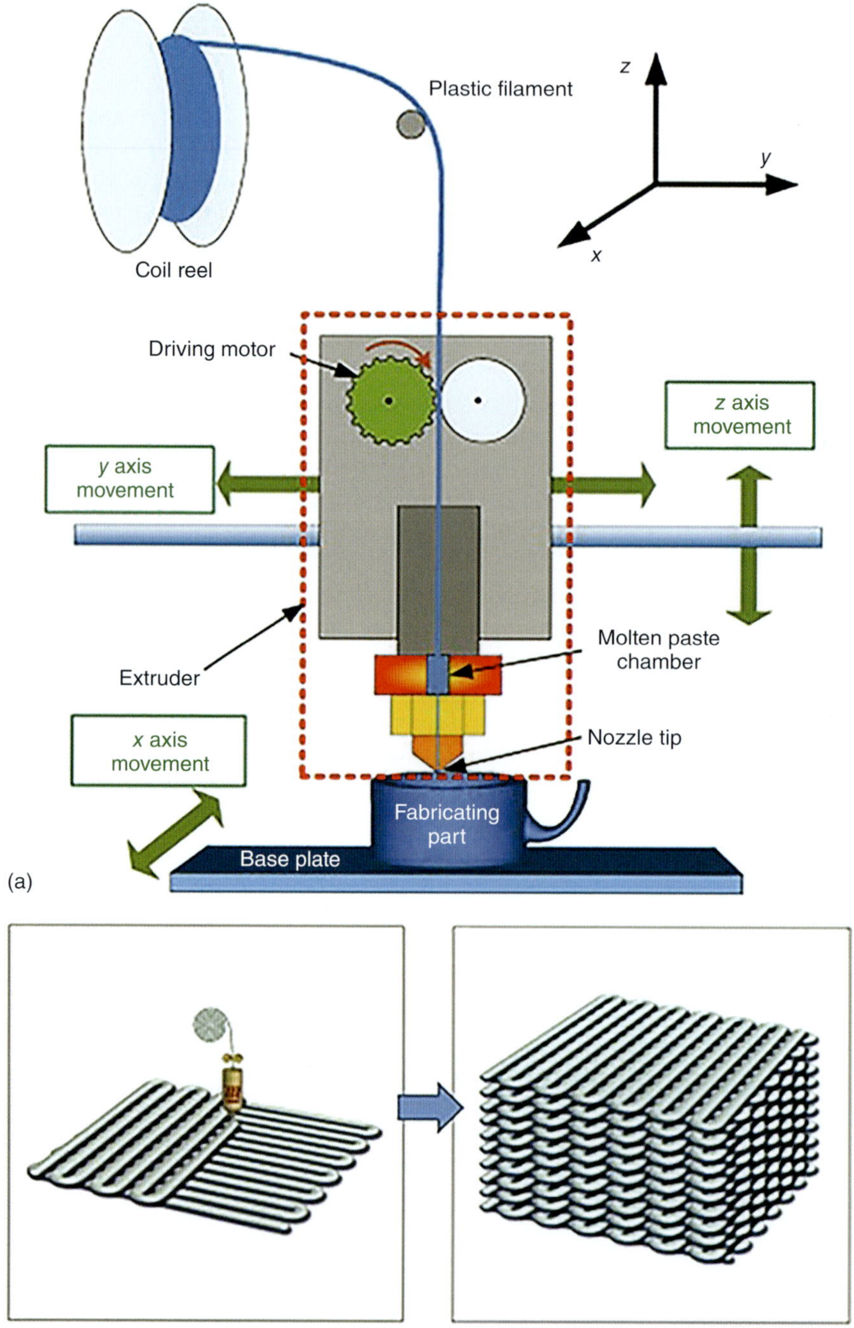

Figure 11.3 (a) Schematic of FDM 3D printing. Source: Reproduced with permission of [31]. (b) Illustration of FDM layer by layer building of a printed object. Source: Mohanty et al. 2015 [32]. Reproduced with permission of Elsevier.

is then followed by a downward movement of the build plate to allow for a second layer to be deposited atop the first, as seen in Figure 11.3b [32]. The feed polymer must have suitable thermal rheological properties to allow it to be extruded through the nozzle [34]. The drug can be incorporated either through impregnation of the polymer filament using drug solutions or by producing the drug–polymer-dispersion-based filament using hot melt extrusion before the FDM printing (detailed in Section 11.3.3). The distribution and dose of the drug can be changed by changing the infill and geometry (e.g. hollow, multilayer, and coated) of the printing, which allows the inclusion of flexible dosing and drug combinations in the 3D printed dosage forms [35]. However, there are three main specific challenges associated with FDM printing. These are ensuring the suitable mechanical properties of heated materials to build the 3D dosage forms, the quality of the printed product in terms of reproducible precision and accuracy, and appropriate drug incorporation into the printed or printing materials without degradation or drug loss.

11.3.2 Printing Parameter Control

There are a number of critical process parameters (CPPs) controlling the quality of FDM 3D printing. Huang et al. grouped them into three categories: machine-specific parameters, operation-specific parameters, and material-specific parameters, as summarized in Table 11.2 [36]. It should be noted that Huang et al. discussed FDM from an industrial point of view, relating strictly to its prototype production but nonpharmaceutical applications. It does not account for special requirements that are undoubtedly necessary for robust pharmaceutical adaptations, nor for the fact that, in the context of pharmaceutical formulation. FDM has been constrained to being explored only

Table 11.2 Summary of the CPPs that could affect a FDM 3D printing process.

	Types of parameters		
	Materials related	**Operation related**	**Machine related**
3D printing process parameters	Thermal stability	Layer resolution	Nozzle inner diameter
	Viscosity and fluidity	Infill percentage	Filament diameter i.e. 1.75 or 2.85 mm
	Stiffness and bending	Fill pattern	Roller and filament feed rate
	Flexibility	Print head speed	Flow of the feedstock materials
	Mechanical properties	Operation temperature	
		Number of extruder tips used	
		Print bed temperature	

Source: Huang et al. 2015 [36]. Modified with permission of IEEE.

by repurposing commercial 3D printers (with printers from the Makerbot® Replicator™ series being the most popular) for the application of FDM in pharmaceutical formulations. However, in terms of the printing control, so far, there is no literature that has systematically studied the effects of printing parameters on pharmaceutical filaments. Such studies are urgently needed to allow better design of pharmaceutical filaments. More importantly, these commercial printers that are primarily designed for fabricating porotypes do not allow the user to control some of the process parameters mentioned previously; parameters such as nozzle diameter (road width), filament diameter, filament feeding speed (melt deposition rate), are preinstalled into the printer's software, firmware, or hardware preventing the user from controlling these easily, at best offering the user the choice between a number of preprogrammed presets [35]. In order to fully utilize the advantages of 3D printing in personalized medicine, pharmaceutically complied printer should be further developed, and these preset parameters should become adjustable to allow the adequate printing of pharmaceutical materials.

A number of pharmaceutical studies have investigated the relationships between the few adjustable CPPs such as infill density, head speed, layer thickness, nozzle, and build plate temperature [8, 20, 37]. The infill density can be defined as the quantity of materials filled into hollow object, which reveals the porosity of the 3D printed tablets. It can range from 0% to 100% infill density, where the 0% is a completely hollow shell and 100% is a solid, nonporous object [32]. As seen in Figure 11.4, by adjusting the infill, the graphic design demonstrates that disk-shaped objects can be printed either as solid nonporous materials or as highly porous materials with adjustable pore size. The height/thickness of the printed layer, which is ranged within 100–400 μm for most commercially available printers and filaments, can be affected by the printing speed and the nozzle diameter. Most of the commercial 3D printers have preinstalled printing settings (i.e. low resolution printing, standard resolution printing, high resolution printing) that manipulate the printing speed and layer thickness to compensate printing time for surface roughness of the printed object [8]. The extrusion temperature and build plate temperature can also affect the printed layer thickness, and their effects are highly dependent on the material used. Using commercial filament acrylonitrile butadiene styrene (ABS) as an example, as seen in Figure 11.5, an increase in the head speed of the printing leads to an unwanted increase of variation in dimensions between the strips

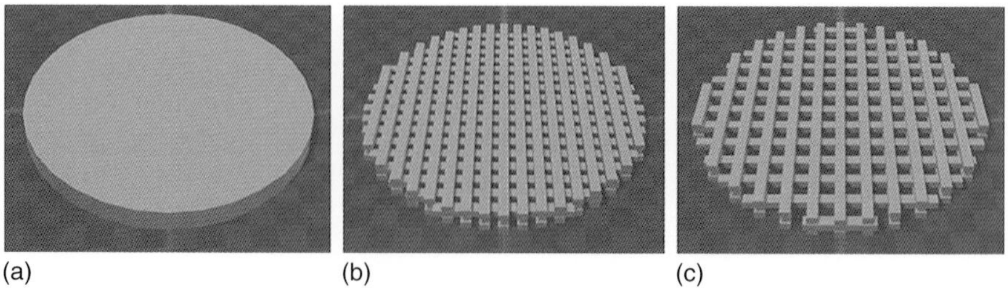

Figure 11.4 The 3D design of layered disks FDM 3D printed using different levels of infills.

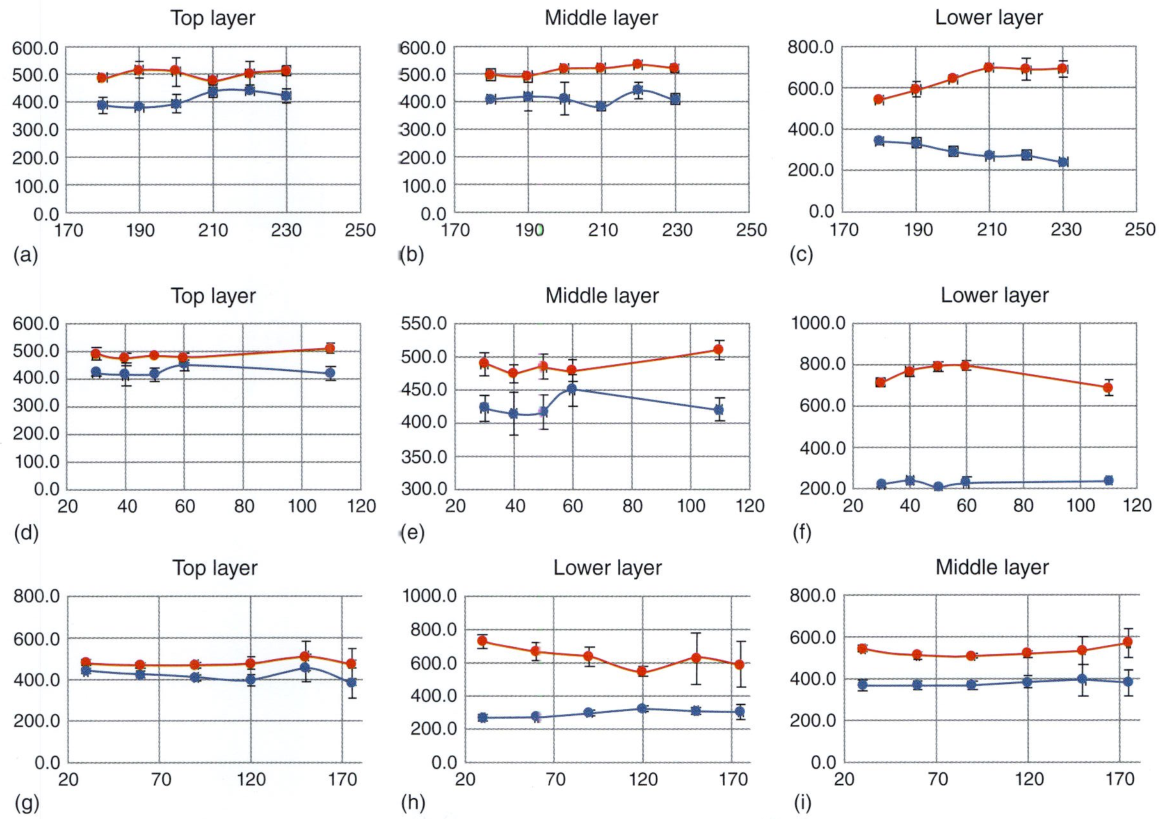

Figure 11.5 The effects of printing temperature (a–c), build plate temperature (d–f), and the head speed (g–i) on the layer thickness of triple-layered FDM 3D printed disks. *x*-Axes are temperature (°C) for a–f and head speed (mm s^{-1}) for g–i; *y*-axes are the printed layer thickness (μm). The blue data points are the height of the printed strips and the red data points are the width of the printed strips (A. Brochard, M. Alhijjaj, S. Qi, unpublished lab report data, 2015).

of a same layer. A decrease of the extrusion temperature leads to more defined cylinders with less difference between the height and the width of the printed strips and the layers that are close to the preset 400 μm thickness. The lower layer obtained the more accurate results with an average printed strip width of 537 μm and an average thickness of 339 μm for the low temperature of printing 180 °C. Although trying to print at a temperature lower than 180 °C, the material did not stick to the heated building plate properly and generated a high level of printing defects. In the case of ABS which has a Tg at approximately 110 °C, changing the building plate temperature between 30 and 100 °C showed little effects on the printed strip thickness. This is likely because of the high printing temperature required for ABS. Other materials requiring lower printing temperature may experience more significant effect of the building plate temperature on the layer thickness.

The optimization of these parameters will allow improvement in the quality of printing. As with other 3D printing technologies, FDM printing relies on the formation and stacking of layers to fabricate the object; the layer thickness and the uniformity of the thickness of each layer are the key factors affecting the overall printing quality. The uniformity of the layer thickness has been discussed earlier. The absolute thickness of each layer is often regarded to be highly correlated with the "resolution" of the printing, which indicates how faithful the fabricated object is to the computer-generated image (CGI) design. Printers capable of printing thinner layers produce higher resolution prints, as thinner layers minimize what is known as the "staircase effect" of which 3D printed objects suffer from [38, 39]. As can be seen in Figure 11.6, thin layers produce a far more faithful approximation of a circle than thick layers, resulting in a "staircase" that is not as steep, and thus a finer surface and edge of the printed object.

Most of the commercial FDM 3D printers available have a minimum layer thickness of approximately 100 μm [10], with more recently, some manufacturers claiming a minimum layer thickness of 50 μm [40]. The most extensively investigated FDM-based 3D printers for pharmaceutical applications are variants

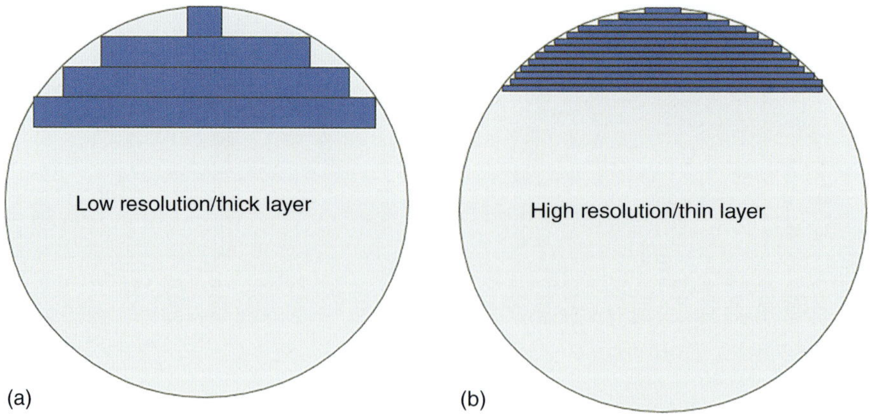

Figure 11.6 Illustration of the effect of printing resolution on the precision of the edges of the printed products.

of the Makerbot Replicator series, which have a minimum layer thickness of 100 μm. From a printing quality point of view, higher printing resolution would allow better printing precision and minimize the chances of producing defects. However, the impact of printing resolution on the performance of FDM-manufactured pharmaceutical dosage forms remains unexplored. It is likely that the effects would be related to the surface area differences in the solid dosage forms printed with different resolutions.

11.3.3 Drug-Loading Methods

Two common methods of preparing drug-loaded filaments have been reported in the pharmaceutical literature, impregnation (FDMi) and extrusion (FDMe). Impregnation can be applied by soaking a polymer filament in a concentrated drug solution [41]. The selection of the solvent to prepare the drug solution is critical in FDMi. The solvent selected for FDMi should neither dissolve nor alter the physical properties of the polymer but still provide a good drug solubilization [8]. In a study by Goyanes et al., polyvinyl alcohol (PVA) filaments were immersed in concentrated ethanolic solutions of a range of model drugs including fluorescein, 4-aminosalicylic acid (ASA), and 5-ASA [20, 37, 41]. PVA is soluble in water but not in ethanol. Despite all the printed tablets fulfilled pharmacopeia specifications in terms of friability (<1%) and weight uniformity (±5%), the drug loading for all the model drugs are disappointingly low. This could also be contributed by the thermal degradation of the drug during the FDM printing process. Similarly, Skowyra et al. impregnated PVA filaments in a saturated solution of prednisolone in methanol at 30 °C for 24 hours [22]. Although prednisolone is thermally stable at the used printing temperature, the drug loading by FDMi in this case was still as low as 1.9% w/w. These early studies indicated that in FDMi, the passive diffusion of the drug molecules from the drug solution into the polymer filament was slow and inefficient and would lead to low drug loadings. Therefore, a more efficient drug-loading method is needed.

To address the previous drawbacks associated with FDMi, an alternative drug-loading method combining hot-melt extrusion (HME) and FDM 3D printing has been developed and is now widely used. HME is used to prepare the solid-dispersion-based drug-loaded polymer filaments, where the raw materials (e.g. API, polymer, plasticizer, and other additives) are melted, mixed, and homogenized at a raised temperature before being extruded via a die as a filament with a uniform diameter [41]. Pietrzak et al. used four different pharmaceutical polymers, Eudragit RL, RS, E, and hydroxypropyl cellulose (HPC), to demonstrate the applicability of the FDMe method for drug-loading and pharmaceutical FDM printing (Figure 11.7) [34]. Using theophylline as the model drug, the drug-loaded filaments were extruded using triethyl citrate as a plasticizer. In comparison to FDMi method, the theophylline loading by FDMe was significantly increased and the theophylline content in the printed tablet was within 91–95% of the targeted dose [35]. From a practical point of view, it was found that the FDM printing temperature needs to be 20–50 °C greater than the HME temperature used to prepare the drug-loaded filaments. This ensures that the melt viscosity of drug polymer mixture is suitably decreased and avoids issues such as nozzle

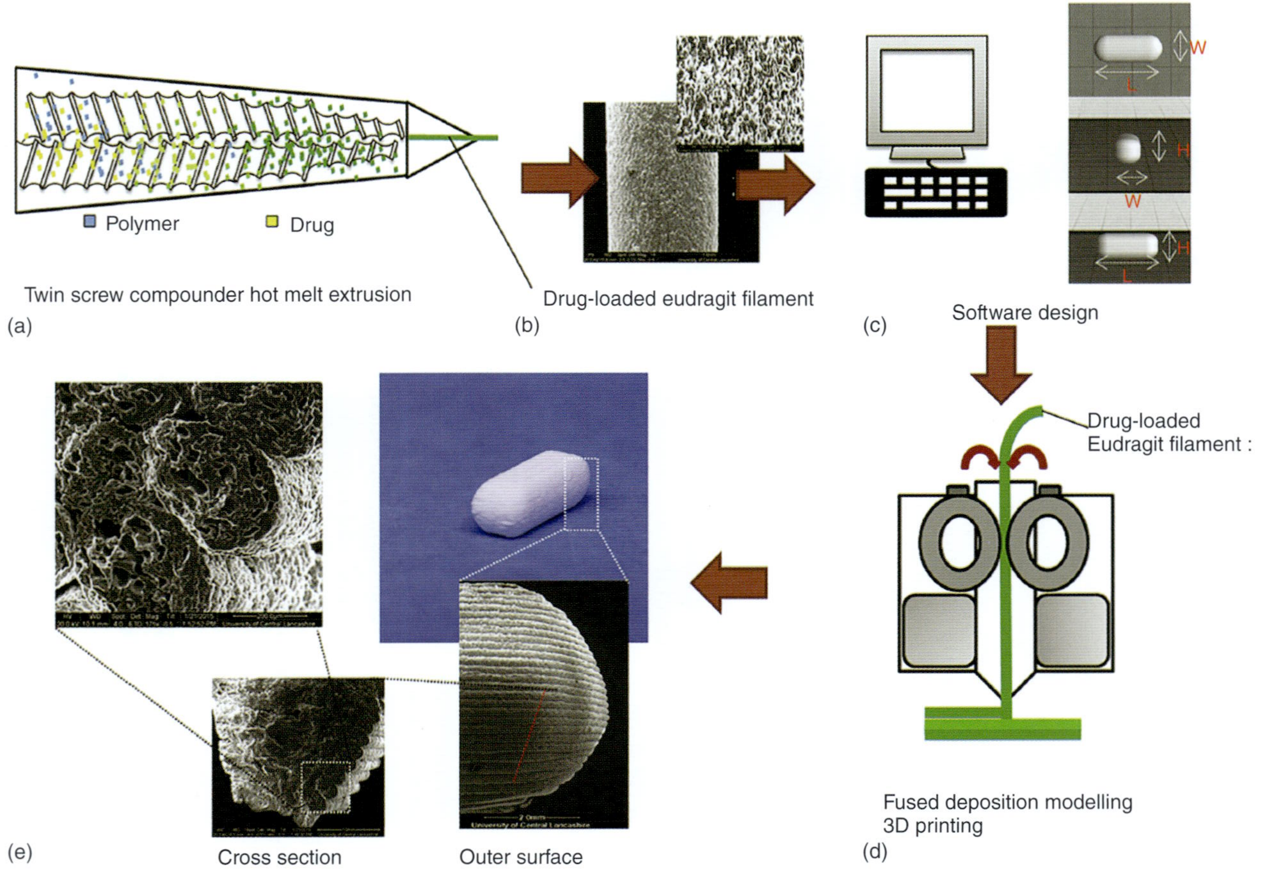

Figure 11.7 Schematic illustration of FDMe fabrication process of solid dosage forms. Source: Pietrzak et al. 2015 [34]. Reproduced with permission of Elsevier.

blockage [7]. However, FDMe requires the materials being processed twice by thermally based methods, and this leads to the great risk of thermal degradation of the loaded drug, which is one of the major limitations of FDM printing [35].

11.4 Key Challenges in the Development of FDM 3D Printed Personalized Polypills

Although the working principle of FDM 3D printing is relatively simple, many technical hurdles are still present in order to truly transform it into a pharmaceutical process that can reliably produce safe solid dosage forms. Norman et al. have summarized some of the unconventional risks associated with FDM printing and discuss controllable factors that could be used to mitigate these risks [1]. Here, we focus the discussion on the technical challenges arisen from both raw materials and process parameters, which could be potentially improved or even resolved by re-engineering the printer. Research on redesigning the pharmaceutical FDM 3D printer is essential for advancing the technique and will play a key role in bringing the printing closer to the clinical use. In addition to technical challenges, currently, there is no regulatory framework established for personalized 3D printed products.

11.4.1 Printable Pharmaceutical Materials

Currently, pharmaceutical FDM 3D printing is restricted to print low-dose, thermostable drugs embedded within dosage forms mainly consisting of a limited number of thermoplastic polymers such as ABS, poly (L-lactic acid) (PLA), and polycaprolactone (PCL). This is because even the materials are thermally stable under both HME and FDM printing temperatures; the currently available commercial FDM printers require the materials to have the suitable thermal rheological properties at the printing temperature to allow the accurate deposition and printing. As the extrusion mechanisms of the commercial FDM printer is not effective in "pushing" the molten/soften materials out of the printing nozzle, many hot melt extrudable materials are not FDM 3D printable. Most of the pharmaceutical materials that can be made into filaments using hot melt extrusion process are either not feedable and/or printable by FDM printing. Feedability describes whether the filament can be successfully fed into the printing head. This feeding process requires the materials to have the suitable mechanical properties (such as elastic modulus, yield strength, and toughness) in order to avoid any breakage or coiling. For printability, the thermal–rheological properties of the materials are critical. This is because in commercial FDM 3D printers, the main function of the extruder (liquefier) is melting and pushing the materials to the nozzle, which rheologically requires materials exhibit a low melt viscosity at the printing temperature. In addition, rapid deposition after extrusion requires a rapid solidification behavior, which can only be provided by highly thermoplastic polymers. However, most thermoplastic pharmaceutical polymers do not possess adequate properties to allow for robust FDM [42]. Moreover, as many drugs can act as plasticizers or anti-plasticizers after being incorporated into the polymers as solid dispersions, using a high drug loading in the extrusion could affect the thermoplastic characteristics of the polymer, hence the printability of the dosage forms [8]. As there are temperature gradients during the FDM printing process (i.e. feeding zone has lower temperature than the printing zone and the nozzle has different temperature to the building plate), the overall viscoelastic properties of the filament are significant, and the rheological properties need to be understood at multiple temperatures [43, 44]. There is currently a lack of such understanding for many pharmaceutical materials in the literature.

In order to overcome the printability issue, researchers have been increasingly reporting the use of polymer blends as carrier materials for FDM printed formulations. Table 11.3 summarizes the blends reported in the literature that can be used to prepare drug delivery systems. Blends offer flexibility to tune the thermoplasticity of the materials and buffer the effect of drug loading. However, such formulations are often complex. For example, Alhijjaj et al. described the fabrication of a five-component matrix containing either Eudragit® E PO, Soluplus®, or PVA, plasticized with polysorbate (Tween®) 80, polyethylene oxide (PEO) N 10, and polyethylene glycol (PEG) 4000 with felodipine as a model drug and the fifth component of the matrix [42].

Systematic studies on developing a rationalized approach to understand and implement printability into pharmaceutical polymers are urgently needed by the pharmaceutical FDM printing research field. With such understanding, formulation-screening tools could allow rapid screening and development of feedable and printable filament formulations and link into the quality by design (QbD) approach of pharmaceutical product development.

11.4.2 Printing Precision and Printer Redesign

In association with materials' responses during different stages of printing discussed in Section 11.4.1, defects (often unavoidable) in FDM printed products are another concern and technical barriers need to be understood, addressed, and regulated. Defects in FDM 3D printing can be minimized by optimizing the processing parameters including infill, extrusion speed and printing, and build plate temperature. In theory, the layer-by-layer building of a FDM 3D printed object should have the same layer thickness and road width of each strip of the laid materials. However, in practice, this is extremely difficult to achieve. As seen in Figure 11.8, the printed layers with 50% infill using ABS filaments had different layer thickness and road width of the printed stripes for each layer. This was caused by the inappropriate pairing of the build plate and printing temperature. Such defects may not affect the drug-loading uniformity, but they would affect the release profile of the API from the matrices. Within the pharmaceutical 3D printing field, defects have been barely mentioned and investigated in the literature. However, to move the technology toward commercial operation, such fundamental aspects of technical problems should be thoroughly investigated and reported.

Table 11.3 Summary of polymer blends using pharmaceutical grand excipients.

Polymers/blends used	API	Additives	References
Eudragit EPO + PEG 4000 + PEO WSR Soluplus + PEG 4000 + PEO WSR PVA + PEG 4000 + PEO WSR	Felodipine	Tween 80	[42]
HPMC E5 + EC N14 HPMC E5 + HPC EF HPMC E5 + HPC LF HPMC E5 + Soluplus HPMC E5 + Eudragit L100 HPC LF + EC N14 EC N14 + Eudragit L100	Paracetamol	Kollidon CL-F	[45]
PVP	Theophylline Dipyridamole	Triethyl citrate Talc	[46]
Eudragit EPO	Theophylline Captopril Prednisolone 5-ASA	Triethyl citrate Tri-calcium phosphate Talc MCC Lactose	[47]
Eudragit E	Hydrochlorothiazide	Triethyl citrate Tri calcium phosphate Sodium starch glycolate Croscarmellose Polyplasdone-XL	[48]
Kollidon® VA64 Kollicoat® IR HPMC 15 CP HPMCAS-MG HPMCAS-MG + Kollidon VA64 Kollidon VA64 + HPMC 15 CP	Haloperidol		[49]

More importantly, as discussed earlier, the commercially available FDM printers have poor extrusion mechanism for feeding the molten/soften materials through the printing nozzle. Therefore, the current printing is heavily relying on the filament having the suitable melt flow properties to enable printing. This significantly limits the number of pharmaceutical materials that are FDM printable. In addition, although the current configuration of FDM printers would cope with printing personalized pills at the point of care on a small scale using preprepared filaments. The quality of HME-prepared filament is still

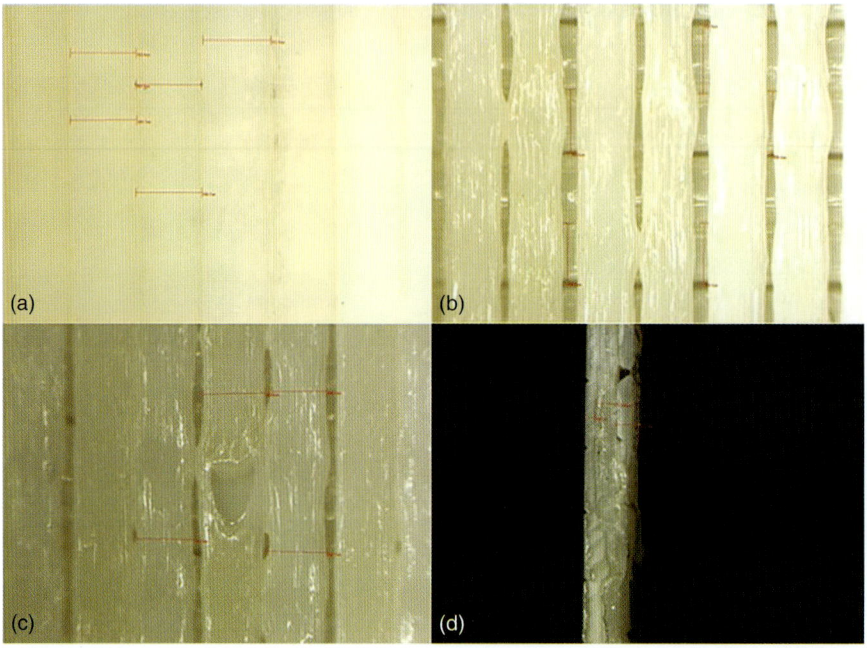

Figure 11.8 The light microscopy images of FDM 3D printed triple-layered patches with 50% infill using ABS. (a) Bottom layer, (b) middle layer, (c) top layer, and (d) the cross section of the patch containing all the three layers (A. Brochard, M. Alhijjaj, S. Qi, unpublished lab report data, 2015).

one of the major limitations to obtain good printing quality with controllable level of defects. Therefore, the development of special filament-making HME instrument is extremely important to allow a good level of control of the printing quality of the personalized products. If FDM 3D printing was used for the large-scale production of FDC products (not personalized product), the filament-making process by HME should be integrated into the printer to allow continuous manufacturing. Therefore, in order to fit the purpose of pharmaceutical manufacturing and process control, the current FDM printers that are used in research need to be re-engineered. Significant research efforts are needed in this area requiring close collaboration between manufacturing engineers and pharmaceutical scientists.

11.4.3 Regulatory Barriers for Personalized Polypill Printing

Another challenge that stands in the way of successful commercialization of personalized 3D printing lies in the barrier posed by regulatory bodies [50]. Regulatory bodies, such as FDA and the Medicines and Healthcare Products Regulatory Agency (MHRA), require demonstration of consistency and reliability during manufacturing; the tried-and-tested methods that have been employed by the pharmaceutical industry for decades are considered to be more trusted, and by extension, more safe, than a newcomer technology such as 3DP [1]. However, the introduction of process exploratory paradigms such as pharmaceutical quality risk management (QRM) and QbD by the FDA provides pharmaceutical manufacturers with a margin of freedom to explore innovative approaches to drug formulation and development, provided that they have demonstrated a satisfactory understanding of the CPPs that govern a particular industrial process. A suitable in-line process analytical technology (PAT) should be developed to monitor the layer-by-layer quality for the FDM 3D printing process in order to minimize the level of defects discussed in Section 11.4.2 [51, 52]. Analytical tools with high-throughput potentials such as imaging-based methods could be used to serve this purpose [53]. Manufacturing environment monitoring and control will minimize the possible risk of material properties altering because of changes in humidity. As discussed earlier, the temperature gradient between the printing nozzle and the environment and the building plate could affect the solidification process of the extruded material, and thus their layering behavior [7]. Therefore, the temperature of the printing chamber can also play an important role in defining the quality of the printed products.

In term of the finished product, a set of QC criteria should be set up to ensure that the quality of the product meets the standards. However, it is challenging and requires innovative approaches for this aspect if the printing will be performed at the point of care. It means that the in-line processing monitoring becomes far more important than conventional large batch production of the solid oral dosage forms. Nevertheless, quality attributes including identity, appearance, assay, content uniformity, drug release, impurity level, hardness, friability, crystallinity, and API polymorphic form should still be closely monitored and controlled [54].

Currently, from the regulatory bodies' prospective, they generally consider that additive manufacturing processes such as 3D printing should be able to follow the same regulatory pathway and manufacturing requirements as conventional processes such as tableting. However, if the final products were produced at the point of care, significant developments in regulation are needed for quality and processing assurance protocols. Future guidelines for defining the responsibility of the industrial partners and the health care professionals are needed for manufacturing individualized medicine. This will require the close collaboration between the regulatory bodies, the industry, and the health care professionals.

In addition to the new regulatory guidelines for validating the printing process, personalized dose and drug combinations face even tougher challenges from the regulatory perspective. The management of the risk of drug–drug interactions and the efficacy of using personalized dosing and drug combinations need to be demonstrated before this treatment approach could be brought into practice. Large clinical trials are likely to be needed to demonstrate the efficacy and viability of personalized treatment methods. With accelerated developments in printing technology and associated materials sciences, the clinical data collection will become the rate-limiting step for bringing 3D printed personalized medicine closer to patients.

11.5 Conclusions and Future Remarks

It has been widely recognized that FDM 3D printing has unique advantages for fabricating personalized polypills at the point of care. Personalization using unconventional geometry design for co-delivery of multiple drugs from a single dosage unit has been the main focus of this pharmaceutical research field. Although the enthusiasm for the unique potentials of this technology from the research community is important for pushing the technology forward, some fundamental and critical technical issues need to be addressed. These include developing a rational formulation approach for producing printable blends as drug-loaded filaments, printer re-engineering to meet clinical use requirements, and the development of in-line processing control tools for monitoring the printing quality of FDM printing. Despite the positive and supportive gestures from the regulatory bodies, significant regulatory barriers need to be overcome before the personalized medicine approach can be brought to clinical practice. 3D printed personalized polypills are still at the conceptual stage, and any possible clinical success of the technology will require tremendous joint efforts from a wide range of cross-disciplinary researchers such as manufacturing engineers, pharmaceutical scientists, process scientists, regulatory, health economics, and health care professionals.

References

1 Norman, J., Madurawe, R.D., Moore, C.M.V. et al. (2017). A new chapter in pharmaceutical manufacturing: 3D-printed drug products. *Advanced Drug Delivery Reviews* 108: 39–50.
2 Pan, F., Chernew, M.E., and Mark Fendrick, A. (2008). Impact of fixed-dose combination drugs on adherence to prescription medications. *Journal of General Internal Medicine* 23 (5): 611–614.
3 US-FDA (2013). *Paving the Way for Personalized Medicine: FDA's Role in a New Era of Medical Product Development*. Washington, DC: US Department of Health and Human Services.
4 Ursan, I.D., Chiu, L., and Pierce, A. (2013). Three-dimensional drug printing: a structured review. *Journal of the American Pharmacists Association (2003)* 53 (2): 136–144.
5 Kolakovic, R., Viitala, T., Ihalainen, P. et al. (2013). Printing technologies in fabrication of drug delivery systems. *Expert Opinion on Drug Delivery* 10 (12): 1711–1723.
6 Yu, D.G., Zhu, L.-M., Branford-White, C.J., and Yang, X.L. (2008). Three-dimensional printing in pharmaceutics: promises and problems. *Journal of Pharmaceutical Sciences* 97 (9): 3666–3690.
7 Prasad, L.K. and Smyth, H. (2016). 3D Printing technologies for drug delivery: a review. *Drug Development and Industrial Pharmacy* 42 (7): 1019–1031.
8 Goole, J. and Amighi, K. (2016). 3D printing in pharmaceutics: A new tool for designing customized drug delivery systems. *International Journal of Pharmaceutics* 499 (1–2): 376–394.

9 Sandler, N. and Preis, M. (2016). Printed drug-delivery systems for improved patient treatment. *Trends in Pharmacological Sciences* 37 (12): 1070–1080.

10 Alhnan, M.A., Okwuosa, T.C., Sadia, M. et al. (2016). Emergence of 3D printed dosage forms: opportunities and challenges. *Pharmaceutical Research* 33 (8): 1817–1832.

11 de Gans, B.J., Duineveld, P.C., and Schubert, U.S. (2004). Inkjet printing of polymers: state of the art and future developments. *Advanced Materials* 16 (3): 203–213.

12 Sandler, N., Määttänen, A., Ihalainen, P. et al. (2011). Inkjet printing of drug substances and use of porous substrates-towards individualized dosing. *Journal of Pharmaceutical Sciences* 100 (8): 3386–3395.

13 Aronson, J.K. (2006). Polypharmacy, appropriate and inappropriate. *British Journal of General Practice* 56 (528): 484–485.

14 WHO (2002). Secondary prevention of noncommunicable diseases in low-and middle-income countries through community-based and health service interventions: World Health Organization-Wellcome Trust meeting report, 1–3 August 2001. Geneva: World Health Organization.

15 Wald, N.J. and Law, M.R. (2003). A strategy to reduce cardiovascular disease by more than 80%. *BMJ* 326 (7404): 1419.

16 Bramlage, P., Sims, H., Minguet, J., and Ferrero, C. (2017). The polypill: an effective approach to increasing adherence and reducing cardiovascular event risk. *European Journal of Preventive Cardiology* 24 (3): 297–310.

17 Wening, K. and Breitkreutz, J. (2011). Oral drug delivery in personalized medicine: unmet needs and novel approaches. *International Journal of Pharmaceutics* 404 (1–2): 1–9.

18 Tamargo, J., Castellano, J.M., and Fuster, V. (2015). The Fuster-CNIC-Ferrer cardiovascular polypill: a polypill for secondary cardiovascular prevention. *International Journal of Cardiology* 201: S15–S23.

19 Guglietta, A. and Guerrero, M. (2009). Issues to consider in the pharmaceutical development of a cardiovascular polypill. *Nature Clinical Pratice Cardiovascular Medicine* 6 (2): 112.

20 Goyanes, A., Buanz, A.B., Basit, A.W., and Gaisford, S. (2014). Fused-filament 3D printing (3DP) for fabrication of tablets. *International Journal of Pharmaceuticals* 476 (1–2): 88–92.

21 Ventola, C.L. (2014). Medical applications for 3D printing: current and projected uses. *Pharmacy and Therapeutics* 39 (10): 704.

22 Skowyra, J., Pietrzak, K., and Alhnan, M.A. (2015). Fabrication of extended-release patient-tailored prednisolone tablets via fused deposition modelling (FDM) 3D printing. *European Journal of Pharmaceutical Sciences* 68: 11–17.

23 Raijada, D., Genina, N., Fors, D. et al. (2013). A step toward development of printable dosage forms for poorly soluble drugs. *Journal of Pharmaceutical Sciences* 102 (10): 3694–3704.

24 Markarian J. (2016). FDA and the emerging technology of 3D printing, vol. 36. http://www.pharmtech.com/fda-and-emerging-technology-3d-printing (accessed 09 July 2018).

25 Khaled, S.A., Burley, J.C., Alexander, M.R. et al. (2015). 3D printing of tablets containing multiple drugs with defined release profiles. *International Journal of Pharmaceutics* 494 (2): 643–650.

26 Vakili, H., Kolakovic, R., Genina, N. et al. (2015). Hyperspectral imaging in quality control of inkjet printed personalised dosage forms. *International Journal of Pharmaceutics* 483 (1–2): 244–249.

27 Katstra, W., Palazzolo, R., Rowe, C. et al. (2000). Oral dosage forms fabricated by Three Dimensional Printing™. *Journal of Controlled Release* 66 (1): 1–9.

28 Yu, D.-G., Branford-White, C., Yang, Y.-C. et al. (2009). A novel fast disintegrating tablet fabricated by three-dimensional printing. *Drug Development and Industrial Pharmacy* 35 (12): 1530–1536.

29 Hsiao, W.-K., Lorber, B., Reitsamer, H., and Khinast, J. (2018). 3D printing of oral drugs: a new reality or hype? *Expert Opinion on Drug Delivery* 15 (1): 1–4.

30 Crump, S.S. (1992). Apparatus and method for creating three-dimensional objects. Google Patents US5121329A.

31 Jin, Y.-a., Li, H., He, Y., and Fu, J.-z. (2015). Quantitative analysis of surface profile in fused deposition modelling. *Additive Manufacturing* 8: 142–148.

32 Mohanty, S., Larsen, L.B., Trifol, J. et al. (2015). Fabrication of scalable and structured tissue engineering scaffolds using water dissolvable sacrificial 3D printed moulds. *Materials Science and Engineering: C* 55: 569–578.

33 Chia, H.N. and Wu, B.M. (2015). Recent advances in 3D printing of biomaterials. *Journal of Biological Engineering* 9 (1): 4.

34 Pietrzak, K., Isreb, A., and Alhnan, M.A. (2015). A flexible-dose dispenser for immediate and extended release 3D printed tablets. *European Journal of Pharmaceutics and Biopharmaceutics* 96: 380–387.

35 Melocchi, A., Parietti, F., Maroni, A. et al. (2016). Hot-melt extruded filaments based on pharmaceutical grade polymers for 3D printing by fused deposition modeling. *International Journal of Pharmaceutics* 509 (1–2): 255–263.

36 Huang, T., Wang, S., and He, K. (eds.) (2015). Quality control for fused deposition modeling based additive manufacturing: current research and future trends. In: *Reliability Systems Engineering (ICRSE), 2015 First International Conference on 2015*. Beijing: IEEE.

37 Goyanes, A., Buanz, A.B., Hatton, G.B. et al. (2015). 3D printing of modified-release aminosalicylate (4-ASA and 5-ASA) tablets. *European Journal of Pharmaceutics and Biopharmaceutics* 89: 157–162.

38 Vasudevarao, B., Natarajan, D.P., Henderson, M., and Razdan, A. (eds.) (2000). Sensitivity of RP surface finish to process parameter variation. In: *Solid Freeform Fabrication Proceedings*. Austin, TX: University of Texas.

39 Van Weeren, R., Agarwala, M., Jamalabad, V. et al. (eds.) (1995). Quality of parts processed by fused deposition. In: *Proceedings of the Solid Freeform Fabrication Symposium*. Austin, TX: University of Texas.

40 Zhang, B., Seong, B., Nguyen, V., and Byun, D. (2016). 3D printing of high-resolution PLA-based structures by hybrid electrohydrodynamic and fused deposition modeling techniques. *Journal of Micromechanics and Microengineering* 26 (2): 025015.

41 Goyanes, A., Chang, H., Sedough, D. et al. (2015). Fabrication of controlled-release budesonide tablets via desktop (FDM) 3D printing. *International Journal of Pharmaceutics* 496 (2): 414–420.

42 Alhijjaj, M., Belton, P., and Qi, S. (2016). An investigation into the use of polymer blends to improve the printability of and regulate drug release from pharmaceutical solid dispersions prepared via fused deposition modeling (FDM) 3D printing. *European Journal of Pharmaceutics and Biopharmaceutics* 108: 111–125.

43 Aho, J., Boetker, J.P., Baldursdottir, S., and Rantanen, J. (2015). Rheology as a tool for evaluation of melt processability of innovative dosage forms. *International Journal of Pharmaceutics* 494 (2): 623–642.

44 Shah, S., Maddineni, S., Lu, J., and Repka, M.A. (2013). Melt extrusion with poorly soluble drugs. *Internatinal Journal of Pharmaceutics* 453 (1): 233–252.

45 Zhang, J., Feng, X., Patil, H. et al. (2017). Coupling 3D printing with hot-melt extrusion to produce controlled-release tablets. *Inteernational Journal of Pharmaceutics* 519 (1-2): 186–197.

46 Okwuosa, T.C., Stefaniak, D., Arafat, B. et al. (2016). A lower temperature FDM 3D printing for the manufacture of patient-specific immediate release tablets. *Pharmaceutical Research* 33 (11): 2704–2712.

47 Sadia, M., Sośnicka, A., Arafat, B. et al. (2016). Adaptation of pharmaceutical excipients to FDM 3D printing for the fabrication of patient-tailored immediate release tablets. *International Journal of Pharmaceutics* 513 (1–2): 659–668.

48 Sadia, M., Arafat, B., Ahmed, W. et al. (2018). Channelled tablets: an innovative approach to accelerating drug release from 3D printed tablets. *Journal of Controlled Release* 269: 355–363.

49 Solanki, N.G., Tahsin, M., Shah, A.V., and Serajuddin, A.T. (2018). Formulation of 3D printed tablet for rapid drug release by fused deposition modeling: screening polymers for drug release, drug-polymer miscibility and printability. *Journal of Pharmaceutical Sciences* 107 (1): 390–401.

50 Lipson, H. and Kurman, M. (2013). *Fabricated: The New World of 3D Printing*. Wiley.

51 US-FDA (2004). Guidance for Industry: PAT — A Framework for Innovative Pharmaceutical Development, Manufacturing, and Quality Assurance.

52 Cerniglia, D., Scafidi, M., Pantano, A., and Rudlin, J. (2015). Inspection of additive-manufactured layered components. *Ultrasonics* 62: 292–298.

53 Alhijjaj, M., Reading, M., Belton, P., and Qi, S. (2015). Thermal analysis by structural characterization as a method for assessing heterogeneity in complex solid pharmaceutical dosage forms. *Analytical Chemistry* 87 (21): 10848–10855.

54 US-FDA (2000). *Q6A Specifications: Test Procedures and Acceptance Criteria for New Drug Substances and New Drug Products: Chemical Substances*, 83041–83063.

12

3D Printing of Metallic Cellular Scaffolds for Bone Implants

Xipeng Tan[1] and Yu Jun Tan[2]

[1] *Nanyang Technology University, Singapore Centre for 3D Printing, 50 Nanyang Avenue, 639798, Singapore*
[2] *National University of Singapore, Biomedical Institute for Global Health Research and Technology, 21 Lower Kent Ridge Road, 119077, Singapore*

12.1 Introduction

The term scaffold has different meanings in various fields, but it is generally defined as a temporary frame or a support structure. Particularly, in tissue engineering, a scaffold is a highly cellular or porous artificial framework that serves as a mimicry of extracellular matrix (ECM) for cell adhesion, migration, proliferation, and tissue regeneration in three dimensions [1]. To facilitate and guide the formation and maturation of new tissues or organs in tissue regeneration or repair, biodegradable scaffolds are typically employed with a desirable balance between temporary mechanical integrity and mass transport [2]. In recent years, the concept of scaffold in the biomedical field has been extended to the open cellular structures with cell-growth-favorable microarchitectures in which materials are bioinert.

Many different types of materials have been used for tissue-scaffolding materials, such as polymers, hydrogels, metals, ceramics, and hybrid materials. There are some necessary requirements that have to be fulfilled for a hard tissue scaffold, including excellent biocompatibility (nontoxic), high porosity, sufficient corrosion and wear resistance, suitable mechanical properties, and osseointegration (in the case of bone implant) [3], as shown in Figure 12.1. The scaffold materials have become an increasingly important category of biomedical engineering materials. Although the scaffold-free tissue engineering methods, such as the cell sheet technology or scaffold-free bioprinting, are very promising, they still require a long-term research before maturation and realization of the technology. Nowadays, the scaffold-based biomedical strategies remain as the most popular, reliable, and achievable solutions to medical care of human beings.

With the drastic increase in aging populations and longevity of human beings, it is experiencing an exceedingly high demand in functional bone grafting or implantation globally. In addition, the worldwide incidence of bone disorders and conditions has trended steeply upward, especially in populations where aging is

3D and 4D Printing in Biomedical Applications: Process Engineering and Additive Manufacturing,
First Edition. Edited by Mohammed Maniruzzaman.

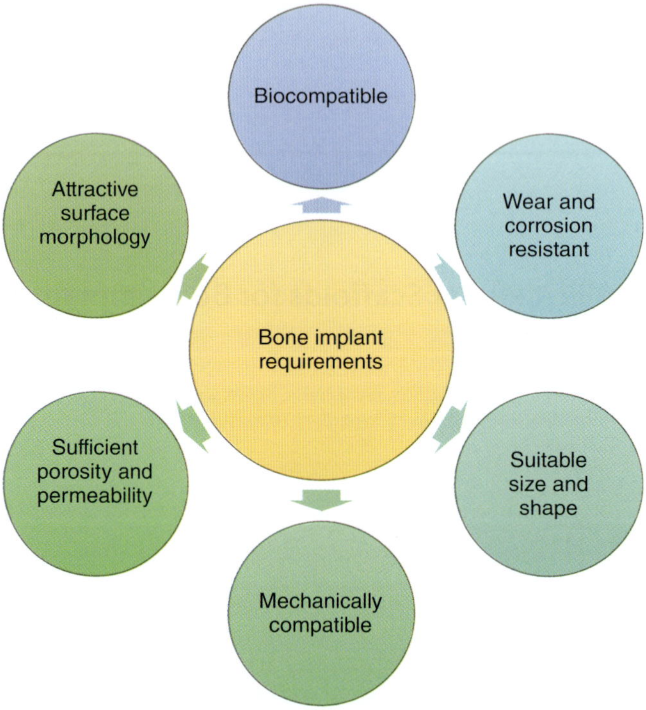

Figure 12.1 Essential requirements for bone implants.

accompanied with increased obesity and poor physical activity. An authorized market survey indicates that the global bone grafts and substitutes market is poised to grow at a compound annual growth rate (CAGR) of around 5.6% over the next decade to reach approximately $ 3.96 billions by 2025 [4]. Undoubtedly, all these facts will boost the consecutive development of bone implant materials and technologies in the long run. Engineered bone implant has been viewed as a potential alternative to the conventional use of bone grafts because of their limitless supply and safe implantation. Bone tissue engineering aims to induce new functional bone regeneration via the synergistic combination of biomaterials, cells, and factor therapy [5]. However, for both bone implant and bone tissue engineering, the fabrication of highly porous scaffolds using the available biomaterials is the primary concern. Thanks to the rapid advance in three-dimensional (3D) printing technique, which is also known as additive manufacturing (AM), or rapid prototyping (RP), or solid freeform fabrication (SFFF), especially for the latest emergence of powder bed fusion (PBF) technology, the fabrication of intricate, customized cellular scaffolds for bone implant can be achieved with relative ease since the past decade.

The ability to fabricate porous biomaterials with a controlled microarchitecture is crucial to the successful bone implantation. Traditional methods, such as direct metal foaming and powder metallurgy, can generally produce random pore distribution and nonhomogeneous porosity; others, such as vapor deposition, can only produce an almost uniform and homogeneous arrangement of unit cells [6]. However, only limited freedom is offered in generating complex cellular geometries, with favorable unit cell topology that can best fulfill specific biological and mechanical requirements. Typically, it is very difficult or impossible to rely on traditional manufacturing methods to craft a cellular structure throughout an orthopedic implant. 3D printing makes this relatively easy as it builds up in a layer-by-layer form, including the internal cellular cross sections. It is hence possible to produce intricate cellular implants tailored to biomedical applications. To produce a useful implant, factors such as topological design of pores, porosity, mechanical properties, and *in vivo* response to natural bone tissues must be taken into account [7].

In this chapter, the emerging PBF 3D printing techniques for producing metallic cellular scaffolds for bone implants as well as the available metallic biomaterials (including nondegradable and biodegradable) will be mainly introduced. Moreover, the recent advances in 3D printing of metallic cellular scaffolds with various unit cell topologies will be reviewed in detail. The main focus of this chapter will be placed on the newly developed 3D printing technology and its applications in cellular scaffolds using a variety of metallic biomaterials.

12.2 Metal 3D Printing Techniques for Bone Implants

3D printing is a process of joining materials (in powder, wire, or sheet forms) to create objects from 3D models usually layer by layer. It is a digital, automated, and layer-wise manufacturing process, which makes it distinct from the traditional subtractive and formative processes. The American Society for Testing and Materials (ASTM) group ASTM F42 has classified 3D printing or AM into seven process categories [8]. The categories pertaining to metal 3D printing include PBF, directed energy deposition (DED), and sheet lamination (SL). Furthermore, based on the material form and the manufacturing manner, metal 3D printing technology can be subdivided into the following systems: PBF, powder-fed fusion, wire-fed fusion, SL, and hybrid manufacturing. The detailed comparison among these metal 3D printing methods is given in Table 12.1. It is noted that PBF 3D printing process owns the best build resolution as compared to others, which enables itself to be the only currently available technique that can print metallic cellular scaffolds with microarchitectures. There are some key advantages of using 3D printing technique to fabricate bone scaffolds, such as customization or personalization, high design freedom, save materials, and reduced buy-to-fly time. Another benefit of 3D printing is the high flexibility of its builds. Various properties, such as porosity, strength, and ductility, can be adjusted to achieve the optimum in replicating the function of bone in a body [7].

Selective laser melting (SLM) and selective electron beam melting (SEBM) are the representative PBF 3D printing techniques for metals and alloys nowadays. In both processes, high-energy beams (either laser or electron beams) are utilized to melt cross-sectional shapes into layers of metal powder, fusing powder particles into a large form, with each layer representing a "slice" of the

Table 12.1 A comparison on various metal 3D printing techniques [8].

System	Energy source	Build environment	Build resolution	Build rate	Densification	Surface finish	Materials saving	Popularity
Powder bed fusion	Laser/electron beam	Inert gas/vacuum	High	Low	High	Good	Good	High
Powder-fed fusion	Laser	Air	Low	High	High	Poor	Good	High
Wire-fed fusion	Laser/electron beam/arc	Air/vacuum	Low	High	High	Poor	Good	Low
Sheet lamination	Ultrasonic	Air	Low	High	Low	Poor	Poor	Low
Hybrid	Laser	Air	Low	High	High	Good	Poor	High

final product. After every slice is formed, the build platform moves downward by the distance equivalent to a layer's thickness and a fresh layer of metal powder is uniformly spread on top of it. The process is repeated so that the cross sections build up cumulatively until the build is finished. At the end, the excess, unmolten powder is removed by high-pressure blowing, leaving behind the near-net-shape build as defined in the computer-aided design (CAD) model. The unused powder can be recycled for subsequent rounds of fabrication. Both processes are capable of fabricating complex designs such as cellular structures with high accuracy. Recent advances in PBF technology have brought versatile layer-by-layer processes and systems that facilitate the fabrication of porous materials and bone replacement implants with controlled unit cell morphology. Additionally, through 3D printing, gradients of porosity and pore size can be introduced into the microarchitecture to enhance mechanobiological performance with respect to anatomical location [9].

12.2.1 Selective Laser Melting

Typically, the setup of an SLM system consists of a laser source, powder containers and delivering and layering apparatus, print platform, and a computer system for process parameter control. Figure 12.2 schematically illustrates an SLM system. The build platform is heated to a temperature usually below ~200 °C and maintained at this temperature throughout the process. The desired material in powder form is loaded into a tank. This powder, of mean particle size ranging between ~20 and 60 μm, is delivered onto the build platform and recoated into layers with the thickness of a few powder particles, typically 30–100 μm. A flat, even layer is crucial to the accuracy and reliability of the SLM process, and hence, the morphology and granulometry of powders used must be scrutinized. This is to ensure that the powder particles are able to be spread effectively

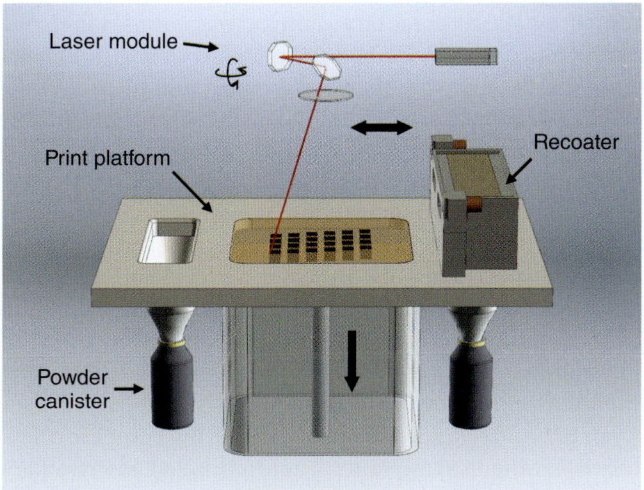

Figure 12.2 Schematic of an SLM system.

and uniformly at operating temperatures to make up the desired thickness. As a result, metal powders produced for SLM and other powder-based 3D printing processes should ideally possess good sphericity and small size distribution range to enhance their overall flowability. Additionally, oxygen must be removed from the system to prevent oxidation. This is usually achieved by introducing either purified argon or nitrogen into the build chamber. After all the above conditions have been met, a laser beam is focused onto the powder bed, tracing the areas to be melted. To achieve an optimal effect depending on the powder material and build specifications, various laser parameters can be altered. A variety of lasers can be used such as CO_2, Nd:YAG, and fiber lasers. These vary in wavelength and energy density and may have different levels of absorptivity with different materials. The laser power, scanning speed, and infill hatch melting pattern could be optimized in order to achieve desirable microstructure and properties [7, 10].

12.2.2 Selective Electron Beam Melting

SEBM technique was developed by a Sweden-based company Arcam AB that then became a GE additive company, who named their commercial machine electron beam melting (EBM). Similar to SLM, EBM involves a system consisting of powder hoppers, rake, print platform, and an energy source used to melt powder as shown in Figure 12.3. However, EBM adopts an electron beam as the energy source. Electrons emitted by a tungsten filament are accelerated by a high voltage of 60 kV to a high velocity before being focused into a high energy beam by electromagnetic lenses. Unlike SLM, EBM must work under a high vacuum because of some inherent features of electron beams. Additionally, certain metals or alloys, being highly reactive, are susceptible to gathering impurities when exposed to air. Impurities afflicted by contact with oxygen or other chemical entities in air can be prevented using a vacuum chamber, and this way, the integrity of a 3D printed titanium part can be ensured. As with SLM, there is a strong correlation between powder quality and the fabrication process. A good powder for metal printing is defined by good flowability, compactness of packing, and heat transfer characteristics, all of which can be maximized by using fine powder with spherical morphology. Powder size distribution must also be taken into consideration as this can have a strong effect on build density, surface finish, and mechanical characteristics [7, 11].

Table 12.2 lists the main indicators for SLM and SEBM processes. The major difference between SLM and SEBM is the energy source, namely, SLM utilizes a laser beam, whereas SEBM employs an electron beam. This leads to a difference in operating environment as it is necessary to use a high vacuum in SEBM to maintain the strength of the electron beam. As compared to SLM, SEBM is faster in producing fully dense builds with a high energy electron beam allowing full melting of powder particles and a faster scanning speed. Completely molten particles are also a key to aid metallurgical bonding between layers. In addition, SEBM has also been proven to be advantageous in manufacturing complex cellular titanium builds suitable as biomedical implants. Titanium, being an extremely reactive metal, is highly susceptible to the formation of impurities simply through contact with oxygen. Such impurities increase the difficulty of manufacturing

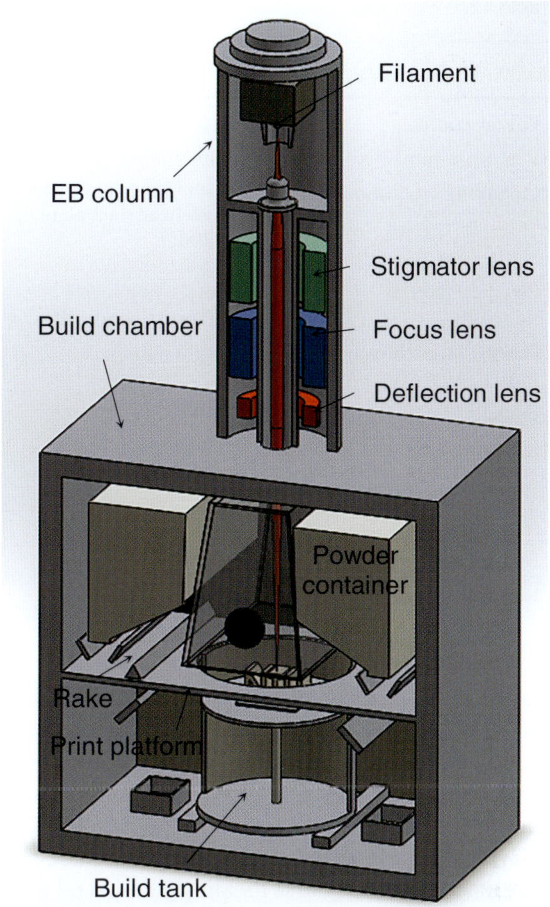

Figure 12.3 Schematic of an Arcam EBM system.

these titanium parts to meet the mechanical properties required of an implant, for example, unsatisfactory ductility. The integrity of titanium implants can be preserved under the high vacuum of the SEBM process. However, because of the high power with a coarse beam spot, SEBM suffers from poor print resolution in comparison with SLM. Also, it is challenging to fully eliminate the metal powders trapped within cellular structures as SEBM works under elevated temperatures. Therefore, SLM is more preferable to print cellular structures with intricate and fine unit cells or pores.

12.3 Biometals for Bone Implants

Metallic biomaterials (also known as biomedical metals and alloys or biometals) have been widely used to produce various types of biomedical implants, ranging from large-scale, load-bearing orthopedic prostheses to small-sized dental implants and cardiovascular stents, owing to their favorable combination of superior properties including high strength, high fracture toughness,

Table 12.2 A comparison between SLM and SEBM processes [7, 8].

	SLM	SEBM
Energy source	Laser beam (up to 1 kW)	Electron beam (3 kW)
Beam size (μm)	~80–100	~150–250
Operating environment	• Inert gas environment • Preheat start plate up to 200 °C	• High vacuum • Chamber temperature maintained at 600–1000 °C
Powder material	Metals, polymers, ceramics, composites, etc.	Metals and metallic-based composites
Powder layer thickness (μm)	20–50	50
Powder particle size (μm)	~20–63	~45–105
Melting method	• Contour melting • Infill hatch melting	• Powder bed preheating • Contour melting • Infill hatch melting
Build rate	Slow	Fast
Build resolution	High	Low
Residual stress	High	Low

outstanding biocompatibility, good wear, and corrosion resistance. Moreover, biometals can be massively processed using the well-established manufacturing techniques such as casting and forging, and recently, AM to fabricate complex personalized human implants. Common biometals include Ti-based alloys (e.g. Ti–6Al–4V ELI), Co-based alloys (e.g. CoCrMo), austenitic stainless steels (e.g. SS316L), Zr–Nb alloys, Ni–Ti shape memory alloys, pure Ta, W-based alloys, precious metals, Mg and its alloys, Fe-based alloys, etc [3, 12]. In terms of their biodegradability in physiological environment, these biometals can be classified into two categories: nondegradable and biodegradable biometals. Currently, one mainstream direction for 3D printing is that of biomedical applications, especially in producing cellular scaffolds for orthopedic implants. Many attempts have been made to print these biometals using PBF techniques and some of them were successful in 3D printing of cellular structures.

12.3.1 Nondegradable Biometals

Nondegradable, inert biometals include titanium and its alloys, surgical stainless steels, cobalt–chromium-based alloys, Ni–Ti shape memory alloy, tantalum, tungsten, precious metals, and so on. The strength of these biometals usually far exceeds the requirements for bone implant. However, the major drawback of these nondegradable biometals is the high Young's modulus. It may attribute to the poor mechanical integrity because of the mismatch between metallic implants and human bones, also called stress-shielding effect, leading to failure of bone implantation. A solution to overcome this issue is to utilize cellular or porous structures with stochastic or regular microarchitectures [13]. Among those inert biometals, Ti-based alloys, Co-based alloys, and stainless steels to date are primarily popular and commercially used for producing orthopedic implants because of its best combination of biocompatibility, corrosion resistance, mechanical properties, and cost. Compared with cobalt alloys and stainless steel, titanium alloys represented by Ti–6Al–4V ELI have been proven to be superior in bone implant applications because of the excellent biocompatibility coupled with lightweight and low modulus [3]. However, titanium and its alloys are usually expensive and hard to manufacture, and their wear resistance is poor. By contrast, stainless steels are cheap and easy to manufacture, and cobalt alloys have outstanding wear properties. Therefore, a new class of high-performance implantable biometals is still highly desired.

12.3.2 Biodegradable Biometals

A key feature of bone scaffolds is to provide temporary mechanical integrity at the defect site until the bone tissue is repaired or regenerated, and normal biomechanical function is restored. Metals that can degrade in physiological environment, namely, biodegradable metals, are proposed as potential materials for hard tissue scaffolding where biodegradable polymers are often considered as having poor strength and bioceramics usually suffer from disastrous brittleness. Biodegradable metal scaffolds have shown interesting mechanical properties that are close to those of human bone with tailored degradation behavior. Hence, biodegradable biometals have been studied as alternative implantable metals in orthopedics for years. Magnesium and its alloys are promising potential candidates, as their density and mechanical properties are similar to human bone tissues. Moreover, they are degradable and resorbable, and Mg ions are essential for cell functions [14]. It is noted that the paradigm of implants must be inert and corrosion resistant and has now been partially changed, owing to the advent of this new class of degradable biomaterials [15].

Table 12.3 lists the key physical and mechanical properties of human bone tissues and some popular nondegradable and degradable biometals. It is seen that in solid form, only NiTi and Mg own similar Young's modulus with natural cortical bone. In particular, it also matches very well between Mg and cortical bone on many other aspects such as density and mechanical strength. Unfortunately, a lot of concerns have been raised about the biocompatibility of NiTi shape memory alloy because of the known toxic, allergenic, and carcinogenic properties of Ni [18] and rapid degradation and release of hydrogen gas for Mg and its alloys, which may seriously inhibit their applications clinically. In addition, there are still some critical challenges regarding the use of implantable biometals. For instance, the long-term presence of nondegradable biometals such as steel, Co–Cr, or Ti alloy in the body is associated with an increased risk of allergic reactions or other side effects, whereas a relatively high Young's modulus of these metals (even in cellular structures) compared to natural bone tissue (especially for cancellous bone) leads to stress shielding and consequent osteopenia. On the other hand, the control over biodegradation kinetics is critical for resorbable metals, such as Mg or Fe, where untimely degradation may lead to premature loss of mechanical strength before the bone tissue or function has been fully restored.

Table 12.3 A comparison of mechanical properties between bone tissue and some popular biometals [3, 12, 14, 16, 17].

Material	Density ($g\,cm^{-3}$)	Young's modulus (GPa)	Yield strength (MPa)	Tensile strength (MPa)	Compression strength (MPa)
Bone tissue					
Cortical bone	1.8–2.0	7–30	104.9–114.3	160–240	100–230
Cancellous bone	1.0–1.4	0.01–3		1.5–38	2–12
Nondegradable biometal					
Ti–6Al–4V ELI	4.43	110	860	930	
CoCrMo	8.2–8.4	200–230	500–1500	900–1540	
SS316L	8.0	200	170–310	540–1000	480–620
Ni–Ti	6.45	38–48	70–140	1240	
Ta	16.69	200	220	310	
Degradable biometal					
Mg	1.74	41	21–100	87–180	40–140
Fe–20Mn	7.73	207	420	700	

12.3.3 3D Printing of Biometals

3D printing provides the opportunities to fabricate engineered cellular metallic scaffolds with tunable mechanical and functional properties via changing the topology design of unit cells and process parameters. Because of the wide applications in many industrial sectors, Ti–6Al–4V, CoCrMo, and SS316L have been extensively printed using various 3D printing techniques such as PBF and DED. Thereinto, SLM and SEBM processing of the medical-grade Ti–6Al–4V, CoCrMo, and SS316L has become relatively mature, and the three commercially available biometals can be manufactured into many different types of high-quality complex functional parts including orthopedic implants [3]. Moreover, the mechanical properties of 3D printed metallic parts are found to be, in general, comparable or even superior to their cast and wrought counterparts. Many other biometals have also been printed into solid or porous samples using SLM and/or SEBM methods, even though the trials are still pending at process parameter optimization. As PBF 3D printing technology is currently the most suitable to fabricate metallic cellular scaffolds with microarchitectures, the state-of-the-art advances in SLM and SEBM printing of some important biometals will be stated in detail below.

12.3.3.1 Ti–6Al–4V ELI Alloy

Ti–6Al–4V is the most popular titanium alloy and one of the most important engineering materials in some key industrial sectors such as biomedicine and aerospace. Medical-grade 23 Ti–6Al–4V ELI is widely used in the biomedical fields because of its lightweight but strong, excellent biocompatibility, bioinertness, and corrosion resistance. Ti and its alloys are expensive and hard to manufacture and machine, so their parts have relatively high added value. Hence, among all the materials, Ti–6Al–4V ELI is the most studied material being used in metal 3D printing field to date. The 3D printing process of Ti–6Al–4V alloy has become very mature [11, 19].

A large number of cellular Ti–6Al–4V scaffolds with various pore shapes and sizes have been fabricated by using both SLM and SEBM. Much work has been conducted on mechanical response, *in vitro* and *in vivo* performance, and clinical trials of those 3D printed Ti–6Al–4V implants [7, 13]. 3D printed Ti–6Al–4V implants have been commercially available in market of some countries like United States and China. However, there is still lack of long-term implantable performance data for the 3D printed Ti–6Al–4V implants.

12.3.3.2 CoCrMo Alloy

CoCrMo alloy was used as a dental and orthopedic implant material since 1940s. It is in general superior to stainless steels in terms of strength, wear, and corrosion resistance, which makes it suitable for a wide range of orthopedic applications including metallic components of all joint replacement, as well as fracture fixation devices. Moreover, its good biocompatibility and bioinertness have been verified based on the long-term clinical use. Although a number of issues are accompanied by the alloy, including failure because of fretting and corrosion fatigue; stress-shielding effect; biological toxicity because of Co, Cr, and Ni ion; or particle release, wrought CoCrMo alloy is still the most popular metallic implant material for load-bearing joint replacement to improve the life quality of human beings worldwide [3]. Similarly, CoCrMo alloy also suffers from difficulties in manufacturing and machining, and high price as compared to stainless steels. It is also one of the available materials for PBF 3D printing process at the earliest period [20]. Its popularity in both academia and industry in the field of metal 3D printing is just secondary to Ti–6Al–4V.

3D printing of cellular CoCrMo scaffolds with different unit cell shapes and sizes has been realized by using both SLM and SEBM [21, 22]. It is worth noting that minor differences in the microstructure of CoCrMo parts fabricated by SLM and SEBM existed because of their different thermal history. SLM process can minimize carbide precipitation and formation of equiaxed ε grains that could enhance corrosion resistance and limit the release of metal ions or carbide particles into surrounding environment, whereby the risk of allergic or toxic reaction will be much reduced [12]. Long-term osseointegration behavior of SEBM-fabricated CoCrMo cellular scaffolds was evaluated by 26-week implantation in adult sheep femora and compared with the Ti–6Al–4V counterpart. The encouraging results showed that bone formation patterns around and inside the open cellular CoCrMo and Ti–6Al–4V implants were comparable [22]. Therefore, it is promising to use 3D printed CoCrMo scaffolds for bone implants.

12.3.3.3 Stainless Steel 316L Alloy

In general, SS316L shows relatively good compatibility, but to a less satisfactory level in comparison with Ti–6Al–4V and CoCrMo alloys, because of its greater corrosion rates. In addition, there are many long-term issues of using SS316L as an implant material, such as stress corrosion cracking, poor resistance to wear and fatigue, and toxicity and carcinogenicity of released Ni and Cr ions. Nowadays, SS316L has been rarely used in any long-term or permanent implant devices. Nevertheless, the low cost of stainless steels has maintained its application in a large quantity of temporary devices, such as internal fracture fixation and traction of the spine, bone screws, bone plates, intramedullary nails, and rods [3, 12].

Numerous research works have been conducted to print SS316L parts by using SLM. SS316L has shown ideal printability under varied process parameters with different printing systems. 3D printing of nearly full density SS316L parts can be achieved under a relatively high build rate [10]. Moreover, SLM-fabricated SS316L samples have shown extraordinary strength and ductility in terms of conventionally processed ones and ASTM standard. Recently, SEBM was also successfully applied to fabricate SS316L parts. Because of the elevated build temperature, the SEBM process promotes microstructural coarsening and the detrimental σ precipitation at columnar grain boundaries, which may result in anisotropic properties and weakening of the printed parts [23].

12.3.3.4 NiTi Shape Memory Alloy

NiTi or nitinol is an attractive alloy because of its unique functional properties (i.e. shape memory effect and superelasticity behavior), low modulus, good biocompatibility, damping characteristics, and corrosion resistance, which make it an excellent candidate in functional implant designs. Cellular NiTi scaffolds are being studied to be used in biomedical implants because they present superelasticity behavior similar to bone, high bone growth through the interconnecting pores, low tunable stiffness (i.e. modulus of elasticity), and capillary properties. One of the advantages of using NiTi implants is that their shape can be further adjusted after implantation. This is possible by initiating the desirable shape memory effect of this material using body temperature. Two primary benefits might be obtained from changes in implant shape: (i) enhanced bone fixation and (ii) minimization of invasive surgery. It, however, suffers from difficulties in manufacturing and machining NiTi alloy because of the high reactivity and high ductility by means of conventional methods [18, 24, 25]. Additionally, release of toxic Ni ions in body fluid is one of the critical concerns for using NiTi implants.

3D printing can resolve the issues of manufacturing and machining of NiTi alloy as it produces net-shaping parts under inert environment or vacuum. There have been a few studies on 3D printing of NiTi lattice structures by using SLM. The cellular NiTi samples were shown to be nontoxic, and the Ni ion release was below cytotoxic concentrations [25]. However, some common issues and concerns on its 3D printing process can be raised. For instance, the optimum process parameters may not be able to be applied between different SLM models or systems; severe geometrical variation may exist between CAD file and the printed lattice structure, and the high residual stress as well as the higher transformation temperature may affect the function of NiTi scaffolds. It is worth noting that very limited works have been reported on SEBM printing of NiTi parts. This is because of the fact that EBM processing may cause poor surface finish, large geometrical deviation, and high consumption in powder material. At the same time, there are many advantages in the fabrication of cellular NiTi scaffolds using SEBM, such as good impurity control, negligible residual stress, and consistent process parameters.

12.3.3.5 Tantalum

Ta is a hard, ductile, and highly chemically inert material with good adherence to human bone. The biocompatibility and nontoxic behavior of Ta in general are excellent in comparison to the more commonly used surgical grade 5 Ti–6Al–4V. However, the wide use of Ta as an implant material is limited because of its high Young's modulus, density, and price, as well as difficulty in manufacturing and machining with precision. Moreover, it is challenging to manufacture dense Ta parts by means of conventional methods, owing to its refractory nature with a very high melting temperature of 3017 °C. High-energy beam-based 3D printing technique is thought to be optimum to process Ta, especially its cellular scaffolds for small implants. SLM has been proven to be able to produce both fully dense Ta solid parts and cellular Ta structures with a pore size of 500 μm and a porosity of 80%. Furthermore, it was found that SLM can be used to fabricate highly porous, pure Ta orthopedic implants with interesting mechanical properties and promising *in vivo* performance [26, 27].

12.3.3.6 Mg and Its Alloy

Mg and its alloys (Mg–Al-based, Mg–Zn-based, and Mg–Ca-based) have been extensively exploited as potential biodegradable scaffold materials for orthopedic implants or bone repair because they own physical and mechanical properties similar to those of natural bone as compared to other biodegradable and nondegradable biometals, ceramics, and polymers. However, the rapid corrosion of Mg-based alloys in physiological environment has become the major obstacle to be used in biomedical applications [14]. One effective solution to overcome this issue is to add alloying elements such as Al, Ca, Zn, and Mn that can react with Mg to form intermetallic phases dissolving in grain matrix or distributing along the grain boundary, thereby improving mechanical properties and corrosion resistance of Mg-based alloys. In addition to rapid corrosion, some critical issues must be addressed before clinical applications including hydrogen generation, infection, and toxicity of any alloying elements in the Mg-based alloys. Because of the inherent characteristics, e.g. high flammability in powder form and low boiling point, it is particularly difficult to carry out 3D printing of Mg and Mg-based alloys. Therefore, there are only several studies reporting the SLM processing of CP Mg [28, 29] and Mg-based alloys [30, 31] and no available reports on SEBM fabrication of them. Moreover, most of the studies were focused on the SLM process parameter optimization and only bulk Mg samples were fabricated. Recently, Fraunhofer ILT, a German institution, developed an SLM process for producing magnesium alloy (WE43) lattice scaffolds, in which both the exact shape and pore size of the scaffold can be designed. The biocompatibility of the implant prototypes has already been demonstrated *in vitro* [32].

12.4 Cellular Structure Design

Bone implants must be able to withstand substantial loading but at the same time have mechanical properties matchable to the stiffness of the local host bone in order to reduce bone resorption induced by stress shielding. In short, scaffolds for bone implants must have high strength but low modulus.

Natural bone tissues constitute open cellular structures with two different types of porosities. Inspired by the native structures of bones, cellular structures are designed using high-strength metallic biomaterials for bone implants. Cellular solids are composites with one phase of solid and the other phase of gas/fluid. The solid phase consists of a lattice, which is a connected network of struts. These cellular structures can significantly decrease the stiffness and at the same time increase the total surface area that encourages bone ingrowth, leading to a sturdier and long-lasting fixation. Cellular structures with a high effective permeability (i.e. with the presence of interconnected pores, optimized topology design of unit cell, and the high porosity in the scaffold [33]) assist in the transportation of cellular nutritional and waste matter. However, the requirements of strength vs stiffness and permeability must be balanced as they depend, in opposite ways, on the structural density (the amount of material) that makes up the scaffold (Figure 12.4). In general, the stiffness and strength of solids decrease with the increasing porosity and in turn the permeability of solids is enhanced.

An optimum design is required to maximize the functionality of cellular scaffold in the body. For example, when comparing between two unit cell shapes, e.g. spherical vs cylindrical, spherical pores were found to be stiffer, whereas cylindrical ones tend to be more permeable. Strut size and density should be optimal. Degradation in mechanical properties can be drastic with the decrease in strut size of the cells. Hence, scaffold design of bone implants includes a rigorous selection of the materials, the optimized cellular topology design (permeability and mechanical properties), and the suitable manufacturing process as illustrated in Figure 12.5.

Cellular solids are characterized by a typical unit cell with certain symmetry elements. The unit cell in millimeter or micrometer scales allows the cellular solids to be viewed both as structures and materials. Hence, the macroscopic

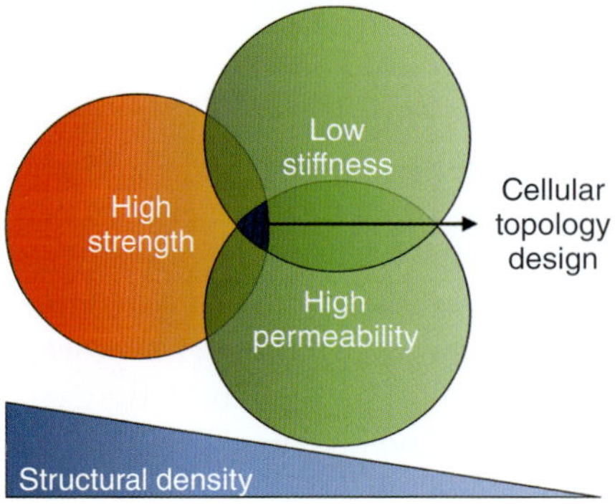

Figure 12.4 Structural density design of cellular topology in scaffolds.

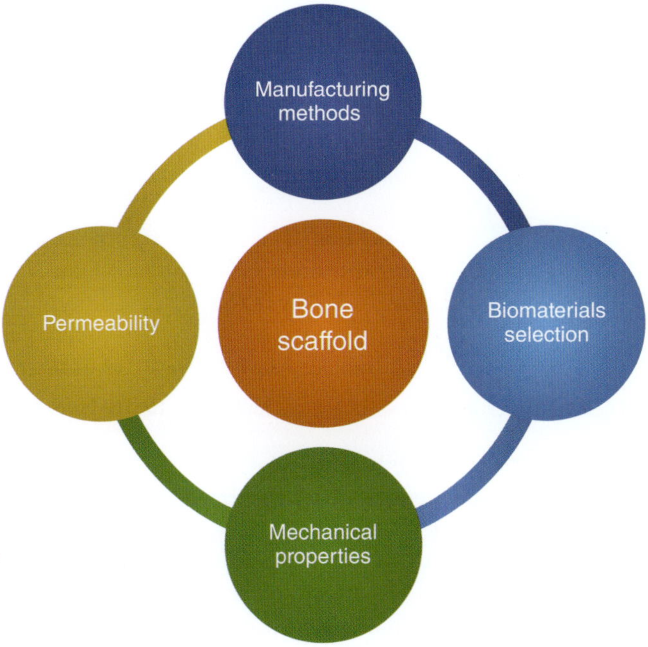

Figure 12.5 Scaffold design for bone implants.

properties of cellular solids, such as the elastic modulus and compressive strength, are governed by both material properties and structural mechanics.

There are two types of cellular solids, namely, the closed and open cellular structures. The closed cellular structures contain closed pores, and the open cellular structures have interconnected pores. This chapter will only focus on open cellular structure designs as only interconnected pores allow transport of cells (and other matters such as nutrients and waste) into and out of the implant. Open cellular structures may originate from many different designs for biomimicry of the macroscopic properties of native human bone structure. These designs may also vary in porosity, pore size, strut thickness, shape, and orientation of unit cells. Figure 12.6 illustrates different cellular designs from varying cellular forms and deformation behavior, which will be discussed in detail in this section.

12.4.1 Stochastic and Reticulated Cellular Design

Cellular structures can be divided into two main groups in terms of unit cell form, i.e. the stochastic (irregular) and the reticulated (regular) structures. The stochastic open cellular foams have pores of random shapes and sizes, and their unit cells do not repeat regularly. On the other hand, reticulated lattice consists of repeating unit cells leading to a highly regular form.

Stochastic open cellular foam structures are often heterogeneous, where the foams are strong in some regions, whereas weak in the others. The weak regions drastically drag down the mechanical performance of the foams. Also, failure of the stochastic foam structure is usually hard to predict, owing to the mechanical

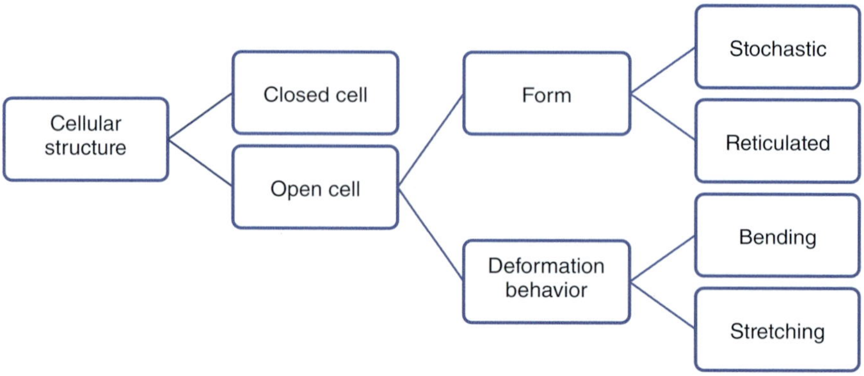

Figure 12.6 Classification of cellular structure in terms of cellular forms and their deformation behavior.

heterogeneity. Their mechanical properties could be improved from structures that contain unit cells arranged in an ordered hierarchy.

Highly regular reticulated lattice structures are preferred over the stochastic foam structures in metallic scaffold as their mechanical properties are predictable. Designing a reticulated cellular scaffold to be 3D printed involves choosing the unit cell design from an archive. To date, many different unit cell designs have been produced and studied, and they vary from the most basic cube or triangular prism, to the more complex octagonal prisms or rhombic dodecahedrons.

12.4.2 Bend- and Stretch-Dominated Cellular Design

Cellular solids can also be classified according to their deformation behavior, into the bend- and the stretch-dominated structures [34]. Most of the cellular structures, especially the stochastic foams, belong to the bend-dominated types, namely, the cell struts bend upon loading. A stretch-dominated unit cell structure is, in general, structurally stronger than the bend-dominated type. The struts of stretch-dominated cellular solids are loaded in tension or compression largely without bending. Fully triangular 3D structures, such as the tetrahedral-truss and octet-truss unit cell, belong to the stretch-dominated type. Designing bone implants using stretch-dominated cellular structures would allow both to exploit the topological advantages and to gain enhanced strength of the scaffold.

12.4.3 Scaffold Design Feasibility

A large variety of cellular designs have been fabricated through various 3D printing techniques. In light of growing interest in the area of 3D printed cellular metallic scaffolds for bone implants, there has been an enormous drive to optimize the structural design to (i) provide mechanical properties similar to natural bone to minimize stress shielding, (ii) to boost load-bearing ability, and (iii) to aid bone cell ingrowth [35]. These can be achieved by tailoring the cellular scaffold's porosity (and gradient in porosity if applicable) as well as the unit cell shape to create a lightweight but strong structure that encourages bone ingrowth.

3D printing technology has greatly favored cellular scaffold design of extremely high porosity and permeability with a periodic microarchitecture. Nonetheless, the design of these cellular scaffolds is limited by the processing-related restrictions. Arabnejad et al. formulated a strategy to identify feasible design limits for different lattice structures fabricated by SLM [36]. A few constraints are considered: (i) bone ingrowth requirements (i.e. pore size and porosity) and (ii) manufacturing limitations (i.e. strut thickness). For bone ingrowth, minimum porosity of the scaffold should be at least 40% for sufficient cell infiltration [37], and pore size should be within 50–800 μm for adequate permeability [38]. The feasible design space of SLM is bigger when compared to SEBM because of its thinner minimum strut thickness [7]. Moreover, stretch-dominated cellular structures have a narrower design space in comparison to bend-dominated ones because they usually have complicated diagonal struts.

12.5 Outlook

3D printing has shown to be able to fabricate most of the biometals into cellular scaffolds for implants. Its ability to produce intricate cellular designs that mimic

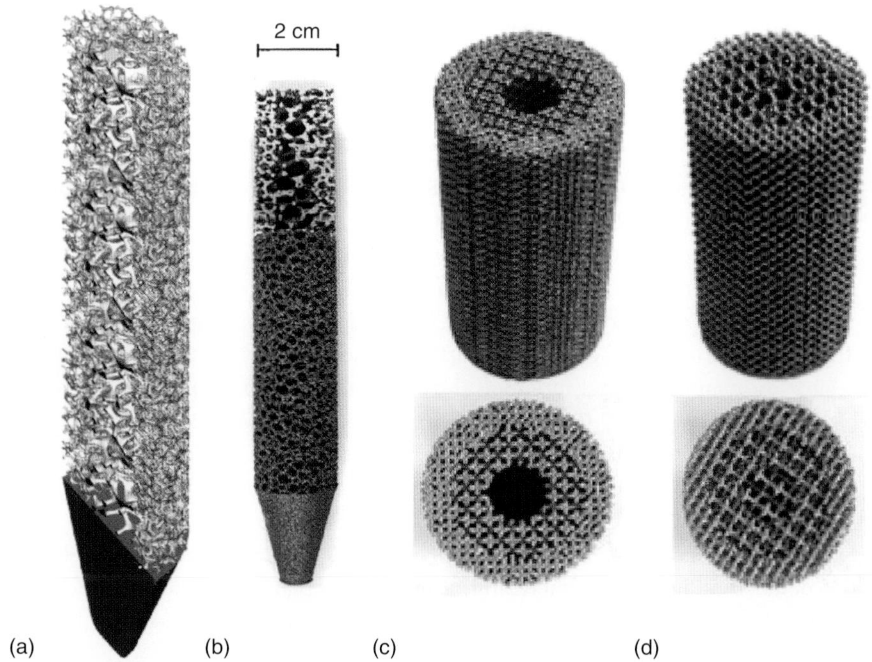

Figure 12.7 CAD models (a–d) of SEBM-fabricated bone implant prototypes with gradient in porosity from inside to outside. Source: Murr et al. 2010 [40] and Surmeneva et al. 2017 [41]. Reproduced with permission of Elsevier.

the properties of naturally occurring structures (in bone) is unmatched by other traditional manufacturing methods. It is envisioned that the next generation of biomedical implants will have complex functional cellular structures fabricated by using metal 3D printing technologies [39]. Biomimicry design of functionally graded structure as shown in Figure 12.7 improves the biological and mechanical properties of bone implants. 3D printing is beneficial in fabricating these structures because of its flexibility to manufacture different pore sizes within the same build. Hence, it is possible to design and manufacture graded cellular structures that optimize both the requirements of mechanical reliability and satisfactory biological performance.

References

1 Yang, S., Leong, K.-F., Du, Z., and Chua, C.-K. (2001). The design of scaffolds for use in tissue engineering. Part I. traditional factors. *Tissue Engineering* 7 (6): 679–689.

2 Yusop, A.H., Bakir, A.A., Shaharom, N.A. et al. (2012). Porous biodegradable metals for hard tissue scaffolds: a review. *International Journal of Biomaterials* 2012: 1–10.

3 Chen, Q. and Thouas, G.A. (2015). Metallic implant biomaterials. *Materials Science and Engineering R Reports* 87: 1–57.

4 Research and Markets (2017). Global Bone Graft and Substitutes Market Analysis & Trends – Industry Forecast to 2025.

5 Amini, A.R., Laurencin, C.T., and Nukavarapu, S.P. (2012). Bone tissue engineering: recent advances and challenges. *Critical Reviews in Biomedical Engineering* 40 (5): 363–408.

6 Melancon, D., Bagheri, Z.S., Johnston, R.B. et al. (2017). Mechanical characterization of structurally porous biomaterials built via additive manufacturing: experiments, predictive models, and design maps for load-bearing bone replacement implants. *Acta Biomaterialia* 63: 350–368.

7 Tan, X.P., Tan, Y.J., Chow, C.S.L. et al. (2017). Metallic powder-bed based 3D printing of cellular scaffolds for orthopaedic implants: a state-of-the-art review on manufacturing, topological design, mechanical properties, and biocompatibility. *Materials Science and Engineering C* 76: 1328–1343.

8 Kok, Y., Tan, X.P., Wang, P. et al. (2018). Anisotropy and heterogeneity of microstructure and mechanical properties in metal additive manufacturing: a critical review. *Materials and Design* 139: 565–586.

9 Sing, S.L., An, J., Yeong, W.Y., and Wiria, F.E. (2016). Laser and electron-beam powder-bed additive manufacturing of metallic implants: a review on processes, materials and designs. *Journal of Orthopaedic Research* 34 (3): 369–385.

10 Sun, Z., Tan, X.P., Tor, S.B., and Yeong, W.Y. (2016). Selective laser melting of stainless steel 316L with low porosity and high build rates. *Materials and Design* 104: 197–204.

11 Tan, X.P., Kok, Y., Tan, Y.J. et al. (2015). Graded microstructure and mechanical properties of additive manufactured Ti-6Al-4V via electron beam melting. *Acta Materialia* 97: 1–16.
12 Prasad, K., Bazaka, O., Chua, M. et al. (2017). Metallic biomaterials: current challenges and opportunities. *Materials (Basel)* 10 (8): 884.
13 Wang, X., Xu, S., Zhou, S. et al. (2016). Topological design and additive manufacturing of porous metals for bone scaffolds and orthopaedic implants: a review. *Biomaterials* 83: 127–141.
14 Agarwal, S., Curtin, J., Duffy, B., and Jaiswal, S. (2016). Biodegradable magnesium alloys for orthopaedic applications: a review on corrosion, biocompatibility and surface modifications. *Materials Science and Engineering C* 68: 948–963.
15 Hermawan, H. (2012). Biodegradable metals: state of the art. In: *Biodegradable Metals*, SpringerBriefs in Materials. Berlin, Heidelberg: Springer-Verlag.
16 Long, M. and Rack, H.J. (1998). Titanium alloys in total joint replacement – a materials science perspective. *Biomaterials* 19 (18): 1621–1639.
17 Murphy, W., Black, J., and Hastings, G. (2016). *Handbook of Biomaterial Properties*, 2e. Springer.
18 Biesiekierski, A., Wang, J., Abdel-Hady Gepreel, M., and Wen, C. (2012). A new look at biomedical Ti-based shape memory alloys. *Acta Biomaterialia* 8 (5): 1661–1669.
19 Tan, X.P., Kok, Y., Toh, W.Q. et al. (2016). Revealing martensitic transformation and α/β interface evolution in electron beam melting three-dimensional-printed Ti-6Al-4V. *Scientific Reports* 6 (26039): 1–10.
20 Tan, X.P., Wang, P., Kok, Y. et al. (2018). Carbide precipitation characteristics in additive manufacturing of Co-Cr-Mo alloy via selective election beam melting. *Scripta Materialia* 143: 117–121.
21 Van Hooreweder, B., Lietaert, K., Neirinck, B. et al. (2017). CoCr F75 scaffolds produced by additive manufacturing: influence of chemical etching on powder removal and mechanical performance. *Journal of the Mechanical Behavior of Biomedical Materials* 70: 60–67.
22 Shah, F.A., Omar, O., Suska, F. et al. (2016). Long-term osseointegration of 3D printed CoCr constructs with an interconnected open-pore architecture prepared by electron beam melting. *Acta Biomaterialia* 36: 296–309.
23 Wang, C.C., Tan, X.P., Liu, E., and Tor, S.B. (2018). Process parameter optimization and mechanical properties for additively manufactured stainless steel 316L parts by selective electron beam melting. *Materials & Design* 147: 157–166.
24 Elahinia, M.H., Hashemi, M., Tabesh, M., and Bhaduri, S.B. (2012). Manufacturing and processing of NiTi implants: a review. *Progress in Materials Science* 57 (5): 911–946.
25 Elahinia, M., Shayesteh Moghaddam, N., Taheri Andani, M. et al. (2016). Fabrication of NiTi through additive manufacturing: a review. *Progress in Materials Science* 83: 630–663.
26 Thijs, L., Montero Sistiaga, M.L., Wauthle, R. et al. (2013). Strong morphological and crystallographic texture and resulting yield strength anisotropy in selective laser melted tantalum. *Acta Materialia* 61 (12): 4657–4668.

27 Wauthle, R., Van Der Stok, J., Yavari, S.A. et al. (2015). Additively manufactured porous tantalum implants. *Acta Biomaterialia* 14: 217–225.
28 Ng, C.C., Savalani, M.M., Lau, M.L., and Man, H.C. (2011). Microstructure and mechanical properties of selective laser melted magnesium. *Applied Surface Science* 257 (17): 7447–7454.
29 Chung Ng, C., Savalani, M., and Chung Man, H. (2011). Fabrication of magnesium using selective laser melting technique. *Rapid Prototyping Journal* 17 (6): 479–490.
30 Taltavull, C., Torres, B., Lopez, A.J. et al. (2012). Selective laser surface melting of a magnesium–aluminium alloy. *Materials Letters* 85: 98–101.
31 Wei, K., Gao, M., Wang, Z., and Zeng, X. (2014). Effect of energy input on formability, microstructure, and mechanical properties of selective laser melted AZ91D magnesium alloy. *Materials Science and Engineering A* 611: 212–222.
32 Jauer, L., Meiners, W., Vervoort, S. et al. (2016). Selective laser melting of magnesium alloys. In: *European Congress and Exhibition on Powder Metallurgy. European PM Conference Proceedings*, 1–6. The European Powder Metallurgy Association.
33 Hollister, S.J. (2005). Porous scaffold design for tissue engineering. *Nature Materials* 4 (July): 518–524.
34 Deshpande, V.S., Ashby, M.F., and Fleck, N.A. (2001). Foam topology: bending versus stretching dominated architectures. *Acta Materialia* 49 (6): 1035–1040.
35 Parthasarathy, J., Starly, B., Raman, S., and Christensen, A. (2010). Mechanical evaluation of porous titanium (Ti6Al4V) structures with electron beam melting (EBM). *Journal of the Mechanical Behavior of Biomedical Materials* 3 (3): 249–259.
36 Arabnejad, S., Burnett Johnston, R., Pura, J.A. et al. (2016). High-strength porous biomaterials for bone replacement: a strategy to assess the interplay between cell morphology, mechanical properties, bone ingrowth, and manufacturing constraints. *Acta Biomaterialia.* 30: 345–356.
37 Bragdon, C.R., Jasty, M., Greene, M. et al. (2004). Biologic fixation of total hip implants. *Journal of Bone and Joint Surgery* 86: 105–117.
38 Harrysson, O.L.A., Cansizoglu, O., Marcellin-Little, D.J. et al. (2008). Direct metal fabrication of titanium implants with tailored materials and mechanical properties using electron beam melting technology. *Materials Science and Engineering C* 28 (3): 366–373.
39 Murr, L.E., Gaytan, S.M., Medina, F. et al. (2010). Next-generation biomedical implants using additive manufacturing of complex, cellular, and functional mesh arrays. *Philosophical Transactions. Series A: Mathematical, Physical and Engineering Sciences* 368 (1917): 1999–2032.
40 Murr, L.E., Gaytan, S.M., Medina, F. et al. (2010). Characterization of Ti-6Al-4V open cellular foams fabricated by additive manufacturing using electron beam melting. *Materials Science and Engineering A* 527 (7–8): 1861–1868.
41 Surmeneva, M.A., Surmenev, R.A., Chudinova, E.A. et al. (2017). Fabrication of multiple-layered gradient cellular metal scaffold via electron beam melting for segmental bone reconstruction. *Materials and Design* 133: 195–204.

13

3D and 4D Scaffold-Free Bioprinting

Chin Siang Ong, Pooja Yesantharao, and Narutoshi Hibino

Johns Hopkins Hospital, Division of Cardiac Surgery, 1800 Orleans Street, Baltimore, MD 21287, USA

13.1 Introduction

Traditional 3D bioprinting (inkjet bioprinting, microextrusion, and laser bioprinting) [1] uses scaffolds (also referred to as hydrogels, bioinks, matrices, biomaterial, biomatrices, etc.) as carriers to embed living cells. The cells, together with the vehicle, are then spatially positioned using 3D bioprinting technologies to create the preprogrammed or intended geometry of the final tissue constructs.

However, scaffold-dependent 3D bioprinting faces multiple limitations [2]. In terms of the scaffold choice, there is no universal scaffold suitable for all cell types. As tissue constructs are frequently made of more than one cell type, the search for an ideal scaffold is even more difficult. In terms of bioprinting, traditional 3D bioprinting methods subject live cells to stresses, such as overheating and pneumatic forces, which may lead to the release of toxic degradation products from a previously biocompatible scaffold. In addition, because of the presence of a scaffold, the cell density within the final construct is lower [3]. Cost-wise, the scaffold itself, usually a bioink, can be proprietary and expensive. In addition, optimizing proprietary scaffolds is difficult from an intellectual property point of view. Finally, for clinical or preclinical applications, after *in vivo* implantation, the scaffold can be immunogenic, degrade, and release toxic degradation products [3], eventually leading to graft failure and loss of biological function. As such, scaffold-free 3D bioprinting was developed [4, 5].

Meanwhile, four-dimensional (4D) printing is a recently developed term encompassing additive manufacturing technologies that produce dynamic, responsive 3D patterned structures that can alter their behavior or shape in response to stimuli. In other words, 4D printing is essentially a family of printing methodologies that allow for the fabrication of "smart" objects that can independently undergo physical or chemical changes over time in response to interaction with human inputs or elements from the environment. Such "smart" behavior, including self-sensing, self-actuating, and shape changing, allows 4D printed structures to have expanded functionality in comparison to

3D and 4D Printing in Biomedical Applications: Process Engineering and Additive Manufacturing,
First Edition. Edited by Mohammed Maniruzzaman.

traditional 3D printed constructs, while also retaining the spatial precision and complex structures that conventional 3D printing allows for. 4D bioprinting is the extension of 4D printing technologies to biomedical applications.

In this chapter, we will discuss 3D scaffold-free bioprinting, 4D bioprinting, and briefly, our recent efforts at 4D scaffold-free bioprinting.

13.2 3D Scaffold-Free Bioprinting

13.2.1 Principles

The basic unit in 3D scaffold-free bioprinting is a multicellular spheroid (Figure 13.1a). Spheroids can be created in a variety of ways [6, 7], such as pellet culture, spinner culture, hanging drop, liquid overlay, rotating wall vessel, external force, cell sheets, microfluidics, and nonadhesive hydrogel micromolds. However, the method of choice for spheroid creation for the first scaffold-free 3D bioprinter (Regenova 3D bioprinter, Cyfuse Biomedical K.K., Tokyo, Japan) is pellet culture without centrifugation, in ultralow attachment 96-well plates [8]. After designing the shape of the intended tissue construct (Figure 13.1b) using the software of the 3D bioprinter, a metal nozzle at the tip of a robotic arm of the 3D bioprinter picks up the individual spheroids from the 96-well plates and deposits them onto a microneedle array (Figure 13.1c) [5]. This process is repeated until the 3D bioprinting of the tissue construct is completed. After giving time for the spheroids to fuse, the tissue construct is removed from the needle array and allowed to mature *in vitro* or *in vivo* (Figure 13.1d).

13.2.2 Spheroid Optimization

In the first step (Figure 13.1a), spheroid characterization and optimization (Figures 13.2 and 13.3) are paramount. Different cell types have different sizes, different growth potentials, and different rates of extracellular matrix (ECM) secretion, and these factors impact on the spheroid strength, dimensions, and circularity.

Not all cell types result in spheroids by pellet culture without centrifugation. For example, using hiPSCs alone, spheroids do not form despite culture for 48 hours (Figure 13.2a). Instead, a soft, gel-like cell clump forms. Similarly, using only human-induced pluripotent stem-cell-derived cardiomyocytes (hiPSC-CMs) without no other cell types, spheroids do not form (Figure 13.2b), instead many small discrete smaller cell clumps form. However, from our experience and others', mesenchymal stem cells (MSCs), human umbilical vein endothelial cells (HUVECs), and fibroblasts (FBs) (Yurie et al. noted as well [10]) form spheroids when cultured by themselves as a single cell type. Moldovan et al. also found that endothelial colony-forming cells (ECFCs) and smooth-muscle forming cells (SMFCs) form spheroids, as a single cell type.

Cell types and their corresponding cell ratio within the spheroids are important considerations. Naturally, the components of the spheroid will determine the cell type and composition of the final construct. Itoh et al. [4] used HUVECs, human

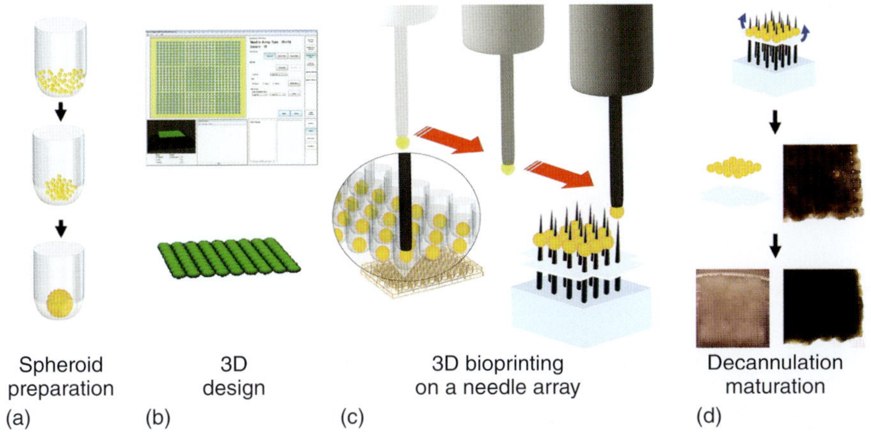

Figure 13.1 Schematic overview of biomaterial-free cardiac 3D bioprinting process. (a) Cells (CMs, FBs, and ECs) are aggregated in ultralow attachment 96-well plates to form cardiospheres. (b) The desired 3D structure to be bioprinted is designed using computer software. (c) The 3D bioprinter picks up individual cardiospheres using vacuum suction and loads them onto a needle array. (d) Cardiospheres are allowed to fuse. The 3D bioprinted cardiac tissue is then removed from the needle array and cultured further to allow the needle holes to be filled in by the surrounding tissue (Reproduced from [5]). Source: https://creativecommons.org/licenses/by/4.0/.

aortic smooth muscle cells, and human normal dermal fibroblasts at a ratio of 4 : 1 : 5 to make their spheroids. We used hiPSC-CMs, human adult ventricular cardiac FBs and HUVECs at a ratio of 14 : 3 : 3 to make our spheroids. The percentage of 15% for the endothelial component of the spheroid was chosen based on the previous published literature [3]. Yanagi et al. [9] used human mature hepatocytes (hHeps), HUVECs, and human MSCs (hMSCs) at 10 : 7 : 3 ratio. Kizawa et al. [11] used primary hepatocytes and mouse fibroblasts at 1 : 1 ratio. Moldovan et al. [12] used ECFCs and SMFCs at 1 : 5 ratio.

As expected, with increasing number of cells in the spheroids, the dimensions of the spheroids increase (Figure 13.2h–i), with a slight variation with a changing spheroid cell composition. Of note, in the method of 3D bioprinting using microneedles (Figure 13.1c), which are positioned 400 μm apart, the ideal spheroid diameter is 500 μm or larger as reported directly in all groups that use scaffold-free 3D bioprinting [5, 8–10, 12], or indirectly by observation of the micrographs of the 3D bioprinted tissue [11]. Failure to create spheroids of these dimensions results in a malformed patch or disintegration and spheroid scattering after removal from the microneedle array (Figure 13.4).

Different cell types, even the same cell type but from different animal species, have different sizes. For example, we found that to obtain a spheroid with a diameter of approximately 500 μm, 33 000 cells (hiPSC-CM:FB:HUVEC ratio of 70 : 15 : 15) are required. For rat MSCs, 60 000 cells are required. Yanagi et al. [9] found that 20 000 cells (hHep:HUVEC:hMSC 10 : 7 : 3) are required. Kizawa et al. [11] used 20 000 cells (primary hepatocytes:mouse FBs 1 : 1). Moldovan et al. used 25 000 cells (ECFC:SMFC 1 : 5).

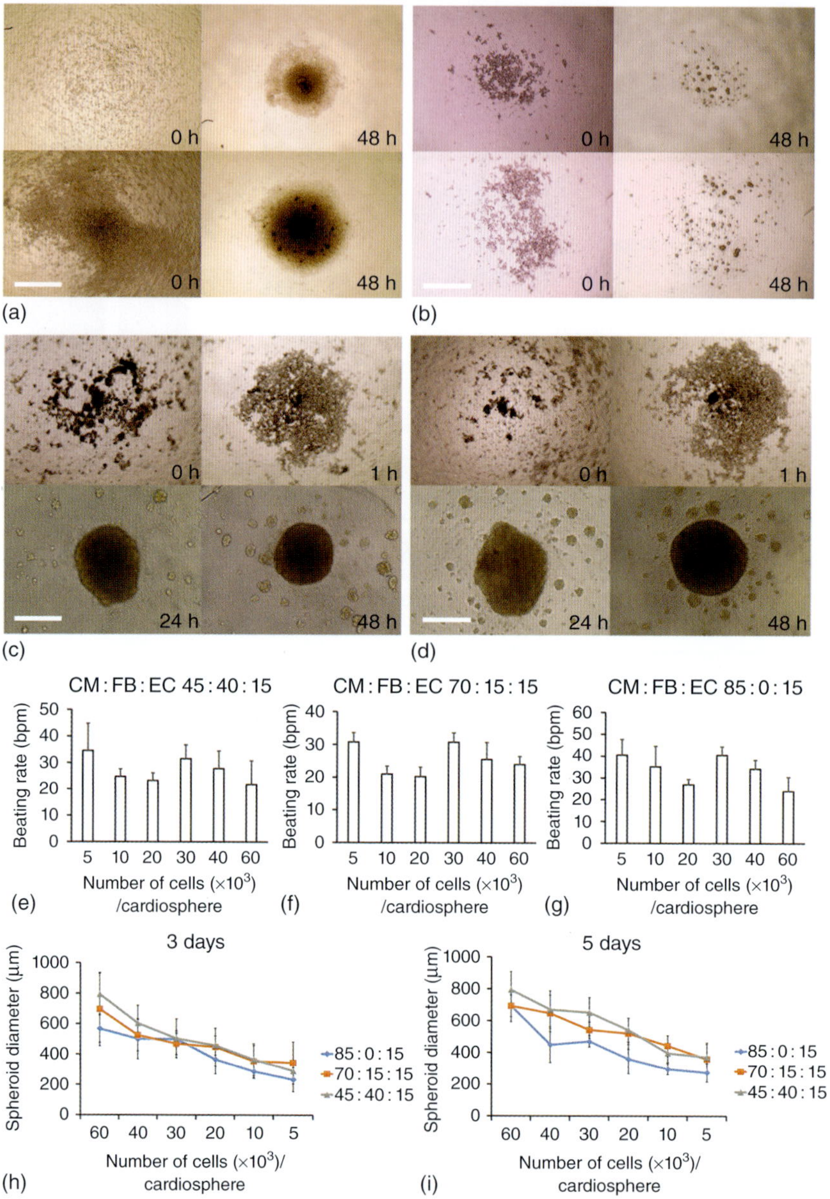

Figure 13.2 Cell type, number, and ratio determine the cardiosphere integrity. (a,b) hiPSCs (100%) (a) and hiPSC-CMs (100%) (b) do not form spheroids after 48 hours in ultralow attachment 96-well plates (top 10 000 cells/well; bottom 40 000 cells/well). (c,d) hiPSC-CM:FB:EC ratio 70 : 15 : 15 (c) and 45 : 45 : 10 (d) cardiospheres form in 24 hours and start beating in 48 hours. Scale bar panels: a–d : 400 μm. (e–g) The beating rate of cardiospheres composed of hiPSC-CM:FB:EC ratio of 45 : 40 : 15 (e), 70 : 15 : 15 (f), and 85 : 0 : 15 (g) at various cell numbers per cardiosphere, in beats per minute (bpm). (h,i) The mean diameter (μm) of cardiospheres of various cell ratios (hiPSC-CM:FB:EC) and various cell numbers, after 3 days (h) and 5 days (i) of culture (Reproduced from [5]). Source: https://creativecommons.org/licenses/by/4.0/.

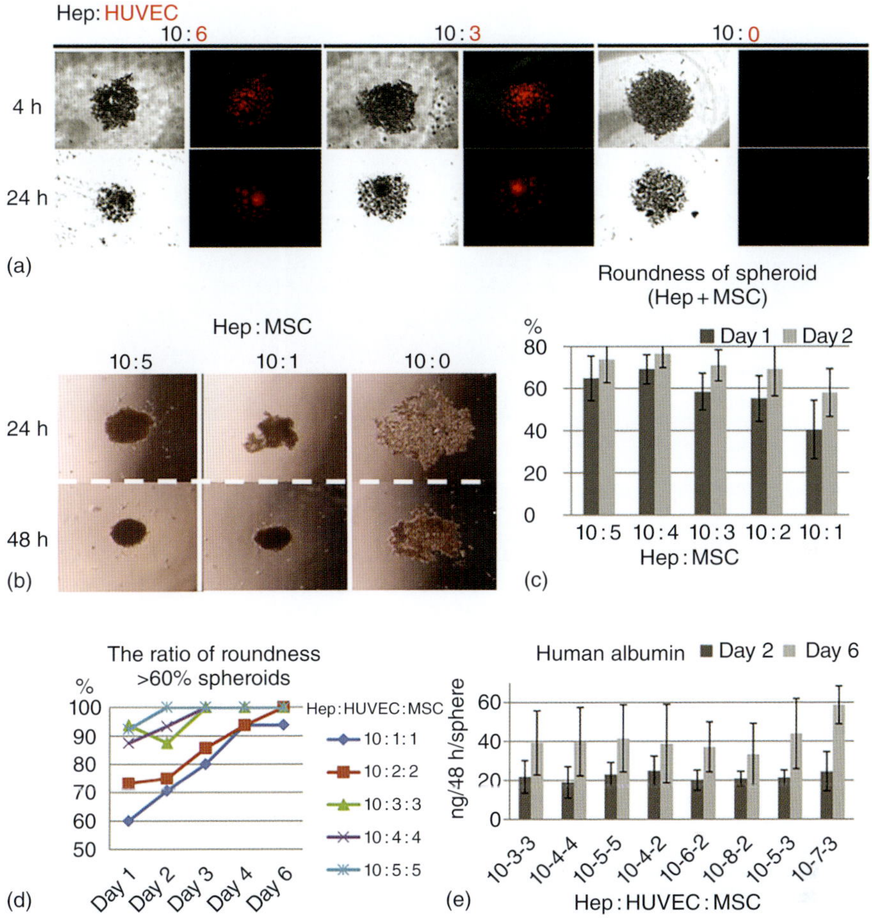

Figure 13.3 The establishment of appropriate culture conditions for generating a large number of liver budlike spheroids. (a) Microscopic observation of spheroid formation at various ratios of hHeps and HUVECs. hHeps, black; HUVECs, red. (b,c) The morphological analysis of heterospheroid formation at various ratios of hHeps and hMSCs. (b) The microscopic observation of spheroid formation. The addition of hMSCs dramatically promoted hepatocyte aggregation and a single spheroid was generated in two days. (c) The roundness of the heterospheroid at various ratios of hHeps and hMSCs. The results represent mean ± SD; $n = 12$. (d) The ratios of the spheroids that had a roundness of ≥ 60% at various ratios of hHeps, HUVECs, and hMSCs; $n = 16$. (e) The secretion of albumin into the culture medium was measured by ELISA. LBSs containing hHeps, HUVECs, and hMSCs at a ratio of 10 : 7 : 3 showed the significantly highest level of albumin secretion. The results represent the mean ± SD; $n = 8$. (f) The profiles of the optimized spheroid (hHeps:HUVECs:hMSCs at $10 \times 10^3 : 7 \times 10^3 : 3 \times 10^3$ cells). Changes in the diameters and roundness of the spheroids and the ratios of the spheroids that had the roundness of ≥ 60%. The results represent the mean ± SD; $n = 16$. (g) The microscopic observation of spheroid formation using mixtures of the three cells types (hHeps: black; HUVECs: red; and MSCs: green). Scale bar, 300 μm (Reproduced from [9]). Source: https://creativecommons.org/licenses/by/4.0/.

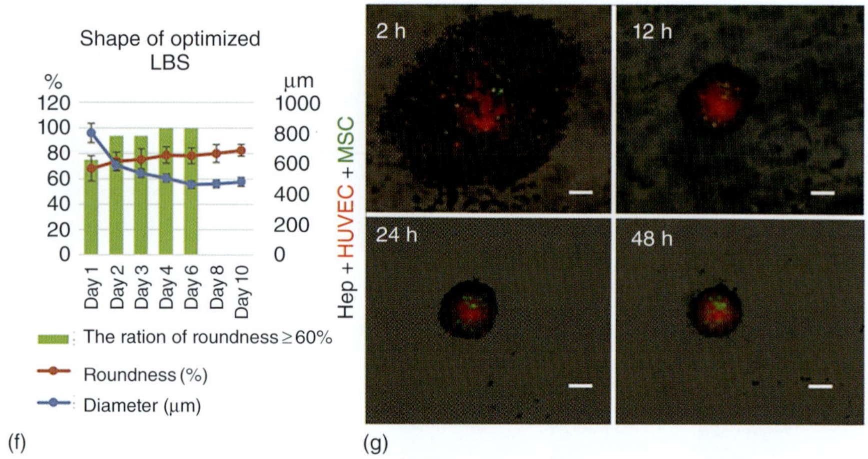

Figure 13.3 (*Continued*)

As the final tissue construct is made of individual spheroids, it is useful to check that the individual spheroids are functional, for the final tissue construct to be functional. In our case of cardiac tissue made of hiPSC-CMs, we had a phenotype to look for, which is spontaneous beating that occurred within 48 hours of spheroid creation (Figure 13.2e–g) [8]. This step is particularly helpful for hiPSC-derived tissue, as there may be failure of differentiation. Itoh et al. [4] used immunostaining to examine the distribution of cells in the spheroids (Figure 13.5). For other groups, this step is presumably omitted as fibroblasts, EPCs, SMFCs, HUVECs, and hepatocytes, etc., do not beat. For all groups, functional and structural tests were performed on the final 3D bioprinted tissue construct to assess the final 3D bioprinted construct.

13.2.3 3D Bioprinting

Before 3D bioprinting, the numbers of spheroids needed should be determined by manual calculation. Taking into account that spheroid creation is not perfect, and transfer losses, a generous margin should be added on top of the calculated margin to ensure that there are enough spheroids for the final 3D bioprinted tissue construct. When the spheroids are ready, the 3D bioprinter is set up, with 3D bioprinting equipment (such as nozzles and microneedle array) autoclaved before use, for sterility.

The principle behind 3D bioprinting on the microneedle array is to assemble spheroids spatially in predetermined Cartesian x–y–z coordinates. In fact, Moldovan [2] and others [13] have argued that this should be called 3D bioassembly, and not 3D bioprinting. The spheroids are skewered onto medical-grade stainless steel needles of 170 μm diameter.

Other than preprogramming spatial positioning, the suction power of the nozzle, the nozzle size, spheroid selection (thresholds can be set for spheroid circularity, diameter, and smoothness), lighting, camera, and other settings can be adjusted. The first step before actual 3D bioprinting is to inspect the nozzle

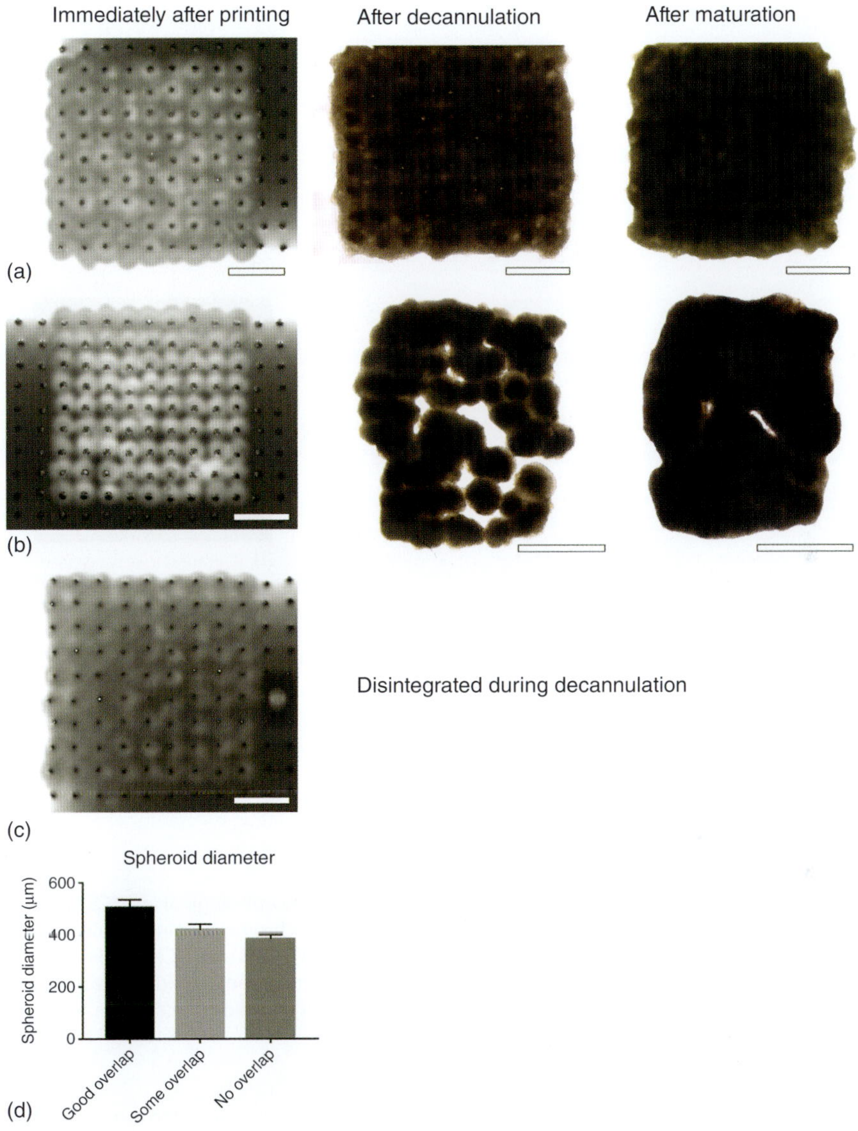

Figure 13.4 Cardiosphere dimensions and the degree of cardiosphere overlap determines the structural properties of 3D bioprinted cardiac patches. (a–c) Representative images of 3D bioprinted cardiac patches (CM:FB:EC 70 : 15 : 15) with overlap at least 50 μm (a), some overlap, 1 to 50 μm (b), and no overlap (c), immediately after printing, after decannulation, and after maturation. Scale bar: 1000 μm. (d) Cardiosphere diameters immediately before printing for 3D bioprinted cardiac patches in panels (a) (504.6 ± 30.6 μm) ($n = 81$), (b) (413.1 ± 21.6 μm) ($n = 81$), (c) (382.4 ± 16.1 μm) ($n = 65$) respectively (Reproduced from [5]). Source: https://creativecommons.org/licenses/by/4.0/.

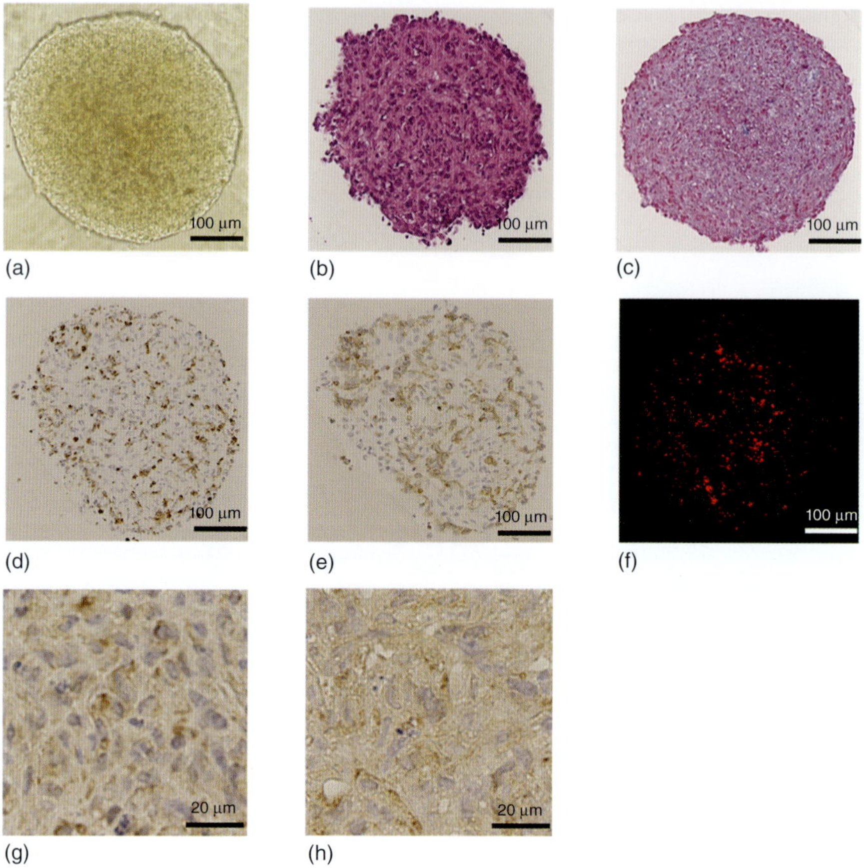

Figure 13.5 A phase-contrast microscopy shows the spheroid morphology of an MCS (a). Spindle to polygonal cells are mixed in the MSC (b). Masson's trichrome staining reveals the extensive collagenous extracellular matrix (ECM) as blue (c). vWF, CD31, and CellTracker Red-positive vascular endothelial cells are distributed to all parts of the MCS (d–f). SMA and desmin-positive HASMC intermingle with the other cell types (g,h) (Reproduced from [4]). Source: https://creativecommons.org/licenses/by/4.0/.

(to ensure the nozzle is not deviated) and then the spheroids (to ensure they are within the thresholds set earlier). There are two cameras to image the microneedle array as well as the wells, live, to show their contents. During 3D bioprinting, if the spheroids cannot be picked up, the spheroids may be too adherent to the well, in which case, tapping the plates or pipetting the media in the wells (spheroid included) may help to dislodge the spheroids. If this fails, the suction can be increased, or the spheroids need to be optimized (as discussed in Section 13.2.2). If the spheroids appear to be successfully picked up but do not appear on the needle array, then the spheroids might have been too small, fragile, or soft and may be aspirated into the nozzle. To troubleshoot this, the nozzle

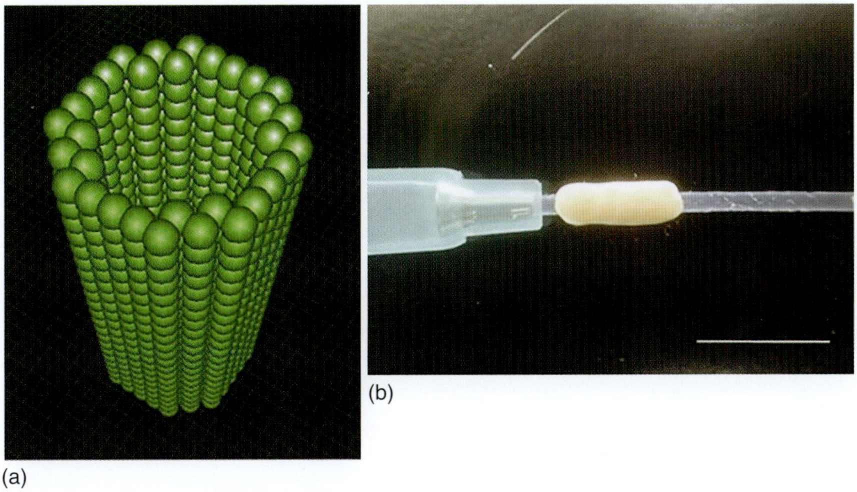

Figure 13.6 (a) The predesigned 3D tubelike structure. The green spheres represent homogeneous multicellular spheroids that were developed using only human normal dermal fibroblasts. (b) The bio 3D conduit according to the predesigned 3D model. The conduit was cannulated via an 18-gauge intravenous catheter (SURFLO: NIPRO, Osaka, Japan). Scale bar = 10 mm (Reproduced from [10]). Source: https://creativecommons.org/licenses/by/4.0/.

size can be decreased and the suction pressure can be decreased. If this fails, the spheroids may need to be optimized again. Problems that may arise include clogging of the nozzle, nozzle bending, plate jamming, presence of bubbles, leaks in the suction system, abnormal resets, microneedle array damage, out-of-focus images, and other technical issues.

In terms of the 3D bioprinting design, it can be a single layer (Figure 13.1b–d), of any shape, or made of multiple layers, again of any shape. Tubular structures (multiple layers) can be bioprinted by stacking rings of tissue (single layer) (Figures 13.6 and 13.7). In our experience, we have 3D bioprinted tissue constructs with complex designs made of multiple layers. The increased bioprinting time may affect the final tissue construct.

13.2.4 Decannulation and Functional Assessment

After allowing a few days for the tissue to mature, the 3D bioprinted tissue construct is removed from the needle array by gently sliding the hard-plastic plate below the tissue upward. This process should be done as gently as possible, with lubrication, to reduce damage to the 3D bioprinted tissue [8]. The holes created by the microneedles (Figure 13.1d) are filled in by the surrounding tissue [11]. The hard-plastic plate is then replaced back onto the microneedle array for the next bioprinting session. The 3D bioprinted tissue is then cultured for further *in vitro* [5, 11] and *in vivo* [4, 5, 9, 10] functional and structural studies, specific to the type of the tissue that was 3D bioprinted.

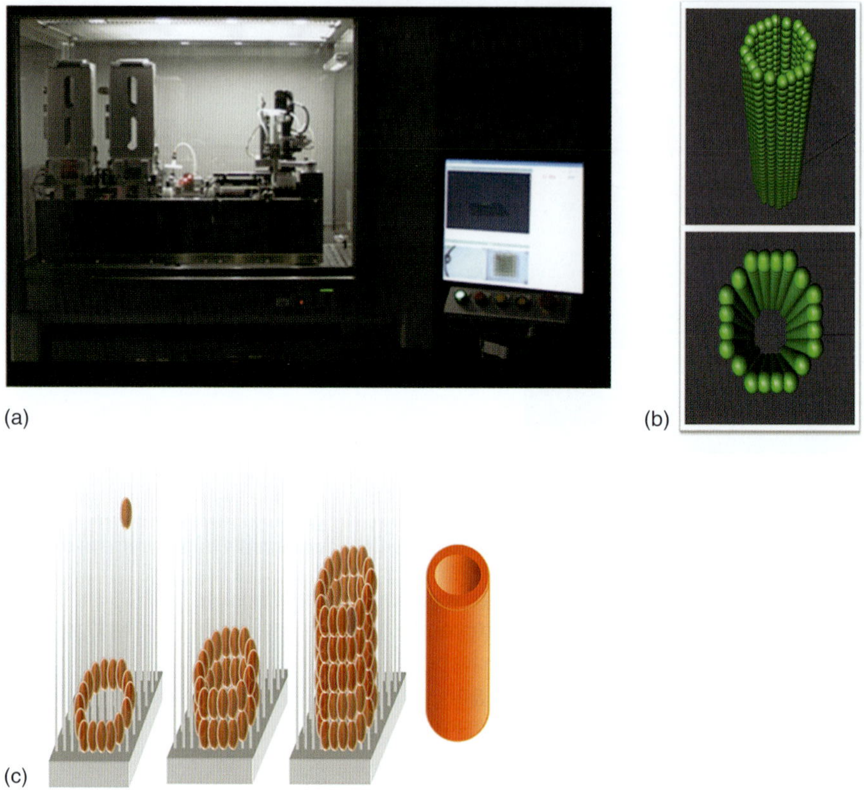

Figure 13.7 (a) The system of lamination of the MCS (Bio-3D printer) that (b) skewers the MCSs into needle array according to a three-dimensional structure predesigned on a computer system. (c) It is possible to design a 3D tube-shaped structure, such as that plotted in green on the workstation computers (Reproduced from Itoh et al., PLOS ONE [4]). Source: https://creativecommons.org/licenses/by/4.0/.

13.3 4D Bioprinting

4D printing encompasses a family of methods and technologies that evolved from 3D printing technologies, but which incorporates the fourth dimension, time, into printed structures through the use of "smart" or reactive materials. In 4D printing, therefore, the printed constructs are no longer static and can dynamically transform in a predefined manner over time based on the interactions with the surrounding environment. Furthermore, current 4D printing techniques allow for the printing of single materials or of combinations of multiple materials, thereby allowing for the creation of complex, dynamic structures. 4D printing technology has only come into existence over the past five years and is still in its infancy. However, 4D printing has already been identified to have diverse applications in fields ranging from biomedical devices to security and optics. As further technologies for 4D printing are developed, applications for this new field of additive manufacturing will continue to broaden.

4D bioprinting involves the use of 4D printing techniques to fabricate the biological structures through the patterning and assembly of biologically relevant materials such as cells, tissues, growth factors, and other molecules and biodegradable scaffolds [14]. In fact, one review suggests that 4D bioprinting is defined as an additive manufacturing methodology encompassing self-assembling and self-accommodating technologies that are man-made, have a programmable structure and design, involves a 2D or 3D printing process, and allows for continued control over the evolution of the printed constructs even after they are printed, through the use of specific biomaterials or external stimuli [15].

Certain key aspects of this definition should be highlighted to best understand 4D bioprinting and differentiate it from 3D bioprinting methods. First, bioprinted structures can only be considered "4D" if there is a preprogrammable component of the printed structure that can change in some biochemical or physical capacity in a predefined manner. This preprogrammed change in the printed structure can be stepwise, for instance, based on the degree of stimulation applied, and the changes can also be reversible [15]. Secondly, the responsive, or "smart," structures must be printed using 2D printing technologies (and subsequently folded into 3D structures after exposure to the predefined stimulus), or using 3D printing technologies (which can then be induced to change into other 3D conformation upon exposure to the predefined stimulus). The "smart" materials used in the construct, therefore, must be printed using some sort of printed technology, and not simply added to a printed construct post-printing. Lastly, the desired, preprogrammed change must be a result of exposure to the predefined stimulus. In other words, the changes should not be those that would occur naturally over time and must be triggered by exposure to the external stimulus. Thus, in 4D bioprinting, it is essential that the fourth dimension, or change over time, is something that occurs as a result of manual manipulation [15].

Just as 4D printing overcomes limitations of 3D printing by incorporating programmable, dynamic "smart" materials, 4D bioprinting allows for printed biological constructs to respond to the microenvironment in which they are to function, just as a native tissue would. Current advances in 3D printing techniques allow for the fabrication of complex 3D constructs using bioinks that encapsulate live cells, using a layer-by-layer approach with controlled positioning of biomaterials to create complex structures [16]. Additionally, 3D printing technologies even allow for stem cell technology [17] that involves patterned printing of growth factors to promote selective differentiation into daughter cell lines. However, these 3D printing technologies can only dictate the initial state of the printed cells, and it is assumed that the cells can automatically carry on any further biological processes such as cell adhesion, sorting, and fusion that are needed to synthesize extracellular matrices and produce tissue constructs with native-like biological, geometric, and mechanical properties [16]. With 4D bioprinting, the integration of the fourth dimension "time" allows for the continued control over the evolution of 3D printed biomaterials and bioinks, so that formation of biomimetic tissues from the printed constructs can be preprogrammed and regulated, in order to optimize outcomes and achieve more native-like results.

13.3.1 Properties of "Smart" Materials

"Smart" materials are defined as those that can sense a particular stimulus from the surrounding environment and create some type of response [18]. It is typically this response that is exploited. "Smart" materials used in 4D bioprinting demonstrate properties or responses that can be divided into four fundamental categories: shape memory, self-assembly, self-actuation, and self-sensing. Materials with shape memory are those that can change their shape in some way upon exposure to an external stimulus [18]. These changes in shape can be stimulated by various factors as previously described in the General Approaches section, such as temperature and electrical stimulation. For instance, Ge et al. demonstrate the fabrication of a flat sheet that can reversibly fold into a predefined 3D shape upon adequate stimulation [19]. As further examples, thermoresponsive shape memory materials have been used to fabricate an endovascular thrombectomy device, and light-induced shape memory materials containing biodegradable segments have also been successfully been used to engineer 3D constructs [20]. Materials that can self-assemble are able to automatically fold for assembly into a predefined structure and have also been used for various biological applications [18]. Lastly, self-actuating materials can automatically actuate or demonstrate some sort of motion, in response to a particular stimulus, and self-sensing materials can automatically detect and sometimes even quantify external stimuli. Self-assembling, self-actuating, and self-sensing materials have all been used for biological applications [18]. For instance, Zhang demonstrated the fabrication of peptides and proteins that can self-assemble into complex tertiary structures (alpha helices and beta pleated sheets) upon exposure to water, replicating protein-folding patterns *in vivo* [21].

13.3.2 General Approaches

Multiple approaches are used in 4D bioprinting to fabricate dynamic, biomimetic tissue constructs [15]. These different approaches employ various strategies to produce responsive 3D printed biomaterials that can exhibit the "smart" behaviors previously described upon stimulation. In general, the 4D bioprinting approaches can be divided into categories based on whether single-material printing is used with just the "smart" material itself or multimaterial printing strategies are used to simultaneously print a combination of multiple "smart" materials or "smart" materials and conventional materials.

13.3.2.1 "Smart" Scaffolds

The first approach to 4D bioprinting involves the 3D printing of a "smart" polymer or hydrogel scaffold that can alter its mechanical or geometric properties and create a precise, predefined shape upon exposure to a particular stimulus [15]. Other "smart" biomaterials have been shown to transform themselves through assembly or disassembly as well [16]. Such changes in the printed scaffolding would subsequently result in alterations to the shape of the concurrently printed biological materials (e.g. cells), thus allowing the entire construct to take on a desired shape. Various "smart" hydrogels have been used in 4D printing processes, including those that respond to changes in electric, ionic, light, magnetic, pH, and temperature inputs from the surrounding environment [16, 22]. Such hydrogels have also demonstrated behaviors such as programmable transition between solid and gel states, self-healing properties, and shape memory [22, 23].

Temperature is most often used to induce conformation changes in bioprinted constructs [16]. Thermoresponsive printed scaffolds can shrink, swell, fold, or undergo phase transitions upon temperature changes. Popular classes of thermoresponsive polymers used in bioprinting applications include poly(*N*-alkylacrylamide)s, poly(*N*-vinyl caprolactam), poly(*N*-ethyl oxazoline), poly(methyl vinyl ether), poly(acrylic acid-*co*-acrylamide), and elastin-like oligo- and polypeptides [24]. These polymers demonstrate diverse properties upon exposure to changes in temperature and have been used in various applications ranging from drug delivery to tissue engineering [24]. Poly(*N*-isopropylacrylamide) (PNIPAAm) is a commonly used thermoresponsive polymer scaffold for bioengineering, as it responds to temperatures near normal body physiologic temperature and it demonstrates rapid, reversible on–off switching between coiled and globule states [24]. Studies have demonstrated the use of PNIPAAm coprinted with poly(ε-caprolactone) (PCL) as well as poly(acrylic acid)-PNIPAAm along with ceramic powder as thermoresponsive platforms for biomedical application [25, 26].

Exposure to aqueous environments is also commonly used to prompt 4D-bioprinted "smart" scaffolds to undergo changes in conformation [16]. Immersion of "smart" scaffolds in aqueous environments can lead to water sorption in the scaffolds, which can lead to selective swelling in particular compartments of a 4D bioprinted scaffold. Control over the rate and extent of swelling is achieved through preprogrammable components of the responsive material used, allowing for controlled changes in conformation driven by water sorption. Cell-laden scaffolds with such water-responsive hydrogels can thus be induced to self-fold into cylinders and other structures of predetermined diameters upon introduction into an aqueous environment and can thus be used to fabricate tissue constructs of customizable and clinically relevant geometries [16]. For instance, one study successfully fabricated a bilayered polyethylene glycol (PEG) "bio-origami" scaffold, whose layers underwent differential swelling in water, thereby introducing scaffold curvature. In this study, the water-induced conformation changes had no substantial impact on cell viability, which remained as high as 90% during the 8-week experimental period [27]. Another study printed composite hydrogel scaffolding that demonstrated preprogrammable, anisotropic swelling upon immersion in water, leading the creation of complex 3D architectures [28]. In this study, the swelling allowed for precise control of curvature in the bilayer constructs, and the use of cellulose fibrils to promote selective swelling of the printed scaffold allows for biocompatibility and creation of biomimetic structures [29]. Thus, this technique holds great promise for use in the development of biomedical devices and tissue engineering.

Similar to aqueous exposure, humidity-responsive materials have also been used in 4D printing of bilayer cellulose films, and the application of such materials to 4D bioprinting is currently under study [29]. Biocompatible humidity-responsive materials can be used in the future to design multilayer constructs whose layers can react differently to changes in ambient humidity. However, the use of humidity-responsive cell-laden materials is limited in that cells often require specific osmotic pressures, which limits the humidity and related osmotic pressures that can be used with the particular constructs in question. The constructs must be chosen such that the humidity-responsive scaffolds can transform or respond within the range of that particular cell type's endurance [29].

Scaffold materials responsive to magnetic fields have also been used to fabricate "smart" 4D bioprinted structures [30]. Magnetic fields have been applied during the printing process itself to direct the orientation of printed particles that contain magnetized stiff aluminum platelets embedded in a poly(urethane acrylate) hydrogel, and they can also be applied after printing to induce conformation changes in the printed tissue structure [16, 29]. Multimaterial printing techniques could allow for the fabrication of heterogeneous tissue constructs composed of pockets of magnetically active biomaterials, thus allowing for pre-programmable changes in the architecture of the 4D printed structures upon exposure to a magnetic field. Such a technique could be used to produce biomaterials that can closely mimic complex structures in nature [16].

Electrical stimulation has also been used as a way to induce changes in the size and/or shape of certain "smart" materials [29]. Electrically conductive materials (polyaniline, polypyrrole, and carbon nanotubes) demonstrate changes in dimension when under applied electrical potential. Biocompatible materials such as carbon nanotubes have important implications for 4D bioprinting, given that electrical stimulation is seen in many native tissues such as cardiomyocytes and neurons. Further work is needed in the field of 4D bioprinting with electrically responsive materials, although challenges such as local changes in pH as a result of electrical stimulation must be addressed through optimization of applied voltages to stimulate transformations in the 4D bioprinted constructs.

Other stimuli have also been used to enhance biomimeticity of 4D printed biological constructs. For instance, one group studied the use of changing osmotic gradients to stimulate precise patterns of folding in printed constructs composed of heterologous picoliter aqueous droplets [31]. These constructs were formed through the self-assembly of the printed droplets into lipid bilayers, and the bilayers were subsequently functionalized with membrane proteins, allowing for electrical communication [32]. Additionally, contents stored in these droplets were released upon changes in pH or temperature of the environment [31]. These "smart" droplet networks have huge potential for use as tissue-engineering substrates, tissue interfacing, or programmable cell membrane constructs [16]. Light-sensitive materials have also been used in 4D printing and have applications for 4D bioprinting, upon combination with cell-laden bioinks. The advantage of these materials lies in the ability for precise and remote control over the light-triggered transformation [29]. However, one challenge to be addressed is the attenuation of light due to biological tissues, which could be mitigated through the use of near-infrared light sources that can penetrate tissue constructs.

13.3.2.2 In Vivo Bioprinting

Another approach to 4D bioprinting is termed "*in vivo* 4D bioprinting," where assembly, organization, and maturation of cells printed in precise microdroplet patterns are induced through some sort of stimulus, whether it comes from an external source or by harnessing the patterns of inter and intracellular communication among the printed cells [15, 16]. Current 3D printing techniques allow for the precise deposition of cell-laden bioinks and growth factors to fabricate tissue constructs that mimic the native environment, but do not allow for the continued control over tissue formation and maturation. The *in vivo* 4D bioprinting approach allows for biological processes post-printing to also be regulated by directing the maturation of the bioprinted tissue constructs. This enables tissue constructs to demonstrate more native-like functionality [16].

Through the *in vivo* 4D bioprinting approach, various cell-induced stimuli can be used to alter the mechanical and biological features of the printed tissue grafts over time. Such stimuli include inter and intracellular communication patterns, cellular organization, and matrix deposition [16]. Beyond intrinsic cellular capacities, various external stimuli, such as added growth factors and other biologically active molecules, have also been used through the *in vivo* 4D bioprinting approach to promote and enhance printed tissue maturation and function in the native microenvironment [14].

Multiple examples of *in vivo* 4D bioprinting have been reported in the literature for various types of tissue engineering. For instance, harnessing control over cell organization in printed tissue constructs through the use of 4D bioprinting techniques can optimize the maturation of engineered tissues by making them more biomimetic, and it can optimize their mechanical properties and function in the native microenvironment as well [16]. The functionality of 3D bioprinted vascular grafts is limited by a lack of adequate tissue maturation and thrombosis. Studies have shown that HUVEC suspensions injected into 3D bioprinted vascular grafts leads to endothelialization, by forming a HUVEC layer on the lumen of the vascular graft. This HUVEC layer acted as a "smart" material, by producing and absorbing various soluble factors from the printed vascular graft, helping the printed cell construct to properly mature, and avoid thrombotic events [33]. Future work should investigate the use of 4D bioprinting technology to coprint HUVECs with vascular cells, to use HUVEC layer as a smart material that stimulates maturation of the printed vascular graft.

Other *in vivo* 4D bioprinting techniques have also harnessed intercellular communication and cellular self-organization capacity as a stimulus for proper maturation of the printed constructs. In one study, discrete cellular spheroids composed of human smooth muscle cells were printed and were stimulated to form a tubular construct through intercellular communication between the cells encapsulated in the spheroids, enabled by multiple signaling molecules [34]. Similarly, another study has demonstrated tubular graft formation from bioprinted cellular toroids containing human ovarian granulosa cells [35]. In both studies, intercellular communication between the printed cells and cell self-organization capacities were harnessed as the "smart" behaviors that promoted post-printing tubular graft formation from the printed sphere or toroid constructs.

Beyond intercellular communication and cellular self-organization capacity, matrix deposition is another cellular process that can serve as a stimulus to promote printed graft maturation through *in vivo* 4D bioprinting approaches. The traditional hydrogels used in 3D bioprinting demonstrate good compatibility with coprinted cells, have good mechanical properties, and are easy to process [16]. However, the rapid degradation of these hydrogels *in vivo* limits printed tissue's mechanical strength and integrity in the native microenvironment. One study demonstrated that human MSCs seeded onto a bioprinted tissue construct were able to preserve printed tissue integrity through matrix deposition and cellular remodeling processes [36]. Future work in the *in vivo* 4D bioprinting approach should also investigate the use of coprinted cells' ability for matrix deposition and scaffold remodeling as a stimulus to help preserve and maintain the robustness of printed tissue.

Cell traction force (CTF) has also been used in 4D *in vivo* bioprinting approaches [29]. This technique, also termed "cell origami," involves harnessing the driving force of cells (through contractile forces exerted by cytoskeletal components and actin–myosin interactions) to move and shape 3D microstructures. CTF is critical to processes such as wound healing, angiogenesis, metastasis, and inflammation, as well as embryonic development (such as by generating critical tensions during gastrulation and neurulation) [29]. CTF can be induced by exposure to a stimulus – for instance, Takeuchi et al. demonstrate the use of CTF to fabricate a 3D microstructure, with the stimulus for the CTF being cellular detachment from a glass substrate [37]. Furthermore, 2D microplates have been successfully folded into 3D cell-laden microstructures by harnessing CTF [29].

13.3.2.3 Hybrid Techniques

A third approach presented in the literature involves "smart" 3D printed polymer constructs that are implanted into a particular tissue microenvironment, such as a postsurgical site [15]. Native cells are stimulated to assemble and grow into specific tissue constructs in the context of the changing polymer framework, which is gradually bioresorbed as the forming tissue or organ gains mechanical strength. In this approach, the growth and organization of the printed cells into organized tissue or organs are seen as the stimulus that promotes degradation of the "smart" printed polymer through the use of mechanical force. Similarly, studies have shown the use of photodegradable hydrogels to promote cell migration along degraded patterns. In fact, MSC and glycosaminoglycan production was shown to be enhanced upon degradation of the "smart" hydrogel, demonstrating that the cells responded to changes in the hydrogel microenvironment [29].

13.3.3 4D Bioprinting Technologies

4D printing technologies have harnessed various advanced solid freeform fabrication techniques, including 3D printing methodologies, to print smart materials either on their own or in combination with other smart or conventional materials. As previously discussed, there are three main types of 3D bioprinting: inkjet bioprinting, microextrusion, and laser bioprinting. Each of these types of

3D bioprinters has been adapted for use in 4D bioprinting. However, although there are many printers currently available for each type of 3D bioprinting, only a few printers in each category allow for multimaterial 4D bioprinting approaches that can fabricate morphologically complex, heterogeneous "smart" structures that have various properties (mechanical, chemical, thermal, etc.). A major multimaterial 3D printing technique that has been adapted to 4D bioprinting includes Polyjet printing, which uses inkjet bioprinting. Furthermore, other solid freeform fabrication techniques, such as fused deposition modeling (FDM) and stereolithography, have also been applied to create multimaterial 4D bioprinted structures [38].

Inkjet printing is a technique that reproduces digital patterns onto a substrate using droplets of ink, typically ejected using heat or mechanical compression. Inkjet printing technologies have been successfully applied in biomedicine to print structures as delicate as DNA molecules. However, there is some well-documented damage to cell membranes and cell lysis after inkjet printing, and generally, thermal inkjet printing is considered to be most biocompatible, producing printed cell viabilities of 90% [39]. Polyjet printers use inkjet bioprinting technology and have the capacity to print multimaterial 4D constructs by serially depositing liquid photopolymers to create 3D structures that incorporate responsive or "smart" materials [14]. Printed layers are instantaneously ultraviolet (UV) cured before the next layer is fabricated. Furthermore, polyjet technology now allows for concurrent deposition and blending of multiple materials, thus allowing for the fabrication of bioprinted structures that are composed of distinct segments or gradients of "smart" and conventional materials. Such technologies have been applied to 4D bioprinting, allowing for the fabrication of biomimetic, responsive tissue constructions such as anisotropic structures that have the ability to change shape upon immersion in water [14]. However, although polyjet printing has the capacity to print multiple materials, the biomaterials that are printed must be of low viscosity to prevent clogging of the nozzle used for printing [38].

Microextrusion 3D printing involves physical extrusion of a bioink using a mechanical or pneumatic dispenser [1]. Multiple groups have developed multinozzle microextrusion systems that can sequentially extrude various types of bioinks and polymer scaffold materials in order to create complex 3D structures [33, 40, 41]. Recent work has also demonstrated the use of a single, multimaterial printhead that can simultaneously print more than one type of bioink to create heterogeneous structures [42–44]. Further work should investigate the use of multimaterial microextrusion printers in 4D bioprinting applications, given that microextrusion offers the advantage of allowing for the use of wider ranges of bioink viscosities and bioinks with higher cell densities, as well as for the continuous extrusion of bioinks instead of discontinuous droplets as is seen in polyjet printing [29].

Other solid freeform fabrication techniques such as FDM have also been used to fabricate 4D bioprinted structures. FDM printers incorporate computer-controlled extrusion and deposition techniques to print more than two materials simultaneously, allowing for the fabrication of customized, complex structures with biologically functional properties [28, 38, 45]. FDM printers allow for the printing of not only solid filaments but also liquids and pastes, polymer melts, and molten materials. The resolution of the printed structures is limited by tip size, which impacts dimensional accuracy and surface quality [38]. FDM printers can be used to print "smart" or responsive biomaterials alongside conventional materials, thus allowing for the fabrication of complex 4D bioprinted structures.

Stereolithography is an additive manufacturing process that allows for the fabrication of complex 3D structures [46]. Stereolithography involves spatially controlled solidification of liquid materials through the use of photopolymerization. Multimaterial printing is possible through stereolithography, but it is more difficult to do than with other techniques such as polyjet printing because this fabrication method uses liquid photopolymers that are easily contaminated, so a rinsing step between material changeover is critical [38]. Stereolithography techniques have been applied to 4D bioprinting through the fabrication of multimaterial "smart" structures. For instance, Ge et al. demonstrate the use of high-resolution microstereolithography to print cardiovascular stents, by utilizing photocurable methacrylate that could undergo preprogrammed alterations in its thermomechanical properties upon exposure to defined changes in environmental temperature [47].

13.3.4 Applications

4D bioprinting technologies allow for greater biomimeticity than 3D bioprinting because of their responsiveness to stimuli and their ability to change in a controlled manner over time. Many native tissues possess complex structures and functions that are achieved through dynamic alterations of tissue conformation, such as peristalsis in the esophagus and intestines used to propel food [29]. 4D bioprinted constructs can thus better replicate and produce native tissues by allowing such changes in printed constructs in response to specific stimuli [16]. Therefore, 4D bioprinting holds great promise for the fields of medicine and biomedical science, through the production of higher resolution, responsive printed constructs that can address unmet biomedical needs in fields such as tissue engineering and drug delivery [14].

4D bioprinting has huge implications for the fields of tissue engineering and organ printing. Given the increasing demand for organ transplants and chronic donor shortage, biofabrication of organs through tissue engineering can potentially alleviate the organ shortage crisis [48]. However, many organs have complex microarchitectures that require combinations of multiple cell types and the fabrication of precise *in vivo* microenvironments to support adequate cell maturation and organization, which current 3D bioprinting techniques have yet to successfully replicate, especially at clinically relevant sizes [49]. Through the fabrication of responsive tissue constructs using 4D bioprinting techniques, it will be possible to construct structures with more complex and precise geometries, such as liver and heart [16]. For instance, *in vivo* 4D bioprinting approaches have been used to print polymeric bone tissue grafts with improved osteoinductive and osteoconductive characteristics, although further work is still needed to improve the mechanical strength of the printed tissue [50].

In vitro tissue modeling, a key application of 3D bioprinting, is also an area in which 4D bioprinting techniques can optimize and improve outcomes [29]. Just as in regenerative medicine applications, 4D bioprinted constructs that are capable of dynamically responding to environmental stimuli can better replicate native tissue. Therefore, 4D bioprinted tissue constructs would allow for more realistic modeling of interactions in biomedical research, such as response of tissues to novel drug therapies. The enhanced compositional complexity of the 4D bioprinted tissue will therefore improve the biomimeticity of the engineered models and respond to research conditions in a way that is more similar to the responses by native tissues themselves.

Beyond organ and tissue printing and *in vitro* modeling, 4D bioprinting also holds huge potential for addressing a critical challenge in tissue engineering: vascularization. Vascularization is needed to engineer tissue constructs of clinically relevant sizes, given that general mass diffusion can only supply tissues over distances of 100–200 μm [16]. Stem cells, fibroblasts, and endothelial cells can be printed into cylindrical structures, which can then be stimulated to mature into vascular-like structures through *in vivo* 4D bioprinting approaches, as previously described [33, 34]. A wide variety of noncellular stimuli have also been used in the 4D printing of vascular structures, prompting the preprogrammed printed constructs to take on precise cylindrical formations upon exposure to the stimulus in question.

3D bioprinting has been used in the fabrication of complex biomedical devices, but these are limited in their capacity to function *in vivo*. The use of 4D bioprinting techniques can greatly expand the functionality of such printed biomedical devices, through the use of "smart" materials that demonstrate environmental adaptive mechanisms [18]. As previously mentioned, one type of "smart" material used in 4D bioprinting is that which has self-sensing capacity and can respond to changes in the environment, sometimes even in a stepwise, quantitative manner. 4D bioprinted devices that are composed of such materials can automatically sense and respond to changes in their microenvironment, thus providing dynamic and instantaneous monitoring of the health state in question [18].

Drug delivery is another area of potential application for 4D bioprinting, as "smart" printed constructs can be programmed to fold or unfold to encapsulate and release biomaterials or cells upon exposure to particular stimuli [16]. Thermoresponsive platforms (such as hydrogels, micelles, films, and particles) that are composed of a "smart" polymer encapsulating cells or other biologically relevant molecules (such as nutrients or growth factors) have long been demonstrated in drug delivery applications [24]. Upon injection into the body, the temperature increases stimulate such "smart" polymers to undergo a phase transition, thus releasing its contents in a targeted manner [24]. However, there are few examples demonstrating the use of bioprinting techniques to fabricate these thermoresponsive platforms. 4D bioprinting has been used to fabricate anisotropic liquid microcapsules using biomimetic polymer films that can demonstrate complex movements such as twisting, folding, and bending upon temperature changes. In this study, PNIPAAm, a thermoresponsive polymer, was coprinted with PCL, forming a bilayered construct that could independently fold or unfold upon temperature changes in the surrounding environment. Thus, this construct was used for the encapsulation and targeted delivery of yeast cells [26]. Future work in this field should continue to investigate the application of 4D bioprinting techniques to fabricate such thermoresponsive platforms, to increase precision and resolution of the engineered constructs. 4D bioprinting techniques using "smart" materials that respond to other stimuli beyond changes in temperature have also been employed in the fabrication of drug delivery constructs. For instance, pH gradients have been used to stimulate encapsulation and release of drugs from printed structures. Osmolarity gradients have also been used to induce differential swelling in a "smart" bioprinted hydrogel scaffold composed of cross-linked poly(methyacrylic acid) (PMAA) and poly(hydroxyethyl methacrylate) (PHEMA) to enable preprogrammable encapsulation and release of therapeutics from a mucoadhesive layer [51]. Folding of this bilayered construct upon swelling of the PMAA layer promoted mucoadhesion, whereas the PHEMA layer served as a barrier to diffusion, to minimize drug leakage. Orange 8 and bovine serum albumin were successfully delivered through mucosal epithelium.

Similar to drug delivery, gene therapy is another area in which 4D bioprinting techniques can fabricate structures that allow for targeted delivery of appropriate therapeutic genes into cells to correct for genetic defects [24]. The delivered genes may replace, repair, or regulate the native, defective gene to ameliorate the disease phenotype. However, delivery of genes is challenging, given that they must pass through the hydrophobic, negatively charged membrane and into the hydrophilic, negatively charged DNA molecule. Thermoresponsive, cationic polymers have been used to circumvent this issue, where alterations in temperature have allowed for polymer carriers to form complexes with the DNA and allow for successful transfection into the target cell. Studies have demonstrated that chitosan and various other materials grafted with polymers such as PNIPAAm allow for the successful transfection of genes into target cell upon temperature changes within the normal body temperature range [24, 52–54]. Further work in gene therapy should investigate the application of 4D bioprinting technologies to the fabrication of such thermoresponsive polymer gene delivery platforms, in order to improve the design and precision of these structures.

13.3.5 Limitations and Future Directions

4D bioprinting techniques are relatively new fields of research that have only emerged within the past decade. Thus, there are still many challenges and limitations that must be addressed before these techniques can be applied to clinically or biomedically relevant applications. Further work should address such limitations in order to continue advancing the field of 4D bioprinting and achieve its human impacts.

Firstly, most "smart" materials that are currently used in 4D bioprinting applications are only responsive to one stimulus type, although native tissues often respond to multiple diverse stimuli in their respective microenvironments [16].

Human tissues are simultaneously regulated by the nervous system, intra- and intercellular communications, and hormonal signals and other biologically active molecules. Thus, to truly recapitulate the function and character of native tissues, 4D bioprinting processes must produce tissues that can respond to multiple stimuli over time.

Secondly, many "smart" materials that can be preprogrammed to respond to certain stimuli and which are used in 4D printing processes are not biocompatible, thus limiting the range of materials that can be used in 4D bioprinting [14]. Making responsive materials that can function in diverse native microenvironments would greatly expand the range of applications for 4D bioprinting techniques. Further work is needed to compile a library of "smart," biocompatible materials that allow for good cell viability, provide appropriate mechanical support, and support biochemical or biophysical signaling between embedded cells in order to allow for preprogrammed deformation or alteration upon response to stimuli [29]. Also, certain "smart" materials currently in use, such as PNIPAAm, lack bioactivity and often require stimuli (such as temperature in the case of PNIPAAm) that are beyond the range of clinical feasibility. Thus, optimization of the current "smart" materials in use is also needed to further improve the field, such as by using copolymers and bioactive molecules alongside PNIPAAm to lower the transformation temperature to normal body temperature ranges and promote cell adhesion, proliferation, and growth [29].

The existing 4D bioprinted constructs can only respond in certain ways to applied stimuli, such as folding and unfolding, or cellular organization. These responses are often made on the macroscale, and further work should evaluate the control over the printed constructs on a more microlevel to improve spatiotemporal control and achieve greater resolution of the engineered tissue [16]. Additionally, the range of stimuli that these "smart" materials can respond to should also be further expanded in order to increase the applicability of these 4D bioprinted structures [18].

13.4 4D Scaffold-Free Bioprinting

As discussed, the majority of 4D bioprinting, mirroring the state of 3D bioprinting (before the development of 3D scaffold-free bioprinting), involves the use of scaffolds. However, the principles of 4D bioprinting can be applied to this nascent field of 3D scaffold-free bioprinting. Instead of having "smart" materials, we have attempted to stimulate live cells in a predictable and preprogrammed way, after they have been bioprinted. In the case of cardiac tissue, we have used electrical stimulation, mechanical loading, and chemical factors to promote maturation of hiPSC-CMs within the bioprinted tissue constructs. In addition, we are in the process of investigating *in vivo* 4D bioprinting through various cell-induced stimuli, such as through paracrine effects following myocardial infarction in a rodent myocardial infarction model and in a neonatal rat model for *in vivo* maturation of 3D cardiac patches.

13.5 Conclusion

In conclusion, the limitations of scaffolds in traditional 3D bioprinting have led to a new field of 3D scaffold-free bioprinting, spurred by technological advances. 4D bioprinting is currently led by the development of "smart" materials that are preprogrammed to respond to a predefined external stimulus in a certain way. A combination of the principles of 4D bioprinting and 3D scaffold-free bioprinting has the potential to open a new field of 4D scaffold-free bioprinting, if cells can be stimulated in a predictable and preprogrammable manner, similar to the manner by which "smart" materials are currently being stimulated. This may lead to scaffold-free bioprinted tissue constructs that continue to develop into more mature native-like tissue, paving the way for the development of improved *in vitro* disease models for drug testing, gene therapy and genome editing, as well as enhanced *in vivo* tissue regeneration, both of which have clinically relevant applications.

Acknowledgments

All figures used in this chapter have been published under Creative Commons Attribution (CC BY) license. Specifically, Figures 13.1, 13.2, and 13.4 are from the authors' own publications [5], Figure 13.3 is from Yanagi et al. [9], both published in Scientific Reports. Figures 13.5 and 13.7 are from Itoh et al. [4] and Figure 13.6 is from Yurie et al. [10], both published in PLOS ONE.

References

1 Murphy, S.V. and Atala, A. (2014). 3D bioprinting of tissues and organs. *Nature Biotechnology* 32: 773–785. doi: 10.1038/nbt.2958.
2 Moldovan, N.I., Hibino, N., and Nakayama, K. (2017). Principles of the Kenzan method for robotic cell spheroid-based three-dimensional bioprinting. *Tissue Engineering Part B, Reviews* 23 (3): 237–244. doi: 10.1089/ten.TEB.2016.0322.
3 Noguchi, R., Nakayama, K., Itoh, M. et al. (2016). Development of a three-dimensional pre-vascularized scaffold-free contractile cardiac patch for treating heart disease. *The Journal of Heart and Lung Transplantation: The Official Publication of the International Society for Heart Transplantation* 35 (1): 137–145. doi: 10.1016/j.healun.2015.06.001.
4 Itoh, M., Nakayama, K., Noguchi, R. et al. (2015). Scaffold-free tubular tissues created by a bio-3D printer undergo remodeling and endothelialization when implanted in rat aortae. *PloS One* 10 (9): e0136681. doi: 10.1371/journal.pone.0136681.
5 Ong, C.S., Fukunishi, T., Zhang, H. et al. (2017). Biomaterial-free three-dimensional bioprinting of cardiac tissue using human induced

pluripotent stem cell derived cardiomyocytes. *Scientific Reports* 7 (1): 4566. doi: 10.1038/s41598-017-05018-4.

6 Achilli, T.-M., Meyer, J., and Morgan, J.R. (2012). Advances in the formation, use and understanding of multi-cellular spheroids. *Expert Opinion on Biological Therapy* 12 (10): 1347–1360. doi: 10.1517/14712598.2012.707181.

7 Fennema, E., Rivron, N., Rouwkema, J. et al. (2013). Spheroid culture as a tool for creating 3D complex tissues. *Trends Biotechnology* 31 (2): 108–115. doi: 10.1016/j.tibtech.2012.12.003.

8 Ong, C.S., Fukunishi, T., Nashed, A. et al. (2017). Creation of cardiac tissue exhibiting mechanical integration of spheroids using 3D bioprinting. *Journal of Visualized Experiments* e55438. doi: 10.3791/55438.

9 Yanagi, Y., Nakayama, K., Taguchi, T. et al. (2017). In vivo and ex vivo methods of growing a liver bud through tissue connection. *Scientific Reports* 7 (1): 14085. doi: 10.1038/s41598-017-14542-2.

10 Yurie, H., Ikeguchi, R., Aoyama, T. et al. (2017). The efficacy of a scaffold-free Bio 3D conduit developed from human fibroblasts on peripheral nerve regeneration in a rat sciatic nerve model. *PloS One* 12 (2): e0171448. doi: 10.1371/journal.pone.0171448.

11 Kizawa, H., Nagao, E., Shimamura, M. et al. (2017). Scaffold-free 3D bio-printed human liver tissue stably maintains metabolic functions useful for drug discovery. *Biochemistry and Biophysics Reports* 10: 186–191. doi: 10.1016/j.bbrep.2017.04.004.

12 Moldovan, L., Barnard, A., Gil, C.H. et al. (2017). iPSC-derived vascular cell spheroids as building blocks for scaffold-free biofabrication. *Biotechnology Journal* doi: 10.1002/biot.201700444.

13 Groll, J., Boland, T., Blunk, T. et al. (2016). Biofabrication: reappraising the definition of an evolving field. *Biofabrication* 8 (1): 013001. doi: 10.1088/1758-5090/8/1/013001.

14 Khoo, Z.X., Teoh, J.E.M., Liu, Y. et al. (2015). 3D printing of smart materials: a review on recent progresses in 4D printing. *Virtual and Physical Prototyping* 10 (3): 103–122. doi: 10.1080/17452759.2015.1097054.

15 An, J., Chua, C.K., and Mironov, V. (2016). A perspective on 4D bioprinting. *International Journal of Bioprinting* 2: doi: 10.18063/ijb.2016.01.003.

16 Gao, B., Yang, Q., Zhao, X. et al. (2016). 4D bioprinting for biomedical applications. *Trends in Biotechnology* 34 (9): 746–756. doi: 10.1016/j.tibtech.2016.03.004.

17 Ong, C.S., Yesantharao, P., Huang, C.Y. et al. (2017). 3D bioprinting using stem cells. *Pediatric Research* doi: 10.1038/pr.2017.252.

18 Li, X., Shang, J., and Wang, Z. (2017). Intelligent materials: a review of applications in 4D printing. *Assembly Automation* 37 (2): 170–185. doi: 10.1108/aa-11-2015-093.

19 Ge, Q., Dunn, C.K., Qi, H.J., and Dunn, M.L. (2014). Active origami by 4D printing. *Smart Materials and Structures* 23 (9): 094007.

20 Wu, L., Jin, C., and Sun, X. (2010). Synthesis, properties, and light-induced shape memory effect of multiblock polyesterurethanes containing biodegradable segments and pendant cinnamamide groups. *Biomacromolecules* 12 (1): 235–241.

21 Zhang, S. (2005). Building from bottom up: Fabrication of materials using peptide motifs. *American Chemical Society* 229: U715–U716.

22 Wang, S., Lee, J.M., and Yeong, W.Y. (2015). Smart hydrogels for 3D bioprinting. *International Journal of Bioprinting* doi: 10.18063/ijb.2015.01.005.

23 Wang, Y., Adokoh, C.K., and Narain, R. (2017). Recent development and biomedical applications of self-healing hydrogels. *Expert Opinion Drug Delivery* doi: 10.1080/17425247.2017.1360865.

24 Gandhi, A., Paul, A., Sen, S.O., and Sen, K.K. (2015). Studies on thermoresponsive polymers: phase behaviour, drug delivery and biomedical applications. *Asian Journal of Pharmaceutical Sciences* 10 (2): 99–107. doi: 10.1016/j.ajps.2014.08.010.

25 Wang, X., Sun, Y., Peng, C. et al. (2015). Transitional suspensions containing thermosensitive dispersant for three-dimensional printing. *ACS Applied Materials and Interfaces* 7 (47): 26131–26136. doi: 10.1021/acsami.5b07913.

26 Zakharchenko, S. and Ionov, L. (2015). Anisotropic liquid microcapsules from biomimetic self-folding polymer films. *ACS Applied Materials and Interfaces* 7 (23): 12367–12372. doi: 10.1021/am505755j.

27 Jamal, M., Kadam, S.S., Xiao, R. et al. (2013). Bio-origami hydrogel scaffolds composed of photocrosslinked PEG bilayers. *Advanced Healthcare Materials* 2 (8): 1142–1150. doi: 10.1002/adhm.201200458.

28 Gladman, A.S., Matsumoto, E.A., Nuzzo, R.G. et al. (2016). Biomimetic 4D printing. *Nature Materials* 15 (4): 413–418. doi: 10.1038/nmat4544.

29 Li, Y.C., Zhang, Y.S., Akpek, A. et al. (2016). 4D bioprinting: the next-generation technology for biofabrication enabled by stimuli-responsive materials. *Biofabrication* 9 (1): 012001. doi: 10.1088/1758-5090/9/1/012001.

30 Kokkinis, D., Schaffner, M., and Studart, A.R. (2015). Multimaterial magnetically assisted 3D printing of composite materials. *Nature Communications* 6: 8643. doi: 10.1038/ncomms9643.

31 Villar, G., Graham, A.D., and Bayley, H. (2013). A tissue-like printed material. *Scientific Reports* 340: 48–52.

32 Villar, G., Heron, A.J., and Bayley, H. (2011). Formation of droplet networks that function in aqueous environments. *Nature Nanotechnology* 6 (12): 803–808. doi: 10.1038/nnano.2011.183.

33 Kolesky, D.B., Truby, R.L., Gladman, A.S. et al. (2014). 3D bioprinting of vascularized, heterogeneous cell-laden tissue constructs. *Advanced Materials* 26 (19): 3124–3130. doi: 10.1002/adma.201305506.

34 Mironov, V., Visconti, R.P., Kasyanov, V. et al. (2009). Organ printing: tissue spheroids as building blocks. *Biomaterials* 30 (12): 2164–2174. doi: 10.1016/j.biomaterials.2008.12.084.

35 Blakely, A.M., Manning, K.L., Tripathi, A., and Morgan, J.R. (2015). Bio-pick, place, and perfuse: a new instrument for three-dimensional tissue engineering. *Tissue Engineering Part C: Methods* 21 (7): 737–746. doi: 10.1089/ten.TEC.2014.0439.

36 Ikegami, T. and Maehara, Y. (2013). Transplantation: 3D printing of the liver in living donor liver transplantation. *Nature Reviews Gastroenterology and Hepatology* 10 (12): 697–698. doi: 10.1038/nrgastro.2013.195.

37 Kuribayashi-Shigetomi, K., Onoe, H., and Takeuchi, S. (2012). Cell origami: self-folding of three-dimensional cell-laden microstructures driven by cell traction force. *PloS One* 7 (12): e51085.

38 Lee, J., Kim, H.-C., Choi, J.-W., and Lee, I.H. (2017). A review on 3D printed smart devices for 4D printing. *International Journal of Precision Engineering and Manufacturing-Green Technology* 4 (3): 373–383. doi: 10.1007/s40684-017-0042-x.

39 Cui, X., Boland, T., D D'Lima, D., and K Lotz, M. (2012). Thermal inkjet printing in tissue engineering and regenerative medicine. *Recent Patents on Drug Delivery and Formulation* 6 (2): 149–155.

40 Chang, R., Nam, J., and Sun, W. (2008). Effects of dispensing pressure and nozzle diameter on cell survival from solid freeform fabrication–based direct cell writing. *Tissue Engineering Part A* 14 (1): 41–48.

41 Kang, H.-W., Lee, S.J., Ko, I.K. et al. (2016). A 3D bioprinting system to produce human-scale tissue constructs with structural integrity. *Nature Biotechnology* 34 (3): 312–319.

42 Colosi, C., Shin, S.R., Manoharan, V. et al. (2016). Microfluidic bioprinting of heterogeneous 3D tissue constructs using low-viscosity bioink. *Advanced Materials* 28 (4): 677–684.

43 Ober, T.J., Foresti, D., and Lewis, J.A. (2015). Active mixing of complex fluids at the microscale. *Proceedings of the National Academy of Sciences* 112 (40): 12293–12298.

44 Hardin, J.O., Ober, T.J., Valentine, A.D., and Lewis, J.A. (2015). Microfluidic printheads for multimaterial 3D printing of viscoelastic inks. *Advanced Materials* 27 (21): 3279–3284.

45 Zein, I., Hutmacher, D.W., Tan, K.C., and Teoh, S.H. (2002). Fused deposition modeling of novel scaffold architectures for tissue engineering applications. *Biomaterials* 23 (4): 1169–1185.

46 Melchels, F.P., Feijen, J., and Grijpma, D.W. (2010). A review on stereolithography and its applications in biomedical engineering. *Biomaterials* 31 (24): 6121–6130.

47 Ge, Q., Sakhaei, A.H., Lee, H. et al. (2016). Multimaterial 4D printing with tailorable shape memory polymers. *Scientific Reports* 6: 31110. doi: 10.1038/srep31110.

48 Huang, Y., Zhang, X.F., Gao, G. et al. (2017). 3D bioprinting and the current applications in tissue engineering. *Biotechnology Journal* 12 (8): doi: 10.1002/biot.201600734.

49 Wang, X., Ao, Q., Tian, X. et al. (2016). 3D bioprinting technologies for hard tissue and organ engineering. *Materials (Basel)* 9 (10): doi: 10.3390/ma9100802.

50 Pati, F., Song, T.-H., Rijal, G. et al. (2015). Ornamenting 3D printed scaffolds with cell-laid extracellular matrix for bone tissue regeneration. *Biomaterials* 37: 230–241.

51 He, H., Guan, J., and Lee, J.L. (2006). An oral delivery device based on self-folding hydrogels. *Journal of Controlled Release* 110 (2): 339–346.

52 Yang, J., Zhang, P., Tang, L. et al. (2010). Temperature-tuned DNA condensation and gene transfection by PEI-g-(PMEO 2 MA-b-PHEMA) copolymer-based nonviral vectors. *Biomaterials* 31 (1): 144–155.
53 Zhou, Y.-M., Ishikawa, A., Okahashi, R. et al. (2007). Deposition transfection technology using a DNA complex with a thermoresponsive cationic star polymer. *Journal of Controlled Release* 123 (3): 239–246.
54 Takeda, N., Nakamura, E., Yokoyama, M., and Okano, T. (2004). Temperature-responsive polymeric carriers incorporating hydrophobic monomers for effective transfection in small doses. *Journal of Controlled Release* 95 (2): 343–355.

14

4D Printing and Its Biomedical Applications

Saeed Akbari, Yuan-Fang Zhang, Dong Wang, and Qi Ge

Singapore University of Technology and Design, Digital Manufacturing and Design Centre, Singapore 487372, Singapore

14.1 Introduction

In recent years, three-dimensional (3D) printing (also known as additive manufacturing) has emerged as a novel technology to fabricate customized biomedical devices such as stents [1, 2], implants [3, 4], prosthetics [5, 6], and surgical tools [7, 8] with remarkable spatial resolution and design freedom. Moreover, 3D bioprinting was adopted from traditional 3D printing technologies to fabricate biological constructs with the capability of directly encapsulating cells into tissues such as liver, heart, and blood vessels [9–13]. The introduction of 3D printing to the biomedical applications enables the personalization and customization of tissues, drugs, and medical products and devices. Additionally, it is less expensive, and faster than the traditional manufacturing methods, which are commonly used in the biomedical field, and often requires molds, dies, or lithographic masks [14].

However, in many biomedical applications, the dynamic shape changes are desirable but hardly achievable by the conventional 3D printing technologies. For instance, in human and other mammals, the heart has four chambers exhibiting a strong expansion and contraction in each cardiac cycle to pump blood through the blood vessels [15]. Thus, the conventional 3D printing techniques that only create 3D objects with static properties cannot meet the functional requirements of the above-mentioned applications. Dynamic behavior and constantly shape-changing features of biological tissues necessitate the development of a more advanced manufacturing technology.

The recent advances in 4D printing, which adds the fourth dimension "time" to the 3D printed complex structures [16, 17], open a new avenue to 3D printing of biological and medical structures and devices with dynamic behaviors. The 4D printing is realized by creatively integrating engineered active materials into high-resolution 3D printing technologies, which enable the precise deposition of the active materials at micron scales. Under environmental stimuli, e.g. temperature, water, and light [18–21], the properties of the active materials change, which subsequently result in the shape changes of the 3D printed structures over time.

3D and 4D Printing in Biomedical Applications: Process Engineering and Additive Manufacturing,
First Edition. Edited by Mohammed Maniruzzaman.

Despite great progress in the field of 4D printing, its further development for biomedical applications requires the development of novel stimuli-responsive 3D printable biomaterials, advanced 3D printing technologies capable of efficient and rapid deposition of multiple active materials, and computational design tools incorporating realistic material models to predict the 3D printed object behavior under environmental stimuli. This chapter will discuss the potentials of 4D printing as the next-generation technology to fabricate complex 3D biomedical devices and biological structures exhibiting dynamic shape change capability. In this chapter, we will first introduce state-of-the-art 3D printing technologies with potential applications to 4D printing. Then, we will review a number of soft active materials (SAMs) that are widely used in 4D printing. Also, some recent applications of 4D printing in biomedical engineering are presented in detail. At the end, future perspectives of 4D printing in creating novel biomedical devices and bioinspired architectures that can be activated by external stimuli to create highly complex shapes are discussed.

14.2 3D Printing Technologies with Potential for 4D Printing

3D printing or additive manufacturing, which allows the digital design and fabrication of complex 3D objects from various materials with a broad range of mechanical and biological properties, has seen a significant progress over the past three decades [22–24]. Figure 14.1 schematically shows four fundamental categories of 3D printing methods appropriate for effective manufacturing of active structures. These variants of 3D printing offer considerable flexibility in the architecture and properties of structures and will be introduced briefly in this section.

In a conventional 3D printer, a pattern-generating device, typically in the form of a printhead connected to an ink reservoir or a light source, is moved by computer-controlled translation stages. The patterned regions of the printable material are then solidified to produce the desired 3D object. Significant research has been conducted in recent years to increase speed, improve resolution and minimum feature size, and integrate multiple materials into these basic 3D printing technologies [14, 25].

14.2.1 Fused Deposition Modeling (FDM)

In fused deposition modeling (FDM) technology, also known as fused filament fabrication (FFF), a thermoplastic filament is extruded from a nozzle by a worm drive at a controlled rate (Figure 14.2). The nozzle is heated to melt or soften the filament and deposit it on a build tray to fabricate 3D structures in a layer-by-layer manner. The deposited material solidifies after its temperature decreases because of convection heat transfer to the ambient temperature. FDM is the most common 3D printing technology because of its simplicity and low cost. However, compared with other 3D printing technologies, FDM has longer operation time and lower resolution. Moreover, only a few thermoplastic polymers such as acrylonitrile butadiene styrene (ABS) and polylactic acid (PLA) are commercially available for FDM 3D printing.

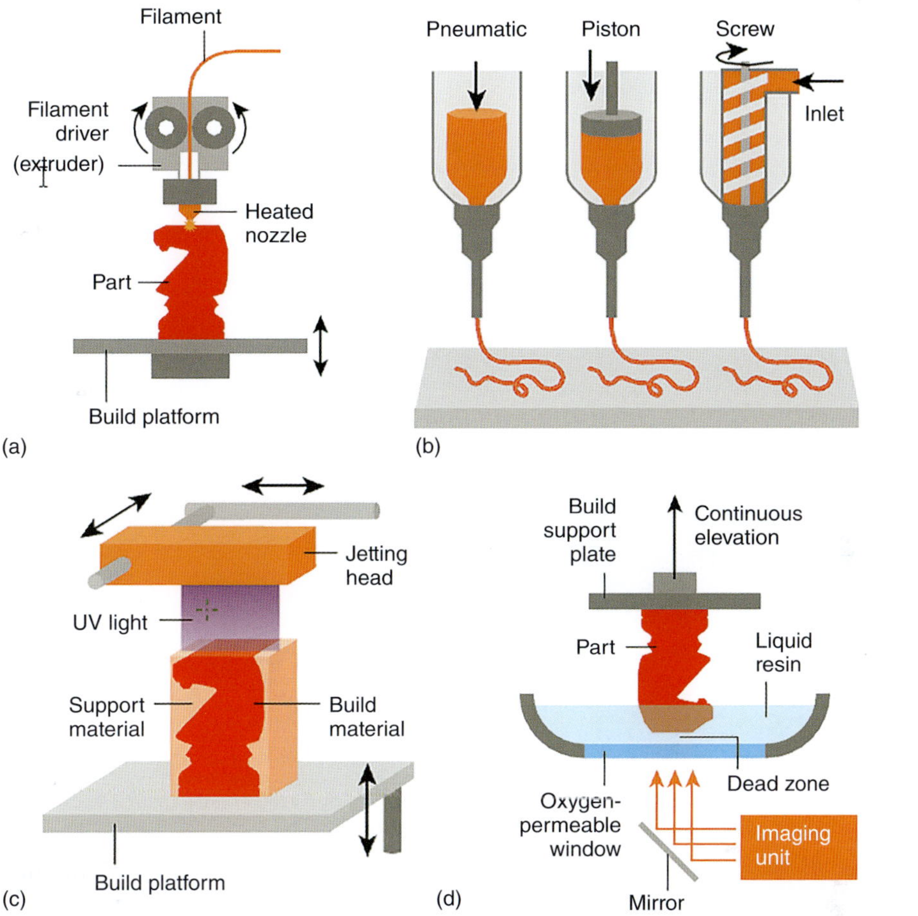

Figure 14.1 Common 3D printing technologies. (a) Fused deposition modeling (FDM), (b) direct ink writing (DIW), (c) inkjet, and (d) projection stereolithography. Source: Truby and Lewis 2016 [14]. Reproduced with permission of Springer Nature.

For specific applications, the initial material should be turned into a filament shape. For example, Yang et al. [26] used a twin-screw extruder to manufacture filaments of a shape memory polymer (SMP) to make it printable using an FDM printer and fabricate a bioinspired robotic finger. Inspired by human finger, each 4D printed finger had three joints printed from the SMP and three rigid segments printed from ABS. Because of the shape memory behavior of the SMP, the finger can be programmed into desired bent shape and then thermally actuated to recover its original shape.

14.2.2 Direct Ink Writing (DIW)

A direct ink writing (DIW) printer uses a printhead similar to an FDM printer. However, instead of a solid thermoplastic, a DIW printer deposits a viscoelastic ink or paste. During printing, the nozzle moves across the build platform and

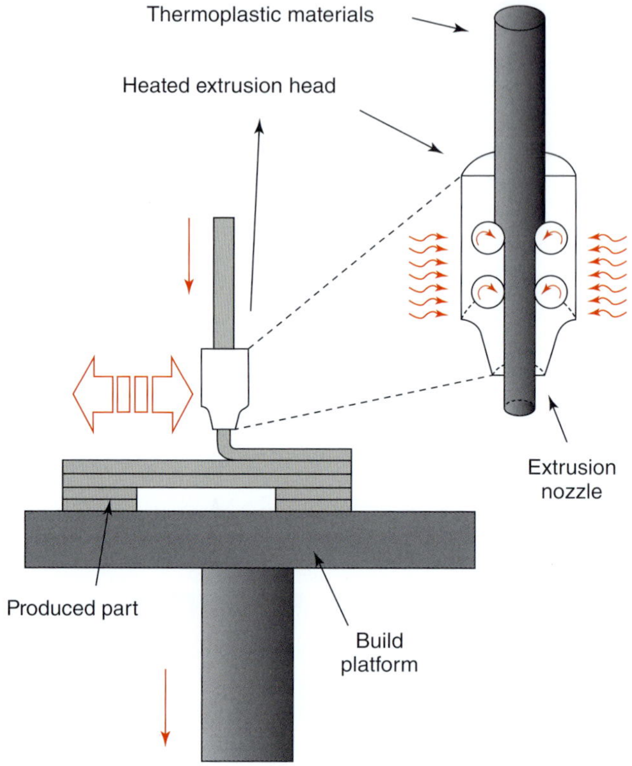

Figure 14.2 Schematics of FDM 3D printing technology. Source: Farahani et al. 2014 [25]. Reproduced with permission of Royal Society of Chemistry.

extrudes the viscoelastic ink under pressure to fabricate the desired 3D object in a layer-by-layer manner. In order to ensure that prohibitively high printing pressures are not required to extrude the viscoelastic ink, it is important to design 3D printable viscoelastic inks that possess significant shear thinning to guarantee efficient extrusion through micro nozzles under ambient conditions. Also, after exiting the nozzle and being deposited on the build tray, the shear thinning materials should have sufficiently high shear elastic modulus and shear yield strength to maintain the desired printed shapes. For this purpose, thickening and thinning agents are typically added to the uncured printing material to modify the respective rheological properties for DIW printing.

Compton and Lewis [27] developed a novel epoxy-based ink to 3D print cellular composites with precise alignment of fiber reinforcement to create hierarchical structures inspired by balsa wood. Although the pure epoxy resin easily flows through the printhead nozzles of a DIW printer, it immediately wets and spreads over the build tray and lacks the ability to support itself. In order to create epoxy inks with desired viscoelastic behavior, nanoclay platelets were used to reinforce the pure epoxy and transform it into a viscoelastic fluid. The resulting epoxy ink exhibited strong shear thinning behavior, with a viscosity of ~20 Pa s at the shear rate of 50 s^{-1}, which is typical of the DIW printing process, and a viscosity of ~10^4 Pa s at the low shear rate (0.01 s^{-1}). In contrast, the pure epoxy resin has a viscosity of 1.5 Pa s that is independent of shear rate. Consequently, the viscosity of the SiC-filled epoxy is only an order of magnitude higher than that of the pure epoxy under the same printing conditions. In order to effectively design bioinspired hierarchical cellular composites, a computational multiscale model was developed to assess the effect of various design parameters at different levels of microstructural hierarchy on the mechanical properties [28].

Spatial arrangement of multiple ingredients in the printed structure is a grand challenge in 3D bioprinting. Most of the existing DIW bioprinters are limited to the use of a single bioink. Multimaterial DIW bioprinting for the fabrication of sophisticated compositional architectures is often realized by assembling multiple printheads on a single DIW bioprinter [29–31]. This strategy, however, reduces the printing speed as it requires successive switching among the printheads. The switch between nozzles for a typical multinozzle bioprinter on average takes approximately 4–20 seconds [29, 31]. To address this issue, Liu et al. [32] developed a multimaterial DIW bioprinter with a single printhead that is able to continuously pattern multiple bioinks simultaneously. The patterned shear-thinning bioinks were prepared by a suspension of synthetic nanosilicates in water. Multiple bioink reservoirs were routed to a single printhead containing seven bundled channels with equal size and independently actuated by programmable pneumatic valves. Using this technology, different organ-like structures such as brain, heart, liver, kidneys, lung, pancreas, stomach, intestines, bladder, and prostate were printed. This printer exhibits much faster fabrication speed than most available multinozzle printers because it functions continuously without any pause to switch among different materials.

14.2.3 Inkjet

Inkjet is a "noncontact" printing technology that deposits tiny droplets of a low-viscosity ink on a build tray using a thermal or piezoelectric technology. Inkjet printers often combine ink nozzles and a UV light source in one platform. For example, photocurable liquid resins are polymerized immediately after being sprayed on the build tray by illumination with a UV light source (Figure 14.3).

Although accurate placement of cells has been realized through inkjet printing [34], a major limitation of this technology is that it has been designed to deposit low-viscosity inks. Therefore, it is not able to deposit viscous extracellular materials, as the viscosity of the ink in this technology should be less than 0.1 Pa s^{-1}. Small droplet size and low viscosity of the ink have been major obstacles to apply this technology to a broader range of clinical contexts [35].

In order to deposit multiple materials, an inkjet printer can be equipped with multiple printheads, with each printhead spraying a different material (Figure 14.3). This technology, referred to as Polyjet, has been widely used as a versatile technology for 4D printing because it allows for the fabrication of complex three-dimensional solids by micron-scale placement of multiple materials. By controlling the composition of various polymers with highly

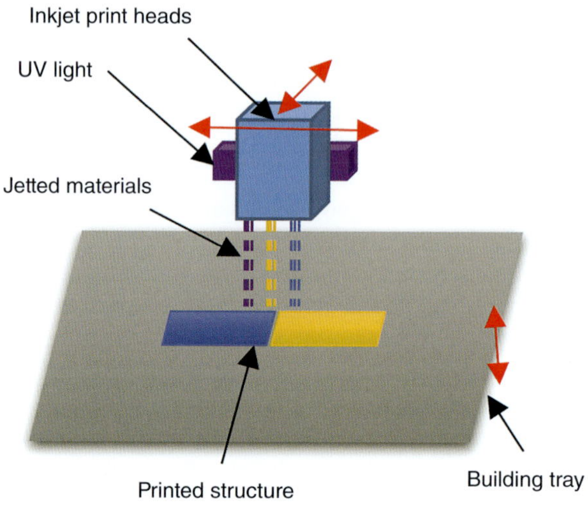

Figure 14.3 Schematics of Polyjet 3D printing technology of liquid photopolymer, which is jetted on a build tray and then UV cured. Source: Adapted from Ding et al. 2017 [33].

different thermomechanical properties as well as changing the spatial variation of constituents over multiple length scales, this technology provides considerable design freedom to produce complex composite structures with controllable anisotropic behavior under thermal and mechanical loadings.

14.2.4 Projection Stereolithography (pSLA)

Stereolithography (SLA), the first 3D printing technology, was developed in 1986 [22]. In this technology, a liquid polymer is selectively polymerized by a rastering laser layer-by-layer to fabricate 3D objects. A major advantage of SLA is that it does not require a nozzle. Thus, the clogging issue commonly observed in other 3D printing technologies could be avoided [36, 37]. However, SLA is relatively slow because it is based on point-source illumination to locally cross-link the polymer.

An alternative method is projection stereolithography (pSLA), also referred to as digital projection lithography (DLP), which is able to polymerize an entire layer using micromirror array devices [38–40]. In pSLA, cross-linking is achieved by projecting a patterned UV light on the surface of a photosensitive polymer resin synergized with upward movement of a linear stage (Figure 14.4). This technology enables the fabrication of complex 3D structures at high resolutions not achievable through the traditional manufacturing technologies. A major limitation of this technology, however, is that it can print only photopolymerizable thermoset resins [14].

In recent years, pSLA has been increasingly used for biofabrication. It allows for precise deposition of biomaterials in relatively small 3D bioconstructs with no adverse effect on cellular viability or functionality [42]. For instance, Ma et al. [41] utilized SLA to print a patient-specific, hydrogel-based model that patterns liver hepatic cells in a physiologically relevant liver-like structure without losing the liver structural and cellular composition (Figure 14.4).

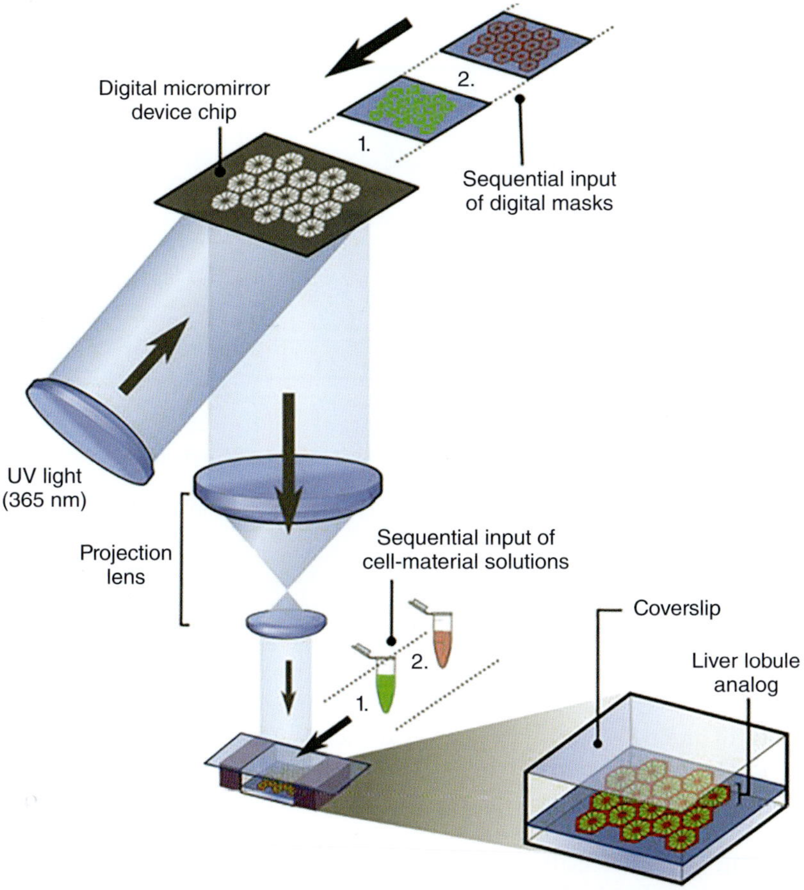

Figure 14.4 pSLA 3D bioprinting of a hydrogel-based hepatic bioconstruct. Source: Ma et al. 2016 [41]. Reproduced with permission of PNAS.

14.3 Soft Active Materials for 4D Printing

The pioneering work advocating the term "4D printing" emerged around 2013 when researchers [16, 17] creatively integrated the SAMs, which exhibit large deformations in response to environmental stimuli, i.e. heat [43, 44], light [45, 46], moisture [47], and others, into 3D printing to create the 3D structures that change shapes over time. Thus, the fourth dimension "time" is added into 3D printing. Despite the rapid development of 4D printing, the most commonly used SAMs for 4D printing are limited to two types: SMPs and hydrogels. In the following, we will introduce these two types of SAMs and the related 4D printing technologies.

14.3.1 Shape Memory Polymers

SMPs are active materials that can recover a predeformed shape under environmental stimuli such as temperature [43, 44, 48–51], light [45, 52–59],

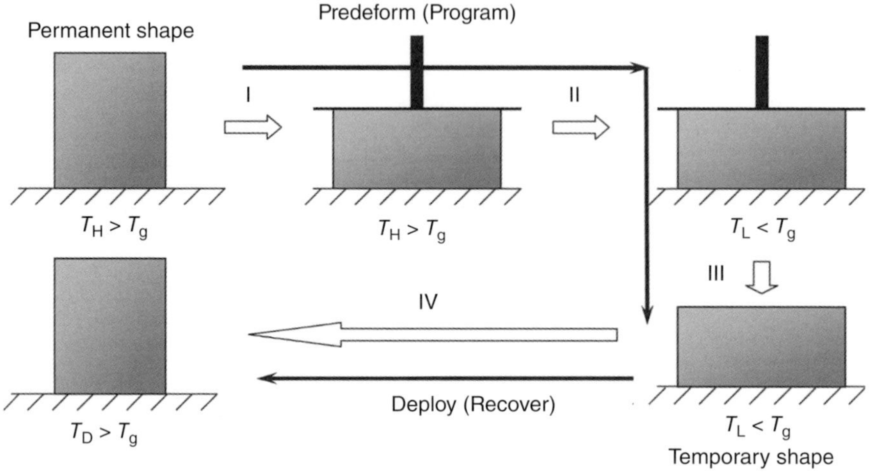

Figure 14.5 A typical thermomechanical loading–unloading cycle in a SMP application. Source: Qi et al. 2008 [67]. Reproduced with permission of Elsevier.

moisture [60], and magnetic field [61]. Compared with the shape memory alloys and ceramics, SMPs possess the advantages of high strain recovery, low density, low cost, easy shape programming procedure, and easy control of recovery. These advantages allow SMPs to be used in many applications such as actuation components in microsystems, biomedical devices, active surface patterning, aerospace deployable structures, and morphing architectures [43, 44, 62–66]. Among the SMPs developed recently, thermally triggered SMPs have been the primary focus in SMP research [65].

For thermally activated SMPs, the most common mechanism used to achieve the shape memory effect is the glass transition that can be obtained following a four-step thermomechanical shape memory cycle (Figure 14.5). At step I, a SMP sample is first deformed at a high temperature T_H ($T_H > T_g$, the phase transition temperature, i.e. glass transition or crystallization temperature) [67]. At step II, the temperature is decreased to a lower one T_L ($T_L < T_g$) while keeping the external constraint. At step III, after removing the external constraint, the SMP sample maintains its deformed shape at T_L. Finally, at step IV, the SMP "remembers" and restores the initial undeformed shape after being heated back to T_H.

The first example of 4D printing with SMPs demonstrated by Ge et al. [16] was printed active composites (PACs) where the prescribed multimaterial microstructures were accurately printed by a multimaterial Polyjet 3D printer (Objet Connex 260, Stratasys, Edina, MN, USA). The base materials loaded onto the printer are a class of elastomers (Tango series) and a class of thermosets (Vero series). A number of digital materials, i.e. composite materials obtained by mixing the two types of base materials at the printing stage, are available with tuned thermomechanical properties ranging from rubbery to glassy at room temperature. Based on the understanding of these thermomechanical behaviors of the SMPs that formed the PACs, a computational design tool was developed to guide the setting of design parameters including geometry, material selection, material placement, and the programming conditions such as deformation and temperature [68]. By precisely defining the different microarchitectures of the PACs, a simple 3D printed strip sample spontaneously deforms into different configurations including coil, helix, wavy-shaped strips (Figure 14.6a) upon a thermomechanical programming. Following the pioneering work of the PAC by 4D printing, researchers also developed a few other SMP-enabled 4D printing examples realized by the Polyjet 3D printer: (i) active origami by 4D printing where the active hinges were carefully designed to achieve different bending angles at different positions of 2D sheets of airplane. (ii) 4D printed structures with the sequential self-folding realized by intelligently placing active hinges with different glass transition temperatures at different positions to control the order of self-bending [19, 70]. The concept of the sequential self-folding is demonstrated in Figure 14.6b where a mailbox self-assembles from a 2D flat

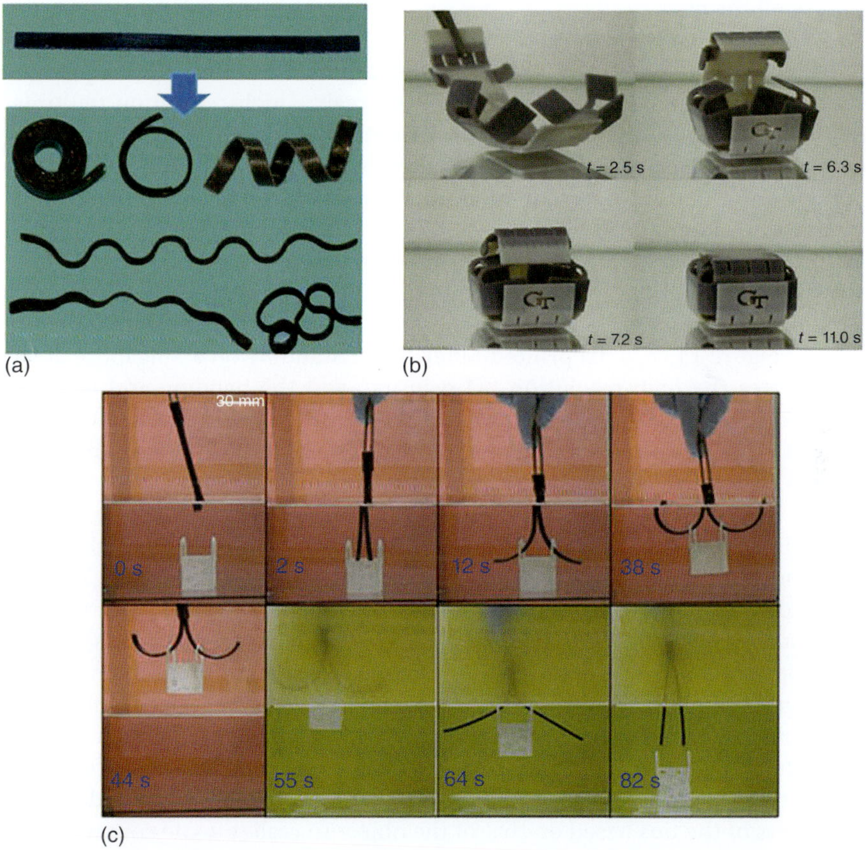

Figure 14.6 (a) A printed strip in its original shape and results of thermomechanical programming with differing fiber architectures. Source: Ge et al. 2013 [16]. Reproduced with permission of AIP. (b) 3D folding structures mimicking the United States Postal Service (USPS) mailbox. Upon heating, the sheet folds into a box with a self-locking mechanism [19]. Source: https://creativecommons.org/licenses/by/4.0/. (c) A smart hook consisting of two SMPs with different T_gs [69]. Source: https://creativecommons.org/licenses/by/4.0/.

sheet into a 3D box with the hinges bending in sequence [19]. (iii) Multishape active composites where SMP fibers with different glass transition temperatures were embedded into an elastomeric matrix to enable a multistep shape memory effect. The recovery of each SMP fiber was triggered only when the ambient temperature exceeded the glass transition temperature [69]. Figure 14.6c demonstrates the multistep recovery in the example of a smart hook that consists of two types of digital SMP fibers embedded in an elastomeric matrix. The glass transition temperatures of the two SMPs are 38 and 57 °C, respectively. The smart hook was first programmed by being stretched by 10%, cooled to 0 °C, and relaxed under 0 °C water for one minute. The first recovery was triggered by placing the smart hook into the warm water with a temperature of 30 °C where the straight strips bend to form two half circular shapes. Using the two half circular strips, a small basket can be lifted from water. To release the box into another position, the temperature of the sample was elevated to be higher than the T_g of all the fibers [69].

Apart from using the commercial Polyjet 3D printer, Ge et al. combined the high-resolution projection microstereolithography (PμSL)-based 3D printing technology with methacrylate-based SMPs to realize high-resolution multimaterial 4D printing [71]. The thermomechanical properties (i.e. glass transition temperature, rubbery modulus, and failure strain) of the SMPs are highly tailorable by choosing different monomers, cross-linkers, and mixing ratios between the monomer and the cross-linker. The methacrylate SMPs exhibit large deformation, which allows the printed structures to accommodate up to 300% large deformation. In addition, the PμSL-based multimaterial 3D printing technology enables the fabrication of 4D printing structures with different SMPs. In Figure 14.7a, a 3D printed Eiffel Tower standing on a Singapore dollar demonstrates the high-resolution 3D printing, and the bending and recovery behaviors of the Eiffel Tower showcase 4D printing with large deformation. Figure 14.7b demonstrates a potential biomedical application, a cardiovascular stent, which was realized by 4D printing with the methacrylate SMPs, and can undergo large local deformation. Figure 14.7c–e demonstrates the examples of 3D printed structures with multiple SMPs: multimaterial grippers that have the potential to function as microgrippers [72] that can grab objects or drug delivery devices [73, 74] that can release objects. Figure 14.7c shows a number of multimaterial grippers with different designs. In Figure 14.7d, an as-printed closed (open) gripper was opened (closed) after programming and the functionality of grabbing (releasing) objects was triggered upon heating. Figure 14.7e shows time-lapsed images of a gripper grabbing an object. The capability of multimaterial fabrication enables the fabrication of the tips of the grippers with the materials different from the SMPs constructing the joints and the design of the stiffness of the tips based on that of the object to realize a safe contact.

Other than UV curing-based SMPs that are thermosets with covalent chemical cross-links, thermoplastic SMPs that are networks with physical cross-links have also been applied to 4D printing. Recently, Yang et al. [21] utilized FDM 3D printing to fabricate 3D structures made of active polymer polyurethane (PU) reinforced with carbon black to introduce the photoresponse. Because of the excellent photothermal conversion efficiency of carbon black (92 ± 3%) upon illumination,

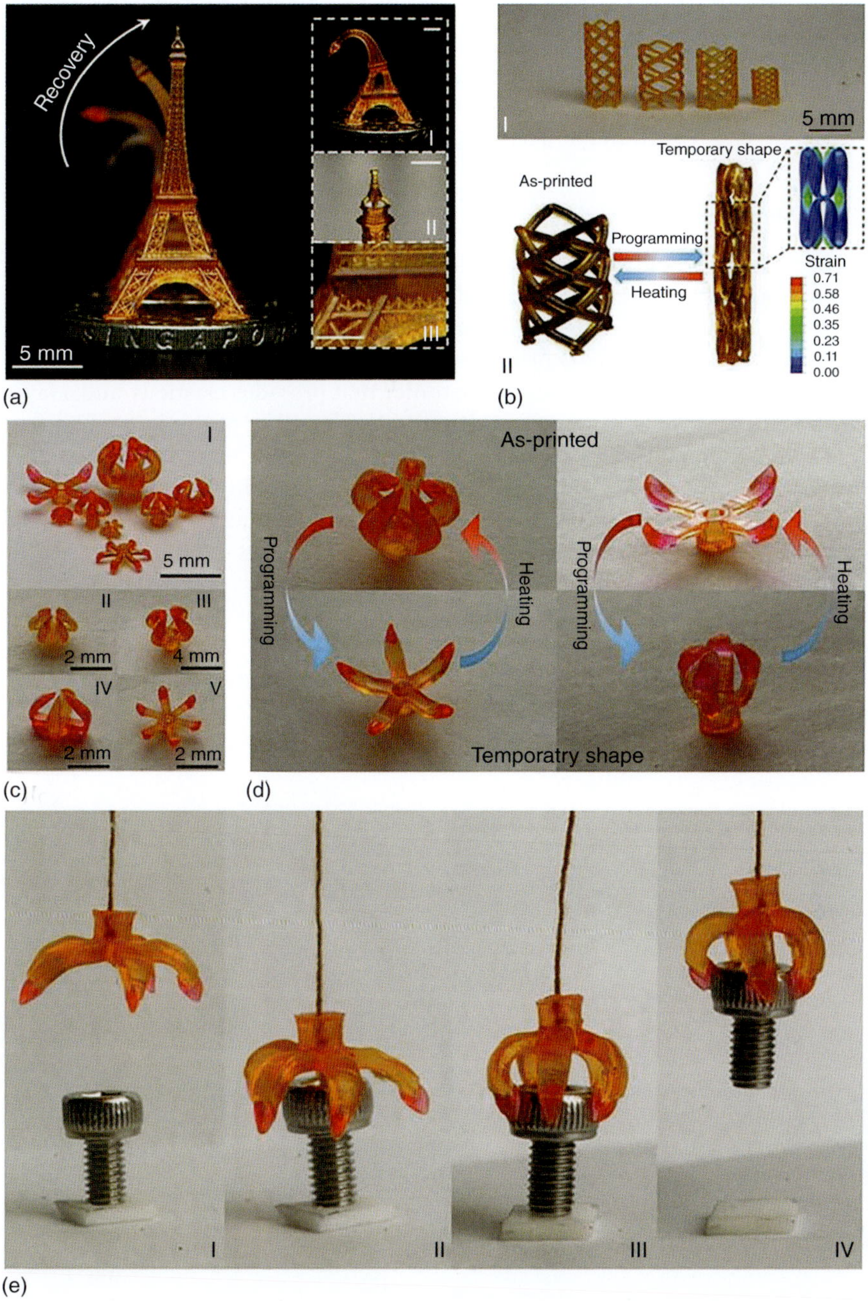

Figure 14.7 (a) 3D printed shape memory (SM) Eiffel Tower. (b) 3D printed SM stents. (c) Multimaterial grippers were fabricated with different designs. (d) Demonstration of the transition between as-printed shape and temporary shape of multimaterial grippers. (e) Snapshots of the process of grabbing an object [71]. Source: https://creativecommons.org/licenses/by/4.0/.

the temperature of the printed structure increases by 20 °C within 10 minutes. The concept of this 4D printing technology was demonstrated via the blooming of a 3D printed sunflower. The petals of the flower gradually opened during illumination under 87 mW cm^{-2} of light intensity, and the temperature increased from 0.4 to 34.4 °C within 280 seconds.

14.3.2 Hydrogels

Hydrogels are another type of SAMs that have been widely applied to 4D printing. They are formed with loosely cross-linked networks of long polymer chains that swell into a larger volume after water or other solvent diffuses into them [75–77]. The essence of designing a 4D printed structure with hydrogels is to design multilayer joints consisting of a layer of elastomer that provides elasticity and a layer of hydrogel that swells in water resulting in the bending of the joint. The first 4D printing example with hydrogels was realized by a Stratasys multimaterial Polyjet 3D printing [17]. It was demonstrated by a straight strand evolved into the "MIT" pattern in the aqueous environment. This self-assembly was realized by precise printing of the hydrogel/elastomer smart hinges. Following this spirit of design, numerous structures exhibiting complex self-evolving deformations were created including sinusoidal waves, hyperbolic surfaces, double-curvature surfaces, time-varying curve (Figure 14.8) [20], etc.

Other than using the Polyjet 3D printing technology, 4D printing with hydrogels can also be realized by employing the DIW 3D printing technology. In Figure 14.9, Gladman et al. [18] developed a bioinspired shape-changing

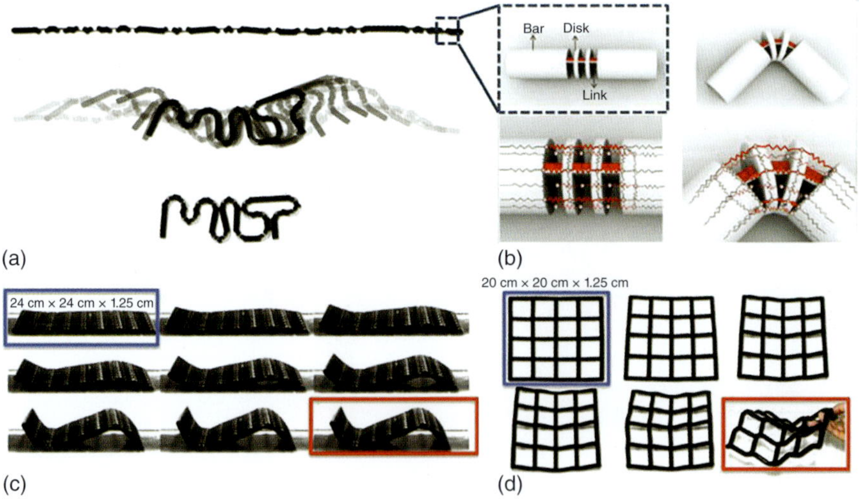

Figure 14.8 Demonstrations of 4D printing presented by the Skyler Tibbits' research group [20]. (a) The self-assembly process from a strand to the "MIT" pattern. (b) The design of a folding joint consisting of a layer of white elastomer providing elasticity and a layer of hydrophilic material that swells after water diffusion and causes the bending of the joint. (c) Deformation of a grid into a sinusoidal wave. (d) Deformation of a grid into a hyperbolic surface [20]. Source: https://creativecommons.org/licenses/by/4.0/.

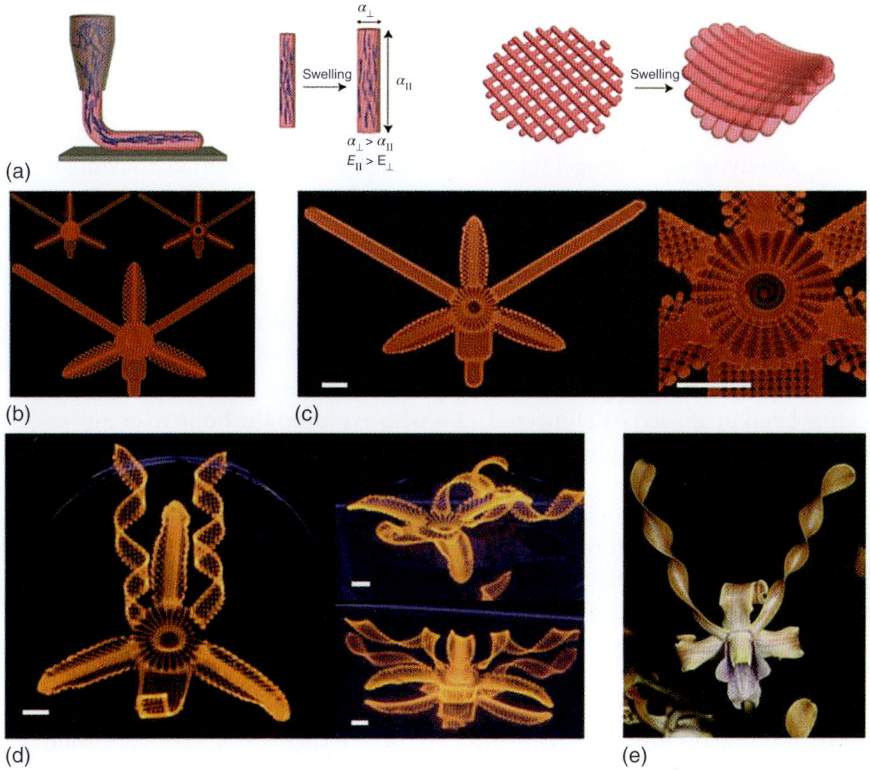

Figure 14.9 Biomimetic 4D printing (Gladman et al.) [18]. (a) Schematic of the shear-induced alignment of cellulose fibrils during direct ink writing and subsequent effects on anisotropic stiffness E and swelling strain α. Print path (b), printed structure (c), and resulting swollen structure (d) of a flower demonstrating a range of morphologies inspired by a native orchid. (e) Based on the print path, this orchid architecture exhibits four different configurations: bending, twisting, and ruffling corolla surrounding the central funnel-like domain (scale bars, 5 mm). Source: Gladman et al. 2016 [18]. Reproduced with permission of Springer Nature.

structure based on 4D printing. Using the DIW technology, they printed a hydrophilic composite material with local control of the orientation of cellulose fibrils within a hydrogel matrix. During printing, fibrils were arranged to induce anisotropic swelling behavior in the longitudinal direction compared to the transverse direction (Figure 14.9a). By following the printing path (Figure 14.9b), generated by a mathematical model that combines thermal expansion in bilayers with a tailored metric-driven approach that employs anisotropic swelling to control the embedding of a complex surface, the printed patterns mimic complex flower morphologies (Figure 14.9c–e) with different configurations.

Another level of actuation can be added to certain hydrophilic materials through the volume change because of the variation of the swelling ratio as a function of temperature [78]. The interpenetrating network of alginate and poly(N-isopropylacrylamide) (PNIPAAm) was used for DIW-based 3D printing. PNIPAAm is a thermally responsive hydrogel that was synthesized from *N*-isopropylacrylamide, a type of commercially available monomer. Upon heating in water above its lower critical solution temperature (LCST) (32–35 °C), PNIPAAm undergoes a coil–globule transition causing a large reversible volume transition. In this process, the material loses its hydrophilicity and hence experiences a large decrease in the water content. A smart valve is created by printing the dynamic hydrogel ink alongside other static materials to control the flow of water in response to the temperature of the water.

14.3.3 Other SAMs

The current 3D printing technologies being costly and time-consuming, some recent advances in 4D printing address these intrinsic drawbacks of 3D printing. For example, a new approach to 4D printing, named direct 4D printing, has recently been proposed by Ding et al. [33] who explored the built-in compressive strain generated during photopolymerization and the mismatch of coefficients of thermal expansion between the elastomer and the SMP in a composite laminate to achieve bending that turns a printed 2D flat sheet into a 3D structure (Figure 14.10a). A structure in its as-printed (temporary) configuration can be deformed to a new shape after removal from the build tray and upon heating that causes the SMP to soften and hence to release the residual strain on the elastomer layer. Cooling down the structure results in a significant increase in the stiffness of the SMP, rendering the new shape permanent. This method makes the fabrication more economical both in terms of time and material consumption compared with the same structure directly printed in three dimensions. The "blooming" of a 4D-printed flower upon heating has been reported elsewhere [33]. The curvatures are merely controlled by varying the layer printing time. A theoretical model is developed to guide the design of structures for direct 4D printing [33, 80]. Ding et al. went further to reduce the dimension of the printed structures to rod elements [79]. Figure 14.10b shows how a 3D buckyball is created by deforming a mesh of 1D rods. This example shows the significant reduction of both printing time (beyond 70%) and material consumption (beyond 90%) that direct 4D printing of rod elements offers in comparison with the same structure printed using conventional 3D printing (Figure 14.10b–d).

Similarly, Huang et al. [81] developed a technique to print thin planar layers with nonuniform cross-linking density distribution by controlling the exposure time at pixel level, requiring, in the meantime, less printing time than the conventional layer-by-layer 3D printing technology. Immersion of the printed structures in water will trigger differential swelling that generates stress causing the structures to deform into 3D objects.

Some other stimulus-response combinations among active materials are less developed for 3D printing, but open interesting perspectives. One example is the 3D printing of pH-responsive polymer to which a postprinting functionalization is applied to form a hydrogel with a tuned degree of swelling [82]. Another example is a computational approach to study the distinct behaviors of thermoresponsive gels and embedded photoresponsive fibers forming a composite in response to heat and illumination [83].

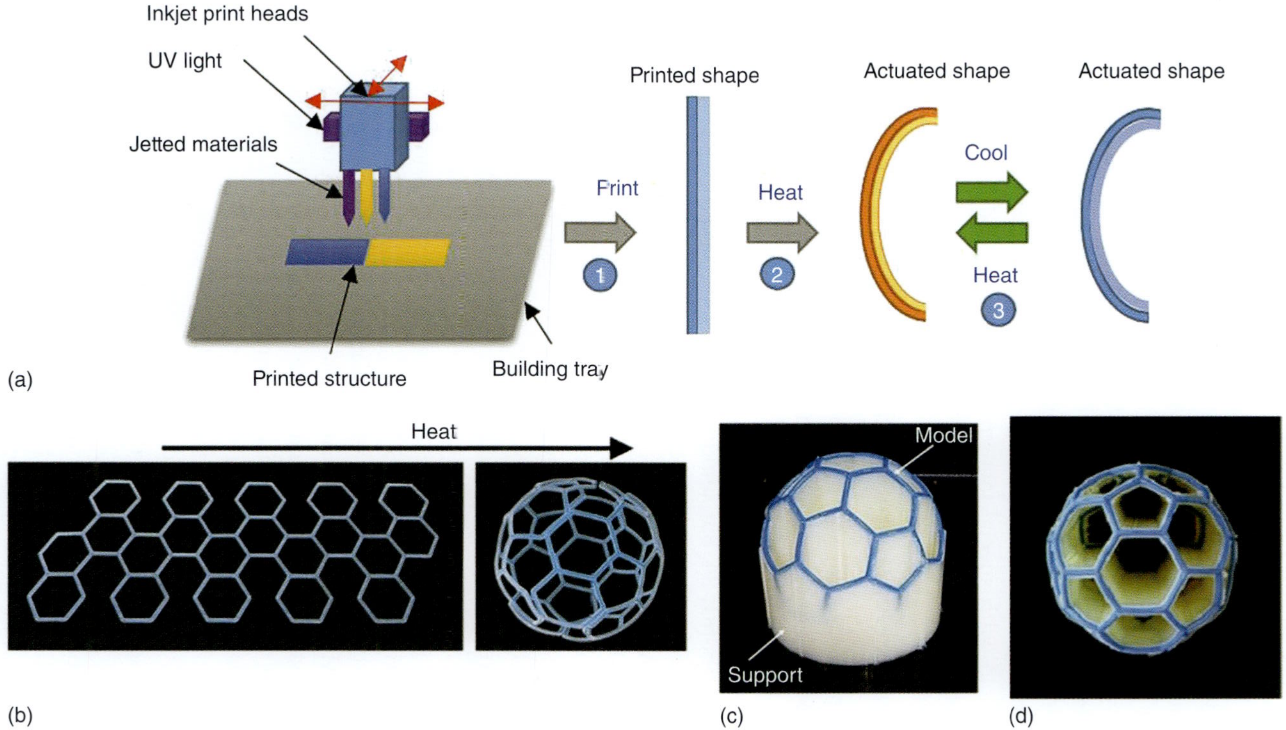

Figure 14.10 (a) The direct 4D printing approach exploits the ability to print controlled multimaterial composites to integrate the five steps into a single one (modified from [33]). (b) A buckyball directly transformed from a planar rod mesh upon heating [79]. The buckyball in conventional 3D printing from (c) isometric view and (d) top view needs to be supported by a large amount of the sacrificial material [79].

14.4 Biomedical Applications of 4D Printing

4D printing has emerged as a versatile technology to fabricate biomedical and bioinspired constructs from active materials. The unique advantage of 4D printing is that it produces structures capable of reshaping under environmental stimuli. For example, oxygen and nutrients are supplied to cells through blood vessels, which are the main channels for vascularization of tissues. Vascularization is a major challenge in fabricating functional tissues. 4D bioprinting is a promising approach to address this issue. Self-folding polymer architectures can be 4D printed to create blood vessels capable of encapsulating blood cells, which subsequently deform into tubelike structures in the presence of water [83]. The following section discusses several examples of the application of 4D printing in biomedical field, with special focus on temperature and humidity-sensitive biostructures.

14.4.1 Temperature-Actuated 4D Printing

The most common external stimulus to actuate the 4D printed structures is temperature. SMPs, including shape memory thermoplastics and shape memory thermosets, are the most commonly used temperature-sensitive active materials in the biomedical field, demonstrate a broad range of mechanical, chemical, and biological properties [81, 84, 85], and have been widely used to fabricate stents [65, 86], scaffolds [87–89], and grippers [90, 91] (Figure 14.11). For biomedical applications, it is important to develop 3D printable materials with activation temperature, e.g. glass transition temperature (T_g) in thermosets or melting temperature (T_m) in thermoplastics, close to physiological temperature.

Compared with shape memory thermoplastics, shape memory thermosets show better shape fixity and recovery, and their thermomechanical properties can be easily tailored to favorable values, but processing shape memory thermoplastics is easier, which makes them more appropriate for some medical applications. Although shape memory thermoset polymers are mostly printed by SLA or inkjet 3D printing methods [16, 20, 68, 71, 93], FDM is the most common technology to print shape memory thermoplastics [94]. For example, Senatov et al. [87] used FDM technology to fabricate a scaffold from PLA with sufficiently high porosity as well as an appropriate pore size necessary for spreading of cells and nutrients throughout the printed structure (Figure 14.11b). The 4D printed structure demonstrated self-fitting effect, resulting from shape memory feature, and has potential to be used for small bone defect replacement. The printing material, PLA, is a biodegradable and bioactive thermoplastic derived from renewable resources. Presently, PLA is one of the world's most consumed biopolymers.

Shape memory behavior of 4D printed structures has the potential to act as a mechanical stimulation bioreactor for cell culture in tissue engineering applications and can be used in various clinical contexts. Active scaffolds can help regenerate tissues, such as cardiovascular tissues, bones, and muscles, that exhibit dynamic mechanical properties. 4D printing enables the customization of the architecture and shape of the active scaffold to ensure that it fits the specific anatomical defect of each patient. This is achieved by combining advanced 3D

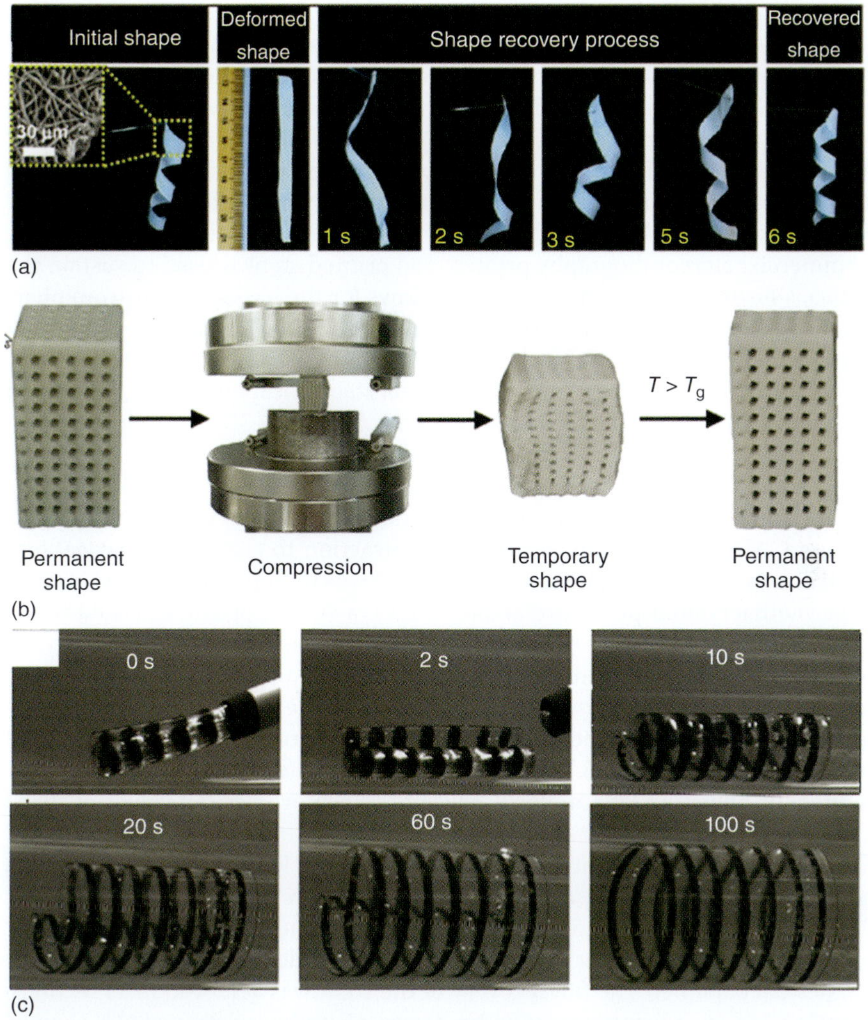

Figure 14.11 Examples of biomedical applications of shape memory polymers. (a) Shape recovery behavior of a poly(D,L-lactide-*co*-trimethylene carbonate) (PLMC) scaffold fabricated by electrospinning. Source: Bao et al. 2014 [92]. Reproduced with permission of ACS. (b) Compression–heating–compression cycles of a PLA-based hydroxyapatite (HA) porous scaffold fabricated by FDM 3D printing technology. Source: Senatov et al. 2016 [87]. Reproduced with permission of Elsevier. (c) Recovery of a cardiovascular stent fabricated from PEGDMA by photopolymerization. Source: Yakacki et al. 2007 [65]. Reproduced with permission of Elsevier.

printing technologies with patient-specific clinical 3D computer-aided design (CAD) images obtained from magnetic resonance imaging (MRI) or computer tomography [95].

Hendrikson et al. [89] used shape memory PU with a T_g of 32 °C to fabricate active scaffolds by FDM technology. The printed scaffolds were loaded at 65 °C using a custom-made stretcher, cooled down to 4 °C, and unloaded to maintain a temporary shape and then seeded by cells cultured at 30 °C to allow for cell adhesion and proliferation. Finally, the temperature was increased to 37 °C and the permanent shape was recovered. After the shape recovery, cells were significantly more elongated and aligned. In order to facilitate cell activity, the training cycle could be further optimized to apply multiple mechanical stimuli to the cells seeded to the active scaffolds.

One of the potential applications of 4D printing is fabrication of self-deploying stents inserted into lumens, such as coronary artery and trachea, to sustain or keep them open. Zarek et al. [96] designed and printed a smart airway stent using a commercial stereolithography printer. The printed stent is used to sustain the trachea, a flexible, tubular structure that allows for the passage of air from pharynx and larynx to the lungs. It typically has a diameter of 2 cm and a length of approximately 12 cm in adult males.

A common reason for the failure of tracheal stent is migration [97], which can be reduced by customized design approaches based on 3D printing technologies. In order to demonstrate that personalized smart stents can be fabricated with a desired geometry, a digital model of the trachea of a middle-aged male was obtained from an MRI scan. The digital model of the trachea was modified in a CAD software using a Boolean subtraction to convert the model into a shell and to remove the dorsal wall. The stent was then printed from semicrystalline methacrylated polycaprolactone (PCL), a thermoplastic actuated based on the melting temperature (T_m) of the crystalline phase. Terminal methacrylate groups were added to PCL diols to prepare them for stereolithography printing. The printed PCL behaved as a rigid polymer below the T_m and as a flexible elastomer above the T_m. Thermomechanical characterization showed that the T_m of the crystalline phase was 55 °C. However, the T_m of the printed structure can be modified by adjusting PCL molecular weight to achieve desired values of shape fixity and recovery.

The fabricated smart stent was approximately 7 cm long and had a maximum and minimum thickness of almost 3.5 and 1.5 mm, respectively. The thickness of each layer was around 150 μm, and the entire printing process took almost six hours. The shape recovery of the smart stent was demonstrated by heating it in a thermal chamber. The full recovery of the stent was achieved in 14 seconds. After three cycles, a printed stent with a degree of methacrylation of 88% showed a shape fixity of 99% and a shape recovery of 98% [96].

The main source of polymers used to fabricate the biostructures and biomedical devices are conventionally petroleum products. Because these resources are rapidly depleting, the use of plant oils for biopolymer fabrication has attracted significant attention in recent years [98–100]. Plant oil polymers are renewable and exhibit excellent biocompatibility [101]. Miao et al. [93] used a stereolithography apparatus to fabricate smart biocompatible scaffolds from soybean oil epoxidized acrylate, which is a renewable liquid resin for biomedical applications. pSLA has rarely been used to fabricate shape memory scaffolds from liquid resins. Differential scanning calorimetry (DSC) analysis showed that the printed samples had a glass transition temperature (T_g) of 20 °C.

As shown in Figure 14.12, shape memory behavior of the scaffold was examined by bending it into a U shape at human body temperature (37 °C), then fixing the temporary shape by reducing temperature to −18 °C, and finally recovering to its original shape by increasing the temperature to 37 °C. Therefore, the printed

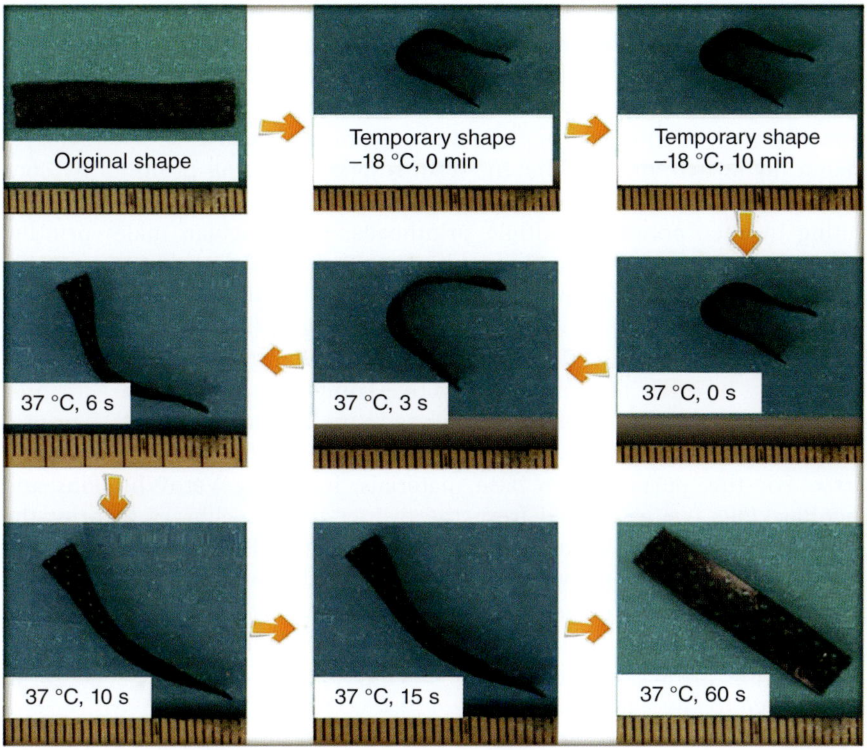

Figure 14.12 Shape memory cycle of the smart scaffold printed from soybean-oil-epoxidized acrylate [93]. Source: https://creativecommons.org/licenses/by/4.0/.

scaffold can be deformed into its temporary shape, embedded in human body, and returned back to its permanent shape with the aid of the body temperature.

Polyethylene glycol diacrylate (PEGDA) resins have been extensively studied to fabricate biomedical scaffolds. However, they are inherently bioinert, and most types of cells are not able to attach to and grow on their surface [102]. Alternatively, PCL and PLA are highly biocompatible polymers, but they lack photosensitive chemical groups and cannot be used for pSLA [103].

In order to evaluate the biocompatibility of the scaffolds printed from soybean oil epoxidized acrylate, Miao et al. studied the attachment of human bone marrow mesenchymal stem cells (hMSCs) on the scaffolds and compared the results with PEGDA, PCL, and PLA [93]. Cytotoxicity analysis indicated that the novel scaffolds printed by stereolithography had better proliferation and adhesion than PEGDA. Moreover, the difference between these new scaffolds, PLA and PCL, was statistically insignificant.

Biomedical devices and natural organs often possess different materials with a broad range of mechanical, chemical, and other functional properties. For instance, human finger consists of rigid bone segments connected together by using relatively flexible muscle joints and ligaments. At least two materials representing these rigid and flexible segments are required to fabricate a bioinspired finger [26, 104, 105].

In order to mimic complex natural structures, printing different materials together, known as multimaterial 3D printing, with high spatial and compositional resolution presents a great potential to the biomedical field [30, 31, 106]. However, light-based printing technologies such as SLA and pSLA are not essentially well adopted for multimaterial printing because it is difficult to dynamically switch among different resin reservoirs during printing. By contrast, FDM, DIW, and inkjet technologies can be easily utilized for multimaterial printing by incorporating multiple printheads. In particular, inkjet printing allows for voxel-by-voxel control of multiple materials composition to smoothly vary mechanical properties over different length scales [107, 108].

Bodaghi et al. [109] used Polyjet printing technology to fabricate multimaterial self-deploying mechanisms that can be used as tubular grippers for biomedical applications. In order to fabricate an active composite mechanism with programmable shape change capability, they used a multimaterial Polyjet 3D printer (Objet 500 Connex, Stratasys, Ltd) to directly print SMP fibers in an elastomeric matrix. For this purpose, two base materials, including VeroWhitePlus and TangoBlackPlus, were selected from the printer material library to print the composite actuator. VeroWhitePlus is a hard and stiff SMP, whereas TangoBlackPlus is a soft and flexible polymer. During the fabrication process, the printer is able to combine these two base materials with various ratios to get a new digital material. Three different digital materials, known as DM9850, DM9885, and DM8510 in the printer material library, with a respective T_g of 18, 31, and 64 °C, were selected to fabricate the actuator.

The actuator unit consisted of two beams; each beam had five layers with a total thickness of 1 mm. Three of these layers consisted of pure matrix (DM9850), whereas the other two were reinforced with SMP fibers (DM9885 and DM8510). Sequential thermal activation of eccentrically positioned SMP fibers with different T_g's controlled the shape change of the actuator.

Programming and actuating included these steps: The printed actuator was first heated up to 100 °C; stretched uniaxially under various nominal strains of 5%, 6.3%, 8%, and 9.25%; cooled down to 0 °C; unloaded; and finally heated up again to 90 °C. With an increase of temperature, the DM9885 SMP fibers with a low T_g of 31 °C were actuated first and applied an axial compressive force to return to their original length. The force generated by eccentrically positioned DM9885 SMP fibers produced a bending moment and resulted in an ellipsoidal shape. A further increase of temperature activated DM8510 SMP fibers with a T_g of 64 °C, which created a bending moment in reverse direction. Consequently, the actuator returned to its original straight shape at a temperature beyond the T_g's of both SMPs. Therefore, the designed actuator exhibits a two-way planar actuation.

The developed actuator unit was then used to fabricate a tubular structure with self-expanding and self-shrinking capabilities. The actuator units were arranged periodically to create a tubular shape with a length of 95 mm, a diameter of 7.7 mm, and a thickness of 1.6 mm. With an increase of temperature, the tubular actuator first expanded, then shrank, and returned back to its original shape. Self-expanding and self-shrinking capability of this actuator, resulting from the strong anisotropy introduced into the structure during the printing process, can be exploited to grip and carry an object.

14.4.2 Humidity-Actuated 4D Printing

Hydrogels are widely used to fabricate scaffolds for tissue engineering applications because they exhibit the features of a natural matrix that creates highly hydrated permeable microenvironments with tunable chemical and mechanical properties appropriate for cell culture [110–112]. They provide uniform and effective cell seeding; are able to effectively transfer physical, chemical, and biological signals; and can be formed into various 3D architectures [35].

Humidity can deform hydrogel structures through different levels of water absorption in different constituents and layers. For instance, Jamal et al. [113] used conventional photolithography to fabricate a two-layered hydrogel scaffold containing photopatterned polyethylene glycol (PEG) of different molecular weights, which exhibited self-folding capability realized by differential swelling of the two PEG bilayers. Insulin-secreting cells were encapsulated in the hydrogel scaffolds. Under humidity, the cell-laden scaffolds self-folded into cylinders. During a period of eight weeks, the scaffold exhibited a robust insulin production with a cell viability as high as 90%.

Gelation is necessary for a 3D printed hydrogel structure to maintain its shape. It is achieved through physical or chemical cross-linking or a combination of both [35]. Physical cross-linking consists of nonchemical reversible mechanisms such as entanglements of polymer chains achieved through shear-thinning property of the bioink, whereas chemical cross-linking is induced by the formation of covalent bonds. Overall, physically cross-linked hydrogels exhibit lower mechanical properties that may result in stability issues of the printed structure.

Shape-morphing architectures found in nature have many potential applications in tissue engineering and biomedical devices. One example of such structures is complex three-dimensional morphologies of plants resulting from local swelling and hydration-induced changes in their shapes caused by directional distribution of cellulose fibrils within plant cell walls. Developing an efficient one-step fabrication process to create bioinspired structures with the capability to achieve certain target shapes based on a predictive model is of great value. Gladman et al. [18] developed a plant-inspired shape-changing structure based on 4D printing. Using DIW technology, they printed a hydrophilic composite material with local control of the orientation of cellulose fibrils within a hydrogel matrix.

Hydrogel can also be used to fabricate bioinspired soft robots [114, 115]. Wehner et al. [116] used DIW technology to fabricate an entirely soft, autonomous robot inspired by octopus (Figure 14.13). The fuel reservoirs, catalytic reaction chambers, actuator networks, and vent orifices were printed from hydrogel-based inks, and the body and microfluidic logic of the robot were fabricated using soft lithography and molding. By incorporating the programmable assembly of multiple materials and parts within this structure, the developed approach allows for the design and fabrication of completely soft, autonomous robots.

Moreover, self-healing hydrogels have attracted significant attention over the past decade [117, 118]. Self-healing is one of the most remarkable features of natural tissues and organs such as skin, bones, and wood [119] and is defined as the ability of a material to self-mend damage, heal cracks, and recover its original mechanical properties so as to have a longer lifetime. Self-healing

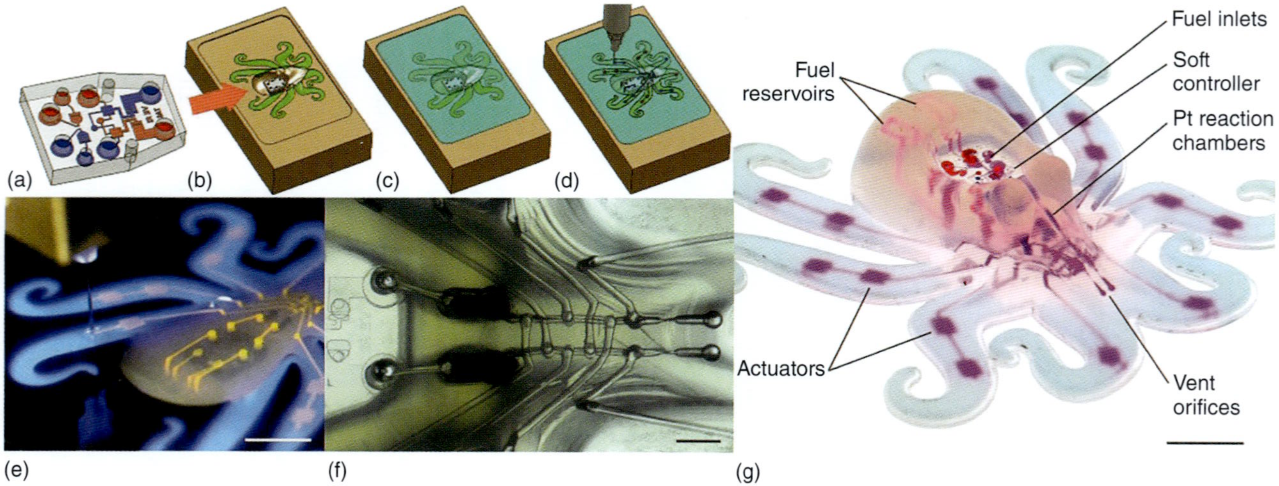

Figure 14.13 Soft robot assembly printed by DIW technology. (a–f) Fabrication steps of the assembly, including loading a prefabricated microfluidic soft controller into a mold and printing hydrogel inks. (g) The fabricate robot is removed from the mold and inverted to reveal a fully soft, autonomous robot. Source: Wehner et al. 2016 [116]. Reproduced with permission of Springer Nature.

hydrogels have the capability to heal damaged tissues and organs intrinsically and automatically, without the intervention of an external signal or stimulus, through reconstructive covalent dangling side chains or noncovalent hydrogen bonding [117]. Although there is no report of 4D printing of self-healing hydrogels in the literature, the dynamic behavior of these active materials offers a great potential for 4D printing of self-healing hydrogels for tissue and organ regeneration applications.

14.5 Conclusion and Outlook

Over the past decade, the 3D printing technologies capable of fabricating complex objects from biomaterials and cellular species have seen significant progress. However, a major drawback of the current 3D printing technologies is that they are only able to create static objects that permanently maintain their original as-printed shape. With the incorporation of various stimuli-responsive biomaterials into advanced 3D printing methodologies and integration of an additional dimension of time, 4D printing is believed to be the future technology to produce transformable objects, e.g. scaffolds used in tissue engineering, which continue to reshape after being fabricated. It is a promising technology to fabricate customized 3D biostructures with myriad applications in the biomedical field. The most common environmental stimuli used to deform 4D printed bioconstructs and medical devices include temperature, which is mostly used to actuate SMPs, and humidity, which is often used to activate hydrophilic materials.

Despite recent significant progress in the area of 4D printing, notable scientific barriers remain that prevent most transformative applications of this new technology in medicine. For further development of this new field, novel stimulus-responsive printable biomaterials with appropriate rheological behavior, identified by the viscosity, surface tension, shear yield stress, and shear elastic and loss moduli, should be optimized for rapid and effective 4D printing. The currently used SMPs used for 4D printing are "one-way," i.e. after recovering to their original as-printed shape, they cannot change their configuration any more after being cooled to a temperature lower than transition temperature, unless the loading–unloading programming step is repeated. This intrinsic irreversibility of shape memory transformation limits the application of SMPs and necessitates the incorporation of novel two-way SMPs into advanced 3D printing modalities. Alternatively, reversible shape change of bioprinted structures can be achieved by combining various types of active materials with different actuation mechanisms.

Furthermore, the current active materials used for 4D printing are only sensitive to one type of external stimulus. Active printable materials and structures that are able to respond to multiple environmental signals are highly desirable for biomedical applications because in the living organisms complicated biostructures are constantly influenced by various regulatory stimuli.

The currently used 4D printed structures show simple uniform deformations, such as twisting and bending, at the macroscale. Precise control of deformation of the 4D printed biostructures at the microscale provides new opportunities to mimic complex deformation of natural organs. To this end, developing appropriate constitutive material models, which are able to capture time- and temperature-dependent nonlinear behavior of SAMs, is of great importance to locally control the morphology of the printed object. These models can be incorporated into computational design tools, such as finite element analysis, to predict deformation of the 4D printed objects and their stress and strain distribution under various environmental stimuli during programming and actuation and optimize the programming conditions such as deformation and temperature to achieve the desired performance.

References

1 Morrison, R.J., Hollister, S.J., Niedner, M.F. et al. (2015). Mitigation of tracheobronchomalacia with 3D-printed personalized medical devices in pediatric patients. *Science Translational Medicine* 7 (285): 285ra64.
2 Park, S.A., Lee, S.J., Lim, K.S. et al. (2015). In vivo evaluation and characterization of a bio-absorbable drug-coated stent fabricated using a 3D-printing system. *Materials Letters* 141: 355–358.
3 Bergmann, C., Lindner, M., Zhang, W. et al. (2010). 3D printing of bone substitute implants using calcium phosphate and bioactive glasses. *Journal of the European Ceramic Society* 30 (12): 2563–2567.
4 Bose, S., Vahabzadeh, S., and Bandyopadhyay, A. (2013). Bone tissue engineering using 3D printing. *Materials Today* 16 (12): 496–504.
5 Gross, B.C., Erkal, J.L., Lockwood, S.Y. et al. (2014). *Evaluation of 3D Printing and its Potential Impact on Biotechnology and the Chemical Sciences.* ACS Publications.
6 Herbert, N., Simpson, D., Spence, W.D., and Ion, W. (2005). A preliminary investigation into the development of 3-D printing of prosthetic sockets. *Journal of Rehabilitation Research and Development* 42 (2): 141.
7 Ventola, C.L. (2014). Medical applications for 3D printing: current and projected uses. *Pharmacy and Therapeutics* 39 (10): 704.
8 Rengier, F., Mehndiratta, A., von Tengg-Kobligk, H. et al. (2010). 3D printing based on imaging data: review of medical applications. *International Journal of Computer Assisted Radiology and Surgery* 5 (4): 335–341.
9 Zhang, Y.S., Yue, K., Aleman, J. et al. (2017). 3D bioprinting for tissue and organ fabrication. *Annals of Biomedical Engineering* 45 (1): 148–163.
10 Zhang, Y.S., Duchamp, M., Oklu, R. et al. (2016). Bioprinting the cancer microenvironment. *ACS Biomaterials Science and Engineering* 2 (10): 1710.
11 Murphy, S.V. and Atala, A. (2014). 3D bioprinting of tissues and organs. *Nature Biotechnology* 32 (8): 773–785.
12 Xu, F., Celli, J., Rizvi, I. et al. (2011). A three-dimensional in vitro ovarian cancer coculture model using a high-throughput cell patterning platform. *Biotechnology Journal* 6 (2): 204–212.
13 Norotte, C., Marga, F.S., Niklason, L.E., and Forgacs, G. (2009). Scaffold-free vascular tissue engineering using bioprinting. *Biomaterials* 30 (30): 5910–5917.
14 Truby, R.L. and Lewis, J.A. (2016). Printing soft matter in three dimensions. *Nature* 540 (7633): 371–378.

15 Lu, L., Mende, M., Yang, X. et al. (2012). Design and validation of a bioreactor for simulating the cardiac niche: a system incorporating cyclic stretch, electrical stimulation, and constant perfusion. *Tissue Engineering Part A* 19 (3–4): 403–414.
16 Ge, Q., Qi, H.J., and Dunn, M.L. (2013). Active materials by four-dimension printing. *Applied Physics Letters* 103 (13): 131901.
17 Tibbits, S. (2013). *The Emergence of "4D Printing"*. Long Beach, FL: TED Conference.
18 Gladman, A.S., Matsumoto, E.A., Nuzzo, R.G. et al. (2016). Biomimetic 4D printing. *Nature Materials* 15 (4): 413–418.
19 Mao, Y., Yu, K., Isakov, M.S. et al. (2015). Sequential self-folding structures by 3D printed digital shape memory polymers. *Scientific Reports* 5: 13616.
20 Raviv, D., Zhao, W., McKnelly, C. et al. (2014). Active printed materials for complex self-evolving deformations. *Scientific Reports* 4: 7422.
21 Yang, H., Leow, W.R., Wang, T. et al. (2017). 3D printed photoresponsive devices based on shape memory composites. *Advanced Materials* (33): 29, 1701627.
22 Hull, C.W. (1986). Apparatus for production of three-dimensional objects by stereolithography. Google Patents.
23 Lee, M.P., Cooper, G.J., Hinkley, T. et al. (2015). Development of a 3D printer using scanning projection stereolithography. *Scientific Reports* 5: 9875.
24 Tumbleston, J.R., Shirvanyants, D., Ermoshkin, N. et al. (2015). Continuous liquid interface production of 3D objects. *Science* 347 (6228): 1349–1352.
25 Farahani, R.D., Chizari, K., and Therriault, D. (2014). Three-dimensional printing of freeform helical microstructures: a review. *Nanoscale* 6 (18): 10470–10485.
26 Yang, Y., Chen, Y., Wei, Y., and Li, Y. (2016). Novel design and three-dimensional printing of variable stiffness robotic grippers. *Journal of Mechanisms and Robotics* 8 (6): 061010.
27 Compton, B.G. and Lewis, J.A. (2014). 3D-printing of lightweight cellular composites. *Advanced Materials* 26 (34): 5930–5935.
28 Malek, S., Raney, J.R., Lewis, J.A., and Gibson, L.J. (2017). Lightweight 3D cellular composites inspired by balsa. *Bioinspiration and Biomimetics* 12 (2): 026014.
29 Kolesky, D.B., Truby, R.L., Gladman, A. et al. (2014). 3D bioprinting of vascularized, heterogeneous cell-laden tissue constructs. *Advanced Materials* 26 (19): 3124–3130.
30 Campbell, J., McGuinness, I., Wirz, H. et al. (2015). Multimaterial and multiscale three-dimensional bioprinter. *Journal of Nanotechnology in Engineering and Medicine* 6 (2): 021005.
31 Kang, H.-W., Lee, S.J., Ko, I.K. et al. (2016). A 3D bioprinting system to produce human-scale tissue constructs with structural integrity. *Nature Biotechnology* 34 (3): 312–319.
32 Liu, W., Zhang, Y.S., Heinrich, M.A. et al. (2017). Rapid continuous multimaterial extrusion bioprinting. *Advanced Materials* (3): 29, 1604630.
33 Ding, Z., Yuan, C., Peng, X. et al. (2017). Direct 4D printing via active composite materials. *Science Advances* 3 (4): e1602890.

34 Nakamura, M., Kobayashi, A., Takagi, F. et al. (2005). Biocompatible inkjet printing technique for designed seeding of individual living cells. *Tissue Engineering* 11 (11–12): 1658–1666.

35 Malda, J., Visser, J., Melchels, F.P. et al. (2013). 25th anniversary article: engineering hydrogels for biofabrication. *Advanced Materials* 25 (36): 5011–5028.

36 Zhang, Z., Xiong, R., Corr, D.T., and Huang, Y. (2016). Study of impingement types and printing quality during laser printing of viscoelastic alginate solutions. *Langmuir* 32 (12): 3004–3014.

37 Xu, C., Zhang, M., Huang, Y. et al. (2014). Study of droplet formation process during drop-on-demand inkjetting of living cell-laden bioink. *Langmuir* 30 (30): 9130–9138.

38 Patel, D.K., Sakhaei, A.H., Layani, M. et al. (2017). Highly stretchable and UV curable elastomers for digital light processing based 3D printing. *Advanced Materials* 29 (15): 1606000.

39 Zheng, X., Deotte, J., Alonso, M.P. et al. (2012). Design and optimization of a light-emitting diode projection micro-stereolithography three-dimensional manufacturing system. *Review of Scientific Instruments* 83 (12): 125001.

40 Sun, C., Fang, N., Wu, D., and Zhang, X. (2005). Projection micro-stereolithography using digital micro-mirror dynamic mask. *Sensors and Actuators A: Physical* 121 (1): 113–120.

41 Ma, X., Qu, X., Zhu, W. et al. (2016). Deterministically patterned biomimetic human iPSC-derived hepatic model via rapid 3D bioprinting. *Proceedings of the National Academy of Sciences* 113 (8): 2206–2211.

42 Guillotin, B. and Guillemot, F. (2011). Cell patterning technologies for organotypic tissue fabrication. *Trends in Biotechnology* 29 (4): 183–190.

43 Lendlein, A. and Kelch, S. (2005). Shape-memory polymers as stimuli-sensitive implant materials. *Clinical Hemorheology and Microcirculation* 32 (2): 105–116.

44 Lendlein, A. and Kelch, S. (2002). Shape-memory polymers. *Angewandte Chemie* 41 (12): 2034–2057.

45 Long, K.N., Scott, T.F., Qi, H.J. et al. (2009). Photomechanics of light-activated polymers. *Journal of the Mechanics and Physics of Solids* 57 (7): 1103–1121.

46 Ryu, J., D'Amato, M., Cui, X. et al. (2012). Photo-origami – bending and folding polymers with light. *Applied Physics Letters* 100 (16): 161908.

47 Choi, J., Kwon, O.-C., Jo, W. et al. (2015). 4D printing technology: a review. *3D Printing and Additive Manufacturing* 2 (4): 159–167.

48 Liu, C., Qin, H., and Mather, P. (2007). Review of progress in shape-memory polymers. *Journal of Materials Chemistry* 17 (16): 1543–1558.

49 Luo, X. and Mather, P.T. (2009). Preparation and characterization of shape memory elastomeric composites. *Macromolecules* 42 (19): 7251–7253.

50 Xie, T. (2010). Tunable polymer multi-shape memory effect. *Nature* 464 (7286): 267–270.

51 Ge, Q., Luo, X., Rodriguez, E.D. et al. (2012). Thermomechanical behavior of shape memory elastomeric composites. *Journal of the Mechanics and Physics of Solids* 60 (1): 67–83.

52 Li, M.H., Keller, P., Li, B. et al. (2003). Light-driven side-on nematic elastomer actuators. *Advanced Materials* 15 (7–8): 569–572.

53 Koerner, H., Price, G., Pearce, N.A. et al. (2004). Remotely actuated polymer nanocomposites – stress-recovery of carbon-nanotube-filled thermoplastic elastomers. *Nature Materials* 3 (2): 115–120.

54 Lendlein, A., Jiang, H., Jünger, O., and Langer, R. (2005). Light-induced shape-memory polymers. *Nature* 434 (7035): 879–882.

55 Scott, T.F., Schneider, A.D., Cook, W.D., and Bowman, C.N. (2005). Photoinduced plasticity in cross-linked polymers. *Science* 308 (5728): 1615–1617.

56 Jiang, H., Kelch, S., and Lendlein, A. (2006). Polymers move in response to light. *Advanced Materials* 18 (11): 1471–1475.

57 Scott, T.F., Draughon, R.B., and Bowman, C.N. (2006). Actuation in crosslinked polymers via photoinduced stress relaxation. *Advanced Materials* 18 (16): 2128–2132.

58 Chen, S., Hu, J., Zhuo, H., and Zhu, Y. (2008). Two-way shape memory effect in polymer laminates. *Materials Letters* 62 (25): 4088–4090.

59 Chen, S., Hu, J., and Zhuo, H. (2010). Properties and mechanism of two-way shape memory polyurethane composites. *Composites Science and Technology* 70 (10): 1437–1443.

60 Huang, W., Yang, B., An, L. et al. (2005). Water-driven programmable polyurethane shape memory polymer: demonstration and mechanism. *Applied Physics Letters* 86 (11): 114105.

61 Mohr, R., Kratz, K., Weigel, T. et al. (2006). Initiation of shape-memory effect by inductive heating of magnetic nanoparticles in thermoplastic polymers. *Proceedings of the National Academy of Sciences of the United States of America* 103 (10): 3540–3545.

62 Tobushi, H., Hara, H., Yamada, E., and Hayashi, S. (1996). Thermomechanical properties in a thin film of shape memory polymer of polyurethane series. *Smart Materials and Structures* 5 (4): 483.

63 Liu, Y., Gall, K., Dunn, M.L., and McCluskey, P. (2004). Thermomechanics of shape memory polymer nanocomposites. *Mechanics of Materials* 36 (10): 929–940.

64 Liu, Y., Gall, K., Dunn, M.L. et al. (2006). Thermomechanics of shape memory polymers: uniaxial experiments and constitutive modeling. *International Journal of Plasticity* 22 (2): 279–313.

65 Yakacki, C.M., Shandas, R., Lanning, C. et al. (2007). Unconstrained recovery characterization of shape-memory polymer networks for cardiovascular applications. *Biomaterials* 28 (14): 2255–2263.

66 Wang, Z., Hansen, C., Ge, Q. et al. (2011). Programmable, pattern-memorizing polymer surface. *Advanced Materials* 23 (32): 3669–3673.

67 Qi, H.J., Nguyen, T.D., Castro, F. et al. (2008). Finite deformation thermo-mechanical behavior of thermally induced shape memory polymers. *Journal of the Mechanics and Physics of Solids* 56 (5): 1730–1751.

68 Ge, Q., Dunn, C.K., Qi, H.J., and Dunn, M.L. (2014). Active origami by 4D printing. *Smart Materials and Structures* 23 (9): 094007.

69 Wu, J., Yuan, C., Ding, Z. et al. (2016). Multi-shape active composites by 3D printing of digital shape memory polymers. *Scientific Reports* 6: 24224.
70 Yu, K., Dunn, M.L., and Qi, H.J. (2015). Digital manufacture of shape changing components. *Extreme Mechanics Letters* 4: 9–17.
71 Ge, Q., Sakhaei, A.H., Lee, H. et al. (2016). Multimaterial 4D printing with tailorable shape memory polymers. *Scientific Reports* 6: 31110.
72 Leong, T.G., Randall, C.L., Benson, B.R. et al. (2009). Tetherless thermobiochemically actuated microgrippers. *Proceedings of the National Academy of Sciences of the United States of America* 106 (3): 703–708.
73 Malachowski, K., Breger, J., Kwag, H.R. et al. (2014). Stimuli-responsive theragrippers for chemomechanical controlled release. *Angewandte Chemie* 53 (31): 8045–8049.
74 Yoon, C., Xiao, R., Park, J. et al. (2014). Functional stimuli responsive hydrogel devices by self-folding. *Smart Materials and Structures* 23 (9): 094008.
75 Lee, H., Xia, C., and Fang, N.X. (2010). First jump of microgel; actuation speed enhancement by elastic instability. *Soft Matter* 6 (18): 4342–4345.
76 Jamal, M., Zarafshar, A.M., and Gracias, D.H. (2011). Differentially photo-crosslinked polymers enable self-assembling microfluidics. *Nature Communications* 2: 527.
77 Malachowski, K., Breger, J., Kwag, H.R. et al. (2014). Stimuli-responsive theragrippers for chemomechanical controlled release. *Angewandte Chemie* 126 (31): 8183–8187.
78 Bakarich, S.E., Gorkin, R., and Spinks, G.M. (2015). 4D printing with mechanically robust, thermally actuating hydrogels. *Macromolecular Rapid Communications* 36 (12): 1211–1217.
79 Ding, Z., Weeger, O., Qi, H.J., and Dunn, M.L. (2018). 4D rods: 3D structures via programmable 1D composite rods. *Materials and Design* 137: 256–265.
80 Yuan, C., Roach, D.J., Dunn, C.K. et al. (2017). 3D printed reversible shape changing soft actuators assisted by liquid crystal elastomers. *Soft Matter* 13 (33): 5558–5568.
81 Huang, W.M., Song, C., Fu, Y.Q. et al. (2013). Shaping tissue with shape memory materials. *Advanced Drug Delivery Reviews* 65 (4): 515–535.
82 Nadgorny, M., Xiao, Z., Chen, C., and Connal, L.A. (2016). Three-dimensional printing of pH-responsive and functional polymers on an affordable desktop printer. *ACS Applied Materials and Interfaces* 8 (42): 28946–28954.
83 Gao, B., Yang, Q., Zhao, X. et al. (2016). 4D bioprinting for biomedical applications. *Trends in Biotechnology* 34 (9): 746–756.
84 Meng, H. and Li, G. (2013). A review of stimuli-responsive shape memory polymer composites. *Polymer* 54 (9): 2199–2221.
85 Hager, M.D., Bode, S., Weber, C., and Schubert, U.S. (2015). Shape memory polymers: past, present, and future developments. *Progress in Polymer Science* 49: 3–33.
86 Agrawal, D. and Habr, F.G. (2009). Removable self-expandable plastic stent to treat postphotodynamic therapy esophageal stricture. *Gastrointestinal Endoscopy* 69 (4): e27–e30.

87 Senatov, F., Niaza, K., Zadorozhnyy, M.Y. et al. (2016). Mechanical properties and shape memory effect of 3D-printed PLA-based porous scaffolds. *Journal of the Mechanical Behavior of Biomedical Materials* 57: 139–148.

88 Neuss, S., Blomenkamp, I., Stainforth, R. et al. (2009). The use of a shape-memory poly(ε-caprolactone)dimethacrylate network as a tissue engineering scaffold. *Biomaterials* 30 (9): 1697–1705.

89 Hendrikson, W.J., Rouwkema, J., Clementi, F. et al. (2017). Towards 4D printed scaffolds for tissue engineering: exploiting 3D shape memory polymers to deliver time-controlled stimulus on cultured cells. *Biofabrication* 9 (3): 031001.

90 Behl, M., Kratz, K., Zotzmann, J. et al. (2013). Reversible bidirectional shape-memory polymers. *Advanced Materials* 25 (32): 4466–4469.

91 Hearon, K., Wierzbicki, M.A., Nash, L.D. et al. (2015). A processable shape memory polymer system for biomedical applications. *Advanced Healthcare Materials* 4 (9): 1386–1398.

92 Bao, M., Lou, X., Zhou, Q. et al. (2014). Electrospun biomimetic fibrous scaffold from shape memory polymer of PDLLA-co-TMC for bone tissue engineering. *ACS Applied Materials and Interfaces* 6 (4): 2611–2621.

93 Miao, S., Zhu, W., Castro, N.J. et al. (2016). 4D printing smart biomedical scaffolds with novel soybean oil epoxidized acrylate. *Scientific Reports* 6: 27226.

94 Yang, Y., Chen, Y., Wei, Y., and Li, Y. (2016). 3D printing of shape memory polymer for functional part fabrication. *The International Journal of Advanced Manufacturing Technology* 84 (9–12): 2079–2095.

95 Tuan, H.S. and Hutmacher, D.W. (2005). Application of micro CT and computation modeling in bone tissue engineering. *Computer-Aided Design* 37 (11): 1151–1161.

96 Zarek, M., Mansour, N., Shapira, S., and Cohn, D. (2017). 4D printing of shape memory-based personalized endoluminal medical devices. *Macromolecular Rapid Communications* 38 (2): 1600628.

97 Strausz, J.B.C.T. (ed.) (2010). *Interventional Pulmonology*, 247. European Respiratory Society Journals Ltd.

98 Liu, Z., Xu, Y., Cao, L. et al. (2012). Phosphoester cross-linked vegetable oil to construct a biodegradable and biocompatible elastomer. *Soft Matter* 8 (21): 5888–5895.

99 Luo, C., Grigsby, W.J., Edmonds, N.R., and Al-Hakkak, J. (2013). Vegetable oil thermosets reinforced by tannin-lipid formulations. *Acta Biomaterialia* 9 (2): 5226–5233.

100 Garrison, T.F., Kessler, M.R., and Larock, R.C. (2014). Effects of unsaturation and different ring-opening methods on the properties of vegetable oil-based polyurethane coatings. *Polymer* 55 (4): 1004–1011.

101 Biermann, U., Bornscheuer, U., Meier, M.A. et al. (2011). Oils and fats as renewable raw materials in chemistry. *Angewandte Chemie* 50 (17): 3854–3871.

102 Liu, S.Q., Tay, R., Khan, M. et al. (2010). Synthetic hydrogels for controlled stem cell differentiation. *Soft Matter* 6 (1): 67–81.

103 Melchels, F.P., Feijen, J., and Grijpma, D.W. (2009). A poly(D,L-lactide) resin for the preparation of tissue engineering scaffolds by stereolithography. *Biomaterials* 30 (23): 3801–3809.
104 Wang, W., Rodrigue, H., Kim, H.-I. et al. (2016). Soft composite hinge actuator and application to compliant robotic gripper. *Composites Part B: Engineering* 98: 397–405.
105 Rodrigue, H., Wang, W., Kim, D.-R., and Ahn, S.-H. et al. (2017). Curved shape memory alloy-based soft actuators and application to soft gripper. *Composite Structures* 176 (15): 398–406.
106 Ober, T.J., Foresti, D., and Lewis, J.A. (2015). Active mixing of complex fluids at the microscale. *Proceedings of the National Academy of Sciences* 112 (40): 12293–12298.
107 Bartlett, N.W., Tolley, M.T., Overvelde, J.T. et al. (2015). A 3D-printed, functionally graded soft robot powered by combustion. *Science* 349 (6244): 161–165.
108 Doubrovski, E., Tsai, E.Y., Dikovsky, D. et al. (2015). Voxel-based fabrication through material property mapping: a design method for bitmap printing. *Computer-Aided Design* 60: 3–13.
109 Bodaghi, M., Damanpack, A., and Liao, W. (2016). Self-expanding/shrinking structures by 4D printing. *Smart Materials and Structures* 25 (10): 105034.
110 Nguyen, K.T. and West, J.L. (2002). Photopolymerizable hydrogels for tissue engineering applications. *Biomaterials* 23 (22): 4307–4314.
111 Drury, J.L. and Mooney, D.J. (2003). Hydrogels for tissue engineering: scaffold design variables and applications. *Biomaterials* 24 (24): 4337–4351.
112 Ifkovits, J.L. and Burdick, J.A. (2007). Photopolymerizable and degradable biomaterials for tissue engineering applications. *Tissue Engineering* 13 (10): 2369–2385.
113 Jamal, M., Kadam, S.S., Xiao, R. et al. (2013). Bio-origami hydrogel scaffolds composed of photocrosslinked PEG bilayers. *Advanced Healthcare Materials* 2 (8): 1142–1150.
114 Palleau, E., Morales, D., Dickey, M.D., and Velev, O.D. (2013). Reversible patterning and actuation of hydrogels by electrically assisted ionoprinting. *Nature Communications* 4: 2257.
115 Ionov, L. (2013). Biomimetic hydrogel-based actuating systems. *Advanced Functional Materials* 23 (36): 4555–4570.
116 Wehner, M., Truby, R.L., Fitzgerald, D.J. et al. (2016). An integrated design and fabrication strategy for entirely soft, autonomous robots. *Nature* 536 (7617): 451–455.
117 Taylor, D.L. and In Het Panhuis, M. (2016). Self-healing hydrogels. *Advanced Materials* 28 (41): 9060–9093.
118 Tuncaboylu, D.C., Sari, M., Oppermann, W., and Okay, O. (2011). Tough and self-healing hydrogels formed via hydrophobic interactions. *Macromolecules* 44 (12): 4997–5005.
119 Amendola, V. and Meneghetti, M. (2009). Self-healing at the nanoscale. *Nanoscale* 1 (1): 74–88.

15

Current Trends and Challenges in Biofabrication Using Biomaterials and Nanomaterials: Future Perspectives for 3D/4D Bioprinting

Luciano P. Silva

Embrapa Genetic Resources and Biotechnology (Cenargen), Laboratory of Nanobiotechnology (LNANO), Brasília 70770-917, DF, Brazil
University of Brasília (UnB), Postgraduate Programs in Nanoscience and Nanobiotechnology and in Molecular Biology, Brasília 70910-900, DF, Brazil

15.1 Introduction

Biofabrication is a multidisciplinary (and maybe interdisciplinary) research field combining principles, protocols, and practice from engineering, biology, and material sciences through the use of manufacturing processes to build models and create biomimetics, bioprototypes, and bioproducts at the cutting edge of bioengineering innovation (Figure 15.1). Within the last decade, biofabrication emerged as a new paradigm and potentially dominant technological platform for the twenty-first century manufacturing [1], particularly for those considering new industries and modes of production as the driving forces of the so-called Fourth Industrial Revolution. More than an isolated scientific approach, this term encompasses multiple techniques and methods that produce or employ biological assets and innovative materials to generate proof-of-concepts that are beyond current knowledge and thus labeled as science fiction or visionary concepts by nonspecialists.

In fact, the scope of biofabrication is still increasing continuously and expanding fast, thereby leading to unpredictable scientific scenarios and trends (Figure 15.2). The current literature reports a variety of strategies and concepts for bioengineering of novel constructs with improved features and functions [2–13]. Now, the production of organs for transplantation or artificial meat has only highlighted the examples of an area of great interest in bioengineering and related areas. Probably, the only certainty about the future of biofabrication is that it will constantly be undergoing changes and be offering opportunities next years. Indeed, this is also evidenced by the already high number of professionals working in the field at academic institutions and companies worldwide, and the estimates of global investments are around multi-billion dollars by 2030 [14]. For example, simple searches on professional networks such as LinkedIn display thousands of people citing biofabrication (or bioprinting) among their own skills/interests and also there are hundreds of companies, particularly startups and spin-offs developing research, development, and innovation in the field.

3D and 4D Printing in Biomedical Applications: Process Engineering and Additive Manufacturing,
First Edition. Edited by Mohammed Maniruzzaman.

Figure 15.1 A word cloud representing the 100 most common terms in the titles and journal names of the 500 works cited in the references section of this chapter.

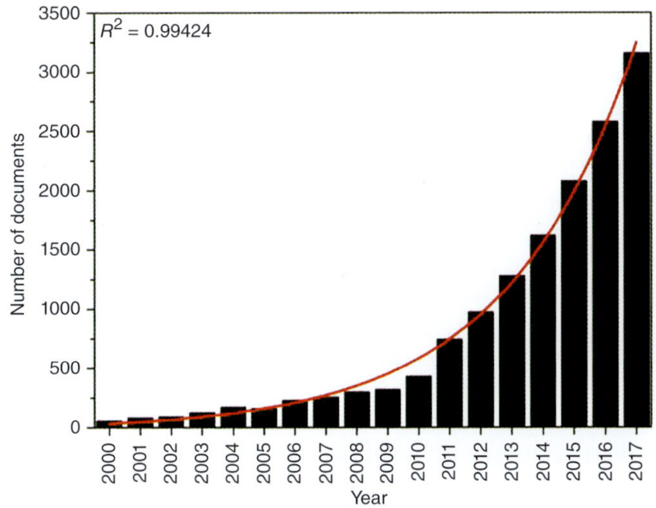

Figure 15.2 Growth in the number of documents available in Google Scholar (including papers, patents, and citations) that were published with the topic query "biofabrication" for 18 years. The line-fitting shows an exponential growth.

Professionals and institutions dealing with biofabrication are focused on the development of software (e.g. computer-aided design – CAD), machines (e.g. bioprinters and bioreactors), and advanced materials (e.g. bioinks) for the production of innovative bioconstructs. Undoubtedly, an increasing interest has been directed to the search for new ways to produce the essential building blocks for biofabrication processes, and thus biomaterials [15–30] and nanomaterials [31–39] are among the most widely used and discussed inputs for many approaches because of their impressive properties. This chapter provides an overview of current trends and challenges in biofabrication, particularly related to the use of biomaterials and nanomaterials as essential components of bioconstructs. However, efforts will also be made to present poorly explored ideas and perceptions so that it could not be just another review chapter on biofabrication but instead a benchmark for those interested in hearing new ways to contextualize and develop this field.

15.2 Biofabrication as a Multidisciplinary to Interdisciplinary Research Field

It is evident that professionals from different disciplines must share thoughts and work interactively together to develop novel technologies and to shape the future of biofabrication field. The cooperation among life scientists, material scientists, and engineers of different specializations is absolutely essential for speeding up this emerging field. Furthermore, multidisciplinary research groups that include scholars and researchers not only from biology [40], biotechnology [41], biomedical engineering [42], regenerative medicine [43], health care [44], chemistry [45], and materials sciences [46] but also from computer science [47, 48], automation engineering [49], and mechatronics [50] are those seeked to move beyond the frontiers of biofabrication.

In addition to the multidisciplinarity research groups, individuals from universities, companies, governmental agencies, and nonprofit organizations must interact to create a proactive innovation culture for biofabrication. Of course, the development of joint efforts among professionals with different skills, attitudes, and values is a challenging issue as it involves a multitude of stakeholders often distributed across both public and private organizations. Indeed, coordination and cooperation are often prerequisites on biofabrication teams surpassing the individual as the fundamental unit. Thus, interdisciplinarity [51] or even transdisciplinarity [52] seems to be the next step in biofabrication despite it still requires the members of teams to learn how to share their knowledge and protocols across the disciplinary boundaries and move beyond disciplines.

Although formal training programs in biofabrication are rare with no more than few examples of postgraduate programs (e.g. the international consortium among Utrecht's University Medical Center in the Netherlands; the Queensland University of Technology and the University of Wollongong, both in Australia; the University of Würzburg in Germany; and more recently the new Master's Program in Biofabrication at the University of Bayreuth in Germany with international partners in Australia, Thailand, France, United States, Netherlands, and Spain), hundreds of institutions worldwide have research groups, research centers, and institutes working in the field. Despite the importance of formal training in biofabrication is noticeable by researchers in the field, it is undeniable that countless other disciplines partially cover the content of the biofabrication curriculum if oriented to natural, engineering, and materials sciences. This may justify the relatively high number of professionals working in the field, although biofabrication is still in its childhood. Undoubtedly, this critical mass of professionals working in biofabrication is essential to consolidate and expand the frontiers of knowledge in the field to the next level.

Anyway, bioengineering professionals developing projects in biofabrication field will have to keep in mind that it will be extremely difficult to develop a functional bioprototype that could be more advanced (or even similar) than the design that nature took millions of years to develop by natural selection. Thus, there will be many mishaps, barriers, and obstacles along the way to make some of the experimental results feasible from practical viewpoint.

Additionally, open-minded professionals are essential in these efforts because they should be able to transfer knowledge, skills, and technologies between scientific entities and entrepreneurs and thus bridging the gap between academic and industrial research. Indeed, new business models must be created and explored offering new capabilities, approaches, and organizational structures without borders or bureaucratic limits that hamper creativity. Of course, the most important question is whether the worldwide people, institutions, and companies are truly ready to accept and promote these ideas and concepts related to biofabrication. There is no answer to this question yet, but it is clear that the forthcoming years will offer more opportunities and experiences for everybody who is interested in breaking paradigms and mindsets.

15.3 Biofabrication as a Multifaceted Approach

Biofabrication is a multifaceted process that commonly encompasses several stages and steps, and each one of them can be methodized and guided focusing on the production of bioconstructs that mimic in some extent the architecture of living systems or their components. In fact, the most fascinating aspect of biofabrication is that a huge diversity of materials, tools, and protocols is currently available and a multitude of new possibilities emerges on a daily basis, making it difficult to keep up-to-date with all recent progresses. Indeed, a literature survey presented in a book chapter or a review paper related to current facts associated with biofabrication would become obsolete after no more one or two years. Surely, this is not the goal of this chapter and it will not dedicate to systematically explore and detail all the current technical possibilities and data, but instead share a general guide about biofabrication so that the readers can themselves search for some specific subject of their interest included in the references section or related literature.

Because of the wide range of techniques and methods available for biofabrication, a general workflow is unfeasible and even impossible to conceive in practice because each type of bioconstruct has specific requirements and steps. However, there are shared aspects and terminologies common to the field, which help in understanding biofabrication and that was recently consensualized by several prominent researchers [53]. Overall, this unprecedented initiative of paramount importance will help people working in the field to avoid or at least minimize generalizations and inappropriate use of interchangeable terms that commonly lead to undesired inconsistencies to the literature.

Another contribution to this scenario could be the herein introduced criteria or premises to be considered when performing biofabrication processes. Accordingly, biofabricated systems should meet at least one (if not all) of the following four bio's criteria: biobased, i.e. to be intentionally composed, in whole or in part, by cells or naturally occurring molecules; bioinspired, i.e. to be intentionally designed to mimic the structure, properties, or function of natural biostructures or biosystems; biofriendly, i.e. to be intentionally nontoxic or minimally toxic for living systems and preferably biodegradable; biotransformable, i.e. to be intentionally conceived to exhibit transformative properties that emerge from the final bioconstruct.

Likewise, other aspects rooted in the core fundamentals of biofabrication processes is that they are overall highly automated [49, 54–59], customized [60–65], personalized [66–69], optimized [70–75], and finally tailored [76–80] solutions. At the same time, these are supported by performance parameters that include reproducibility, accuracy, precision, and speed.

15.4 Biofabrication Beyond Biomedical Pharmaceutical Applications

Undoubtedly, biofabrication debuted in scientific scenario with potential strategies and advanced techniques for human tissue engineering and regenerative medicine aiming at the production of tissues for reconstruction [81–83], organs for transplantation [31, 44, 69, 84–92], and tissue-engineered organoid models for diseases investigation [93–96], drug discovery [95–102], and cosmetics testing [103–105]. However, agriculture (e.g. artificial or synthetic seeds) [106–110], veterinary (e.g. animal transplants) [89, 111], and food industry (e.g. animal-free food and artificial meat) [112–118] are sectors that can directly benefit from the use of biofabrication processes.

Curiously, biofabrication applications to agriculture (e.g. production of alginate spheroids entrapping plant somatic embryos) predate in decades the first reports of applications related to medical field, but this fact is simply neglected by almost all the authors because of unexplained reasons. This is still more incomprehensible because of the fact that historically it is easier to a new technology to achieve the human health sector if it has been previously used and validated in agriculture or veterinary. Such applications may represent predecessors of further pharmaceutical, biomedical, and clinical applications with potential to expand the biofabrication scope to other manufacturing industries and also stimulate the development of more advanced and robust technologies.

15.5 The Diversity of Techniques Used in Biofabrication

From simple labware and low-cost materials available in any laboratory (e.g. plastic pipettes useful for the production of alginate spheroids) to machines of hundreds of thousands of dollars (e.g. advanced 3D bioprinting platforms), there is a wide range of possibilities for routes and protocols of biofabrication. In fact, even among the machines dedicated to biofabrication applications, such as 3D bioprinters, there are examples from big and expensive machines to benchtop and low-cost do-it-yourself (DIY) options [50, 119–125]. Consequently, patented and open-source technologies share the same space to speed up innovation in this emerging field. In fact, few research and development themes have so high diversity of possibilities and options like biofabrication and related disruptive technologies.

For now, there is not an entirely accepted consensus regarding the classification of biofabrication strategies and techniques. Indeed, there are only proposals based on ideas and perceptions from different researchers and groups. In fact, the most elegant, coherent, and less tendentious proposal was that presented by a group composed by very active researchers in the field in a very recently published paper [53]. There, the authors cluster the major biofabrication techniques into 15 different groups and present the main features of each one, particularly related to the fabrication rate, minimum feature width, and common limitations. Probably that paper will represent a milestone in the field and all biofabrication community wishes that it could finally pragmatize this complex subject. However, the existence of so many different groups of technologies could still be difficult to overall people working in the field rationalize their knowledge and thinking.

Thus, perhaps the best choice would be to take advantage of the relatively well-accepted classifications of digital manufacturing processes (Industry 4.0) and after the assignment within different techniques as proposed by those

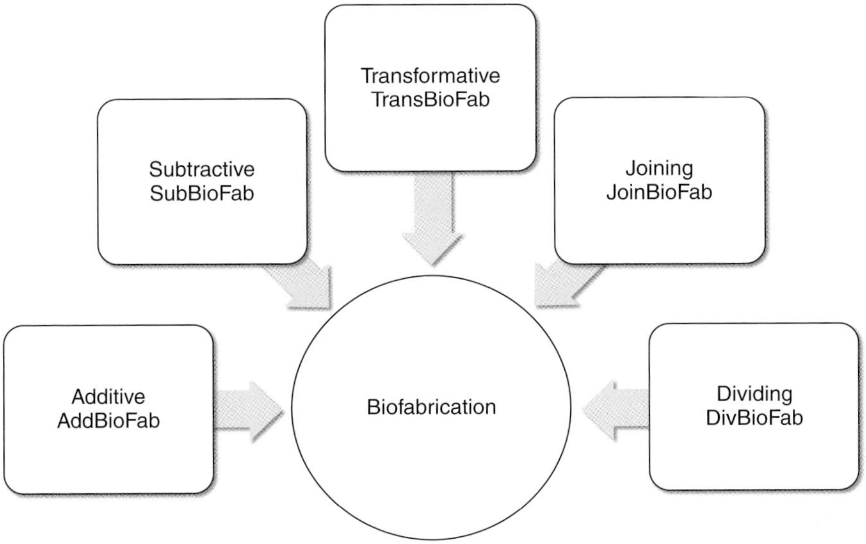

Figure 15.3 New proposals of categorized strategies to perform biofabrication processes using different techniques.

authors [53], split them into simplified classifications. Therefore, it is possible to split all the techniques (and others omitted by those authors [53] like magnetic 3D bioprinting of spheroids and rings [126–130]) into the following categories: additive biofabrication (AddBioFab), subtractive biofabrication (SubBioFab), transformative biofabrication (TransBioFab), joining biofabrication (JoinBioFab), and dividing biofabrication (DivBioFab) (Figure 15.3).

The AddBioFab term can be attributed to several bottom-up manufacturing approaches based on joining, accumulating, building, or assembling biomaterials and cells to produce patterns and spatially oriented objects designed, conceived, and constructed to mimic living systems. Among the vast array of techniques developed for biofabrication purposes [53], AddBioFab-based techniques gained an increasing interest in the last few years for their suitability in producing complex and detailed shapes offering high potential for developing processes with automation, reproducibility, precision, and accuracy. In addition, the possibility of quickly customizing, prototyping, and delivering solutions matching specific criteria, in terms of shape, size, chemical composition, and physical properties as well as biological features, makes AddBioFab techniques a remarkable way of materializing bioconstructs. Established and novel methods based on AddBioFab techniques such as 3D bioprinting [13, 20, 131–137], electrospinning [80, 138–141], bioplotting [142–146], and several others, including variants and hybrids of all techniques, are under constant application and investigation, and on a weekly (if not daily) basis result in new advances to the field in laboratories around the globe.

SubBioFab can be defined as the opposite strategy when compared to AddBioFab because this term refers to top-down processes of removing components or constituents (commonly biomaterials and cells) from a bulk solid to produce a final shape for the modeled biomimetics. The technologies based on SubBioFab include physical (e.g. heat and cold), chemical (e.g. solvents and salts), or biological (e.g. degrading microorganisms and enzymes) processes for milling [147], drilling [148], engraving [149–151], polishing [152], grinding [153], carving [154], or leaching [155, 156] of work pieces containing biomaterials and cells to desired organizations and shapes. The potential of SubBioFab approaches to the biofabrication field is relatively neglected by researchers, but they are now being realized throughout the world.

TransBioFab strategy encompasses those approaches aiming at the creation of a biomimetic by modulating the shape or behavior of scaffolds produced with biomaterials and cells. TransBioFab techniques include molding [157–165], casting [166–171], and forming [172, 173]. Combinatorial approaches are possible, and thus, TransBioFab strategies can be easily combined with AddBioFab [165, 174–178] and/or SubBioFab [179–182] strategies. In fact, there is an increasing trend for the development of multifunctional biofabrication platforms based on a multitude of techniques and instrumental setups. Thus, multistrategy biofabrication approaches have allowed the combination of various cells and biomaterials to construct complex biomimetics with tailorable mechanostructural, physicochemical, and biological properties [124, 183, 184].

JoinBioFab is still a relatively unexplored strategy aiming at the manufacturing of biomimetics. In the processes based on JoinBioFab, two or more biofabricated pieces are joined to each other or broken parts of bioconstructs are repaired or joined together. Gluing [185–187], screw-fastening [188], welding [189], brazing [190], cementing [191], and soldering [192] represent some of the techniques of JoinBioFab and they use biomaterials and cells for attaching biofabricated pieces. Modularization and parallelization that will be discussed further are concepts in biofabrication that require JoinBioFab approaches.

DivBioFab strategy consists in fragmenting bioconstructs into two or more blocks of manufacturing, and it is probably the least explored among the five categories discussed here. However, it is expected that it will gain attention mainly with the advances related to maturation processes because functional pieces derived from larger grown tissue mimetics can be used as representative parts aiming at, e.g., the screening of drugs and cosmetics.

15.6 Natural Resources as Sources of Biomaterials Useful for Biofabrication

Despite the lack of consensus related to several aspects, the fundamental importance of biomaterials for biofabrication processes is unequivocal and deserves attention from stakeholders [15, 18–30]. Biomaterials can be defined from different viewpoints, perspectives, and interpretations. From biofabrication viewpoint, biomaterials are pure substances or mixtures, whether natural or engineered, alive, or lifeless; able to serve as building blocks; interact; and undergo desired modifications in biological systems, particularly to repair [3, 193–196], restore [197], reconstruct [81, 198–201], restructure [202], replace [78, 89, 203, 204], re-establish [205], reconstitute [206, 207], recover [208], regenerate [3, 33, 39, 72, 82, 147, 191], remodel [209], redesign [210], or recreate [211, 212] structural or functional components. Additionally, from biological host (receptor) perspective, a biomaterial can be an autobiomaterial (from the same individual), an isobiomaterial (from an individual sharing the same genetics), an allobiomaterial (from another individual of the same species), a xenobiomaterial (from other species), or a synthetic biomaterial (from an artificial source). From chemical source, biomaterials can be divided into natural (from animals, plants, or microorganisms), synthetic (from nondegradable materials such as metals to man-made biodegradable polymers), or semisynthetic/hybrid (from natural sources but with enhanced performance). Finally, from host interaction and bioactivity viewpoint, biomaterials can be classified into bioinert (illicit minimal or no interaction and immune response), bioresorbable (dissolve and are replaced by endogenous biocomponents), or bioactive (promote biological responses).

Evidently, there is a huge diversity of biomaterials, and their inherent features represent outstanding opportunities for scientists and engineers looking for developing their studies in biofabrication. According to their architecture and macroscopic patterns, biomaterials useful to fabricate bioconstructs can be divided into several categories including beads [213–218], microspheres [219–222], pellets [223, 224], blends [77, 225–228], films [229–231], foams [232–235], sponges [236–238], fibers [239–245], sheets [9, 246, 247], porous scaffolds [167, 248–255], and mainly hydrogels [17, 56, 57, 68, 122, 123, 137, 142, 145, 172, 176, 178, 187, 191, 193, 204, 211, 218, 220, 256–269].

Probably, hydrogels represent the most important building blocks for the production of biomaterial-based scaffolds aiming at biofabrication processes over the past years [256, 270]. Most of the hydrogels are biocompatible materials formed by hydrophilic polymers cross-linked into 3D structures, which absorb from a small percentage up to hundreds of times their dry weights in water [271–274]. Highly varied kinds of hydrogels are used to produce scaffolds for biofabrication and they derive from a variety of origins including synthetic [256, 274–276], recombinant [277, 278], and natural (autologous and heterologous) [279, 280] polymers. Interestingly, natural polymers derived from agricultural, forestry, and livestock products and mainly their by-products are the most largely used raw materials used for the production of hydrogels because they are abundant, inexpensive, renewable, biodegradable, biocompatible, and considered appropriate from structural and mechanical standpoints [281–288].

Examples of these biopolymers for the production of hydrogels applicable to biofabrication include collagen [288–293], gelatin [56, 142, 155, 294–298], keratin [221, 299], silk fibroin [299–302], elastin [299, 303], resilin [299], laminin [2], fibronectin [304], gellan gum [305–307], alginate [106, 122, 132, 142, 163, 213, 215–219, 221, 224–226, 244, 251, 308–314], chitin [315, 316], chitosan [316–322], hyaluronan [289, 291, 323], agarose [324–326], dextran [327–329], κ-carrageenan [330, 331], pectin [332, 333], xanthan gum [334], cellulose [335–338], lignin [283], hemicellulose [283], and starch [339–342], which are chemically or physically cross-linked as homopolymers [56, 265, 284, 292, 293, 299, 306, 308–314, 318, 320–327, 332, 334, 339, 340, 343] or copolymers [77, 142, 219, 225, 343–360] to form network lattices. In addition, most of them is nontoxic or has low cytotoxicity upon contact and incubation with different cell types even at high concentrations being suitable for biofabrication and thus termed biopapers [184, 361–363]. In fact, on the beginning of biofabrication processes, the term biopaper was routinely employed in order to refer to biocompatible, bioprocessable, and biomimetic hydrogels, in contrast to the term bioink that was used originally to refer to cell blocks and tissue spheroids used for biofabrication processes [364]. However, now there is a higher generalization of the term bioink, and for almost all situations, it refers either to the hydrogel containing the materials necessary for the scaffolds' production and also the living cells under single-cell suspensions or small multicellular aggregate (spheroids) configurations [46, 365–385].

Therefore, it is foreseeable that bioinks are the most critical inputs in biofabrication processes, as they must represent appropriate compositional [373], mechanical [373, 384], and rheological [366, 383] properties required not only for the bioconstruction but also to allow an excellent surrounding for cells entrapment, surviving, proliferation, migration, sometimes even differentiation and angiogenesis, and always metabolic function. Indeed, many researchers and companies worldwide are currently developing innovative bioinks because these are probably the most important and the only indispensable consumables for all biofabrication processes. Commonly, an optimal bioink must be biocompatible (e.g. nontoxic), cell-friendly (e.g. allow cell attachment, proliferation, migration, differentiation, and also angiogenesis), immunocompatible (e.g. acceptable by the immune system), affordable (e.g. cost-effective), plentiful (e.g. scalable production), customizable (e.g. organ/tissue-specific), and eco-friendly (e.g. environmentally sustainable) in order to obtain truly viable biomimetic constructs. In addition, such bioink must have enough viscoelasticity to achieve the stability (to maintain the overall shape) of the produced scaffold during (short-term) and after (long-term) biofabrication, particularly following the curing process.

15.7 Nanomaterials as Much More Than Just New Building Blocks for Biofabrication

In recent years, in addition to biomaterial approaches, nanomaterial and nanotechnology-based approaches have also attracted great attention for biofabrication processes because of their unique properties that arise from nanoscale [31–39, 386–389]. Typical nanomaterials such as carbon nanotubes [390–397], graphene [398, 399], polymeric nanofibers [400–408], polymeric nanoparticles [409], micelles [410], emulsions [411], ceramic nanomaterials [397, 412], silica nanoparticles [413, 414], quantum dots [415, 416], magnetic nanoparticles [126–130, 417], noble metal nanoparticles [418–420], and others nanomaterials [421–425] are currently being used for innovative biofabrication processes.

Curiously and coincidentally, the term biofabrication (or biosynthesis or green synthesis) also refers to the use of biological resources aiming at the production of nanomaterials including those from noble metals such as silver, gold, platinum, and among others [426–433], and for such situations, they are included in the emerging area called green nanotechnology [434–436]. It is probably more one case of double use of the same term. However, this fact helps researchers which are working in the field of biofabrication (as discussed in this chapter) remember that biosystems are undoubtedly the grand masters in designing and developing well-ordered and well-controlled structures. Indeed, research, development, and innovation initiatives and concepts such as NanoBioFabLab (https://youtu.be/jBNzMZGmTqU) emerge focusing on 3D biofabrication using nanomaterials as essential building blocks for biomimetic production with sustainability based on green nanotechnology principles and opening opportunities to the development of the green biofabrication related to the development of sustainable alternatives aiming at the production of bioconstructs.

In addition to the use of nanomaterials as building blocks for biofabrication, may be the most promising use of nanomaterials in biofabrication process is related to the so-called bioactive nanoparticles. Thus, several nanomaterials may serve as nanocarriers or vehicles for transport and targeted delivery of bioactive substances to the final products of biofabrication processes [388, 437–442]. Particularly, the smart nanomaterials, such as stimuli-responsive nanotechnology-based materials engineered to understand the surrounding environment and elicit a response according to the demands [443, 444], are going to have the large impact in biofabrication processes leading to the upcoming nanobiofabrication area. Thus, it is expected that the development of the new so-called nanobioinks will probably revolutionize the way we perform biofabrication in the very near future. Particularly, 3D bioprinting and mainly 4D bioprinting strategies may directly benefit from such upcoming nanotechnological advances because they match the technical requirements for using such impressive and innovative materials.

15.8 3D Bioprinting as the New Gold Standard for Biofabrication

Despite 3D bioprinting is only one among various biofabrication techniques, its high reliability, spatial precision, dimensional accuracy, repeatability, compatibility with multiple other techniques, and model compliance with native biostructures and biosystems contribute in becoming almost a synonym of biofabrication [10–15, 43, 44, 83, 85, 92, 133, 184, 445]. Indeed, recent trends have progressively empowered 3D bioprinting technologies to be the reference in biofabrication strategies to the point of being considered by stakeholders the gold standard technique. Surely, 3D bioprinting is emerging as an essential biofabrication technology based on AddBioFab and thus on automated deposition of successive layers of biologically relevant and targetable materials, including biomaterials and nanomaterials, in a precise 3D space by potentially combining multiple methods [446].

There is a huge number of recent book chapters and review papers that provide systematic reviews about 3D bioprinting, and thus this subject will not be the focus of discussion in this chapter [10, 445–448]. However, some facts (not rumors) must be cited and considered. There are now almost hundreds

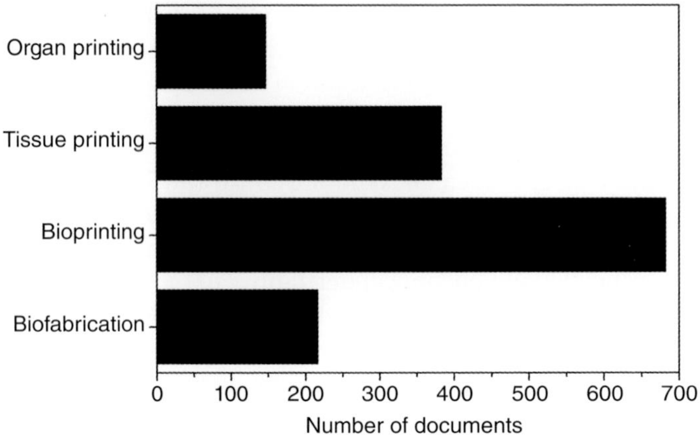

Figure 15.4 Number of documents available until 5 May 2018 in Google Patents (excluding nonpatent literature) with the topic queries "biofabrication" or "bioprinting" or "tissue printing" or "organ printing."

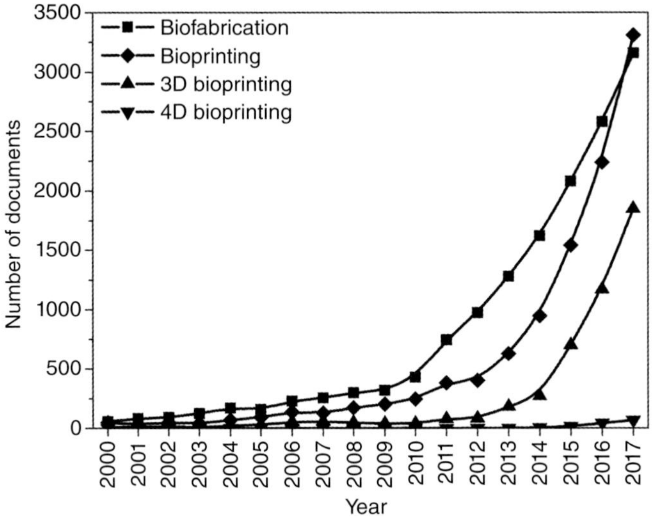

Figure 15.5 Growth in the number of documents available in Google Scholar (including papers, patents, and citations) that were published with the topic queries "biofabrication" or "bioprinting" or "3D bioprinting" or "4D bioprinting" for 18 years.

of companies spread over dozens of countries developing 3D bioprinters, compatible bioinks, and add-ons. Not surprisingly, there are substantially more patents with the term bioprinting than biofabrication or other related terms (Figure 15.4). Nevertheless, even the number of published articles containing the term bioprinting is very similar to the number of articles with the term biofabrication (in 2017, it was higher for the first time); and the term biofabrication is also used in several other senses and contexts (Figure 15.5). In fact, new 3D bioprinting methods and techniques are being introduced to the scientific and technological community continually and fast-growing biofabrication industry market. It is possibly envisioned that the next trend in bioprinting will be an increasing demand for smart materials that can introduce new possibilities and dimensions to the biofabrication processes and lead this shift [449]. Particularly, 4D bioprinting approaches are now presenting the first stimuli-responsive hierarchical self-morphing and active shape-changing biofabricated structures using smart materials.

15.9 When 3D Bioprinting Is Not Sufficient for Bioconstruction: 4D Bioprinting

The goal of 4D biofabrication (bioprinting) approaches is to construct and fine-tune 3D structures through dynamic processes of self-assembly that could programmably modulate their morphologies or functionalities over time, particularly when a certain chemical (e.g. pH, ionic strength due to salts, and other solution components), biological (biomolecules and other endogenous organic compounds), or physical (e.g. temperature, light, magnetic field, ultrasound, electric field, or osmotic pressure) stimulus is applied or a cell/tissue post-processing self-organization occurs. In such respect, new strategies of 4D bioprinting probably represent no more than 100 published research papers, but they are probably among the most remarkable advances and opportunities in the area of biofabrication [450–462].

Despite 4D bioprinting has emerged only last few years allowing that the first smart 3D bioconstructs have debuted, researchers in the area are already finding the first gaps. New software solutions related to predictability of the temporal and dynamical events are essential; new machines for the whole process including not only bioprinters but also bioreactors and real-time monitoring systems should be developed; new smart materials (biomaterials and nanomaterials) with enhanced capabilities are fundamental to achieve reliable bioconstructs; and new facile, precise, and efficient methods must be developed to establish these unprecedented biofabrication processes. There is certainty that new challenges and obstacles will arise from the researches, but the multidisciplinary aspect of this area will help in solving the issues.

15.10 An Overview About Current Bottlenecks in Biofabrication

Despite the recent advances, several challenging issues are still bottlenecks of some practical applications in biofabrication. Thus, the building of reliable 3D models; the definition of feasible biofabrication conditions by considering size, time, cell type, maturogen availability, and incubation parameters; the selection of standard and robust methods for characterization; and the evaluation of potential economical, social, and ethical impacts are all essential aspects to be considered during the strategic planning and design of biofabrication initiatives before real applications can emerge.

15.10.1 Does 3D Model Matter in Biofabrication?

The intricate structures of biological systems that serve as an inspiration to biofabricate the mimetics are commonly examined and structurally decoded by 3D imaging techniques such as laser scanning [199], microcomputer tomography (micro-CT) [164, 357], and magnetic resonance imaging (MRI) [164, 463]. This step is necessary for the generation of biological 3D models because *ab initio* approaches commonly fail in consistency and trustworthiness. Currently, public (free and premium) and private repositories of 3D models are growing, and several organizations, including companies, are currently in the process of generating and storing CAD and computer-aided manufacturing (CAM) models (blueprints) for biological objects ranging from small tissue fragments to human organs. However, a lot of knowledge and hard work are still necessary to create representative models for several biological structures unavailable to date.

15.10.2 Does Size and Time Matter in Biofabrication?

Most human organs have several thousands of cubic millimeters and this volume is relatively prohibitive not only for all current biofabrication techniques but even for traditional 3D printers working with thermoplastics through fused deposition modeling (FDM) or resins through stereolithography (SLA). Thus, the speed of execution of biofabrication processes is an essential aspect to be considered during the experimental planning because biologicals including biomaterials and cells would not be stable over all this required time. From these perceptions, the concepts of parallelization and modularization in the construction of large and complex tissues and organs have emerged.

The concept of 3D printer's farms, or clusters, is relatively well established as a collection of networked 3D printers that someone can run on-site or remotely to industrial or serial production [464]. Several visionary people in the digital fabrication field argue that the future of 3D printing lies in parallelization – the use of many machines working together to additively manufacture parts rapidly. However, this scenario of production line was never tested for bioprinting, but it should only be a matter of time before the first bioprinter farms can be in operation in some laboratories around the world. At that moment, strategies of the type JoinBioFab will be fundamental to carry out the coupling between different parts for the building of larger bioconstructions.

In addition to parallelization, modularization is another innovative idea to be explored during the biofabrication of complex structures for managing complexity and final construct heterogeneity. From engineering viewpoint, a module is a functional unit, a collection of components, or parts that harmoniously interact together to perform a specific function, but it is not easily discernible the representation of a module from biological viewpoint. Life scientists will have to work together with engineers trying to establish which modules are significant in each organ or any biological system and then select the best tools to produce each of them. Thus, each specialized technique/machine could produce those parts or modules for which they are more reliable and efficient (i.e. faster, cheaper, or with higher performance). The final goal would be to optimize the biofabrication processes and to maximize the productivity.

15.10.3 Do Choice Materials and Cells Matters in Biofabrication?

As exhaustively presented throughout this chapter, the range of possibilities for materials useful for biofabrication dramatically increases over the years. Therefore, choosing the adequate materials is not an easy task, despite there are several options available. It is possible that the development of databases and computational tools based on mathematical modeling and statistics will be essential and imperative in order to rationalize that information in the new era of information technology applied to biofabrication [465].

In addition, cells are recognized by always dynamically reorganize the surrounding environment and cell environment (e.g. extracellular matrix and scaffolds) interactions are regulated spatially and temporally. Thus, independent of the use of which biomaterial or nanomaterial during a biofabrication process, it is consensus that each cell type has specific requirements and also elicits specific responses. In fact, the diversity and plasticity of biological cells must be considered and explored in forthcoming studies. The application of either differentiated, stem, or fetal cells as well as their progenitors and derivatives is essential to better understand the role of each cell type in a variety of biofabrication contexts [75, 84, 97, 143, 215, 219, 288, 292, 335, 336, 358, 376, 391, 397, 398, 404, 407, 441, 466].

Large-scale and industrial production of different cell types and spheroids is also a very important challenging aspect to be addressed because they are one of the most important consumables of biofabrication processes [467].

15.10.4 Does Maturation of the Bioconstructs Matter in Biofabrication?

Usually, even though a bioconstruct is biofabricated by some technique, it still requires to be maturated before be considered a truly functional or at least usable biomimetics. However, the number of research papers describing the maturation step related to the biofabrication of a mimetics is inconceivably low. A high number of studies simply neglects that the process is not final when a bioprinted object is obtained. Indeed, mechanical, thermal, chemical, and biological stabilities should be assessed over time under incubation conditions as close as possible to those of the native biosystem in order to gain the maximum information and to determine potential applications.

To this end, specific culture media and maturogens including growth factors [438, 468–471] and morphogens [472, 473] play a pivotal role in the preservation of the functional viability of the biological structures produced by biofabrication because they sustain cell viability and also accelerate and enhance bioconstruct maturation achieving the desired level of cell subpopulations, the proliferative responses, and the effective vascularization [474].

However, for some situations, maturogens are the most limiting factors for biofabrication because of the inexistence of commercially available options or simply because they are not known for several tissues and organs. In this context, prospection of signaling pathways of morphogenesis and the use of molecules prospected from the biodiversity may take several advantages over synthetic compounds. Therefore, the use of endogenous compounds [475] and tissue-specific peptide pools [476] may represent the next frontier related to the maturation step associated with biofabrication. Additionally, the use of recombinant growth factors produced in large scale and at a lower cost is expected to have an expressive and positive impact on biofabrication of tissues and organs [477].

Some efforts are also necessary for the development of bioreactors specifically designed for the maturation of biofabricated mimetics. These systems must allow promoting maturation of biomimetic under conditions suitable for maintaining viability and stimulating the proliferation capacity of the cells. Additionally, the use of microsystem technologies such as microfluidic components, microactuators, and microsensors has expanded as a growing need for biofabrication processes, in particular *in situ* and real-time monitoring technologies [478].

15.10.5 Do Characterization Methods Matters in Biofabrication?

The use of classical and high-throughput methods for characterization of biofabricated constructs should be considered an essential step to be evaluated during the rational planning of a biofabrication project. Simply put, researchers should test their bioconstructs for assessing the biological activity, functionality, and other characteristics necessary to validate the biomimetic from the biochemical, morphological, and physiological viewpoints. However, it is surprising that several published studies in this field do not explore in detail the comparison of chemical, physical, and biological characteristics of the biofabricated mimetics even by classical techniques such as histology, clinical chemistry, and molecular diagnostics. Accordingly, the use of high-throughput methods such as transcriptomics, proteomics, lipidomics, and metabolomics to characterize biofabricated constructs must be part of the next frontier in this emerging field.

Another aspect is that although direct comparisons between native biostructures and biofabricated constructs sound as a fundamental step to be considered and if possible performed during the development of biofabrication projects, there is a surprisingly limited number of studies specifically addressing such similarities and dissimilarities. Even when some correlation is established, most of the studies are restricted to morphological comparisons and more rarely some classical biochemical investigation.

15.10.6 Does Economic and Social Impact Matter Biofabrication?

Current estimates vary according to the sources, but all of them agree and suggest that biofabrication, in particular 3D bioprinting, will represent a global market of several billion USD over the next decade. Another aspect to be considered from economic perspective is that the costs related to biofabrication processes will fall gradually and proportionally with the increase in offering solutions to the market.

On the social aspect, biofabrication technologies allow the promotion of activities related to teaching of bioengineering that creates a playful and tangible experience for students at all levels. Not to mention the unique approaches involving 3D bioprinting combined with other areas such as synthetic biology would truly enable people to construct mimetics of biological structures with characteristics that resemble those observed in living organisms. Some other social aspects encompass political, cultural, and religious concerns that must be accounted for the success of this field.

15.10.7 Does Ethical and Legal Issues Matter in Biofabrication?

Ethical and legal concerns are part of the core aspects of the biofabrication field since the first insights [479–482]. For drug discovery and cosmetics testing, biofabrication models contribute to minimize the use of animals while assessing the safety, efficacy, and security. Furthermore, for medical applications, although biofabrication can minimize some ethical dilemmas and moral distress associated with clinical organ transplantation and xenotransplantation, it is not devoid of its own restrictions, ethical and regulatory issues, particularly because of the fact that some protocols propose the use of stem cells from different sources, including embryonic stem cells. In addition, regulatory issues related

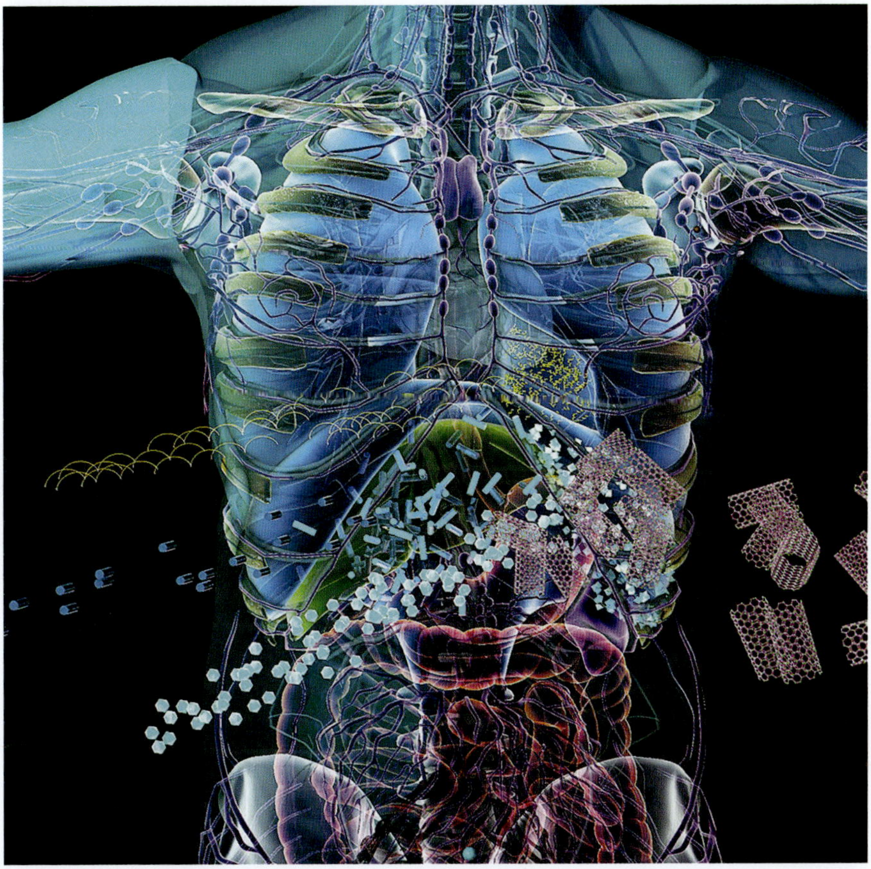

Figure 15.6 Illustration representing the use of biomaterials and nanomaterials for the biofabrication of human organs. Image: Ella Maru Studio.

to biofabrication are currently under discussion by worldwide agencies and organizations. Furthermore, it is possible that the use of previously approved and safety-certified materials (e.g. biomaterials, nanomaterials, and cells) will accelerate the application of biofabrication technologies to solve real-world challenging issues.

15.11 Conclusion

There is no exaggeration to say that biofabrication is the research and development field in that the cell culture meets the innovation culture. Afterward, it is absolutely consensus that organ printing will become a reality in the near future. To do so, it is necessary that all biofabrication community can understand that collaborative efforts are necessary to take this area to the next stage. Today, there is an ecosystem of opportunities for researchers, students, entrepreneurs, and anyone interested to move forward this emerging field. Similar to a manufacturing process for the construction of a 3D object, people from different areas are sculpting, molding, and fabricating the future of biofabrication. In addition, these people are using their creativity, resources, tools, protocols, and strategies to generate innovative outcomes that were unbelievable some years ago. In addition, researchers must develop truly innovative and practical approaches instead of fictional and sometimes surreal proposals when considering achievable applications. To achieve this, nature is inspiring us toward the development of new bioinspired machines and advanced materials such as biomaterials and nanomaterials that will fabricate the future of the bioindustry. Finally, it is expected that biofabrication will continue to break our scientific paradigms and revolutionize our thinking process (Figure 15.6).

References

1 Mironov, V., Trusk, T., Kasyanov, V. et al. (2009). Biofabrication: a 21st century manufacturing paradigm. *Biofabrication* 1 (2): 022001.

2 Bajaj, P., Schweller, R.M., Khademhosseini, A. et al. (2014). 3D biofabrication strategies for tissue engineering and regenerative medicine. *Annual Review of Biomedical Engineering* 16 (1): 247–276.

3 Pereira, R.F., Barrias, C.C., Granja, P.L., and Bartolo, P.J. (2013). Advanced biofabrication strategies for skin regeneration and repair. *Nanomedicine* 8 (4): 603–621.

4 Bártolo, P.J., Domingos, M., Patrício, T. et al. (2011). Biofabrication strategies for tissue engineering. *Computational Methods in Applied Sciences Advances on Modeling in Tissue Engineering* 137–176.

5 Riggs, B., Dias, A., Schiele, N. et al. (2011). Matrix-assisted pulsed laser methods for biofabrication. *MRS Bulletin* 36 (12): 1043–1050.

6 Schiele, N.R., Corr, D.T., Huang, Y. et al. (2010). Laser-based direct-write techniques for cell printing. *Biofabrication* 2 (3): 032001.

7 Ferris, C.J., Gilmore, K.G., Wallace, G.G., and Panhuis, M.I.H. (2013). Biofabrication: an overview of the approaches used for printing of living cells. *Applied Microbiology and Biotechnology* 97 (10): 4243–4258.

8 Patra, S. and Young, V. (2016). A review of 3D printing techniques and the future in biofabrication of bioprinted tissue. *Cell Biochemistry and Biophysics* 74 (2): 93–98.

9 Hannachi, I.E., Yamato, M., and Okano, T. (2009). Cell sheet technology and cell patterning for biofabrication. *Biofabrication* 1 (2): 022002.

10 Murphy, S.V. and Atala, A. (2014). 3D bioprinting of tissues and organs. *Nature Biotechnology* 32 (8): 773–785.

11 Li, J., Chen, M., Fan, X., and Zhou, H. (2016). Recent advances in bioprinting techniques: approaches, applications and future prospects. *Journal of Translational Medicine* 14 (1): doi: https://doi.org/10.1186/s12967-016-1028-0.

12 Hong, N., Yang, G.-H., Lee, J., and Kim, G. (2017). 3D bioprinting and its in vivo applications. *Journal of Biomedical Materials Research Part B: Applied Biomaterials* 106 (1): 444–459.

13 Giannitelli, S., Mozetic, P., Trombetta, M., and Rainer, A. (2015). Combined additive manufacturing approaches in tissue engineering. *Acta Biomaterialia* 24: 1–11.

14 3D Bioprinting Market 2014–2030 (2014). Research Report. Roots Analysis Private Limited, Business Research and Consulting.

15 Nakamura, M., Iwanaga, S., Henmi, C. et al. (2010). Biomatrices and biomaterials for future developments of bioprinting and biofabrication. *Biofabrication* 2 (1): 014110.

16 Mironov, V., Visconti, R.P., Kasyanov, V. et al. (2009). Organ printing: tissue spheroids as building blocks. *Biomaterials* 30 (12): 2164–2174.

17 Shim, J.-H., Kim, J.Y., Park, M. et al. (2011). Development of a hybrid scaffold with synthetic biomaterials and hydrogel using solid freeform fabrication technology. *Biofabrication* 3 (3): 034102.

18 Skardal, A. and Atala, A. (2014). Biomaterials for integration with 3-D bioprinting. *Annals of Biomedical Engineering* 43 (3): 730–746.

19 Atala, A. (2015). *Essentials of 3D Biofabrication and Translation*. Elsevier.

20 Zadpoor, A.A. and Malda, J. (2016). Additive manufacturing of biomaterials, tissues, and organs. *Annals of Biomedical Engineering* 45 (1): 1–11.

21 Zhu, W., Ma, X., Gou, M. et al. (2016). 3D printing of functional biomaterials for tissue engineering. *Current Opinion in Biotechnology* 40: 103–112.

22 Chua, C., Leong, K., and An, J. (2014). Introduction to rapid prototyping of biomaterials. In: *Rapid Prototyping of Biomaterials*, 1–15. Elsevier.

23 O'brien, F.J. (2011). Biomaterials & scaffolds for tissue engineering. *Materials Today* 14 (3): 88–95.

24 Wu, L.-Q. and Payne, G.F. (2004). Biofabrication: using biological materials and biocatalysts to construct nanostructured assemblies. *Trends in Biotechnology* 22 (11): 593–599.

25 Bandyopadhyay, A., Bose, S., and Das, S. (2015). 3D printing of biomaterials. *MRS Bulletin* 40 (2): 108–115.
26 Chia, H.N. and Wu, B.M. (2015). Recent advances in 3D printing of biomaterials. *Journal of Biological Engineering* 9 (1): 22.
27 Subia, B., Kundu, J., and Kundu, S.C. (2010). Biomaterial scaffold fabrication techniques for potential tissue engineering applications. In: *Tissue Engineering* (ed. D. Eberli), 141–158. IntechOpen.
28 Ravi, S. and Chaikof, E.L. (2010). Biomaterials for vascular tissue engineering. *Regenerative Medicine* 5 (1): 107–120.
29 Kundu, J., Pati, F., Jeong, Y.H., and Cho, D.-W. (2013). Biomaterials for biofabrication of 3D tissue scaffolds. *Biofabrication* 23–46.
30 Rollo, J., Raghunath, J., Sales, K.M. et al. (2007). Biomaterials and scaffold design: key to tissue-engineering cartilage. *Biotechnology and Applied Biochemistry* 46 (2): 73.
31 Rezende, R.A., Azevedo, F.D.S., Pereira, F.D. et al. (2012). Nanotechnological strategies for biofabrication of human organs. *Journal of Nanotechnology* 2012: 1–10.
32 Zhang, L.G., Fisher, J.P., and Leong, K.W. (2015). *3D Bioprinting and Nanotechnology in Tissue Engineering and Regenerative Medicine*. Elsevier/AP.
33 O'brien, C.M., Holmes, B., Faucett, S., and Zhang, L.G. (2015). Three-dimensional printing of nanomaterial scaffolds for complex tissue regeneration. *Tissue Engineering Part B: Reviews* 21 (1): 103–114.
34 Mironov, V., Kasyanov, V., and Markwald, R.R. (2008). Nanotechnology in vascular tissue engineering: from nanoscaffolding towards rapid vessel biofabrication. *Trends in Biotechnology* 26 (6): 338–344.
35 Rana, D., Kumar, T.S.S., and Ramalingam, M. (2017). Impact of nanotechnology on 3D bioprinting. *Journal of Bionanoscience* 11 (1): 1–6.
36 Rezende, R.A., Sabino, M.A., de Andréa Dernowsek, J. et al. (2018). From nano to macro: enabling nanotechnologies for human organ biofabrication (electrospun nanofibers and hybrid technique). *International Journal of Advances in Medical Biotechnology – IJAMB* 1 (1): 8–14.
37 Castro, N.J., O'brien, J., and Zhang, L.G. (2015). Integrating biologically inspired nanomaterials and table-top stereolithography for 3D printed biomimetic osteochondral scaffolds. *Nanoscale* 7 (33): 14010–14022.
38 Thrivikraman, G., França, C.M., Athirasala, A. et al. (2018). Nanomaterials in 3D bioprinting: current approaches and future possibilities. In: *Nanotechnologies in Preventive and Regenerative Medicine: An Emerging Big Picture*, 349. Elsevier.
39 Zhang, G. (2016). 3D bioprinting nanocomposite scaffolds for complex tissue regeneration. *The FASEB Journal* 30 (1Supplement): 13–11.
40 Groll, J., Boland, T., Blunk, T. et al. (2016). Biofabrication: reappraising the definition of an evolving field. *Biofabrication* 8 (1): 013001.
41 Nguyen, A. and Pavlovich, M. (2018). Engineering a strong bond between stem cells and biotechnology. *Cell Stem Cell* 22 (3): 279–280.
42 Raman, R., Mitchell, M., Perez-Pinera, P. et al. (2016). Design and integration of a problem-based biofabrication course into an undergraduate

biomedical engineering curriculum. *Journal of Biological Engineering* 10 (1): doi: https://doi.org/10.1186/s13036-016-0032-5.

43 Collins, S.F. (2014). Bioprinting is changing regenerative medicine forever. *Stem Cells and Development* 23 (S1): 79–82.

44 Mir, T.A. and Nakamura, M. (2017). Three-dimensional bioprinting: toward the era of manufacturing human organs as spare parts for healthcare and medicine. *Tissue Engineering Part B: Reviews* 23 (3): 245–256.

45 Holmes, A.M., Charlton, A., Derby, B. et al. (2017). Rising to the challenge: applying biofabrication approaches for better drug and chemical product development. *Biofabrication* 9 (3): 033001.

46 Chimene, D., Lennox, K.K., Kaunas, R.R., and Gaharwar, A.K. (2016). Advanced bioinks for 3D printing: a materials science perspective. *Annals of Biomedical Engineering* 44 (6): 2090–2102.

47 Baumann, F.W., Wellekötter, J., Roller, D., and Bonten, C. (2016). Software-aided measurement of geometrical fidelity for 3D printed objects. *Computer-Aided Design and Applications* 14 (4): 486–497.

48 Sklare, S.C., Richey, W., Vinson, B.T., and Chrisey, D.B. (2017). Directed self-assembly software for single cell deposition. *International Journal of Bioprinting* 3 (2): 100–108.

49 Mekhileri, N.V., Lim, K.S., Brown, G.C.J. et al. (2018). Automated 3D bioassembly of micro-tissues for biofabrication of hybrid tissue engineered constructs. *Biofabrication* 10 (2): 024103.

50 Ozbolat, I.T., Chen, H., and Yu, Y. (2014). Development of 'multi-arm bio-printer' for hybrid biofabrication of tissue engineering constructs. *Robotics and Computer-Integrated Manufacturing* 30 (3): 295–304.

51 Cuevas-Garcia, C.A. (2017). Understanding interdisciplinarity in its argumentative context: thought and rhetoric in the perception of academic practices. *Interdisciplinary Science Reviews* 43 (1): 54–73.

52 Brookes, W. (2017). Transdisciplinary learning in technology degrees. 2017 16th International Conference on Information Technology Based Higher Education and Training (ITHET).

53 Moroni, L., Boland, T., Burdick, J.A. et al. (2018). Biofabrication: a guide to technology and terminology. *Trends in Biotechnology* 36 (4): 384–402.

54 Schon, B.S., Hooper, G.J., and Woodfield, T.B.F. (2016). Modular tissue assembly strategies for biofabrication of engineered cartilage. *Annals of Biomedical Engineering* 45 (1): 100–114.

55 Prendergast, M.E., Montoya, G., Pereira, T. et al. (2018). Microphysiological systems: automated fabrication via extrusion bioprinting. *Microphysiological Systems* 2: 3–3.

56 Nawroth, J.C., Scudder, L.L., Halvorson, R.T. et al. (2018). Automated fabrication of photopatterned gelatin hydrogels for organ-on-chips applications. *Biofabrication* 10 (2): 025004.

57 Yang, Y., Liu, X., Wei, D. et al. (2017). Automated fabrication of hydrogel microfibers with tunable diameters for controlled cell alignment. *Biofabrication* 9 (4): 045009.

58 Othman, R., Morris, G.E., Shah, D.A. et al. (2015). An automated fabrication strategy to create patterned tubular architectures at cell and tissue scales. *Biofabrication* 7 (2): 025003.

59 Momtahan, N., Poornejad, N., Struk, J.A. et al. (2015). Automation of pressure control improves whole porcine heart decellularization. *Tissue Engineering Part C: Methods* 21 (11): 1148–1161.

60 Costa, P.F., Vaquette, C., Baldwin, J. et al. (2014). Biofabrication of customized bone grafts by combination of additive manufacturing and bioreactor knowhow. *Biofabrication* 6 (3): 035006.

61 Lee, J.-Y., Choi, B., Wu, B., and Lee, M. (2013). Customized biomimetic scaffolds created by indirect three-dimensional printing for tissue engineering. *Biofabrication* 5 (4): 045003.

62 Castilho, M., Dias, M., Vorndran, E. et al. (2014). Application of a 3D printed customized implant for canine cruciate ligament treatment by tibial tuberosity advancement. *Biofabrication* 6 (2): 025005.

63 Forrestal, D.P., Klein, T.J., and Woodruff, M.A. (2017). Challenges in engineering large customized bone constructs. *Biotechnology and Bioengineering* 114 (6): 1129–1139.

64 Zhao, M.-L., Li, X.-H., Fu, F. et al. (2017). Magnetic resonance imaging-three-dimensional printing technology fabricates customized scaffolds for brain tissue engineering. *Neural Regeneration Research* 12 (4): 614.

65 Kuo, C.-Y., Wilson, E., Fuson, A. et al. (2018). Repair of tympanic membrane perforations with customized bioprinted ear grafts using chinchilla models. *Tissue Engineering Part A* 24 (5–6): 527–535.

66 Lim, S.H., Ng, J.Y., and Kang, L. (2017). Three-dimensional printing of a microneedle array on personalized curved surfaces for dual-pronged treatment of trigger finger. *Biofabrication* 9 (1): 015010.

67 Neves, L.S., Rodrigues, M.T., Reis, R.L., and Gomes, M.E. (2016). Current approaches and future perspectives on strategies for the development of personalized tissue engineering therapies. *Expert Review of Precision Medicine and Drug Development* 1 (1): 93–108.

68 Liaw, C.-Y., Ji, S., and Guvendiren, M. (2018). Human tissue models: engineering 3D hydrogels for personalized in vitro human tissue models. *Advanced Healthcare Materials* 7 (4): 1870021.

69 Radenkovic, D., Solouk, A., and Seifalian, A. (2016). Personalized development of human organs using 3D printing technology. *Medical Hypotheses* 87: 30–33.

70 Morrison, K.A., Cohen, B.P., Asanbe, O. et al. (2016). Optimizing cell sourcing for clinical translation of tissue engineered ears. *Biofabrication* 9 (1): 015004.

71 Soffer-Tsur, N., Shevach, M., Shapira, A. et al. (2014). Optimizing the biofabrication process of omentum-based scaffolds for engineering autologous tissues. *Biofabrication* 6 (3): 035023.

72 Duan, B., Cheung, W.L., and Wang, M. (2011). Optimized fabrication of Ca–P/PHBV nanocomposite scaffolds via selective laser sintering for bone tissue engineering. *Biofabrication* 3 (1): 015001.

73 Hendriks, J., Visser, C.W., Henke, S. et al. (2015). Optimizing cell viability in droplet-based cell deposition. *Scientific Reports* 5 (1): doi: https://doi.org/10.1038/srep11304.
74 Han, X., Courseaus, J., Khamassi, J. et al. (2018). Optimized vascular network by stereolithography for tissue engineered skin. *International Journal of Bioprinting* 4 (2): doi: https://doi.org/10.18063/ijb.v4i2.134.
75 Baptista, L., Kronemberger, G., Côrtes, I. et al. (2018). Adult stem cells spheroids to optimize cell colonization in scaffolds for cartilage and bone tissue engineering. *International Journal of Molecular Sciences* 19 (5): 1285.
76 Neves, S.C., Mota, C., Longoni, A. et al. (2016). Additive manufactured polymeric 3D scaffolds with tailored surface topography influence mesenchymal stromal cells activity. *Biofabrication* 8 (2): 025012.
77 Köpf, M., Campos, D.F.D., Blaeser, A. et al. (2016). A tailored three-dimensionally printable agarose–collagen blend allows encapsulation, spreading, and attachment of human umbilical artery smooth muscle cells. *Biofabrication* 8 (2): 025011.
78 Uden, S.V., Silva-Correia, J., Correlo, V.M. et al. (2015). Custom-tailored tissue engineered polycaprolactone scaffolds for total disc replacement. *Biofabrication* 7 (1): 015008.
79 Luca, A.D., Longoni, A., Criscenti, G. et al. (2016). Toward mimicking the bone structure: design of novel hierarchical scaffolds with a tailored radial porosity gradient. *Biofabrication* 8 (4): 045007.
80 Rogers, C.M., Morris, G.E., Gould, T.W.A. et al. (2014). A novel technique for the production of electrospun scaffolds with tailored three-dimensional micro-patterns employing additive manufacturing. *Biofabrication* 6 (3): 035003.
81 Otto, I.A., Melchels, F.P.W., Zhao, X. et al. (2015). Auricular reconstruction using biofabrication-based tissue engineering strategies. *Biofabrication* 7 (3): 032001.
82 Chhaya, M.P., Melchels, F.P.W., Holzapfel, B.M. et al. (2015). Sustained regeneration of high-volume adipose tissue for breast reconstruction using computer aided design and biomanufacturing. *Biomaterials* 52: 551–560.
83 Jessop, Z.M., Al-Sabah, A., Gardiner, M.D. et al. (2017). 3D bioprinting for reconstructive surgery: principles, applications and challenges. *Journal of Plastic, Reconstructive & Aesthetic Surgery* 70 (9): 1155–1170.
84 Murphy, S.V. and Atala, A. (2012). Organ engineering – combining stem cells, biomaterials, and bioreactors to produce bioengineered organs for transplantation. *BioEssays* 35 (3): 163–172.
85 Ozbolat, I.T. (2015). Bioprinting scale-up tissue and organ constructs for transplantation. *Trends in Biotechnology* 33 (7): 395–400.
86 Shin'oka, T., Imai, Y., and Ikada, Y. (2001). Transplantation of a tissue-engineered pulmonary artery. *New England Journal of Medicine* 344 (7): 532–533.
87 Arenas-Herrera, J.E., Ko, I.K., Atala, A., and Yoo, J.J. (2013). Decellularization for whole organ bioengineering. *Biomedical Materials* 8 (1): 014106.

88 Munoz-Abraham, A.S., Rodriguez-Davalos, M.I., Bertacco, A. et al. (2016). 3D printing of organs for transplantation: where are we and where are we heading? *Current Transplantation Reports* 3 (1): 93–99.
89 Watanabe, M., Shin'oka, T., Tohyama, S. et al. (2001). Tissue-engineered vascular autograft: inferior vena cava replacement in a dog model. *Tissue Engineering* 7 (4): 429–439.
90 Ravnic, D.J., Leberfinger, A.N., Koduru, S.V. et al. (2017). Transplantation of bioprinted tissues and organs. *Annals of Surgery* 266 (1): 48–58.
91 Lysaght, M.J. and Hazlehurst, A.L. (2004). Tissue engineering: the end of the beginning. *Tissue Engineering* 10 (1–2): 309–320.
92 Mironov, V., Reis, N., and Derby, B. (2006). Review: bioprinting: a beginning. *Tissue Engineering* 12 (4): 631–634.
93 Desrochers, T.M., Palma, E., and Kaplan, D.L. (2014). Tissue-engineered kidney disease models. *Advanced Drug Delivery Reviews* 69–70: 67–80.
94 Schneeberger, K., Spee, B., Costa, P. et al. (2017). Converging biofabrication and organoid technologies: the next frontier in hepatic and intestinal tissue engineering? *Biofabrication* 9 (1): 013001.
95 Nguyen, D.G. and Pentoney, S.L. (2017). Bioprinted three dimensional human tissues for toxicology and disease modeling. *Drug Discovery Today: Technologies* 23: 37–44.
96 Skardal, A., Devarasetty, M., Rodman, C. et al. (2015). Liver-tumor hybrid organoids for modeling tumor growth and drug response in vitro. *Annals of Biomedical Engineering* 43 (10): 2361–2373.
97 Murrow, L.M., Weber, R.J., and Gartner, Z.J. (2017). Dissecting the stem cell niche with organoid models: an engineering-based approach. *Development* 144 (6): 998–1007.
98 Peng, W., Unutmaz, D., and Ozbolat, I.T. (2016). Bioprinting towards physiologically relevant tissue models for pharmaceutics. *Trends in Biotechnology* 34 (9): 722–732.
99 Skardal, A., Shupe, T., and Atala, A. (2016). Organoid-on-a-chip and body-on-a-chip systems for drug screening and disease modeling. *Drug Discovery Today* 21 (9): 1399–1411.
100 Nam, K.-H., Smith, A.S.T., Lone, S. et al. (2015). Biomimetic 3D tissue models for advanced high-throughput drug screening. *Journal of Laboratory Automation* 20 (3): 201–215.
101 Kimlin, L., Kassis, J., and Virador, V. (2013). 3D in vitro tissue models and their potential for drug screening. *Expert Opinion on Drug Discovery* 8 (12): 1455–1466.
102 Santo, V.E., Rebelo, S.P., Estrada, M.F. et al. (2016). Drug screening in 3D in vitro tumor models: overcoming current pitfalls of efficacy read-outs. *Biotechnology Journal* 12 (1): 1600505.
103 Vijayavenkataraman, S., Lu, W.F., and Fuh, J.Y.H. (2016). 3D bioprinting of skin: a state-of-the-art review on modelling, materials, and processes. *Biofabrication* 8 (3): 032001.
104 Chau, D.Y.S., Johnson, C., Macneil, S. et al. (2013). The development of a 3D immunocompetent model of human skin. *Biofabrication* 5 (3): 035011.

105 Mori, N., Morimoto, Y., and Takeuchi, S. (2017). Skin integrated with perfusable vascular channels on a chip. *Biomaterials* 116: 48–56.

106 Tay, L.-F., Khoh, L.-K., Loh, C.-S., and Khor, E. (1993). Alginate–chitosan coacervation in production of artificial seeds. *Biotechnology and Bioengineering* 42 (4): 449–454.

107 Lechelt-Kunze, C., Simon, J., Zitzmann, W. et al. (2000). US Patent 6, 164,012. Washington, DC: U.S. Patent and Trademark Office.

108 Sakhanokho, H.F., Pounders, C.T., and Blythe, E.K. (2013). Alginate encapsulation of begonia microshoots for short-term storage and distribution. *The Scientific World Journal* 2013: 1–7.

109 Hartle, J.E. (2010). US Patent 7,654,037. Washington, DC: U.S. Patent and Trademark Office.

110 McKersie, B.D., Senaratna, T., Bowley, S.R. et al. (1989). Application of artificial seed technology in the production of hybrid alfalfa (*Medicago sativa* L.). *In vitro Cellular & Developmental Biology* 25 (12): 1183–1188.

111 Tatman, P.D., Gerull, W., Sweeney-Easter, S. et al. (2015). Multiscale biofabrication of articular cartilage: bioinspired and biomimetic approaches. *Tissue Engineering Part B: Reviews* 21 (6): 543–559.

112 Langelaan, M.L., Boonen, K.J., Polak, R.B. et al. (2010). Meet the new meat: tissue engineered skeletal muscle. *Trends in Food Science and Technology* 21 (2): 59–66.

113 Bhat, Z. and Bhat, H. (2011). Animal-free meat biofabrication. *American Journal of Food Technology* 6 (6): 441–459.

114 Bonny, S.P.F., Gardner, G.E., Pethick, D.W., and Hocquette, J.-F. (2015). What is artificial meat and what does it mean for the future of the meat industry? *Journal of Integrative Agriculture* 14 (2): 255–263.

115 Stephens, D.N., Dunsford, I., Di Silvio, L. et al. (2018). Bringing cultured meat to market: technical, socio-political, and regulatory challenges in cellular agriculture. *Trends in Food Science and Technology* 78: 155–166.

116 Chiles, R.M. (2013). If they come, we will build it: in vitro meat and the discursive struggle over future agrofood expectations. *Agriculture and Human Values* 30 (4): 511–523.

117 Bhat, Z., Bhat, H., and Pathak, V. (2014). Prospects for in vitro cultured meat – a future harvest. In: *Principles of Tissue Engineering*, 1663–1683. Elsevier.

118 Bhat, Z.F., Kumar, S., and Fayaz, H. (2015). In vitro meat production: challenges and benefits over conventional meat production. *Journal of Integrative Agriculture* 14 (2): 241–248.

119 Mielczarek, J., Gazdowicz, G., Kramarz, J. et al. (2015). A prototype of a 3D bioprinter. *Solid State Phenomena* 237: 221–226.

120 Ozbolat, I.T. (2017). Bioprinter technologies * *with contributions by Hemanth Gudupati and Kazim Moncal, The Pennsylvania State University. In: *3D Bioprinting*, 199–241. Elsevier.

121 Pepper, M., Parzel, C., Burg, T. et al. (2009). Design and implementation of a two-dimensional inkjet bioprinter. 2009 Annual International Conference of the IEEE Engineering in Medicine and Biology Society.

122 Gasperini, L., Maniglio, D., Motta, A., and Migliaresi, C. (2015). An electrohydrodynamic bioprinter for alginate hydrogels containing living cells. *Tissue Engineering Part C: Methods* 21 (2): 123–132.
123 Nishiyama, Y., Nakamura, M., Henmi, C. et al. (2009). Development of a three-dimensional bioprinter: construction of cell supporting structures using hydrogel and state-of-the-art inkjet technology. *Journal of Biomechanical Engineering* 131 (3): 035001.
124 Hesuani, Y.D., Pereira, F.D., Parfenov, V. et al. (2016). Design and implementation of novel multifunctional 3D bioprinter. *3D Printing and Additive Manufacturing* 3 (1): 64–68.
125 Campbell, J., Mcguinness, I., Wirz, H. et al. (2015). Multimaterial and multiscale three-dimensional bioprinter. *Journal of Nanotechnology in Engineering and Medicine* 6 (2): 021001.
126 Tseng, H., Gage, J.A., Shen, T. et al. (2015). A spheroid toxicity assay using magnetic 3D bioprinting and real-time mobile device-based imaging. *Scientific Reports* 5 (1): doi: https://doi.org/10.1038/srep13987.
127 Tseng, H., Gage, J.A., Haisler, W.L. et al. (2016). A high-throughput in vitro ring assay for vasoactivity using magnetic 3D bioprinting. *Scientific Reports* 6 (1): doi: https://doi.org/10.1038/srep30640.
128 Souza, G.R., Molina, J.R., Raphael, R.M. et al. (2010). Three-dimensional tissue culture based on magnetic cell levitation. *Nature Nanotechnology* 5 (4): 291–296.
129 Haisler, W.L., Timm, D.M., Gage, J.A. et al. (2013). Three-dimensional cell culturing by magnetic levitation. *Nature Protocols* 8 (10): 1940–1949.
130 Timm, D.M., Chen, J., Sing, D. et al. (2013). A high-throughput three-dimensional cell migration assay for toxicity screening with mobile device-based macroscopic image analysis. *Scientific Reports* 3 (1): doi: https://doi.org/10.1038/srep03000.
131 Mota, C., Puppi, D., Chiellini, F., and Chiellini, E. (2012). Additive manufacturing techniques for the production of tissue engineering constructs. *Journal of Tissue Engineering and Regenerative Medicine* 9 (3): 174–190.
132 Kundu, J., Shim, J.-H., Jang, J. et al. (2013). An additive manufacturing-based PCL–alginate–chondrocyte bioprinted scaffold for cartilage tissue engineering. *Journal of Tissue Engineering and Regenerative Medicine* 9 (11): 1286–1297.
133 Hafezi, F., Kucukgul, C., Ozler, S., and Koc, B. (2015). Bioprinting: application of additive manufacturing in medicine. In: *Additive Manufacturing*, 197–214. Taylor & Francis Group.
134 Mitchell, M.G. (2017). Additive manufacturing and 3D bioprinting for pharmaceutical testing. In: *Bioprinting*, 113–122. Elsevier.
135 Chang, J., He, J., Mao, M. et al. (2018). Advanced material strategies for next-generation additive manufacturing. *Materials* 11 (1): 166.
136 Shuai, C., Yang, Y., Feng, P. et al. (2018). A multi-scale porous scaffold fabricated by a combined additive manufacturing and chemical etching process. *International Journal of Bioprinting* 4 (2): doi: https://doi.org/10.18063/ijb.v4i1.133.

137 Seidel, J., Ahlfeld, T., Adolph, M. et al. (2017). Green bioprinting: extrusion-based fabrication of plant cell-laden biopolymer hydrogel scaffolds. *Biofabrication* 9 (4): 045011.

138 Hochleitner, G., Jüngst, T., Brown, T.D. et al. (2015). Additive manufacturing of scaffolds with sub-micron filaments via melt electrospinning writing. *Biofabrication* 7 (3): 035002.

139 Chen, F., Hochleitner, G., Woodfield, T. et al. (2015). Additive manufacturing of a photo-cross-linkable polymer via direct melt electrospinning writing for producing high strength structures. *Biomacromolecules* 17 (1): 208–214.

140 Ryan, C.N., Fuller, K.P., Larrañaga, A. et al. (2015). An academic, clinical and industrial update on electrospun, additive manufactured and imprinted medical devices. *Expert Review of Medical Devices* 12 (5): 601–612.

141 Niu, H. and Lin, T. (2012). Fiber generators in needleless electrospinning. *Journal of Nanomaterials* 2012: 1–13.

142 Zehnder, T., Sarker, B., Boccaccini, A.R., and Detsch, R. (2015). Evaluation of an alginate–gelatine crosslinked hydrogel for bioplotting. *Biofabrication* 7 (2): 025001.

143 Mehendale, S.V., Mellor, L.F., Taylor, M.A. et al. (2017). Effects of 3D-bioplotted polycaprolactone scaffold geometry on human adipose-derived stem cell viability and proliferation. *Rapid Prototyping Journal* 23 (3): 534–542.

144 Sheshadri, P. and Shirwaiker, R.A. (2015). Characterization of material–process–structure interactions in the 3D bioplotting of polycaprolactone. *3D Printing and Additive Manufacturing* 2 (1): 20–31.

145 Singh, R., Sarker, B., Silva, R. et al. (2015). Evaluation of hydrogel matrices for vessel bioplotting: vascular cell growth and viability. *Journal of Biomedical Materials Research Part A* 104 (3): 577–585.

146 Warren, P.B., Huebner, P., Spang, J.T. et al. (2016). Engineering 3D-bioplotted scaffolds to induce aligned extracellular matrix deposition for musculoskeletal soft tissue replacement. *Connective Tissue Research* 58 (3–4): 342–354.

147 Biscaia, S., Viana, T., Almeida, H., and Bártolo, P. (2013). Biofabrication of three-dimensional scaffolds of polycaprolactone with eggshell powder for bone regeneration. In: *High Value Manufacturing: Advanced Research in Virtual and Rapid Prototyping*, 171–176. Taylor & Francis Group.

148 Shaoki, A., Xu, J.-Y., Sun, H. et al. (2016). Osseointegration of three-dimensional designed titanium implants manufactured by selective laser melting. *Biofabrication* 8 (4): 045014.

149 Zieber, L., Or, S., Ruvinov, E., and Cohen, S. (2014). Microfabrication of channel arrays promotes vessel-like network formation in cardiac cell construct and vascularization in vivo. *Biofabrication* 6 (2): 024102.

150 Lee, S.-W., Noh, J.-Y., Park, S.C. et al. (2012). Biomimetic engineering of a generic cell-on-membrane architecture by microfluidic engraving for on-chip bioassays. *Langmuir* 28 (20): 7585–7590.

151 Shapira, A., Kim, D.-H., and Dvir, T. (2014). Advanced micro- and nanofabrication technologies for tissue engineering. *Biofabrication* 6 (2): 020301.

152 Singh, A.V., Patil, R., Thombre, D.K., and Gade, W.N. (2013). Micro-nanopatterning as tool to study the role of physicochemical properties on cell-surface interactions. *Journal of Biomedical Materials Research Part A* 101 (10): 3019–3032.
153 Bolay, N.L., Santran, V., Dechambre, G. et al. (2009). Production, by co-grinding in a media mill, of porous biodegradable polylactic acid–apatite composite materials for bone tissue engineering. *Powder Technology* 190 (1–2): 89–94.
154 Liao, C.-J., Chen, C.-F., Chen, J.-H. et al. (2001). Fabrication of porous biodegradable polymer scaffolds using a solvent merging/particulate leaching method. *Journal of Biomedical Materials Research* 59 (4): 676–681.
155 Lee, S.B., Kim, Y.H., Chong, M.S. et al. (2005). Study of gelatin-containing artificial skin V: fabrication of gelatin scaffolds using a salt-leaching method. *Biomaterials* 26 (14): 1961–1968.
156 Oh, S. (2003). Fabrication and characterization of hydrophilic poly(lactic-*co*-glycolic acid)/poly(vinyl alcohol) blend cell scaffolds by melt-molding particulate-leaching method. *Biomaterials* 24 (22): 4011–4021.
157 Gadegaard, N., Mosler, S., and Larsen, N.B. (2003). Biomimetic polymer nanostructures by injection molding. *Macromolecular Materials and Engineering* 288 (1): 76–83.
158 Wu, L., Jing, D., and Ding, J. (2006). A "room-temperature" injection molding/particulate leaching approach for fabrication of biodegradable three-dimensional porous scaffolds. *Biomaterials* 27 (2): 185–191.
159 Sousa, R.A., Kalay, G.R., Reis, R.L. et al. (2000). Injection molding of a starch/EVOH blend aimed as an alternative biomaterial for temporary applications. *Journal of Applied Polymer Science* 77 (6): 1303–1315.
160 Hiebl, B., Lützow, K., Lange, M. et al. (2010). Cytocompatibility testing of cell culture modules fabricated from specific candidate biomaterials using injection molding. *Journal of Biotechnology* 148 (1): 76–82.
161 Koh, W.-G. and Pishko, M. (2003). Photoreaction injection molding of biomaterial microstructures. *Langmuir* 19 (24): 10310–10316.
162 Kramschuster, A. and Turng, L.-S. (2009). An injection molding process for manufacturing highly porous and interconnected biodegradable polymer matrices for use as tissue engineering scaffolds. *Journal of Biomedical Materials Research Part B: Applied Biomaterials* 92 (2): 366–376.
163 Chang, S.C.N., Rowley, J.A., Tobias, G. et al. (2001). Injection molding of chondrocyte/alginate constructs in the shape of facial implants. *Journal of Biomedical Materials Research* 55 (4): 503–511.
164 Ballyns, J.J., Gleghorn, J.P., Niebrzydowski, V. et al. (2008). Image-guided tissue engineering of anatomically shaped implants via MRI and micro-CT using injection molding. *Tissue Engineering Part A* 14 (7): 1195–1202.
165 Lee, K.-W., Wang, S., Lu, L. et al. (2006). Fabrication and characterization of poly(propylene fumarate) scaffolds with controlled pore structures using 3-dimensional printing and injection molding. *Tissue Engineering* 12 (10): 2801–2811.

166 Kasyanov, V.A., Hodde, J., Hiles, M.C. et al. (2008). Rapid biofabrication of tubular tissue constructs by centrifugal casting in a decellularized natural scaffold with laser-machined micropores. *Journal of Materials Science Materials in Medicine* 20 (1): 329–337.

167 Thadavirul, N., Pavasant, P., and Supaphol, P. (2013). Development of polycaprolactone porous scaffolds by combining solvent casting, particulate leaching, and polymer leaching techniques for bone tissue engineering. *Journal of Biomedical Materials Research Part A* 102 (10): 3379–3392.

168 Mironov, V., Kasyanov, V., Markwald, R.R., and Prestwich, G.D. (2008). Bioreactor-free tissue engineering: directed tissue assembly by centrifugal casting. *Expert Opinion on Biological Therapy* 8 (2): 143–152.

169 Deville, S., Saiz, E., and Tomsia, A.P. (2006). Freeze casting of hydroxyapatite scaffolds for bone tissue engineering. *Biomaterials* 27 (32): 5480–5489.

170 Ramay, H.R. and Zhang, M. (2003). Preparation of porous hydroxyapatite scaffolds by combination of the gel-casting and polymer sponge methods. *Biomaterials* 24 (19): 3293–3302.

171 Miller, J.S., Stevens, K.R., Yang, M.T. et al. (2012). Rapid casting of patterned vascular networks for perfusable engineered three-dimensional tissues. *Nature Materials* 11 (9): 768–774.

172 Tan, H., Chu, C.R., Payne, K.A., and Marra, K.G. (2009). Injectable in situ forming biodegradable chitosan–hyaluronic acid based hydrogels for cartilage tissue engineering. *Biomaterials* 30 (13): 2499–2506.

173 Madihally, S.V. and Matthew, H.W. (1999). Porous chitosan scaffolds for tissue engineering. *Biomaterials* 20 (12): 1133–1142.

174 Rezende, R., Pereira, F., Kasyanov, V. et al. (2013). Scalable biofabrication of tissue spheroids for organ printing. *Procedia CIRP* 5: 276–281.

175 Park, J.H., Jung, J.W., Kang, H.-W., and Cho, D.-W. (2014). Indirect three-dimensional printing of synthetic polymer scaffold based on thermal molding process. *Biofabrication* 6 (2): 025003.

176 Yang, L., Shridhar, S.V., Gerwitz, M., and Soman, P. (2016). An in vitro vascular chip using 3D printing-enabled hydrogel casting. *Biofabrication* 8 (3): 035015.

177 Snyder, J.E., Hunger, P.M., Wang, C. et al. (2014). Combined multi-nozzle deposition and freeze casting process to superimpose two porous networks for hierarchical three-dimensional microenvironment. *Biofabrication* 6 (1): 015007.

178 Hinton, T.J., Jallerat, Q., Palchesko, R.N. et al. (2015). Three-dimensional printing of complex biological structures by freeform reversible embedding of suspended hydrogels. *Science Advances* 1 (9): e1500758.

179 Gonsalves, K.E., He, W., Poker, D.B. et al. (2001). A versatile approach for biomaterial patterning: masked ion beam lithography. *MRS Proceedings* 705: doi: https://doi.org/10.1557/proc-705-y4.6.

180 Park, D.Y., Mun, C.H., Kang, E. et al. (2014). One-stop microfiber spinning and fabrication of a fibrous cell-encapsulated scaffold on a single microfluidic platform. *Biofabrication* 6 (2): 024108.

181 Vecchione, R., Pitingolo, G., Guarnieri, D. et al. (2016). From square to circular polymeric microchannels by spin coating technology: a low cost platform for endothelial cell culture. *Biofabrication* 8 (2): 025005.

182 Yost, M.J. and Rodriguez-Rivera, V. (2017). US Patent Application No. 15/514,091.

183 Daly, A.C., Cunniffe, G.M., Sathy, B.N. et al. (2016). 3D bioprinting: 3D bioprinting of developmentally inspired templates for whole bone organ engineering. *Advanced Healthcare Materials* 5 (18): 2353–2362.

184 Mironov, V., Kasyanov, V., Drake, C., and Markwald, R.R. (2008). Organ printing: promises and challenges. *Regenerative Medicine* 3 (1): 93–103.

185 Wang, D.-A., Varghese, S., Sharma, B. et al. (2007). Multifunctional chondroitin sulphate for cartilage tissue–biomaterial integration. *Nature Materials* 6 (5): 385–392.

186 Passage, J., Jalali, H., Tam, R.K. et al. (2002). BioGlue surgical adhesive – an appraisal of its indications in cardiac surgery. *The Annals of Thoracic Surgery* 74 (2): 432–437.

187 Phuc, D.H., Nguyen, T.-H., Toi, V.V., and Tien, P.V. (2015). Fabrication of hyaluronan–chitosan–polyvinyl phosphonic acid hydrogel for bioglue applications. IFMBE Proceedings 5th International Conference on Biomedical Engineering in Vietnam, pp. 329–331.

188 Pinczewski, L.A. and Salmon, L.J. (2017). Editorial commentary: the acrid bioscrew in anterior cruciate ligament reconstruction of the knee. *Arthroscopy: The Journal of Arthroscopic & Related Surgery* 33 (12): 2195–2197.

189 Weber, M., Hoheisel, A.L., and Glasmacher, B. (2016). Automated control of the laser welding process of heart valve scaffolds. *Current Directions in Biomedical Engineering* 2 (1): 301–305.

190 Singh, B. (2016). Artificial bone – an overview. *International Journal of Pharmaceutical & Biological Archive* 7 (1): 26–32.

191 Raja, N. and Yun, H.-S. (2016). A simultaneous 3D printing process for the fabrication of bioceramic and cell-laden hydrogel core/shell scaffolds with potential application in bone tissue regeneration. *Journal of Materials Chemistry B* 4 (27): 4707–4716.

192 Mushaben, M., Urie, R., Flake, T. et al. (2017). Spatiotemporal modeling of laser tissue soldering using photothermal nanocomposites. *Lasers in Surgery and Medicine* 50 (2): 143–152.

193 Klotz, B.J., Gawlitta, D., Rosenberg, A.J. et al. (2016). Gelatin-methacryloyl hydrogels: towards biofabrication-based tissue repair. *Trends in Biotechnology* 34 (5): 394–407.

194 Cohen, D.L., Lipton, J.I., Bonassar, L.J., and Lipson, H. (2010). Additive manufacturing for in situ repair of osteochondral defects. *Biofabrication* 2 (3): 035004.

195 Sawkins, M.J., Mistry, P., Brown, B.N. et al. (2015). Cell and protein compatible 3D bioprinting of mechanically strong constructs for bone repair. *Biofabrication* 7 (3): 035004.

196 Cui, X., Breitenkamp, K., Finn, M. et al. (2012). Direct human cartilage repair using three-dimensional bioprinting technology. *Tissue Engineering Part A* 18 (11–12): 1304–1312.

197 Grant, G.T., Liacouras, P., Santiago, G.F. et al. (2014). Restoration of the donor face after facial allotransplantation. *Annals of Plastic Surgery* 72: 720–724.

198 Visscher, D.O., Farré-Guasch, E., Helder, M.N. et al. (2016). Advances in bioprinting technologies for craniofacial reconstruction. *Trends in Biotechnology* 34 (9): 700–710.

199 Melchels, F., Wiggenhauser, P.S., Warne, D. et al. (2011). CAD/CAM-assisted breast reconstruction. *Biofabrication* 3 (3): 034114.

200 Nakamura, M. (2010). Reconstruction of biological three-dimensional tissues: bioprinting and biofabrication using inkjet technology. In: *Cell and Organ Printing* (ed. B. Ringeisen, B. Spargo and P. Wu), 23–33. Dordrecht: Springer.

201 Chhaya, M.P., Melchels, F.P., Wiggenhauser, P.S. et al. (2013). Breast reconstruction using biofabrication-based tissue engineering strategies. In: *Biofabrication*, Chapter 10 (ed. G. Forgacs and W. Sun), 183–216. William Andrew Publishing. ISBN: 9781455728527.

202 Malyshkina, S., Lu, C., and Dedukh, N. (2014). Restructuring of bone tissue under filling bone cavities with carbon synthetic biomaterial. *Orthopaedics, Traumatology and Prosthetics* 3: 30–37.

203 Nimeskern, L., Ávila, H.M., Sundberg, J. et al. (2013). Mechanical evaluation of bacterial nanocellulose as an implant material for ear cartilage replacement. *Journal of the Mechanical Behavior of Biomedical Materials* 22: 12–21.

204 Fedorovich, N.E., Schuurman, W., Wijnberg, H.M. et al. (2012). Biofabrication of osteochondral tissue equivalents by printing topologically defined, cell-laden hydrogel scaffolds. *Tissue Engineering Part C: Methods* 18 (1): 33–44.

205 Akins, R., Boyce, R., Madonna, M. et al. (1999). Cardiac organogenesis in vitro: reestablishment of three-dimensional tissue architecture by dissociated neonatal rat ventricular cells. *Tissue Engineering* 5 (2): 103–118.

206 Kato, Y., Christiansen, D.L., Hahn, R.A. et al. (1989). Mechanical properties of collagen fibres: a comparison of reconstituted and rat tail tendon fibres. *Biomaterials* 10 (1): 38–42.

207 Lu, S.-H., Sacks, M.S., Chung, S.Y. et al. (2005). Biaxial mechanical properties of muscle-derived cell seeded small intestinal submucosa for bladder wall reconstitution. *Biomaterials* 26 (4): 443–449.

208 Lv, S., Dudek, D.M., Cao, Y. et al. (2010). Designed biomaterials to mimic the mechanical properties of muscles. *Nature* 465 (7294): 69–73.

209 Hung, J., Solis, J., Guerrero, J.L. et al. (2008). A novel approach for reducing ischemic mitral regurgitation by injection of a polymer to reverse remodel and reposition displaced papillary muscles. *Circulation* (14 Suppl. 1): 118, S263–S269.

210 Aguib, H. and Lueth, T.C. (2008). Absorbable biomaterials for medical devices: a redesign of endoscopic rivets for biological tissue. 2008 Cairo International Biomedical Engineering Conference.

211 Steimberg, N. and Mazzoleni, G. (2016). From the macroscale to nanostructures: can tissue engineering recreate bone features? In: *Nanomaterials and Regenerative Medicine*, Chapter 12, 289–332. International Association of Physical Chemists (IAPC).

212 Budhwani, K.I., Thomas, V., and Sethu, P. (2016). Lab-on-a-brane: nanofibrous polymer membranes to recreate organ–capillary interfaces. *Journal of Micromechanics and Microengineering* 26 (3): 035013.

213 Nakamura, M., Nishiyama, Y., Henmi, C. et al. (2008). Ink jet three-dimensional digital fabrication for biological tissue manufacturing: analysis of alginate microgel beads produced by ink jet droplets for three dimensional tissue fabrication. *Journal of Imaging Science and Technology* 52 (6): 060201.

214 Matsunaga, Y.T., Morimoto, Y., and Takeuchi, S. (2011). 3D tissue architectures: molding cell beads for rapid construction of macroscopic 3D tissue architecture. *Advanced Materials* (12): 23. doi: https://doi.org/10.1002/adma.201190040.

215 Phamduy, T., Dias, A., Raof, N.A. et al. (2012). Laser direct-write of embryonic stem cells and cells encapsulated in alginate beads for engineered biological constructs. *MRS Proceedings* 1418: doi: https://doi.org/10.1557/opl.2012.798.

216 Yu, L., Ni, C., Grist, S.M. et al. (2015). Alginate core–shell beads for simplified three-dimensional tumor spheroid culture and drug screening. *Biomedical Microdevices* 17 (2): doi: https://doi.org/10.1007/s10544-014-9918-5.

217 Lehmann, R., Gallert, C., Roddelkopf, T. et al. (2015). 3 dimensional cell cultures: a comparison between manually and automatically produced alginate beads. *Cytotechnology* 68 (4): 1049–1062.

218 Fan, C. and Wang, D.-A. (2015). Potential use of alginate beads as a chondrocyte delivery vehicle and stepwise dissolving porogen in a hydrogel scaffold for cartilage tissue engineering. *RSC Advances* 5 (98): 80688–80697.

219 Yao, R., Zhang, R., Luan, J., and Lin, F. (2012). Alginate and alginate/gelatin microspheres for human adipose-derived stem cell encapsulation and differentiation. *Biofabrication* 4 (2): 025007.

220 Yao, R., Zhang, R., Lin, F., and Luan, J. (2012). Injectable cell/hydrogel microspheres induce the formation of fat lobule-like microtissues and vascularized adipose tissue regeneration. *Biofabrication* 4 (4): 045003.

221 Zhou, W. and Ma, G. (2013). Microspheres for cell culture. In: *Microspheres and Microcapsules in Biotechnology*, 49–83. Pan Stanford Publishing.

222 Ghosh, P., Gruber, S.M.S., Lin, C.-Y., and Whitlock, P.W. (2018). Microspheres containing decellularized cartilage induce chondrogenesis in vitro and remain functional after incorporation within a poly(caprolactone) filament useful for fabricating a 3D scaffold. *Biofabrication* 10 (2): 025007.

223 Acharya, C., Adesida, A., Zajac, P. et al. (2011). Enhanced chondrocyte proliferation and mesenchymal stromal cells chondrogenesis in coculture pellets

mediate improved cartilage formation. *Journal of Cellular Physiology* 227 (1): 88–97.
224 Bernstein, P., Dong, M., Corbeil, D. et al. (2009). Pellet culture elicits superior chondrogenic redifferentiation than alginate-based systems. *Biotechnology Progress* 25 (4): 1146–1152.
225 Schütz, K., Placht, A.-M., Paul, B. et al. (2015). Three-dimensional plotting of a cell-laden alginate/methylcellulose blend: towards biofabrication of tissue engineering constructs with clinically relevant dimensions. *Journal of Tissue Engineering and Regenerative Medicine* 11 (5): 1574–1587.
226 Detsch, R., Sarker, B., Grigore, A., and Boccaccini, A.R. (2013). Alginate and gelatine blending for bone cell printing and biofabrication. Biomedical Engineering.
227 Jung, J.W., Lee, H., Hong, J.M. et al. (2015). A new method of fabricating a blend scaffold using an indirect three-dimensional printing technique. *Biofabrication* 7 (4): 045003.
228 Ma, L., Jiang, W., and Li, W. (2014). Solvent-free fabrication of tissue engineering scaffolds with immiscible polymer blends. *International Journal of Polymeric Materials and Polymeric Biomaterials* 63 (10): 510–517.
229 Terrell, J.L., Gordonov, T., Cheng, Y. et al. (2012). Integrated biofabrication for electro-addressed in-film bioprocessing. *Biotechnology Journal* 7 (3): 428–439.
230 Hald, E.S., Steucke, K.E., Reeves, J.A. et al. (2014). Long-term vascular contractility assay using genipin-modified muscular thin films. *Biofabrication* 6 (4): 045005.
231 Prigipaki, A., Papanikolopoulou, K., Mossou, E. et al. (2017). Laser processing of protein films as a method for accomplishment of cell patterning at the microscale. *Biofabrication* 9 (4): 045004.
232 Zhou, C., Ma, L., Li, W., and Yao, D. (2011). Fabrication of tissue engineering scaffolds through solid-state foaming of immiscible polymer blends. *Biofabrication* 3 (4): 045003.
233 Hadlock, T., Sundback, C., Hunter, D. et al. (2000). A polymer foam conduit seeded with Schwann cells promotes guided peripheral nerve regeneration. *Tissue Engineering* 6 (2): 119–127.
234 Ock, J. and Li, W. (2014). Fabrication of a three-dimensional tissue model microarray using laser foaming of a gas-impregnated biodegradable polymer. *Biofabrication* 6 (2): 024110.
235 Xu, C., Bai, Y., Yang, S. et al. (2018). A versatile three-dimensional foam fabrication strategy for soft and hard tissue engineering. *Biomedical Materials* 13 (2): 025018.
236 Müller, W.E.G., Wang, X., Cui, F.-Z. et al. (2009). Sponge spicules as blueprints for the biofabrication of inorganic–organic composites and biomaterials. *Applied Microbiology and Biotechnology* 83 (3): 397–413.
237 Zhang, Y. and Zhang, M. (2002). Three-dimensional macroporous calcium phosphate bioceramics with nested chitosan sponges for load-bearing bone implants. *Journal of Biomedical Materials Research* 61 (1): 1–8.

238 Nguyen, V.C., Nguyen, V.B., and Hsieh, M.-F. (2013). Curcumin-loaded chitosan/gelatin composite sponge for wound healing application. *International Journal of Polymer Science* 2013: 1–7.
239 Gaharwar, A.K., Nikkhah, M., Sant, S., and Khademhosseini, A. (2014). Anisotropic poly(glycerol sebacate)-poly(ε-caprolactone) electrospun fibers promote endothelial cell guidance. *Biofabrication* 7 (1): 015001.
240 Jeffries, E.M. and Wang, Y. (2013). Incorporation of parallel electrospun fibers for improved topographical guidance in 3D nerve guides. *Biofabrication* 5 (3): 035015.
241 Wong, H.K., Lam, C.R.I., Wen, F. et al. (2016). Novel method to improve vascularization of tissue engineered constructs with biodegradable fibers. *Biofabrication* 8 (1): 015004.
242 Blackwood, K.A., Ristovski, N., Liao, S. et al. (2015). Improving electrospun fibre stacking with direct writing for developing scaffolds for tissue engineering for non-load bearing bone. IFMBE Proceedings 5th International Conference on Biomedical Engineering in Vietnam, pp. 125–128.
243 Nam, J., Huang, Y., Agarwal, S., and Lannutti, J. (2007). Improved cellular infiltration in electrospun fiber via engineered porosity. *Tissue Engineering* 13 (9): 2249–2257.
244 Dai, X., Kathiria, K., and Huang, Y.-C. (2014). Electrospun fiber scaffolds of poly(glycerol-dodecanedioate) and its gelatin blended polymers for soft tissue engineering. *Biofabrication* 6 (3): 035005.
245 Li, W.-J., Jiang, Y.J., and Tuan, R.S. (2006). Chondrocyte phenotype in engineered fibrous matrix is regulated by fiber size. *Tissue Engineering* 060706073730053.
246 Shimizu, T., Yamato, M., Kikuchi, A., and Okano, T. (2001). Two-dimensional manipulation of cardiac myocyte sheets utilizing temperature-responsive culture dishes augments the pulsatile amplitude. *Tissue Engineering* 7 (2): 141–151.
247 Hannachi, I.E., Itoga, K., Kumashiro, Y. et al. (2009). Fabrication of transferable micropatterned-co-cultured cell sheets with microcontact printing. *Biomaterials* 30 (29): 5427–5432.
248 Shimizu, K., Ito, A., and Honda, H. (2006). Enhanced cell-seeding into 3D porous scaffolds by use of magnetite nanoparticles. *Journal of Biomedical Materials Research Part B: Applied Biomaterials* 77B (2): 265–272.
249 Yang, S., Leong, K.-F., Du, Z., and Chua, C.-K. (2001). The design of scaffolds for use in tissue engineering. Part I. Traditional factors. *Tissue Engineering* 7 (6): 679–689.
250 Lode, A., Meissner, K., Luo, Y. et al. (2012). Fabrication of porous scaffolds by three-dimensional plotting of a pasty calcium phosphate bone cement under mild conditions. *Journal of Tissue Engineering and Regenerative Medicine* 8 (9): 682–693.
251 Li, Z., Gunn, J., Chen, M.H. et al. (2008). On-site alginate gelation for enhanced cell proliferation and uniform distribution in porous scaffolds. *Journal of Biomedical Materials Research Part A* 86A (2): 552–559.

252 Tang, W., Lin, D., Yu, Y. et al. (2016). Bioinspired trimodal macro/micro/nano-porous scaffolds loading rhBMP-2 for complete regeneration of critical size bone defect. *Acta Biomaterialia* 32: 309–323.

253 Li, J.L., Cai, Y.L., Guo, Y.L. et al. (2013). Fabrication of three-dimensional porous scaffolds with controlled filament orientation and large pore size via an improved E-jetting technique. *Journal of Biomedical Materials Research Part B: Applied Biomaterials* 102 (4): 651–658.

254 Kang, H.-W., Seol, Y.-J., and Cho, D.-W. (2008). Development of an indirect solid freeform fabrication process based on microstereolithography for 3D porous scaffolds. *Journal of Micromechanics and Microengineering* 19 (1): 015011.

255 Wu, H., Wan, Y., Dalai, S., and Zhang, R. (2010). Response of rat osteoblasts to polycaprolactone/chitosan blend porous scaffolds. *Journal of Biomedical Materials Research Part A* 92A (1): 238–245.

256 Malda, J., Visser, J., Melchels, F.P. et al. (2013). 25th anniversary article: engineering hydrogels for biofabrication. *Advanced Materials* 25 (36): 5011–5028.

257 Huang, G.Y., Zhou, L.H., Zhang, Q.C. et al. (2011). Microfluidic hydrogels for tissue engineering. *Biofabrication* 3 (1): 012001.

258 Huang, G., Wang, L., Wang, S. et al. (2012). Engineering three-dimensional cell mechanical microenvironment with hydrogels. *Biofabrication* 4 (4): 042001.

259 Wang, S., Lee, J.M., and Yeong, W.Y. (2015). Smart hydrogels for 3D bioprinting. *International Journal of Bioprinting* 1 (1): 3–14.

260 Nicodemus, G.D. and Bryant, S.J. (2008). Cell encapsulation in biodegradable hydrogels for tissue engineering applications. *Tissue Engineering Part B: Reviews* 14 (2): 149–165.

261 Geckil, H., Xu, F., Zhang, X. et al. (2010). Engineering hydrogels as extracellular matrix mimics. *Nanomedicine* 5 (3): 469–484.

262 Xu, F., Finley, T.D., Turkaydin, M. et al. (2011). The assembly of cell-encapsulating microscale hydrogels using acoustic waves. *Biomaterials* 32 (31): 7847–7855.

263 Boere, K.W.M., Blokzijl, M.M., Visser, J. et al. (2015). Biofabrication of reinforced 3D-scaffolds using two-component hydrogels. *Journal of Materials Chemistry B* 3 (46): 9067–9078.

264 Liu, J., Zheng, H., Poh, P. et al. (2015). Hydrogels for engineering of perfusable vascular networks. *International Journal of Molecular Sciences* 16 (7): 15997–16016.

265 Altomare, L., Cochis, A., Carletta, A. et al. (2016). Thermo-responsive methylcellulose hydrogels as temporary substrate for cell sheet biofabrication. *Journal of Materials Science Materials in Medicine* 27 (5): doi: https://doi.org/10.1007/s10856-016-5703-8.

266 Shi, Z., Gao, X., Ullah, M.W. et al. (2016). Electroconductive natural polymer-based hydrogels. *Biomaterials* 111: 40–54.

267 Slaughter, B.V., Khurshid, S.S., Fisher, O.Z. et al. (2009). Hydrogels in regenerative medicine. *Advanced Materials* 21 (32–33): 3307–3329.

268 Loo, Y. and Hauser, C.A.E. (2015). Bioprinting synthetic self-assembling peptide hydrogels for biomedical applications. *Biomedical Materials* 11 (1): 014103.

269 Kageyama, T., Kakegawa, T., Osaki, T. et al. (2014). Rapid engineering of endothelial cell-lined vascular-like structures in in situ crosslinkable hydrogels. *Biofabrication* 6 (2): 025006.

270 Murphy, S.V., Skardal, A., and Atala, A. (2012). Evaluation of hydrogels for bio-printing applications. *Journal of Biomedical Materials Research Part A* 101A (1): 272–284.

271 Omidian, H., Rocca, J.G., and Park, K. (2005). Advances in superporous hydrogels. *Journal of Controlled Release* 102 (1): 3–12.

272 Barbucci, R. and Leone, G. (2004). Formation of defined microporous 3D structures starting from cross-linked hydrogels. *Journal of Biomedical Materials Research* 68B (2): 117–126.

273 Hu, B.-H., Su, J., and Messersmith, P.B. (2009). Hydrogels cross-linked by native chemical ligation. *Biomacromolecules* 10 (8): 2194–2200.

274 Chen, J., Park, H., and Park, K. (1999). Synthesis of superporous hydrogels: hydrogels with fast swelling and superabsorbent properties. *Journal of Biomedical Materials Research* 44 (1): 53–62.

275 Lutolf, M., Raeber, G., Zisch, A. et al. (2003). Cell-responsive synthetic hydrogels. *Advanced Materials* 15 (11): 888–892.

276 Ratner, B.D. and Hoffman, A.S. (1976). Synthetic hydrogels for biomedical applications. In: *Hydrogels for Medical and Related Applications*, ACS Symposium Series, 1–36. ACS Publications.

277 Connon, C. and Hamley, I. (2014). *Hydrogels in Cell-Based Therapies*, RSC Soft Matter Series, vol. 2. Royal Society of Chemistry.

278 Benitez, P.L. and Heilshorn, S.C. Recombinant protein hydrogels for cell injection and transplantation. In: *Hydrogels in Cell-Based Therapies*, Soft Matter Series, Chapter 3, 48–72. RSC Publishing.

279 Iqbal, M.S., Akbar, J., Saghir, S. et al. (2011). Thermal studies of plant carbohydrate polymer hydrogels. *Carbohydrate Polymers* 86 (4): 1775–1783.

280 Tibbitt, M.W. and Anseth, K.S. (2009). Hydrogels as extracellular matrix mimics for 3D cell culture. *Biotechnology and Bioengineering* 103 (4): 655–663.

281 Joanitti, G. and Silva, L. (2014). The emerging potential of by-products as platforms for drug delivery systems. *Current Drug Targets* 15 (5): 478–485.

282 Gómez-Guillén, M., Pérez-Mateos, M., Gómez-Estaca, J. et al. (2009). Fish gelatin: a renewable material for developing active biodegradable films. *Trends in Food Science and Technology* 20 (1): 3–16.

283 Li, X. and Pan, X. (2010). Hydrogels based on hemicellulose and lignin from lignocellulose biorefinery: a mini-review. *Journal of Biobased Materials and Bioenergy* 4 (4): 289–297.

284 Hamed, I., Özogul, F., and Regenstein, J.M. (2016). Industrial applications of crustacean by-products (chitin, chitosan, and chitooligosaccharides): a review. *Trends in Food Science and Technology* 48: 40–50.

285 Luckachan, G.E. and Pillai, C.K.S. (2011). Biodegradable polymers – a review on recent trends and emerging perspectives. *Journal of Polymers and the Environment* 19 (3): 637–676.
286 Fertier, L., Koleilat, H., Stemmelen, M. et al. (2013). The use of renewable feedstock in UV-curable materials – a new age for polymers and green chemistry. *Progress in Polymer Science* 38 (6): 932–962.
287 Gorgieva, S. and Kokol, V. (2011). Collagen- vs. gelatine-based biomaterials and their biocompatibility: review and perspectives. In: *Biomaterials* (ed. R. Pignatello). IntechOpen.
288 Bosnakovski, D., Mizuno, M., Kim, G. et al. (2006). Chondrogenic differentiation of bovine bone marrow mesenchymal stem cells (MSCs) in different hydrogels: influence of collagen type II extracellular matrix on MSC chondrogenesis. *Biotechnology and Bioengineering* 93 (6): 1152–1163.
289 Park, J.Y., Choi, J.-C., Shim, J.-H. et al. (2014). A comparative study on collagen type I and hyaluronic acid dependent cell behavior for osteochondral tissue bioprinting. *Biofabrication* 6 (3): 035004.
290 Chen, C.-H., Shyu, V.B.-H., Chen, J.-P., and Lee, M.-Y. (2014). Selective laser sintered poly-ε-caprolactone scaffold hybridized with collagen hydrogel for cartilage tissue engineering. *Biofabrication* 6 (1): 015004.
291 Liao, E., Yaszemski, M., Krebsbach, P., and Hollister, S. (2007). Tissue-engineered cartilage constructs using composite hyaluronic acid/collagen I hydrogels and designed poly(propylene fumarate) scaffolds. *Tissue Engineering* 13 (3): 537–550.
292 Yang, K., Sun, J., Wei, D. et al. (2017). Photo-crosslinked mono-component type II collagen hydrogel as a matrix to induce chondrogenic differentiation of bone marrow mesenchymal stem cells. *Journal of Materials Chemistry B* 5 (44): 8707–8718.
293 Yuan, T., Zhang, L., Li, K. et al. (2013). Collagen hydrogel as an immunomodulatory scaffold in cartilage tissue engineering. *Journal of Biomedical Materials Research Part B: Applied Biomaterials* 102 (2): 337–344.
294 Hu, X., Ma, L., Wang, C., and Gao, C. (2009). Gelatin hydrogel prepared by photo-initiated polymerization and loaded with TGF-β1 for cartilage tissue engineering. *Macromolecular Bioscience* 9 (12): 1194–1201.
295 Zhao, X., Lang, Q., Yildirimer, L. et al. (2015). Photocrosslinkable gelatin hydrogel for epidermal tissue engineering. *Advanced Healthcare Materials* 5 (1): 108–118.
296 Boere, K.W., Visser, J., Seyednejad, H. et al. (2014). Covalent attachment of a three-dimensionally printed thermoplast to a gelatin hydrogel for mechanically enhanced cartilage constructs. *Acta Biomaterialia* 10 (6): 2602–2611.
297 Billiet, T., Gevaert, E., Schryver, T.D. et al. (2014). The 3D printing of gelatin methacrylamide cell-laden tissue-engineered constructs with high cell viability. *Biomaterials* 35 (1): 49–62.
298 Bellan, L.M., Pearsall, M., Cropek, D.M., and Langer, R. (2012). A 3D interconnected microchannel network formed in gelatin by sacrificial shellac microfibers. *Advanced Materials* 24 (38): 5187–5191.

299 Costa, F., Silva, R., and Boccaccini, A. (2018). Fibrous protein-based biomaterials (silk, keratin, elastin, and resilin proteins) for tissue regeneration and repair. In: *Peptides and Proteins as Biomaterials for Tissue Regeneration and Repair*, 175–204. Elsevier.

300 Lee, J., Sultan, M., Kim, S. et al. (2017). Artificial auricular cartilage using silk fibroin and polyvinyl alcohol hydrogel. *International Journal of Molecular Sciences* 18 (8): 1707.

301 Jindal, S., Kiamehr, M., Sun, W., and Yang, X. (2014). Silk scaffolds for dental tissue engineering. In: *Silk Biomaterials for Tissue Engineering and Regenerative Medicine*, 403–428. Elsevier.

302 Wang, H.-Y. and Zhang, Y.-Q. (2015). Processing silk hydrogel and its applications in biomedical materials. *Biotechnology Progress* 31 (3): 630–640.

303 Rodríguez-Cabello, J.C., de Torre, I.G., Ibañez-Fonseca, A., and Alonso, M. (2018). Bioactive scaffolds based on elastin-like materials for wound healing. *Advanced Drug Delivery Reviews* 129: 118–133.

304 Chan, V., Collens, M.B., Jeong, J.H. et al. (2012). Directed cell growth and alignment on protein-patterned 3D hydrogels with stereolithography. *Virtual and Physical Prototyping* 7 (3): 219–228.

305 Lozano, R., Stevens, L., Thompson, B.C. et al. (2015). 3D printing of layered brain-like structures using peptide modified gellan gum substrates. *Biomaterials* 67: 264–273.

306 Silva-Correia, J., Oliveira, J.M., Caridade, S.G. et al. (2010). Gellan gum-based hydrogels for intervertebral disc tissue-engineering applications. *Journal of Tissue Engineering and Regenerative Medicine* 5 (6): e97–e107.

307 Ferris, C.J., Gilmore, K.J., Wallace, G.G., and Panhuis, M.I.H. (2013). Modified gellan gum hydrogels for tissue engineering applications. *Soft Matter* 9 (14): 3705.

308 Tabriz, A.G., Hermida, M.A., Leslie, N.R., and Shu, W. (2015). Three-dimensional bioprinting of complex cell laden alginate hydrogel structures. *Biofabrication* 7 (4): 045012.

309 Song, S.-J., Choi, J., Park, Y.-D. et al. (2011). Sodium alginate hydrogel-based bioprinting using a novel multinozzle bioprinting system. *Artificial Organs* 35 (11): 1132–1136.

310 Cao, N., Chen, X.B., and Schreyer, D.J. (2012). Influence of calcium ions on cell survival and proliferation in the context of an alginate hydrogel. *ISRN Chemical Engineering* 2012: 1–9.

311 Kuo, C.K. and Ma, P.X. (2008). Maintaining dimensions and mechanical properties of ionically crosslinked alginate hydrogel scaffolds in vitro. *Journal of Biomedical Materials Research Part A* 84A (4): 899–907.

312 Mosahebi, A., Simon, M., Wiberg, M., and Terenghi, G. (2001). A novel use of alginate hydrogel as Schwann cell matrix. *Tissue Engineering* 7 (5): 525–534.

313 Luckanagul, J., Lee, L.A., Nguyen, Q.L. et al. (2012). Porous alginate hydrogel functionalized with virus as three-dimensional scaffolds for bone differentiation. *Biomacromolecules* 13 (12): 3949–3958.

314 Williams, S.K., Touroo, J.S., Church, K.H., and Hoying, J.B. (2013). Encapsulation of adipose stromal vascular fraction cells in alginate hydrogel

spheroids using a direct-write three-dimensional printing system. *BioResearch Open Access* 2 (6): 448–454.
315 Liu, H., Liu, J., Qi, C. et al. (2016). Thermosensitive injectable in-situ forming carboxymethyl chitin hydrogel for three-dimensional cell culture. *Acta Biomaterialia* 35: 228–237.
316 Sudha, P.N., Saranya, M., Gomathi, T. et al. (2017). Perspectives of chitin- and chitosan-based scaffolds dressing in regenerative medicine. In: *Chitosan*, 253–269. Wiley.
317 Mahanta, A.K. and Maiti, P. (2015). Chitin and chitosan nanocomposites for tissue engineering. In: *Chitin and Chitosan for Regenerative Medicine*, Springer Series on Polymer and Composite Materials (ed. P. Dutta), 123–149. New Delhi: Springer.
318 Yi, H., Wu, L.-Q., Bentley, W.E. et al. (2005). Biofabrication with chitosan. *Biomacromolecules* 6 (6): 2881–2894.
319 Cheng, Y., Luo, X., Payne, G.F., and Rubloff, G.W. (2012). Biofabrication: programmable assembly of polysaccharide hydrogels in microfluidics as biocompatible scaffolds. *Journal of Materials Chemistry* 22 (16): 7659.
320 Demirtaş, T.T., Irmak, G., and Gümüşderelioğlu, M. (2017). A bioprintable form of chitosan hydrogel for bone tissue engineering. *Biofabrication* 9 (3): 035003.
321 Xu, H. and Matysiak, S. (2017). Effect of pH on chitosan hydrogel polymer network structure. *Chemical Communications* 53 (53): 7373–7376.
322 Tsuda, Y., Hattori, H., Tanaka, Y. et al. (2012). Ultraviolet light-irradiated photocrosslinkable chitosan hydrogel to prevent bone formation in both rat skull and fibula bone defects. *Journal of Tissue Engineering and Regenerative Medicine* 7 (9): 720–728.
323 Skardal, A., Sarker, S.F., Crabbé, A. et al. (2010). The generation of 3-D tissue models based on hyaluronan hydrogel-coated microcarriers within a rotating wall vessel bioreactor. *Biomaterials* 31 (32): 8426–8435.
324 Kinoshita, K., Iwase, M., Yamada, M. et al. (2016). Fabrication of multilayered vascular tissues using microfluidic agarose hydrogel platforms. *Biotechnology Journal* 11 (11): 1415–1423.
325 Roach, B.L., Nover, A.B., Ateshian, G.A., and Hung, C.T. (2016). Agarose hydrogel characterization for regenerative medicine applications: focus on engineering cartilage. In: *Biomaterials from Nature for Advanced Devices and Therapies*, 258–273. Wiley.
326 Zarrintaj, P., Manouchehri, S., Ahmadi, Z. et al. (2018). Agarose-based biomaterials for tissue engineering. *Carbohydrate Polymers* 187: 66–84.
327 Sun, G., Zhang, X., Shen, Y.-I. et al. (2011). Dextran hydrogel scaffolds enhance angiogenic responses and promote complete skin regeneration during burn wound healing. *Proceedings of the National Academy of Sciences of the United States of America* 108 (52): 20976–20981.
328 Lévesque, S.G., Lim, R.M., and Shoichet, M.S. (2005). Macroporous interconnected dextran scaffolds of controlled porosity for tissue-engineering applications. *Biomaterials* 26 (35): 7436–7446.
329 Lévesque, S.G. and Shoichet, M.S. (2006). Synthesis of cell-adhesive dextran hydrogels and macroporous scaffolds. *Biomaterials* 27 (30): 5277–5285.

330 Aminabhavi, T.M. and Deshmukh, A.S. (2015). Polysaccharide-based hydrogels as biomaterials. In: *Polymeric Hydrogels as Smart Biomaterials*, Springer Series on Polymer and Composite Materials, 45–71. Cham: Springer.

331 Popa, E., Reis, R., and Gomes, M. (2012). Chondrogenic phenotype of different cells encapsulated in κ-carrageenan hydrogels for cartilage regeneration strategies. *Biotechnology and Applied Biochemistry* 59 (2): 132–141.

332 Mishra, R.K., Banthia, A.K., and Majeed, A.B.A. (2012). Pectin based formulations for biomedical applications: a review. *Asian Journal of Pharmaceutical and Clinical Research* 5 (4): 1–7.

333 Banks, A., Guo, X., Chen, J. et al. (2017). Novel bioprinting method using a pectin based bioink. *Technology and Health Care* 25 (4): 651–655.

334 Kumar, A., Rao, K.M., and Han, S.S. (2018). Application of xanthan gum as polysaccharide in tissue engineering: a review. *Carbohydrate Polymers* 180: 128–144.

335 Lou, Y.-R., Kanninen, L., Kuisma, T. et al. (2014). The use of nanofibrillar cellulose hydrogel as a flexible three-dimensional model to culture human pluripotent stem cells. *Stem Cells and Development* 23 (4): 380–392.

336 Favi, P.M., Benson, R.S., Neilsen, N.R. et al. (2013). Cell proliferation, viability, and in vitro differentiation of equine mesenchymal stem cells seeded on bacterial cellulose hydrogel scaffolds. *Materials Science and Engineering C* 33 (4): 1935–1944.

337 Sannino, A., Demitri, C., and Madaghiele, M. (2009). Biodegradable cellulose-based hydrogels: design and applications. *Materials* 2 (2): 353–373.

338 Cai, Z. and Kim, J. (2010). Preparation and characterization of novel bacterial cellulose/gelatin scaffold for tissue regeneration using bacterial cellulose hydrogel. *Journal of Nanotechnology in Engineering and Medicine* 1 (2): 021002.

339 Pal, K., Banthia, A.K., and Majumdar, D.K. (2006). Preparation of transparent starch based hydrogel membrane with potential application as wound dressing. *Trends in Biomaterials and Artificial Organs* 20 (1): 59–67.

340 Dong, D., Li, J., Cui, M. et al. (2016). In situ "clickable" zwitterionic starch-based hydrogel for 3D cell encapsulation. *ACS Applied Materials & Interfaces* 8 (7): 4442–4455.

341 Nieuwenhove, I.V., Salamon, A., Peters, K. et al. (2016). Gelatin- and starch-based hydrogels. Part A: Hydrogel development, characterization and coating. *Carbohydrate Polymers* 152: 129–139.

342 Ismail, H., Irani, M., and Ahmad, Z. (2013). Starch-based hydrogels: present status and applications. *International Journal of Polymeric Materials* 62 (7): 411–420.

343 Silva, R., Singh, R., Sarker, B. et al. (2014). Hybrid hydrogels based on keratin and alginate for tissue engineering. *Journal of Materials Chemistry B* 2 (33): 5441–5451.

344 Hadisi, Z., Nourmohammadi, J., and Mohammadi, J. (2015). Composite of porous starch-silk fibroin nanofiber-calcium phosphate for bone regeneration. *Ceramics International* 41 (9): 10745–10754.

345 Ng, W.L., Yeong, W.Y., and Naing, M.W. (2016). Polyelectrolyte gelatin–chitosan hydrogel optimized for 3D bioprinting in skin

tissue engineering. *International Journal of Bioprinting* 2: doi: https://doi.org/10.18063/ijb.2016.01.009.
346 Sarker, B., Singh, R., Silva, R. et al. (2014). Evaluation of fibroblasts adhesion and proliferation on alginate–gelatin crosslinked hydrogel. *PLoS ONE* 9 (9): e107952.
347 Leite, Á.J., Sarker, B., Zehnder, T. et al. (2016). Bioplotting of a bioactive alginate dialdehyde–gelatin composite hydrogel containing bioactive glass nanoparticles. *Biofabrication* 8 (3): 035005.
348 Toba, T., Nakamura, T., Shimizu, Y. et al. (2001). Regeneration of canine peroneal nerve with the use of a polyglycolic acid-collagen tube filled with laminin-soaked collagen sponge: a comparative study of collagen sponge and collagen fibers as filling materials for nerve conduits. *Journal of Biomedical Materials Research* 58 (6): 622–630.
349 Mouser, V.H.M., Melchels, F.P.W., Visser, J. et al. (2016). Yield stress determines bioprintability of hydrogels based on gelatin-methacryloyl and gellan gum for cartilage bioprinting. *Biofabrication* 8 (3): 035003.
350 Huang, J., Fu, H., Wang, Z. et al. (2016). BMSCs-laden gelatin/sodium alginate/carboxymethyl chitosan hydrogel for 3D bioprinting. *RSC Advances* 6 (110): 108423–108430.
351 Campos, F., Bonhome-Espinosa, A.B., García-Martínez, L. et al. (2016). Ex vivo characterization of a novel tissue-like cross-linked fibrin–agarose hydrogel for tissue engineering applications. *Biomedical Materials* 11 (5): 055004.
352 Law, N., Doney, B., Glover, H. et al. (2018). Characterisation of hyaluronic acid methylcellulose hydrogels for 3D bioprinting. *Journal of the Mechanical Behavior of Biomedical Materials* 77: 389–399.
353 Numata, K., Yamazaki, S., Katashima, T. et al. (2014). Silk-pectin hydrogel with superior mechanical properties, biodegradability, and biocompatibility. *Macromolecular Bioscience* 14 (6): 799–806.
354 Birch, N.P., Barney, L.E., Pandres, E. et al. (2015). Thermal-responsive behavior of a cell compatible chitosan/pectin hydrogel. *Biomacromolecules* 16 (6): 1837–1843.
355 Chen, F., Ni, Y., Liu, B. et al. (2017). Self-crosslinking and injectable hyaluronic acid/RGD-functionalized pectin hydrogel for cartilage tissue engineering. *Carbohydrate Polymers* 166: 31–44.
356 Baniasadi, M. and Minary-Jolandan, M. (2015). Alginate–collagen fibril composite hydrogel. *Materials* 8 (2): 799–814.
357 Duan, B., Hockaday, L.A., Kang, K.H., and Butcher, J.T. (2012). 3D bioprinting of heterogeneous aortic valve conduits with alginate/gelatin hydrogels. *Journal of Biomedical Materials Research Part A* 101A (5): 1255–1264.
358 Das, S., Pati, F., Choi, Y.-J. et al. (2015). Bioprintable, cell-laden silk fibroin–gelatin hydrogel supporting multilineage differentiation of stem cells for fabrication of three-dimensional tissue constructs. *Acta Biomaterialia* 11: 233–246.
359 Bertassoni, L.E., Cardoso, J.C., Manoharan, V. et al. (2014). Direct-write bioprinting of cell-laden methacrylated gelatin hydrogels. *Biofabrication* 6 (2): 024105.

360 Skardal, A., Zhang, J., Mccoard, L. et al. (2010). Photocrosslinkable hyaluronan–gelatin hydrogels for two-step bioprinting. *Tissue Engineering Part A* 16 (8): 2675–2685.

361 Lee, W., Bae, C.Y., Kwon, S. et al. (2012). Cellular hydrogel biopaper for patterned 3D cell culture and modular tissue reconstruction. *Advanced Healthcare Materials* 1 (5): 635–639.

362 Ringeisen, B.R., Pirlo, R.K., and Wu, P.K. (2014). US Patent Application No. 14/177,240.

363 Imani, R., Emami, S.H., Sharifi, A.M. et al. (2011). Evaluation of novel "biopaper" for cell and organ printing application: an in vitro study. *Journal of Diabetes & Metabolic Disorders* 10: 1.

364 Jakab, K., Neagu, A., Mironov, V., and Forgacs, G. (2004). Organ printing: fiction or science. *Biorheology* 41 (3–4): 371–375.

365 Pati, F., Jang, J., Ha, D.-H. et al. (2014). Printing three-dimensional tissue analogues with decellularized extracellular matrix bioink. *Nature Communications* 5: 3935.

366 Colosi, C., Shin, S.R., Manoharan, V. et al. (2015). Microfluidic bioprinting of heterogeneous 3D tissue constructs using low-viscosity bioink. *Advanced Materials* 28 (4): 677–684.

367 Jia, J., Richards, D.J., Pollard, S. et al. (2014). Engineering alginate as bioink for bioprinting. *Acta Biomaterialia* 10 (10): 4323–4331.

368 Markstedt, K., Mantas, A., Tournier, I. et al. (2015). 3D bioprinting human chondrocytes with nanocellulose–alginate bioink for cartilage tissue engineering applications. *Biomacromolecules* 16 (5): 1489–1496.

369 Rutz, A.L., Hyland, K.E., Jakus, A.E. et al. (2015). A multimaterial bioink method for 3D printing tunable, cell-compatible hydrogels. *Advanced Materials* 27 (9): 1607–1614.

370 Jia, W., Gungor-Ozkerim, P.S., Zhang, Y.S. et al. (2016). Direct 3D bioprinting of perfusable vascular constructs using a blend bioink. *Biomaterials* 106: 58–68.

371 Melchels, F.P.W., Dhert, W.J.A., Hutmacher, D.W., and Malda, J. (2014). Development and characterisation of a new bioink for additive tissue manufacturing. *Journal of Materials Chemistry B* 2 (16): 2282–2289.

372 Hölzl, K., Lin, S., Tytgat, L. et al. (2016). Bioink properties before, during and after 3D bioprinting. *Biofabrication* 8 (3): 032002.

373 Skardal, A., Devarasetty, M., Kang, H.-W. et al. (2015). A hydrogel bioink toolkit for mimicking native tissue biochemical and mechanical properties in bioprinted tissue constructs. *Acta Biomaterialia* 25: 24–34.

374 Chung, J.H.Y., Naficy, S., Yue, Z. et al. (2013). Bio-ink properties and printability for extrusion printing living cells. *Biomaterials Science* 1 (7): 763–773.

375 Ferris, C.J., Gilmore, K.J., Beirne, S. et al. (2013). Bio-ink for on-demand printing of living cells. *Biomaterials Science* 1 (2): 224–230.

376 Gu, Q., Tomaskovic-Crook, E., Lozano, R. et al. (2016). Functional 3D neural mini-tissues from printed gel-based bioink and human neural stem cells. *Advanced Healthcare Materials* 5 (12): 1429–1438.

377 Armstrong, J.P.K., Burke, M., Carter, B.M. et al. (2016). 3D bioprinting: 3D bioprinting using a templated porous bioink. *Advanced Healthcare Materials* 5 (14): 1681–1681.

378 Hospodiuk, M., Dey, M., Sosnoski, D., and Ozbolat, I.T. (2017). The bioink: a comprehensive review on bioprintable materials. *Biotechnology Advances* 35 (2): 217–239.

379 Yu, Y. and Ozbolat, I.T. (2014). Tissue strands as "bioink" for scale-up organ printing. 2014 36th Annual International Conference of the IEEE Engineering in Medicine and Biology Society.

380 Lee, H., Han, W., Kim, H. et al. (2017). Development of liver decellularized extracellular matrix bioink for three-dimensional cell printing-based liver tissue engineering. *Biomacromolecules* 18 (4): 1229–1237.

381 Mecozzi, L., Gennari, O., Rega, R. et al. (2016). Simple and rapid bioink jet printing for multiscale cell adhesion islands. *Macromolecular Bioscience* 17 (3): 1600307.

382 Panwar, A. and Tan, L. (2016). Current status of bioinks for micro-extrusion-based 3D bioprinting. *Molecules* 21 (6): 685.

383 Catros, S., Guillotin, B., Bačáková, M. et al. (2011). Effect of laser energy, substrate film thickness and bioink viscosity on viability of endothelial cells printed by Laser-Assisted Bioprinting. *Applied Surface Science* 257 (12): 5142–5147.

384 Jang, J., Kim, T.G., Kim, B.S. et al. (2016). Tailoring mechanical properties of decellularized extracellular matrix bioink by vitamin B2-induced photo-crosslinking. *Acta Biomaterialia* 33: 88–95.

385 Khademhosseini, A. and Camci-Unal, G. (2018). *3D Bioprinting in Regenerative Engineering: Principles and Applications.* Taylor & Francis.

386 Elloumi-Hannachi, I., Yamato, M., and Okano, T. (2010). Cell sheet engineering: a unique nanotechnology for scaffold-free tissue reconstruction with clinical applications in regenerative medicine. *Journal of Internal Medicine* 267 (1): 54–70.

387 Das, A.A.K., Fakhrullin, R.F., and Paunov, V.N. Artificial multicellular assemblies from cells interfaced with polymers and nanomaterials. In: *Cell Surface Engineering Smart Materials Series*, 162–184. RSC Publishing.

388 Verma, S., Domb, A.J., and Kumar, N. (2011). Nanomaterials for regenerative medicine. *Nanomedicine* 6 (1): 157–181.

389 Mooney, E., Mackle, J.N., Blond, D.J.-P. et al. (2012). The electrical stimulation of carbon nanotubes to provide a cardiomimetic cue to MSCs. *Biomaterials* 33 (26): 6132–6139.

390 Hopley, E.L., Salmasi, S., Kalaskar, D.M., and Seifalian, A.M. (2014). Carbon nanotubes leading the way forward in new generation 3D tissue engineering. *Biotechnology Advances* 32 (5): 1000–1014.

391 Holmes, B., Castro, N.J., Li, J. et al. (2013). Enhanced human bone marrow mesenchymal stem cell functions in novel 3D cartilage scaffolds with hydrogen treated multi-walled carbon nanotubes. *Nanotechnology* 24 (36): 365102.

392 Ostrovidov, S., Shi, X., Zhang, L. et al. (2014). Myotube formation on gelatin nanofibers – multi-walled carbon nanotubes hybrid scaffolds. *Biomaterials* 35 (24): 6268–6277.

393 Dorj, B., Won, J.-E., Kim, J.-H. et al. (2012). Robocasting nanocomposite scaffolds of poly(caprolactone)/hydroxyapatite incorporating modified carbon nanotubes for hard tissue reconstruction. *Journal of Biomedical Materials Research Part A* 101A (6): 1670–1681.

394 Dolati, F., Yu, Y., Zhang, Y. et al. (2014). In vitro evaluation of carbon-nanotube-reinforced bioprintable vascular conduits. *Nanotechnology* 25 (14): 145101.

395 Zhu, W., O'brien, C., O'brien, J.R., and Zhang, L.G. (2014). 3D nano/microfabrication techniques and nanobiomaterials for neural tissue regeneration. *Nanomedicine* 9 (6): 859–875.

396 An, J., Chua, C.K., Yu, T. et al. (2013). Advanced nanobiomaterial strategies for the development of organized tissue engineering constructs. *Nanomedicine* 8 (4): 591–602.

397 Paul, A., Manoharan, V., Krafft, D. et al. (2016). Nanoengineered biomimetic hydrogels for guiding human stem cell osteogenesis in three dimensional microenvironments. *Journal of Materials Chemistry B* 4 (20): 3544–3554.

398 Chaudhuri, B., Bhadra, D., Moroni, L., and Pramanik, K. (2015). Myoblast differentiation of human mesenchymal stem cells on graphene oxide and electrospun graphene oxide–polymer composite fibrous meshes: importance of graphene oxide conductivity and dielectric constant on their biocompatibility. *Biofabrication* 7 (1): 015009.

399 Lin, D., Jin, S., Zhang, F. et al. (2015). 3D stereolithography printing of graphene oxide reinforced complex architectures. *Nanotechnology* 26 (43): 434003.

400 Gatenholm, P. and Klemm, D. (2010). Bacterial nanocellulose as a renewable material for biomedical applications. *MRS Bulletin* 35 (03): 208–213.

401 Kim, M.S., Lee, B., Kim, H.N. et al. (2017). 3D tissue formation by stacking detachable cell sheets formed on nanofiber mesh. *Biofabrication* 9 (1): 015029.

402 Kai, D., Prabhakaran, M.P., Stahl, B. et al. (2012). Mechanical properties and in vitro behavior of nanofiber–hydrogel composites for tissue engineering applications. *Nanotechnology* 23 (9): 095705.

403 Lim, Y.C., Johnson, J., Fei, Z. et al. (2010). Micropatterning and characterization of electrospun poly(ε-caprolactone)/gelatin nanofiber tissue scaffolds by femtosecond laser ablation for tissue engineering applications. *Biotechnology and Bioengineering* 108 (1): 116–126.

404 Martins, A., Silva, M.L.A.D., Faria, S. et al. (2011). The influence of patterned nanofiber meshes on human mesenchymal stem cell osteogenesis. *Macromolecular Bioscience* 11 (7): 978–987.

405 Baylan, N., Bhat, S., Ditto, M. et al. (2013). Polycaprolactone nanofiber interspersed collagen type-I scaffold for bone regeneration: a unique injectable osteogenic scaffold. *Biomedical Materials* 8 (4): 045011.

406 Lin, H.-Y., Tsai, W.-C., and Chang, S.-H. (2017). Collagen–PVA aligned nanofiber on collagen sponge as bi-layered scaffold for surface cartilage repair. *Journal of Biomaterials Science Polymer Edition* 28 (7): 664–678.

407 Narayanan, L.K., Huebner, P., Fisher, M.B. et al. (2016). 3D-bioprinting of polylactic acid (PLA) nanofiber–alginate hydrogel bioink containing human adipose-derived stem cells. *ACS Biomaterials Science & Engineering* 2 (10): 1732–1742.

408 Liu, X., Xu, S., Kuang, X., and Wang, X. (2017). 3D cardiac cell culture on nanofiber bundle substrates for the investigation of cell morphology and contraction. *Micromachines* 8 (5): 147.

409 Park, C.H., Oh, J.-H., Jung, H.-M. et al. (2017). Effects of the incorporation of ε-aminocaproic acid/chitosan particles to fibrin on cementoblast differentiation and cementum regeneration. *Acta Biomaterialia* 61: 134–143.

410 Müller, M., Becher, J., Schnabelrauch, M., and Zenobi-Wong, M. (2015). Nanostructured Pluronic hydrogels as bioinks for 3D bioprinting. *Biofabrication* 7 (3): 035006.

411 Das, A.A.K., Filby, B.W., Geddes, D.A. et al. (2017). High throughput fabrication of cell spheroids by templating water-in-water Pickering emulsions. *Materials Horizons* 4 (6): 1196–1200.

412 Wang, Z., Wen, F., Lim, P.N. et al. (2017). Nanomaterial scaffolds to regenerate musculoskeletal tissue: signals from within for neovessel formation. *Drug Discovery Today* 22 (9): 1385–1391.

413 Vallet-Regí-, M. (2010). Nanostructured mesoporous silica matrices in nanomedicine. *Journal of Internal Medicine* 267 (1): 22–43.

414 Schmidt, N., Schulze, J., Warwas, D.P. et al. (2018). Long term delivery of brain-derived neurotrophic factor (BDNF) from nanoporous silica nanoparticles improves the survival of spiral ganglion neurons in vitro. *PLoS ONE* 13 (3): e0194778.

415 Tang, A., Liu, Y., Wang, Q. et al. (2016). A new photoelectric ink based on nanocellulose/CdS quantum dots for screen-printing. *Carbohydrate Polymers* 148: 29–35.

416 Ruiz-Palomero, C., Soriano, M.L., Benítez-Martínez, S., and Valcárcel, M. (2017). Photoluminescent sensing hydrogel platform based on the combination of nanocellulose and S,N-codoped graphene quantum dots. *Sensors and Actuators B: Chemical* 245: 946–953.

417 Leonard, F. and Godin, B. (2016). 3D in vitro model for breast cancer research using magnetic levitation and bioprinting method. In: *Breast Cancer*, Methods in Molecular Biology, 239–251. New York: Humana Press.

418 García-Astrain, C., Chen, C., Burón, M. et al. (2015). Biocompatible hydrogel nanocomposite with covalently embedded silver nanoparticles. *Biomacromolecules* 16 (4): 1301–1310.

419 Skardal, A., Zhang, J., Mccoard, L. et al. (2010). Dynamically crosslinked gold nanoparticle – hyaluronan hydrogels. *Advanced Materials* 22 (42): 4736–4740.

420 Blaeser, A., Million, N., Campos, D.F.D. et al. (2016). Laser-based in situ embedding of metal nanoparticles into bioextruded alginate hydrogel tubes enhances human endothelial cell adhesion. *Nano Research* 9 (11): 3407–3427.

421 Shuai, C., Guo, W., Gao, C. et al. (2017). An *n*MgO containing scaffold: antibacterial activity, degradation properties and cell responses. *International Journal of Bioprinting* 4 (1): doi: https://doi.org/10.18063/ijb.v4i1.120.

422 Cao, L., Wu, X., Wang, Q., and Wang, J. (2018). Biocompatible nanocomposite of TiO_2 incorporated bi-polymer for articular cartilage tissue regeneration: a facile material. *Journal of Photochemistry and Photobiology B: Biology* 178: 440–446.

423 Zhu, K., Shin, S.R., Kempen, T.V. et al. (2017). Tissue engineering: gold nanocomposite bioink for printing 3D cardiac constructs. *Advanced Functional Materials* 27 (12): doi: https://doi.org/10.1002/adfm.201770077.

424 Rinoldi, C., Kijeńska, E., Chlanda, A. et al. (2018). Nanobead-on-string composites for tendon tissue engineering. *Journal of Materials Chemistry B* 6 (19): 3116–3127.

425 Roh, H.-S., Lee, C.-M., Hwang, Y.-H. et al. (2017). Addition of MgO nanoparticles and plasma surface treatment of three-dimensional printed polycaprolactone/hydroxyapatite scaffolds for improving bone regeneration. *Materials Science and Engineering C* 74: 525–535.

426 Prasad, T. and Elumalai, E. (2011). Biofabrication of Ag nanoparticles using *Moringa oleifera* leaf extract and their antimicrobial activity. *Asian Pacific Journal of Tropical Biomedicine* 1 (6): 439–442.

427 Suresh, A.K., Pelletier, D.A., Wang, W. et al. (2011). Biofabrication of discrete spherical gold nanoparticles using the metal-reducing bacterium *Shewanella oneidensis*. *Acta Biomaterialia* 7 (5): 2148–2152.

428 Gade, A., Gaikwad, S., Tiwari, V. et al. (2010). Biofabrication of silver nanoparticles by opuntia ficus-indica: in vitro antibacterial activity and study of the mechanism involved in the synthesis. *Current Nanoscience* 6 (4): 370–375.

429 Rajasekharreddy, P. and Rani, P.U. (2014). Biofabrication of Ag nanoparticles using *Sterculia foetida* L. seed extract and their toxic potential against mosquito vectors and HeLa cancer cells. *Materials Science and Engineering C* 39: 203–212.

430 Kotakadi, V.S., Gaddam, S.A., Rao, Y.S. et al. (2014). Biofabrication of silver nanoparticles using *Andrographis paniculata*. *European Journal of Medicinal Chemistry* 73: 135–140.

431 Jafarirad, S., Mehrabi, M., Divband, B., and Kosari-Nasab, M. (2016). Biofabrication of zinc oxide nanoparticles using fruit extract of *Rosa canina* and their toxic potential against bacteria: a mechanistic approach. *Materials Science and Engineering C* 59: 296–302.

432 Verma, V.C., Singh, S.K., Solanki, R., and Prakash, S. (2010). Biofabrication of anisotropic gold nanotriangles using extract of endophytic *Aspergillus clavatus* as a dual functional reductant and stabilizer. *Nanoscale Research Letters* 6 (1): 16.

433 Vázquez, E. and Villaverde, A. (2013). Microbial biofabrication for nanomedicine: biomaterials, nanoparticles and beyond. *Nanomedicine* 8 (12): 1895–1898.

434 Silva, L.P., Bonatto, C.C., and Polez, V.L.P. (2016). Green synthesis of metal nanoparticles by fungi: current trends and challenges. In: *Advances and Applications Through Fungal Nanobiotechnology. Fungal Biology*, 71–89. Cham: Springer.

435 Silva, L.P., Reis, I.G., and Bonatto, C.C. (2015). Green synthesis of metal nanoparticles by plants: current trends and challenges. In: *Green Processes for Nanotechnology*, 259–275. Cham: Springer.

436 Silva, L.P., Silveira, A.P., Bonatto, C.C. et al. (2017). Silver nanoparticles as antimicrobial agents: past, present, and future. In: *Nanostructures for Antimicrobial Therapy*, 577–596. Elsevier.

437 Izadifar, M., Haddadi, A., Chen, X., and Kelly, M.E. (2014). Rate-programming of nano-particulate delivery systems for smart bioactive scaffolds in tissue engineering. *Nanotechnology* 26 (1): 012001.

438 Kang, M.S., Kim, J.-H., Singh, R.K. et al. (2015). Therapeutic-designed electrospun bone scaffolds: mesoporous bioactive nanocarriers in hollow fiber composites to sequentially deliver dual growth factors. *Acta Biomaterialia* 16: 103–116.

439 Sahoo, S., Ang, L.T., Goh, J.C.-H., and Toh, S.-L. (2009). Growth factor delivery through electrospun nanofibers in scaffolds for tissue engineering applications. *Journal of Biomedical Materials Research Part A* 93 (4): 1539–1550.

440 Gu, W., Wu, C., Chen, J., and Xiao, Y. (2013). Nanotechnology in the targeted drug delivery for bone diseases and bone regeneration. *International Journal of Nanomedicine* 8 (1): 2305.

441 Kerativitayanan, P., Carrow, J.K., and Gaharwar, A.K. (2015). Nanomaterials for engineering stem cell responses. *Advanced Healthcare Materials* 4 (11): 1600–1627.

442 Kaliva, M., Chatzinikolaidou, M., and Vamvakaki, M. Applications of smart multifunctional tissue engineering scaffolds. In: *Smart Materials for Tissue Engineering*, Smart Materials Series, Chapter 1, 1–38. RSC Publishing.

443 Koutsopoulos, S. (2012). Molecular fabrications of smart nanobiomaterials and applications in personalized medicine. *Advanced Drug Delivery Reviews* 64 (13): 1459–1476.

444 Choi, S., Tripathi, A., and Singh, D. (2014). Smart nanomaterials for biomedics. *Journal of Biomedical Nanotechnology* 10 (10): 3162–3188.

445 Mandrycky, C., Wang, Z., Kim, K., and Kim, D.-H. (2016). 3D bioprinting for engineering complex tissues. *Biotechnology Advances* 34 (4): 422–434.

446 Jose, R.R., Rodriguez, M.J., Dixon, T.A. et al. (2016). Evolution of bioinks and additive manufacturing technologies for 3D bioprinting. *ACS Biomaterials Science & Engineering* 2 (10): 1662–1678.

447 Zhang, Y.S., Yue, K., Aleman, J. et al. (2016). 3D bioprinting for tissue and organ fabrication. *Annals of Biomedical Engineering* 45 (1): 148–163.

448 Arslan-Yildiz, A., Assal, R.E., Chen, P. et al. (2016). Towards artificial tissue models: past, present, and future of 3D bioprinting. *Biofabrication* 8 (1): 014103.

449 Rutz, A.L. and Shah, R.N. (2015). Protein-based hydrogels. In: *Polymeric Hydrogels as Smart Biomaterials Springer Series on Polymer and Composite Materials*, 73–104. Springer.
450 Gao, B., Yang, Q., Zhao, X. et al. (2016). 4D bioprinting for biomedical applications. *Trends in Biotechnology* 34 (9): 746–756.
451 An, J., Chua, C.K., and Mironov, V. (2016). A perspective on 4D bioprinting. *International Journal of Bioprinting* 2 (1): 3–5.
452 Li, Y.-C., Zhang, Y.S., Akpek, A. et al. (2016). 4D bioprinting: the next-generation technology for biofabrication enabled by stimuli-responsive materials. *Biofabrication* 9 (1): 012001.
453 Kirillova, A., Maxson, R., Stoychev, G. et al. (2017). 4D biofabrication using shape-morphing hydrogels. *Advanced Materials* 29 (46): 1703443.
454 Ionov, L. (2017). 4D biofabrication by shape-morphing polymers. *BioMedical Engineering* 62 (S2): 13–18.
455 Stroganov, V., Pant, J., Stoychev, G. et al. (2018). 4D biofabrication: 3D cell patterning using shape-changing films. *Advanced Functional Materials* 28 (11): 1706248.
456 Ong, C.S., Nam, L., Ong, K. et al. (2018). 3D and 4D bioprinting of the myocardium: current approaches, challenges, and future prospects. *BioMed Research International* 2018: 1–11.
457 Khoo, Z.X., Teoh, J.E.M., Liu, Y. et al. (2015). 3D printing of smart materials: a review on recent progresses in 4D printing. *Virtual and Physical Prototyping* 10 (3): 103–122.
458 Bakarich, S.E., Gorkin, R., Panhuis, M.I.H., and Spinks, G.M. (2015). 4D printing with mechanically robust, thermally actuating hydrogels. *Macromolecular Rapid Communications* 36 (12): 1211–1217.
459 Castro, N.J., Meinert, C., Levett, P., and Hutmacher, D.W. (2017). Current developments in multifunctional smart materials for 3D/4D bioprinting. *Current Opinion in Biomedical Engineering* 2: 67–75.
460 Leist, S.K. and Zhou, J. (2016). Current status of 4D printing technology and the potential of light-reactive smart materials as 4D printable materials. *Virtual and Physical Prototyping* 11 (4): 249–262.
461 Morouço, P., Lattanzi, W., and Alves, N. (2017). Four-dimensional bioprinting as a new era for tissue engineering and regenerative medicine. *Frontiers in Bioengineering and Biotechnology* 5: 61.
462 Gladman, A.S., Matsumoto, E.A., Nuzzo, R.G. et al. (2016). Biomimetic 4D printing. *Nature Materials* 15 (4): 413–418.
463 Cengiz, I., Pitikakis, M., Cesario, L. et al. (2016). Building the basis for patient-specific meniscal scaffolds: from human knee MRI to fabrication of 3D printed scaffolds. *Bioprinting* 1–2: 1–10.
464 Mota, C. (2011). The rise of personal fabrication. Proceedings of the 8th ACM conference on Creativity and cognition – C&C '11, pp. 279–288.
465 Rezende, R.A., Kasyanov, V., Mironov, V., and Silva, J.V.L.D. (2015). Organ printing as an information technology. *Procedia Engineering* 110: 151–158.
466 Yin, X., Mead, B.E., Safaee, H. et al. (2016). Engineering stem cell organoids. *Cell Stem Cell* 18 (1): 25–38.

467 Kumar, A. and Starly, B. (2015). Large scale industrialized cell expansion: producing the critical raw material for biofabrication processes. *Biofabrication* 7 (4): 044103.

468 Ker, E.D., Nain, A.S., Weiss, L.E. et al. (2011). Bioprinting of growth factors onto aligned sub-micron fibrous scaffolds for simultaneous control of cell differentiation and alignment. *Biomaterials* 32 (32): 8097–8107.

469 Miller, E.D., Li, K., Kanade, T. et al. (2011). Spatially directed guidance of stem cell population migration by immobilized patterns of growth factors. *Biomaterials* 32 (11): 2775–2785.

470 Willerth, S.M., Faxel, T.E., Gottlieb, D.I., and Sakiyama-Elbert, S.E. (2007). The effects of soluble growth factors on embryonic stem cell differentiation inside of fibrin scaffolds. *Stem Cells* 25 (9): 2235–2244.

471 Solorio, L., Zwolinski, C., Lund, A.W. et al. (2010). Gelatin microspheres crosslinked with genipin for local delivery of growth factors. *Journal of Tissue Engineering and Regenerative Medicine* 4 (7): 514–523.

472 Zhu, X. and Yang, H. (2017). 4D biofabrication of branching multicellular structures: a morphogenesis simulation based on Turing's reaction–diffusion dynamics. *IOP Conference Series: Materials Science and Engineering* 280: 012018.

473 Payne, G.F., Bentley, W.E., Sun, W., and Forgacs, G. (2013). Biofabrication. In: *Encyclopedia of Biophysics*, 193–194. Berlin, Heidelberg: Springer-Verlag.

474 Malheiro, A., Wieringa, P., Mota, C. et al. (2016). Patterning vasculature: the role of biofabrication to achieve an integrated multicellular ecosystem. *ACS Biomaterials Science & Engineering* 2 (10): 1694–1709.

475 Pillet, F., Gibot, L., Madi, M. et al. (2017). Importance of endogenous extracellular matrix in biomechanical properties of human skin model. *Biofabrication* 9 (2): 025017.

476 Ivanov, V.T., Karelin, A.A., Philippova, M.M. et al. (1997). Hemoglobin as a source of endogenous bioactive peptides: the concept of tissue-specific peptide pool. *Biopolymers* 43 (2): 171–188.

477 Rustad, K.C., Sorkin, M., Levi, B. et al. (2010). Strategies for organ level tissue engineering. *Organogenesis* 6 (3): 151–157.

478 Zhang, Y.S., Aleman, J., Shin, S.R. et al. (2017). Multisensor-integrated organs-on-chips platform for automated and continual in situ monitoring of organoid behaviors. *Proceedings of the National Academy of Sciences of the United States of America* (12): 114, E2293–E2302.

479 Vijayavenkataraman, S., Lu, W., and Fuh, J. (2016). 3D bioprinting – an ethical, legal and social aspects (ELSA) framework. *Bioprinting* 1–2: 11–21.

480 Otto, I.A., Breugem, C.C., Malda, J., and Bredenoord, A.L. (2016). Ethical considerations in the translation of regenerative biofabrication technologies into clinic and society. *Biofabrication* 8 (4): 042001.

481 Gilbert, F., O'Connell, C.D., Mladenovska, T., and Dodds, S. (2017). Print me an organ? Ethical and regulatory issues emerging from 3D bioprinting in medicine. *Science and Engineering Ethics* 24 (1): 73–91.

482 Vijayavenkataraman, S. (2016). A perspective on bioprinting ethics. *Artificial Organs* 40 (11): 1033–1038.

16

Orthopedic Implant Design and Analysis: Potential of 3D/4D Bioprinting

Chang Jiang Wang[1] and Kevin B. Hazlehurst[2]

[1] *University of Sussex, School of Engineering and Informatics, Department of Engineering and Design, Falmer, Brighton, BN1 9RH, UK*
[2] *Conflux Technology, 75 Pigdons Road, Waurn Ponds, VIC 3216, Australia*

16.1 Orthopedic Implant Design with 3D Printing

The use of 3D printing for surgical planning and personalized medical implants are increasing, as 3D printed implants can be used in many types of surgery. Starting from CT or MRI images, personalized implant design and manufacturing can allow surgeons to make customized implants for each patient with an improved anatomical fit. 3D printed cellular or porous orthopedic implants are less rigid when comparing to solid counterparts that may result in bone stress shielding and bone loss. Porous structures can also be designed into the implant for bone ingrowth by varying the pore shape and size.

16.1.1 Bone Properties and Orthopedic Implants

The human skeleton consists of two types of bone, cortical and cancellous bone, as shown in Figure 16.1. Bones form the major component of the human musculoskeletal system. They support body weight, resist mechanical loading, perform motion, and protect the body's internal organs. Apart from this, bones are also important for calcium metabolism and the generation of blood cells.

Osteoarthritis is a degenerative disease common in the older population. The hip and knee joints are commonly affected; however, the disease can also be present in joints such as the shoulder, base of the thumb, and big toe. Trauma to a joint such as a fracture, dislocation, or previous inflammation can also lead to osteoarthritis. Many different joints are routinely replaced in people with advanced arthritis. Joint replacements are sophisticated procedures and have historically been very successful. Total hip arthroplasty (THA) is an orthopedic procedure that removes the head and neck of the proximal femur and the subchondral bone in the acetabulum. THA is performed to reduce pain and to restore the functionality of hip joints that are affected with diseases such as osteoarthritis. A THA typically consists of a metallic femoral stem, acetabulum component, and polyethylene liner. Across England, Wales, Northern Ireland,

3D and 4D Printing in Biomedical Applications: Process Engineering and Additive Manufacturing,
First Edition. Edited by Mohammed Maniruzzaman.

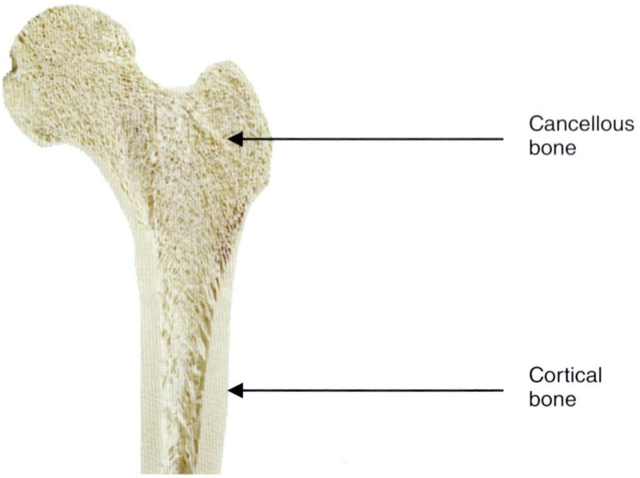

Figure 16.1 Cortical and cancellous bone in the proximal femur.

and the Isle of Man, there were 101 651 cases of THA performed during 2016, as well as 108 713 total knee replacements being reported in this period [1].

When patients have abnormal and excessive motion at a vertebral segment, it results in severe pain and an inability to function. This condition can be treated with a spinal fusion surgery, which is a surgical procedure that was designed to stop the motion at the painful vertebral segment. Often, the spinal fusion procedure includes using a cage, which is placed in the interbody space and filled with bone graft to promote bone growth. Intervertebral cages manufactured from titanium, polyetheretherketone (PEEK), and other biomaterials have historically been used for spinal fusion.

Orthopedic implants are used to replace or provide the fixation of bone or to replace the articulating surfaces of a joint. Orthopedic implants are mechanical devices and are mainly manufactured from stainless steels, titanium alloys, cobalt–chromium alloys, ceramics, and polymers. Materials that are selected for orthopedic applications must be biocompatible and be able to withstand the mechanical loads of humans during daily activities. Functionally graded materials (FGM) can also be utilized for orthopedic applications as they can be beneficial to vary the microstructure of one material to another and thus tailor the properties of the implant or device.

Some basic examples of orthopedic implants are presented in orthopedic plates, nails, and screws. Orthopedic screws are commonly used for bone fracture fixation and can be used as standalone fixators or with other orthopedic implants such as plates. Screws are used to provide the stability of most screw plate fixation devices. However, metallic screws are much stiffer than the adjacent bone and screw loosening can occur through micromotion and stress shielding around the screw threads.

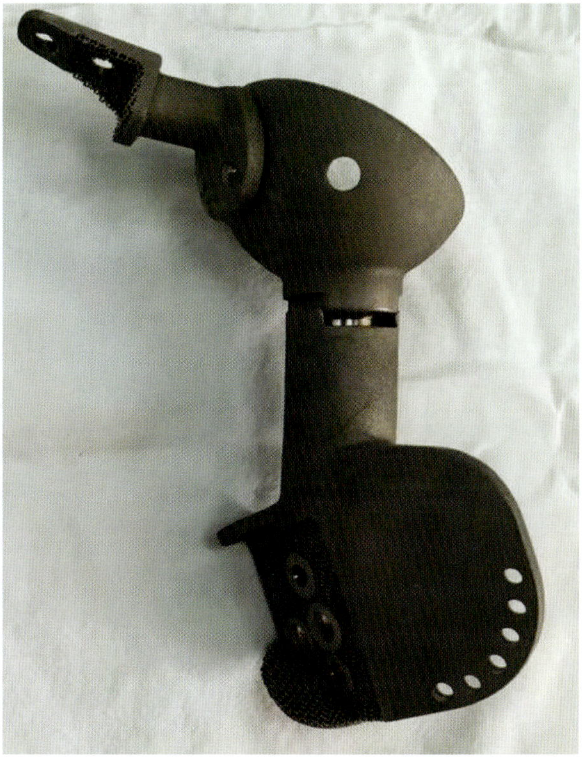

Figure 16.2 A 3D printed customized pelvic implant (Suzhou Kangli Orthopedics Instrument).

Based on patient's medical images, customized implants have been developed for a patient with a pelvic sarcoma, as reported by Chen et al. [2]. The implant was designed using computer-aided design (CAD) software. The 3D printed titanium alloy implant possessed design features such as holes for fixing the implant to the remaining pelvic structures.

"Off-the-shelf" orthopedic implants are normally designed based on a limited number of cadaver bones, which do not necessarily span the whole variability in the global population. Although statistical shape analysis methods have been used to build patient-specific anatomical models for implant design, a model created from a patient's computed tomography (CT) or magnetic resonance image (MRI) data will be more accurate than the statistical shape analysis method. Figures 16.2 and 16.3 show a 3D printed pelvic implant and its installation during operation. This customized implant is designed based on a patient's 3D pelvic model and was used for the treatment of pelvic sarcoma. It can be observed that the implant has predesigned holes for the fixation and the porous surfaces for bone ingrowth.

The load transfer to the proximal femur is reduced following the insertion of metallic femoral stems, which are much stiffer than the bone. This change in the mechanical loading environment is known as stress shielding and often causes a reduction in periprosthetic bone density. Stress shielding can cause the premature failure of THA through aseptic loosening, stem migration, and peripros-

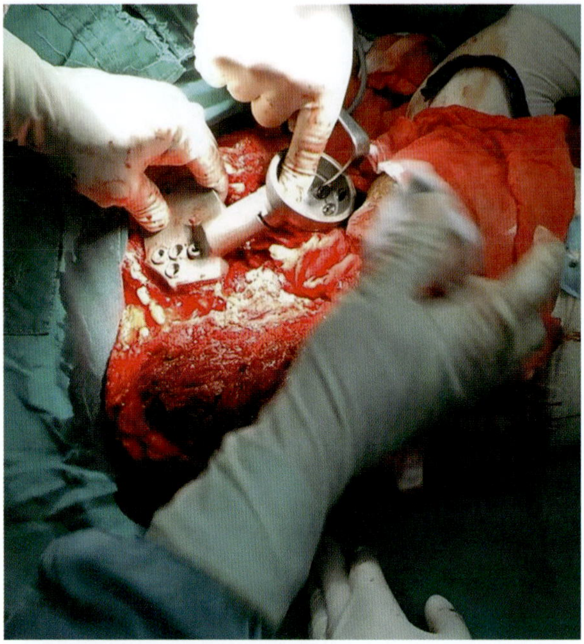

Figure 16.3 Insertion of the 3D printed customized pelvic implant (Suzhou Kangli Orthopedics Instrument).

thetic bone fractures. Therefore, it would be beneficial if femoral stems could be manufactured with characteristics that would assist in the reduction or elimination of stress shielding. For instance, a more flexible femoral stem could alleviate stress shielding and help to preserve periprosthetic bone stock [3]. Orthopedic implant failures are common in total hip and knee arthroplasty, intertrochanteric hip fractures, and long bone fractures. Implant failures can be caused by excessive motion of the implant and bone interfaces, stress concentrations within the implant, and stress shielding of bone. Thus, implant design can affect longevity and marginal bone loss.

16.1.2 3D Printing and Porous Implant Design

Additive manufacturing (AM) technologies, also known as 3D printing, such as selective laser melting (SLM) and electron beam melting (EBM) are capable of manufacturing components from a single alloy with tailored mechanical properties, in a layer-by-layer build process. EBM is a process that was developed by Arcam and uses a high-powered electron beam to selectively melt and fuse together the powder to build parts out of metal powders, whereas SLM utilizes a high-power fiber laser to selectively melt the metal powder. This process was first introduced by EOS GmbH under the name of direct metal laser sintering (DMLS).

Compared with traditional manufacturing technologies, AM can fabricate complex geometries directly from a digital file. When you compare the capabilities of AM against more traditional manufacturing technologies such as machining, casting, and forging, it is clear that AM is more capable of producing porous metallic implants with complex and customized structures. Given this, orthopedic implants can be manufactured with mechanical properties that are much closer to those of human bone. Thus, 3D printing technologies provide a new approach for the clinical treatment and outcomes of orthopedic, maxillofacial, and any other surgery where bone needs to be replaced.

Porous metals are suitable for repairing and replacing bones as the stiffness and porosity can be adjusted. Wang et al. [4] reviewed the topological design and AM of porous metals for bone scaffolds and orthopedic implants, where it was shown that 3D printing provides good opportunities for producing customized medical implants. Topology optimization is also a powerful digital tool for the design of optimal structures. The combined use of 3D printing and topology optimization are an important consideration for the design and manufacturing of customized orthopedic implants.

Cellular structures are composed of an interconnected network of solid struts or plates, which, when combined, produce lightweight porous structures. The use of porous materials for biomedical applications can aid bone growth into the implants' surface. Metallic open cellular structures have been used for the long-term biological fixation of orthopedic implants and tissue scaffolds. Closed cell porous metals can be used to manufacture flexible, lightweight femoral stems for either cemented or cementless fixation. Tan et al. [5] reviewed the 3D printing of cellular scaffolds for orthopedic implants. Three types of existing cellular designs are open foams, reticulated unit cells, and functionally graded structures. The stochastic open cellular foams have random pore shapes and sizes. The reticulated lattice consists of repeating unit cells, leading to a regular foam. Functionally graded structures are designed to vary the porosity of a structure along a required axis. The optimal design of cellular structures for orthopedic implants can boost the affinity between bone tissues and the implant surface, which can ultimately affect bone cell infiltration and bone ingrowth.

Bone ingrowth is important for a range of orthopedic procedures such as the repair of significant bone defects and cementless joint replacement. Wang et al. [6] reviewed factors that influence bone ingrowth into 3D printed porous metal scaffolds. These factors are the material, pore size, porosity, surface modification, and mechanical properties. The porosity of human trabecular bone is between 70% and 90%, and the optimal pore size for bone regeneration still needs to be determined. Values approaching human bone porosity showed the best potential for bone ingrowth by Markhoff et al. [7], whereas a pore size of 600 μm in porous titanium implants showed rapid bone ingrowth by Taniguchi et al. [8]. *In vivo* bone ingrowth studies were performed with rabbits using porous titanium implants that exhibited a constant porosity. The connectivity of the pores is also important. The influence of the network structure and pore throats were studied by Otsuki et al. [9]. It was observed that isolated pores inhibit bone ingrowth.

Simoneau et al. (2017) [10] designed a porous femoral stem to reduce stress shielding and provide implant fixation through bone ingrowth into the pores. The porous material was located in the trochanteric zone where cancellous bone is abundant. The stem was designed to have two dense strips on the medial and lateral sides of the stem, with openings on its dorsal and ventral planes, where porous materials can be accessed for bone ingrowth. The dense strips increased

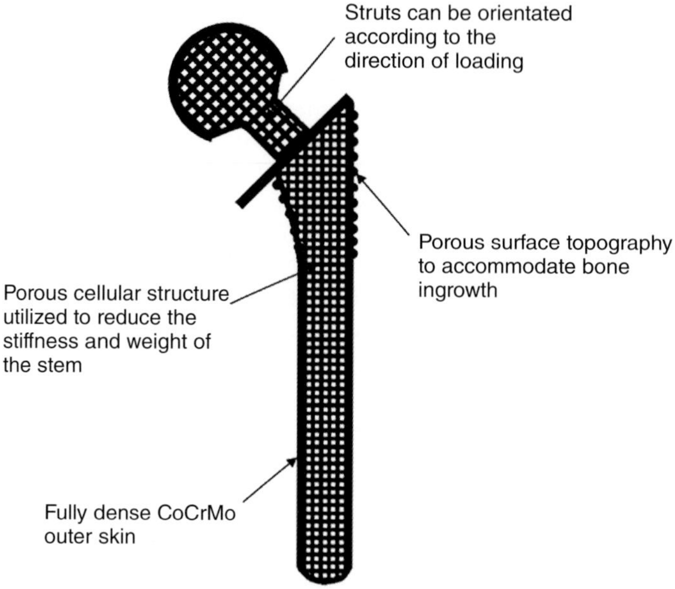

Figure 16.4 Total hip replacement stem with cellular structures. Source: Hazlehurst et al. 2013 [3]. Reproduced with permission of Elsevier.

the global stiffness of the stem and are used to reduce stress concentration at the interface of the porous structure with the dense distal shell. Arabnejad et al. [11] showed that a fully 3D printed porous femoral stem with an optimized material microstructure can reduce the bone loss due to stress shielding by 75% when compared to a fully solid implant.

The femoral stem of a THA can be designed with either a monoblock or a modular design approach. A monoblock femoral stem consists of the stem and head being produced in a single piece, whereas the modular design consists of a separate stem and head that are mechanically interlocked by a taper junction.

The conceptual femoral stem shown in Figure 16.4 was designed by Hazlehurst et al. [3]. The proposed cellular stem possesses a fully dense outer skin and internal cellular structures to reduce the stiffness of the stem. A porous surface topography was applied in the trochanteric region to aid bone ingrowth. 3D printed femoral stems utilizing cellular structures are shown in Figure 16.5a,b. The 3D model shown in Figure 16.5c is a conceptual design of a fully porous implant.

16.2 Analysis of 3D Printed Orthopedic Implants

3D printed orthopedic implants are required to withstand varying loads when implanted. Given this, the mechanical properties of implants, especially implants with porous structures must be evaluated. Yield strength and elastic modulus of 3D printed implants are normally obtained from experimental tests. Finite element models can be used to predict the mechanical performance of

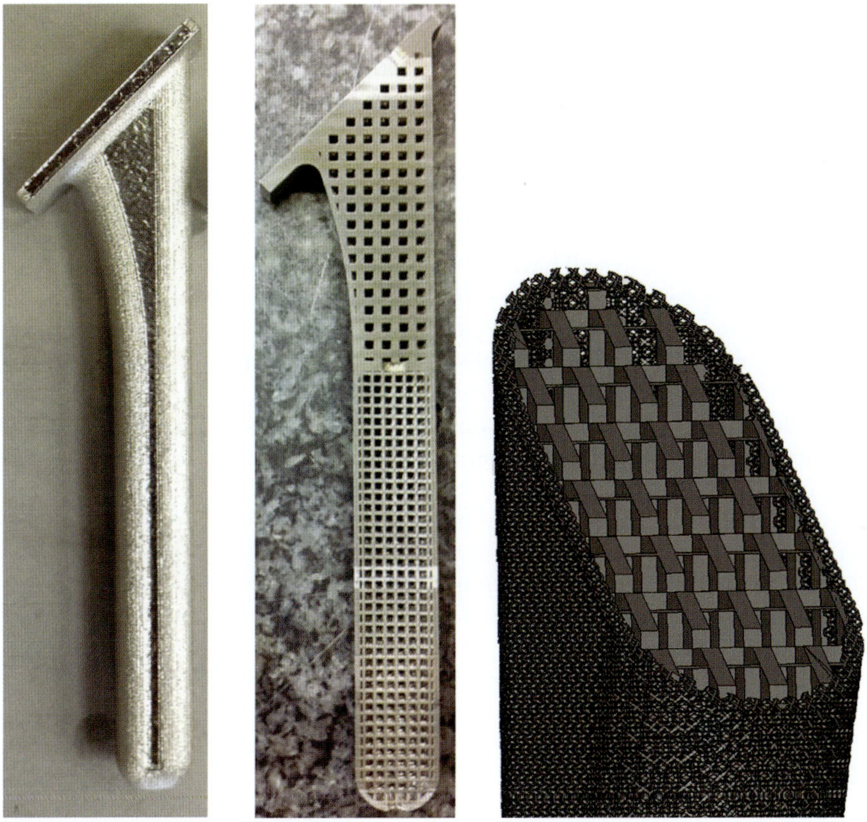

Figure 16.5 (a) 3D printed femoral stem, (b) sectional view showing the internal cellular structures, (c) proximal section of a conceptual fully porous stem [12].

3D printed implants, however due to the nature of 3D printing and porous structures, numerical simulated results should be carefully interpreted.

16.2.1 Mechanical Properties of Porous Structures

The most commonly used orthopedic implant materials are metals and polymers such as stainless steel, cobalt–chromium alloys, titanium alloys, and ultrahigh-molecular-weight polyethylene (UHMWPE). Stainless steel and cobalt–chromium alloys possess a Young' s modulus of around 200 GPa with titanium alloys being approximately 110 GPa. These materials are all much stiffer than the typical cortical bone, which possesses a Young's modulus of approximately 20 GPa. Given this, porous structures can be utilized to reduce the stiffness of implants to be more aligned with the stiffness of the bone. Porosity is defined as the ratio of the volume of voids over the total volume of a material and is governed by the pore size, strut thickness, and pore structure.

The mechanical properties of orthopedic implants are important and cannot be overlooked as failure can occur through common modes such as fatigue and increased wear rate. Therefore, there are many testing standards in place that have been implemented by the US Food and Drug Administration (FDA), American Society for Testing and Materials (ASTM) International, International Organization for Standardization (ISO), and the European Union (EU) to name a few. Croitoru et al. [13] observed that the mechanical strength of 3D printed personalized implants was poor. 3D printed implants should be clinically studied because of the high risk associated with Class III orthopedic implants. Class III orthopedic implants cannot be assessed by comparing with other devices and regulatory agencies. This will have a profound influence on the mechanical testing of 3D printed devices. As specified by the FDA [14], devices manufactured by 3D printing are generally subject to the same regulatory requirements as more traditionally manufactured medical devices. With the advent of 3D printing and the demand of patient-specific devices, the FDA has been working with many stakeholders to develop regulations on the commercial use of 3D printed medical devices. To overcome the shortcomings of conventional implant designs such as nonfusion, biocompatibility issues, subsidence, and migration, 3D printed cellular titanium spinal implants have recently been approved for clinical use by the FDA.

Regarding the technical considerations for additive manufactured medical devices, the FDA produced draft guidelines for industry and FDA staff in May 2016 [14]. In this draft document, it is recommended that the test coupons should be suitable for destructive testing and it should be the worst-case orientation or location in the device. As discussed previously, one of the advantages of 3D printed implants is being able to produce porous structures. The CoCrMo cellular structures shown in Table 16.1 were tested by Hazlehurst [12, 15], and the obtained effective elastic modulus E_{eff} of the square pore CoCrMo cellular structures is presented by Eq. (16.1).

$$E_{\mathrm{eff}} = E_{\mathrm{s}}(1 - \varphi)e^{-2.376\varphi} \tag{16.1}$$

where E_{s} is the elastic modulus of the solid CoCrMo alloy, and φ is the porosity.

From left to right, the cellular structures shown in Figure 16.6 are numbers 3 to 7 in Table 16.1. The micrograph shown in Figure 16.7 of cellular structure 3 with a porosity of 82.5% shows that the strut surfaces are not smooth and potential areas of stress concentration are present within the cellular structure itself.

Table 16.1 Cellular structures for compression testing [12].

Cellular structures	Strut size (mm)	Pore size (mm)	Porosity (%)
1	0.20	1.44	95.14
2	0.35	1.74	90.85
3	0.50	1.57	82.46
4	1.00	1.80	64.80
5	1.50	1.88	50.00
6	2.00	2.33	45.00
7	2.50	1.67	25.93

Figure 16.6 CoCrMo cellular structures for compression tests [12].

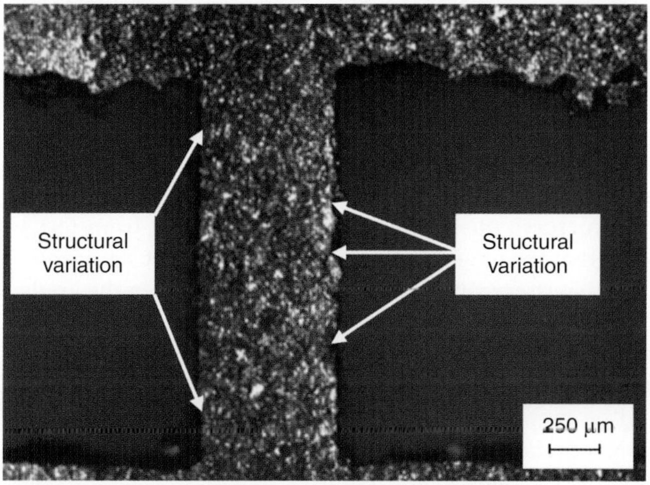

Figure 16.7 Micrograph of 0.5 mm struts of a cellular structure with 82.5% volumetric porosity [12].

During the physical tests, it was observed that the cellular structure with a 1 mm strut size and 64.8% volumetric porosity exhibited the most repeatable mechanical behavior when subjected to uniaxial compression. In contrast to this, cellular structures with strut sizes smaller than 0.5 mm exhibited a more brittle failure mode when compared to the other structures. The compressive stress–strain relationship of the tested cellular structures is shown in Figure 16.8, where the yield strength and slope change significantly as porosity increases from 25.93% to 95.14%.

The effective elastic modulus E_{eff} of square pore CoCrMo cellular structures is also shown in Eq. (16.2) [16]:

$$E_{\text{eff}} = 0.795E_{\text{s}}(1 - \varphi)^{2.159} \tag{16.2}$$

where E_{s} is the elastic modulus of the solid CoCrMo alloy, and φ is the porosity.

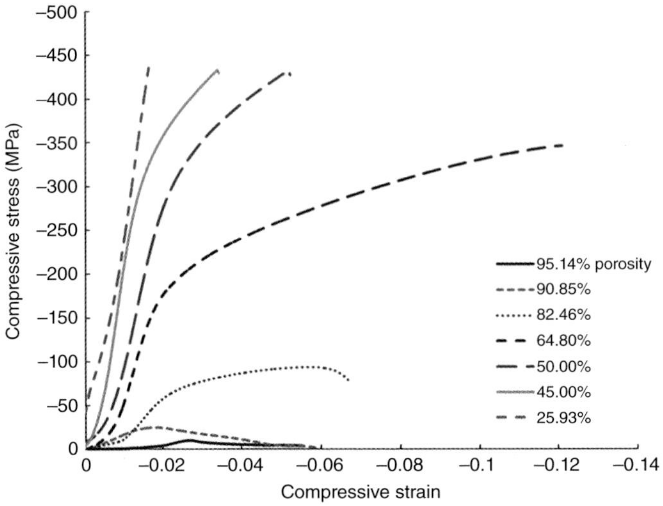

Figure 16.8 Compressive stress and strain curves for cellular structures [12].

Limmahakhun et al. [17] (2017) also conducted experimental tests on cobalt–chromium-graded cellular structures for the stress-shielding reduction of hip implants. The tests showed that the pillar-octahedral-shaped cellular structures with porosity between 41% and 67% were comparable to human bone, with the cellular structures possessing a pore size ranging from 551 to 941 μm. It was also shown that the graded cellular structures do not significantly increase the mechanical properties over cellular structures. However, the graded approach can increase the mechanical strength in the radial orientation. The tested CoCr cellular structures had stiffness values ranging between 2.33 and 3.14 GPa, and the corresponding compressive strengths were in the range of 113–523 MPa. It should be mentioned here that the tested cellular structures were designed with a cylindrical geometry as opposed to the square cellular structures tested by Hazlehurst et al. [15] and Wang et al. [16].

The normalized elastic modulus E of the square pore CoCrMo cellular structures [17] is shown in Eq. (16.3), where the porosity of the tested cellular structures was between 14% and 67%.

$$E = 1.1(1 - \varphi)e^{1.917} \tag{16.3}$$

where the elastic modulus of the solid CoCrMo alloy is 200 GPa, and φ is the porosity.

The hip joint is normally loaded in excess of 1 million gait cycles every year. Therefore, the fatigue strength of femoral stems manufactured from 3D printing is critical for the longevity and reliability of the implant. The mechanical behavior of 3D printed titanium alloy cellular structures when subjected to cyclic loading conditions was tested by Li et al. [18], where the fatigue strength was found to increase with an increasing relative density. Brenne et al. [19] investigated the cyclic behavior of square pore titanium alloy cellular structures under uniaxial and four-point bending loading conditions. It was observed that the fatigue

Table 16.2 Femoral stems with cellular structures for 3D printing.

Femoral stem design	Effective Young's modulus (GPa)	Yield strength (MPa)	Tangent modulus (GPa)
FDS (fully dense)	200.00	600.00	29.41
PC1 (cellular structure 1)	17.98	295.72	4.10
PC2 (cellular structure 2)	13.64	175.15	1.66
PC3 (cellular structure 3)	4.79	65.43	0.05

strength and energy absorption of the structures were improved when the components were heat treated.

16.2.2 Experimental Testing of 3D Printed Femoral Stems

The draft FDA document recommends that the performance testing protocols should be the same for both 3D printed and traditionally manufactured orthopedic implants and devices [14]. Additionally, it specifies that the performance testing should be conducted on fully finished devices. The quality of orthopedic implants from 3D printing should be mechanically tested to evaluate the structural integrity. A selection of femoral stems and their internal cellular structural properties are listed in Table 16.2. Following printing, the excess powder located within the stems was removed. During the 3D printing process, thermally induced stresses are developed in the parts [20]. To alleviate this, the stems were stress relieved in an argon atmosphere at a temperature of 1050 °C for two hours and were left to cool in the furnace. When compared to a fully dense implant, it can be observed that the stems with internal cellular structures were less stiff and possessed a lower yield strength.

A set of 3D printed CoCrMo femoral stems shown in Figure 16.9 were physically tested to measure the stiffness of the implants that were designed and manufactured with internal cellular structures. The stem was clamped at the distal end, and a load was applied at the proximal end. A bending arm length of 140 mm was used and the test setup is shown in Figure 16.10. The stems were loaded to failure by using a constant crosshead displacement speed of 2 mm min^{-1}.

Load and displacement curves are shown in Figure 16.11, where it clearly shows that the stiffness and loading capacity of the porous stems are reduced. Given this, it is imperative to physically test 3D printed orthopedic implants to determine strength, structural integrity, and most importantly to ensure that the device can withstand the loads generated by the patient. As shown in Figure 16.11, the femoral stem (PC2) is 60% more flexible and 48% lighter than the fully dense CoCrMo stem (FDS) [12, 21].

Simoneau et al. [10] designed a porous stem with 33% porosity in the trochanteric region. The porous stem and a dense replica were manufactured using an EOSINT M280 with Ti6Al4V titanium alloy powder. The test stems were potted in an aluminum tube using epoxy resin. The orientation of the stem and embedment length were controlled and achieved using 3D printed jigs.

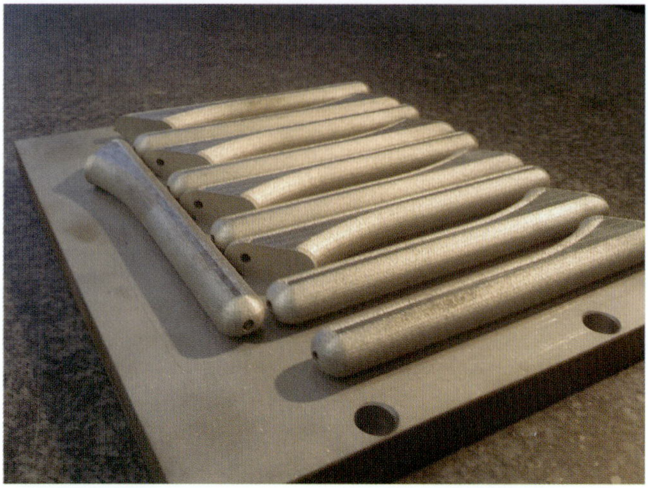

Figure 16.9 3D printed CoCrMo femoral stems manufactured for experimental tests [12].

Figure 16.10 Experimental bending test of a 3D printed femoral stem [12].

The compressive load was applied at a speed of 2 mm min^{-1} at room temperature. During the testing, digital image correlation was performed with a noncontact optical strain measurement system to obtain the displacement and strain fields on the external surfaces of the femoral stem. The test results showed that the flexural stiffness of the porous stem was reduced by 47% when compared to the dense stem. It was reported that there were three sharp sounds during the loading process, indicating that local failures might have occurred. The local failure of 3D printed porous stems should be further studied.

Croitoru et al. [13] designed two personalized hip implants, which were manufactured using an EOSINT M270 extended machine with Ti6Al4V titanium alloy powder. The main feature of the implant was that the fenestration with sizes

Figure 16.11 Load–displacement curves from the bending tests [12].

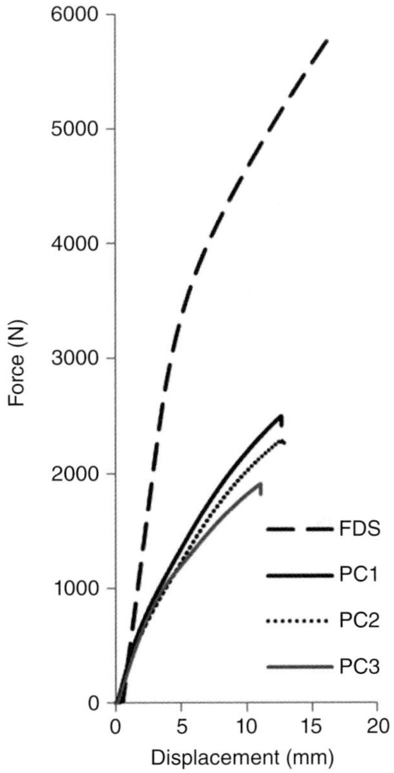

of 6 × 6 and 2 × 2 mm was created in the trochanteric region to enhance the elasticity of the entire femoral stem. For the static test, the compressive load was applied at a speed of 3 mm min^{-1} and stopped at 2700 N. The static results showed that the 3D printed implant with large fenestration had less stiffness, i.e. more elasticity. The load–extension curve of the large fenestration implant had a pronounced nonlinear elastic behavior, similar to rubber. To determine the fatigue limit, the Locati method was employed by Croitoru et al. [13]. Fatigue tests were conducted in a physiological and biochemical environment similar to the human body. The implants were immersed in a saline solution of distilled water with a concentration of 0.91%, at a temperature of 37 °C. Dynamic tests were conducted on the 3D printed implants where it revealed that the fatigue limit for the implant with fenestrations was significantly above the minimum value as specified by the standard. Future work investigating the local failure of 3D printed porous stems is justified and will assist with the accurate predication of strength and structural integrity.

16.2.3 Finite Element Analysis of Porous Stems with 3D Printing

Finite element analysis has been applied in orthopedic biomechanics for more than 40 years, since 1972 [22]. It has been widely used to simulate a full spectrum of orthopedic devices and bone interactions with implants. Finite element analysis can improve our understanding of the mechanical behavior of orthopedic devices. The finite element analysis can solve problems that cannot be analyzed clinically, for example, the development of implants at the preclinical stage. The finite element method can also be used to solve problems that cannot be analyzed experimentally. For example, muscle forces cannot be easily realized in an experimental setup, and the testing of multiple iterations of prototype implants maybe too costly.

Finite element analysis was conducted by Hazlehurst et al. [12, 23] to investigate the load transfer to the periprosthetic femur. The femur was implanted with varying stiffness CoCrMo monoblock femoral stems that incorporated the mechanical properties of square pore cellular structures. A large number of elements are required to mesh cellular structures. Therefore, an alternative method is to model the cellular structures as continuum parts with an effective Young's modulus that represents the stiffness of the cellular structures. It has been shown that this simplified approach is suitable for modeling the behavior of porous femoral stems [23]. However, the finite element modeling of additive manufactured cellular structures should simulate the real cellular structure geometry if computing power is not an issue. The finite element model of a femoral stem, proximal femoral bone, and boundary conditions is shown in Figure 16.12. The stress distribution in the periprosthetic bone is presented in Figure 16.13, and it can be observed that the porous stem resulted in a more evenly distributed stress when compared to the fully dense stem. Therefore, fully porous stems are superior when compared to the fully dense stem in terms of improving stress shielding characteristics.

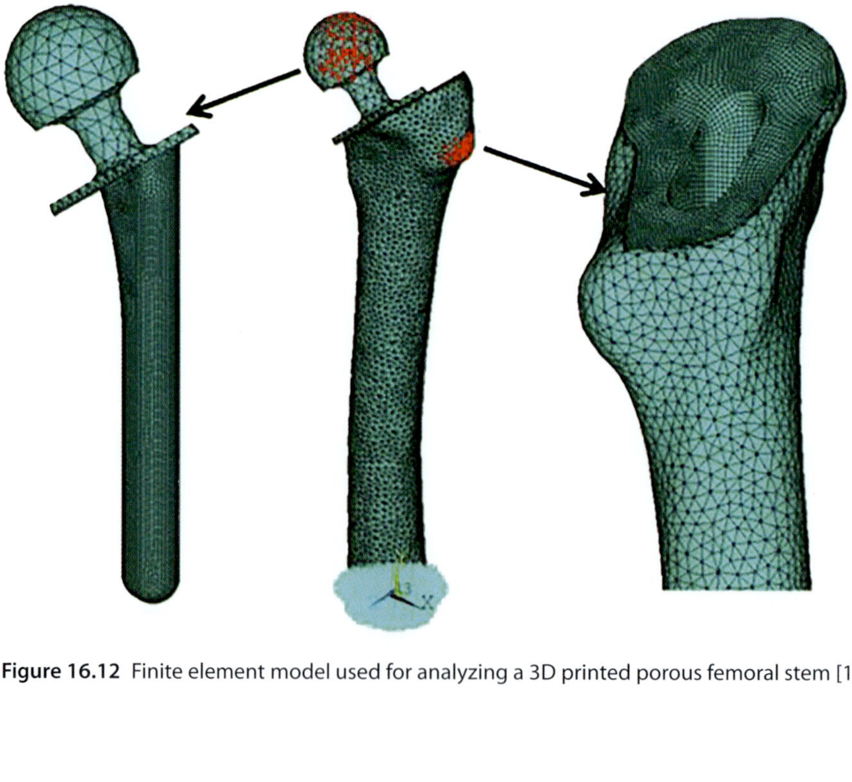

Figure 16.12 Finite element model used for analyzing a 3D printed porous femoral stem [12].

Simoneau et al. (2017) [10] studied a porous metallic biomimetic femoral stem. Both the finite element method and experimental tests were utilized. Digital image correlation technique was used to obtain the displacement and strain fields during the tests. The finite element model of the dense stem was successfully validated; however, the porous stem showed a much higher stiffness from the numerical analysis when compared to the experimental test. It was reported that there were discrepancies between the experimentally measured and numerically targeted porosities.

Although the finite element method is very powerful, there are some limitations that need to be considered when the numerical results are interpreted. For example, the bone and implant material properties may be oversimplified because the properties of biological materials are difficult to obtain. However, to summarize, finite element analysis still provides a very effective tool for the comparative or parametric studies of 3D printed orthopedic implants.

16.3 3D Printed Orthopedic Implant Installation and Instrumentation

3D printing is used to create models of anatomy for preoperative planning. This preoperative 3D modeling and printing process allows the surgeon to effectively diagnose and plan the surgical procedure with an increased level of communication to the patient. For instance, plates can be contoured over the model,

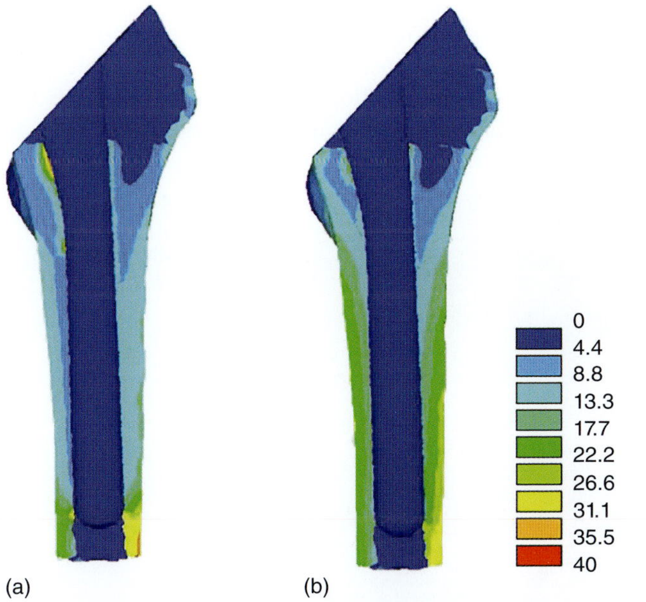

Figure 16.13 Von Mises stress (MPa) in the proximal femur: results from a fully dense femoral stem (a) and a fully porous stem (b). [12].

therefore reducing operative time and increasing the accuracy of the anatomical restoration.

In addition to producing 3D printed customized implants, the technology has been used to print patient-specific instrumentation (PSI) for surgical purposes. Chana-Rodriguez et al. [24] presented a good example of using 3D printing in the preoperative planning of acetabular fractures. A mirror image 3D model of the opposite hemipelvis and precontouring plates over the mold were created, and this led to the achievement of the predefined surgical objective.

Patients' knee anatomies are individual. Given this, the surface geometry of a knee implant affects joint congruence and contact mechanics. To reduce the surgical time and ensure good alignment, Patil et al. [25] suggested that fully customized 3D printed knee replacements, cutting measures, and fitting templates could provide an improved solution. Both MRI and CT data can be used for the creation of patient-specific guides (PSG). Ganapathi [26] suggested using PSG to replace traditional jigs in performing total knee replacement surgery. To produce PSG, computerized 3D models of the distal femur and proximal tibia were created and negative molds of the patient's distal femur and proximal tibia were generated. Ganapathi [26] showed that the advantage of PSG is notable in terms of adequate fit and accuracy of PSG. However, Goyal and Tripathy [27] surveyed the functional outcomes of total knee replacement using PSIs. It was expressed that the PSI is not a patient-matched implant and that the main focus of implant design should be in fact creating the patient-matched implant.

Figures 16.14 and 16.15 show PSI for total knee replacement surgery. These PSI were designed and 3D printed to improve the surgical outcome of the total knee replacement.

Chen et al. [2] indicated that the precision of 3D printed customized implant installation is important and may be difficult to achieve because of anatomic intricacy. The work showed that a surgical navigation system can be used for the tumor resection and to guide the customized implant intraoperative installation.

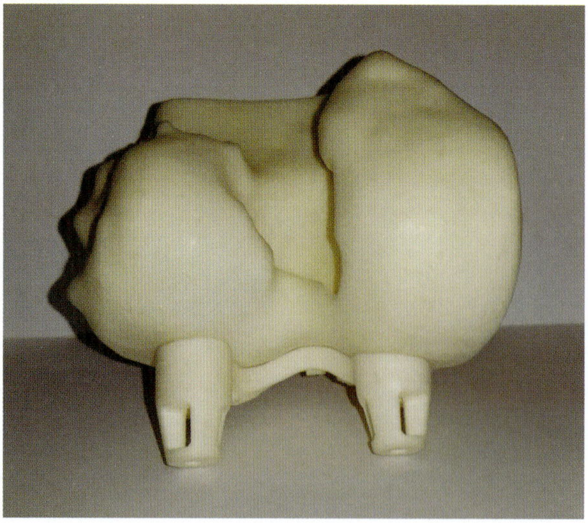

Figure 16.14 Femoral distal end personal specific instrumentations for knee joint replacement.

Figure 16.15 Tibial proximal end personal specific instrumentations for knee joint replacement.

16.4 Orthopedic Implants Manufactured with 4D Printing

Currently, orthopedic implants do not change shape, apart from the wear of the articular surface while in use. However, the host bone and the interface between the implant and bone is constantly remodeling. 4D printing aims to use stimuli-responsive materials to create 3D active structures, which can transform their shapes or behavior under various stimuli such as humidity, temperature, electric and magnetic field, and light. The 3D/4D printing could be used to print bone graft materials used in orthopedics. 4D printing is similar to 3D printing technology but uses more advanced materials and digital designs. Given this, the product is able to change its shape or functionality. For example, by incorporating shape memory polymer fibers into composite materials in 3D printing, when the product is heated or cooled to a certain temperature, it will transform into a different shape. 4D printing could be used to print orthopedic implants using shape memory polymers, which not only fit the anatomy of patients but also apply forces to the host bone through shape changing. Andani et al. [28] studied the mechanical and shape memory properties of porous NiTi alloys manufactured by SLM. It was observed that porous NiTi structures have a lower elastic modulus and density than dense NiTi but still have good shape memory properties, thus creating a promising material for biomedical implants. Superelastic NiTi and resorbable magnesium-based alloys could also be developed for 4D printing and other metamaterials will also have potential in future orthopedic applications.

16.5 Summary

3D printing can be used to create complex orthopedic implants from digital files that are developed from medical images. It has the potential to be used in day-to-day clinics for patient-specific solutions. 3D printed customized implants can be used to improve surgical precision and patient outcomes. However, such products must be subjected to regulatory approval, before commercialization. This may require a different approach or additional testing when compared to traditionally manufactured implants. Although 3D printing of medical devices has advanced recently, the clinical use is still not widespread. This is because 3D printing still possesses complexities that are not yet fully understood. The design and analysis of 3D printed orthopedic implants remains a challenging problem because of the variation of porous structures and the variability of the 3D printing process itself. It will require surgeons, engineers, and radiologists to work together to realize the full potential of 3D and 4D printing in orthopedics and other medical applications.

References

1 National Joint Registry (2017). 14th annual report of the National Joint Registry for England, Wales, Northern Ireland and Isle of Man [online]. http://www.njrreports.org.uk (accessed 10 October 2017).

2 Chen, X., Xu, L., Wang, Y. et al. (2016). Image-guided installation of 3D-printed patient-specific implant and its application in pelvic tumor resection and reconstruction surgery. *Computer Methods and Programs in Biomedicine* 125: 66–78.

3 Hazlehurst, K., Wang, C.J., and Stanford, M. (2013). The potential application of a cobalt chrome molybdenum femoral stem with functionally graded orthotropic structures manufactured using direct metal laser sintering. *Journal of Medical Hypotheses* 81 (6): 1096–1099.

4 Wang, X., Xu, S., Zhou, S. et al. (2016). Topological design and additive manufacturing of porous metals for bone scaffolds and orthopaedic implants: a review. *Biomaterials* 83: 127–141.

5 Tan, X.P., Tan, Y.J., Chow, C.S.L. et al. (2017). Metallic powder-bed based 3D printing of cellular scaffolds for orthopaedic implants: a state-of-art review on manufacturing, topological design, mechanical properties and biocompatibility. *Materials Science and Engineering C* 76: 1328–1343.

6 Wang, Z., Wang, C., Li, C. et al. (2017). Analysis of factors influencing bone ingrowth into three-dimensional printed porous metal scaffolds: a review. *Journal of Alloys and Compounds* 717: 271–285.

7 Markhoff, J., Wieding, J., Weissmann, V. et al. (2005). Influence of different three-dimensional open porous titanium scaffold designs on human osteoblasts behavior in static and dynamic cell investigations. *Materials* 8 (8): 5490–5507.

8 Taniguchi, N., Fujibayashi, S., Takemoto, M. et al. (2016). Effect of pore size on bone ingrowth into porous titanium implants fabricated by additive manufacturing: an in vivo experiment. *Material Science Engineering C* 59: 690–701.

9 Otsuki, B., Takemoto, M., Fujibayashi, S. et al. (2006). Pore throat size and connectivity determine bone and tissue ingrowth into porous implants:

three-dimensional micro-CT based structural analyses of porous bioactive titanium implants. *Biomaterials* 27 (35): 5892–5900.
10 Simoneau, C., Terriault, P., Jette, B. et al. (2017). Development of a porous metallic femoral stem: design, manufacturing, simulation and mechanical testing. *Materials and Design* 114: 546–556.
11 Arabnejad, S., Johnston, B., Tanzer, M., and Pasini, D. (2017). Fully porous 3D printed titanium femoral stem to reduce stress-shielding following total hip replacement. *Journal of Orthopaedic Research* 35 (8): 1774–1783.
12 Hazlehurst, K. (2014). The adoption of laser melting technology for the manufacture of functionally graded cobalt chrome alloy femoral stems. PhD Dissertation. University of Wolverhampton.
13 Croitoru, A.S.M., Pacioga, B.A., and Comsa, C.S. (2017). Personalized hip implants manufacturing and testing. *Applied Surface Science* 417: 256–261.
14 FDA (2016). Technical considerations for additive manufactured devices, draft guidance for industry and food and drug administration staff (draft guidance). https://www.fda.gov/downloads/MedicalDevices/DeviceRegulationandGuidance/GuidanceDocuments/UCM499809.pdf (accessed 25 June 2018).
15 Hazlehurst, K., Wang, C.J., and Stanford, M. (2013). Evaluation of the stiffness characteristics of CoCrMo cellular structures manufactured using laser melting technology for potential orthopaedic applications. *Journal of Materials and Design* 51: 949–955.
16 Wang, L., Kang, J., Sun, C. et al. (2017). Mapping porous microstructures to yield desired mechanical properties for application in 3D printed bone scaffolds and orthopaedic implants. *Materials and Design* 133: 62–68.
17 Limmahakhun, A., Oloyede, A., Sitthiseripratip, K. et al. (2017). Stiffness and strength tailoring of cobalt chromium graded cellular structures for stress-shielding reduction. *Material and Design* 114: 633–641.
18 Li, S.J., Murr, L.E., Cheng, X.Y. et al. (2012). Compression fatigue behaviour of Ti–6Al–4V mesh arrays fabricated by electron beam melting. *Acta Materialia* 60 (3): 793–802.
19 Brenne, F., Niendorf, T., and Maier, H.J. (2013). Additively manufactured cellular structures: impact of microstructure and local strains on the monotonic and cyclic behaviour under uniaxial and bending load. *Journal of Materials Processing Technology* 213 (9): 1558–1564.
20 Roberts, I.A., Wang, C.J., Esterlein, R. et al. (2009). A three-dimensional finite element analysis of temperature field during laser melting of metal powders in additive layer manufacturing. *International Journal of Machine Tools and Manufacture* 49: 916–923.
21 Hazlehurst, K., Wang, C.J., and Stanford, M. (2014). An investigation into the flexural characteristics of functionally graded cobalt chrome femoral stems manufactured using laser melting technology. *Materials and Design* 60: 177–183.
22 Taylor, M. and Prendergast, P.J. (2015). Four decades of finite element analysis of orthopaedic devices: where are we now and what are the opportunities? *Journal of Biomechanics* 48 (5): 767–778.

23 Hazlehurst, K., Wang, C.J., and Stanford, M. (2014). A numerical investigation into the influence of the properties of cobalt chrome cellular structures on the load transfer to the periprosthetic femur following total hip arthroplasty. *Medical Engineering and Physics* 36: 458–466.

24 Chana-Rodriguez, F., Mananes, R.P., Rojo-Manaute, J. et al. (2016). 3D surgical printing and pre-contoured plates for acetabular fractures. *Injury: International Journal of the Care of the Injured* 47: 2507–2511.

25 Patil, S., Bunn, A., Bugbee, W.D. et al. (2015). Patient-specific implants with custom cutting blocks better approximate natural knee kinematics than standard TKA without custom cutting blocks. *The Knee* 22: 624–629.

26 Ganapathi, M. (2014). Patient specific guides for total knee replacements – a review. *Orthopaedic Trauma* 28 (5): 315–321.

27 Goyal, T. and Tripathy, S.K. (2016). Does patient specific instrumentations improve short-term functional outcomes after total knee replacement? A systematic review and meta-analysis. *The Journal of Arthroplasty* 31: 2173–2180.

28 Andani, M.T., Saedi, S., Turabi, A.S. et al. (2017). Mechanical and shape memory properties of Porous Ni50.1Ti49.9 alloys manufactured by selective laser melting. *Journal of the Mechanical Behavior of Biomedical Materials* 68: 224–231.

17

Recent Innovations in Additive Manufacturing Across Industries: 3D Printed Products and FDA's Perspectives

Brett Rust[1], Olga Tsaponina[2], and Mohammed Maniruzzaman[2]

[1] *3D Printing Engineering (3DPE), Sussex Innovation Centre, Science Park Square, Falmer BN1 9SB, UK*
[2] *University of Sussex, School of Life Sciences, Department of Pharmacy/Chemistry, John Maynard Smith (JMS) Building, Falmer, Brighton, BN1 9QJ, UK*

17.1 Introduction

3D printing, also known as additive manufacturing (AM), is a process in which layers of material are combined to produce a three-dimensional object [1–6]. These layers are generated by slicing up a 3D model to create a stereolithograpy (stl) file, which is the native format read and processed by most 3D printers today. Chuck Hull invented it back in 1984, and today, he is still the ever-present vice president and CTO of *3D Systems*, alongside *Stratasys* in the USA, two of the biggest 3D printing companies in the world [7]. The technology, processes, and materials used have come a long way since the initial groundbreaking idea to form objects in layers using UV light to cure material. Additive processes (material on/added) offer different benefits to subtractive processes (material off/removed such as computer-navigated cutting (CNC) and laser cutting), and this will be detailed further later in this chapter. First, we will explain the methods currently employed in the industry today to fabricate 3D objects via 3D printing followed by regulatory aspects.

17.2 Current Widely Used Processes Across Industries

A brief introduction to the most commonly used forms of 3D printing with a particular focus on the practical applications is provided in the following sections.

17.2.1 Fused Deposition Modeling (FDM)

The fused deposition modeling (FDM) works by depositing layers of material filament via extrusion through a nozzle onto a flat printing bed. This is the most economical and accessible technique and makes up the leading portion of the 3D printer/production market. This process was invented and patented by Scott Crump in 1989, and just like Chuck Hull did with 3D systems, Crump jumped out the gates as a pioneer of this new technology and cofounded Stratasys with his

3D and 4D Printing in Biomedical Applications: Process Engineering and Additive Manufacturing,
First Edition. Edited by Mohammed Maniruzzaman.

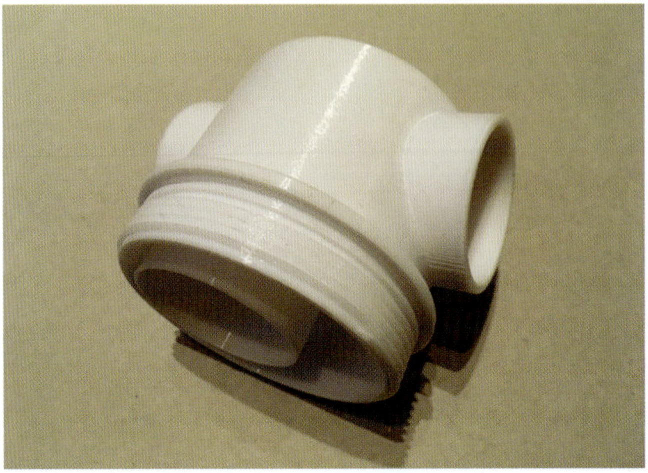

Figure 17.1 PC pipe section made with FDM process.

wife Lisa to lead the way as these two dominant companies are still leading in the industry [8]. FDM processing has become a popular choice due to the functional use of polymers in filaments that allow robust prototypes and end-use parts to be made. Low filament costs and bigger build layers lower the cost of 3D printing and prototyping.

FDM processing yields a slightly grained finish and this is typical of parts built at 200–300 μm. The pipe section shown in Figure 17.1 is a great example of not just a typical FDM part but also used in 3D printing to create a part that could not be made by any other way in small quantities. Due to the complex multi tool design required to mold this part, significant quantities would be required to make it cost viable over 3D printing.

Chuck Hull had already secured a broad patent for stereolithography (SLA) in1986 under "material capable of solidification," which covers many processes currently employed today to create 3D objects with additive manufacturing. Before Hull's breakthrough with FDM, these two innovators continued to expand their businesses in the late 1980s and early 1990s, which lead to the emergence of new materials and processes such as selected laser sintering (SLS) and color jet printing (CJP) .

17.2.2 Stereolithography (SLA) and Digital Light Processing (DLP)

SLA and digital light processing (DLP) use photopolymerization by light to cure liquid resin forming polymers. By linking section data in multiple layers, 3D models can be formed. Machinery is similar to FDM in the way it has led to the development of desktop and professional machines and is a rapidly growing part of the printing market taking a 15% revenue share [9]. This technology and usage has grown steadily over the last five years because of its incredible surface detail and processing speed. Resolution on 3D printed parts using SLA

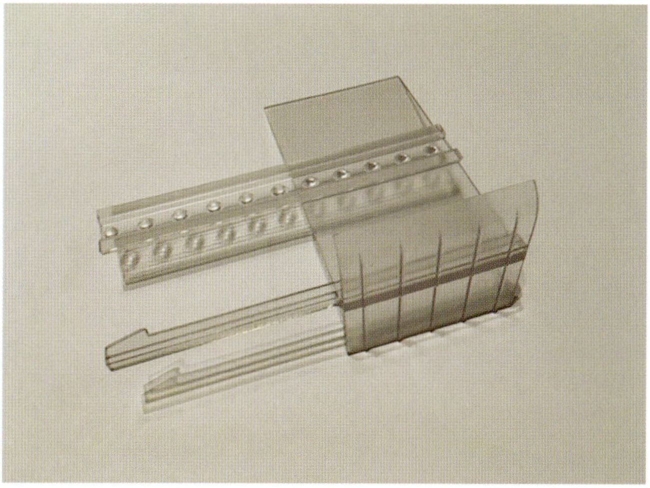

Figure 17.2 SLA tray section, slightly translucent in its natural state.

and DLP processing can go as low as 20–25 μm to provide surface finishing that other 3D printing processing cannot match. Typically, FDM can be processed at 100 μm resolution, but the layering methods increase the build time and therefore costs, making it a less viable choice if you want a better quality surface finish more akin to a CNC machined or cast/molded part. Because of its ability to process fine details, the SLA and DLP processing with resins are used for prototyping and creating patterns and master models for items such as jewelry and figurines. More recently, processing with resins has improved to offer very high-quality translucent (and transparent with polishing) parts such as the tray sample section shown in Figure 17.2.

17.2.3 Selective Laser Sintering (SLS)

SLS forms 3D objects in a chamber of powder where the print layers are fused together by the heat from a high-powered laser. A range of high-performance powders, including nylon, carbon, and graphite, are used in this process; additionally, there are further possibilities with metals using slight variations of the process such as direct metal laser sintering (DMLS) to make parts in aluminum and titanium. The advantage of building 3D objects in this way is the powder bed that acts as a support for additional layers to be built upon. Once the build is complete and the part is dry and set, it can be carefully removed and depowdered with compressed air. FDM and SLA/DLP processes require supports that act as a scaffold to build the model that must be carefully removed. However, in recent years, with professional SLA printers such as the 3D Systems Projet series and now on desktop with machines such as the Ultimaker 3, dissolvable supports remove the printing supports effortless. A good example of this process is the impeller prototype, 100 μm resolution (Figure 17.3).

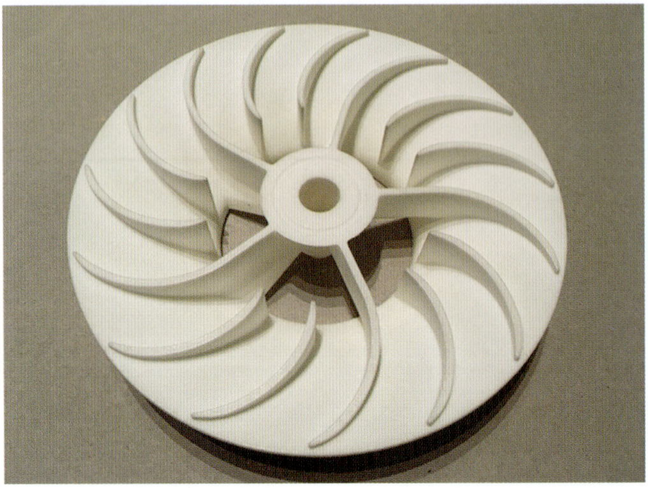

Figure 17.3 Nylon SLS impeller.

17.3 Emerging 3D Printing Processes and Technologies

Some variations to the established 3D printing processes have broken through from newcomers to the market in such as *Carbon 3D* and well known from their printing and computing background *Hewlett Packard* with their eagerly awaited machine debuts this year.

17.3.1 Continuous Liquid Interface Production (CLIP)

Continuous liquid interface production (CLIP) is a breakthrough process that uses "digital light projection, oxygen permeable optics, and programmable liquid resins to produce parts with excellent mechanical properties, resolution, and surface finish" [10]. This process is the first of its kind to build 3D objects continuously as opposed to all the previous processes discussed above built by sliced layers stacked on top of each other. *Carbon 3D*, the inventors of this novel process founded their company in 2014 that brought unique materials to the 3D printing market such as silicone and polyurethane, which was previously formed by traditional casting and molding processes.

17.3.2 Multi Jet Fusion (MJF)

Multi jet fusion (MJF) is a "unique, hybrid methodology that employs printheads to first deposit powdered material onto a build plate. MJF then adds fusing and detailing agents into the mix at the voxel (3D Pixel) level" [11].

A voxel, as opposed to a 2d pixel, contains three-dimensional volumetric information and is a format being used more often for design and preparation of 3D print files to tailor properties for performance and visual characteristics of parts. This will allow, for example, flexible, solid sections similar to Objet technologies with the addition of full color. The *Dunewald's* diagram and summary below further explain the novel process (Figure 17.4):

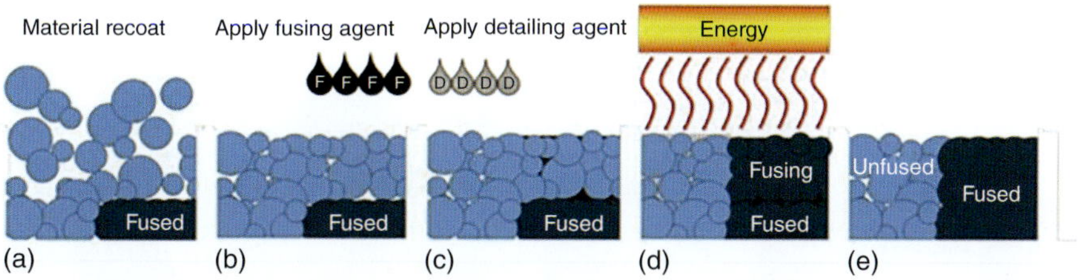

Figure 17.4 MJF process illustration.

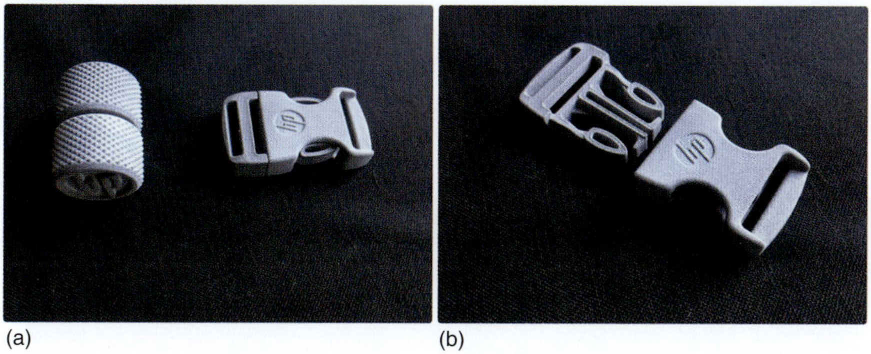

Figure 17.5 Mating part examples built with HP MJF technologies.

Two separate MJF thermal inkjet arrays will work together to build the full-color three-dimensional part. One array will lay down the basic building blocks and structure of the part while the second array combines the coating, color, and fusing steps that will solidify and give the part the desired strength and texture. In our business, we have just been able to see the benefits of this technology with the cost of parts less than SLS process largely because of the increase in build speed but also offering improvement in surface finish with 80 μm resolution. Figure 17.5 shows a snap fit clip and a threaded screw assembled on the left and on the right, the snap fit clip being open.

17.4 Industry Uses of Additive Manufacturing Technologies

3D printing has now established itself, among other traditional manufacturing processes and as highlighted with CLIP and MJF, as a new processing technique emerged from the previous existing technologies that will drive market shares higher. Frost and Sullivan forecast the big three industries, aerospace, automotive, and medical, to account for 51% of the 3D printing market by 2025. The remaining industrial uses and growth rate are detailed in Figure 17.6 [12].

This rise of 3D printing has led to improved offerings from traditionally used methods such as CNC and laser cutting to compete. New hybrid technologies where machinery combines additive and subtractive operations are providing value and engineering solutions, combining the best of both processes. A good

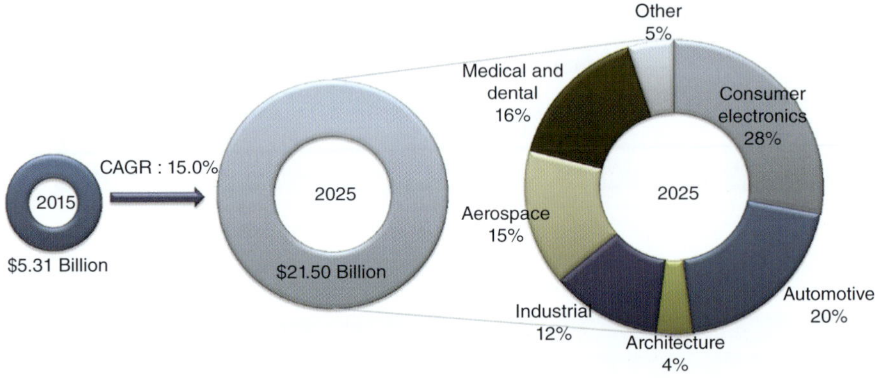

Figure 17.6 Market projection of 3D printing across the industries. Source: Moulton et al. 2014 [6]. Reproduced with permission of Elsevier.

Figure 17.7 The Ethereal Halo and sample component.

example of this is the award-winning Ethereal Halo that offers five-axis 3D printing/CNC milling hybrid manufacturing. Ethereal provided CNC machines but looked to capitalize on weaknesses in 3D printing processes. Within 3D printing, CLIP and MJF have produced counter processes to produce parts with continuous formation. Ethereal's 5D hybrid processing allows building from different angles as opposed to stacked layers making them much stronger. Figure 17.7 shows an example of a 5D processed part on the machine.

This spiral can be built with layers stacking at 45° instead of directly in the Z axis and machined with five-axis capability where it would not be possible without splitting on three-axis CNC.

17.5 Material and Processes for Medical and Motorsport Sectors

In our business (3DPE), through Research and Development (R&D), live projects for customers and with the support from development partners we have utilized the many benefits of 3D AM technologies. Our focus is to develop our business into market sectors that have the best opportunities and where we have our core expertise. Two of these sectors are medical and motorsport, which utilize our materials and process knowledge with thermoplastics and composites. In motorsport, AM supports the rapid continuous development of parts, such as wind tunnel test parts and on-track race car parts. As the surface quality, structure, weight, and cost improve on AM parts, they will become more common on racing cars. Within the medical sector, the big driver for AM is personalized parts such as implants, surgical aids such as jigs and fixtures, or even 3D models to help walk-through complex surgeries. Within 3D printing, the medical devices sector is expected to grow by 23% year-on-year between 2015 and 2025 [12].

We put this to good use recently when 3DPE was contacted by a pharmaceutical company earlier this year looking to produce personalized tablet molds. As part of their product release to the market, they must test products for identification (ID) among other tests. A quality assurance analyst from the company details the requirements and highlights where 3D printing is of great benefit:

'For IT tests of some samples, we conducted a near-infrared (NIR) analysis on the tablets. For this analysis, an NIR beam leaves its source, hits the sample, part of the radiation is absorbed, part is reflected, and the remaining part is scattered. Using the instrument, we have in place, the diffusely reflected beam from the sample is directed onto an internal detector and the components absorbed by the sample can then be measured.

We are in the process of installing a NIR capability, and part of it will be used as a tablet autosampler (TA). The TA is a generic of the shelf accessory. This means the fit between the tablets and the TA was far from perfect, and during testing, the energy throughput was not perfect in addition to the unmeasurable amount of stray light. As a solution for this problem, the TA manufacturer prepared a kit that would allow us to create tablet molds to create a tight seal between the tablet and the TA. This was a very good solution, if you need to make up one, two or three tablet molds. However, when there was a need for 100 molds, the process to make our own molds became time-consuming, and a lack of reproducibility and presentation among the molds was identified.

This is when the option for obtaining 3D print of one of the molds was considered. Working iteratively with the client 3DPE was able to quickly achieve a successful outcome for the client. Essential ingredients to this outcome were good and open communication, trial and error, and a philosophy of continuous improvement.

The tablet molds look very professional and the clients are hopeful that they will deliver reproducible results. This is essential if they want to deploy a robust system in addition to something that will look very good in front of auditors.

(a) (b)

Figure 17.8 Design and prototype prints. Dimensions engraved were the base thicknesses.

(a) (b)

Figure 17.9 Graphite water outlet and carbon SLS brake duct.

Figure 17.8 shows the process of computer-aid design (CAD), engraving, and final use.

In motorsport, carbon fiber composites are still extensively used to deliver the optimum stiffness and strength at the lowest weight possible for parts. Our development partners have developed a number of novel processes and engineering solutions for customers in motorsport. Using SLS processing, they introduce graphite and carbon powder to provide enhanced material properties for the development of functional parts. Two good examples of this (Figure 17.9) are made with graphite and carbon SLS.

The carbon brake duct is a particularly relevant race car component for AM processing because of its complex shape that, if made with tooling, requires multiple sections and careful design to ensure that the part can be laminated in carbon fiber on the tool and demolded after the part is cured in an autoclave. One additional feature to this brake duct is its simulated honeycomb structure that can be customized and is visible here because of the exposed section. Again, foam or honeycomb is encapsulated by carbon fiber layers in conventional molding process to reduce weight further and retain stiffness and strength, but this infill structure can be simulated and tailored in AM processing to achieve

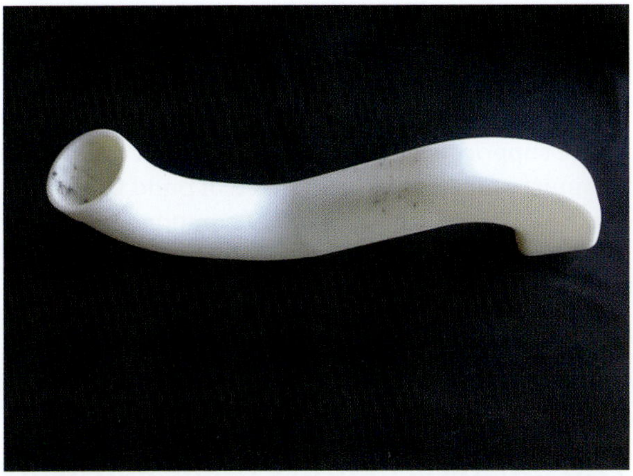

Figure 17.10 SR30 "Washout mandrel".

similar results to make this complex part inexpensive and time-consuming in toolmaking. AM is finding ways to replace and complement conventional manufacturing processes. In some instances, a brake duct may be preferable to be made by molding, if say a batch of 5, 10 parts are required or the part was long and outside optimum build envelope of the machine. In this instance, producing a sacrificial tool that can be wrapped with carbon and dissolved later allows very complex duct-like forms to be created. Another development partner provides these sacrificial tools, known in industry as "washout cores," using FDM processing in SR30 material. Figure 17.10 shows the typical twisty duct sections where washout cores are used.

Tooling for composite manufacturing, particularly for larger parts where it may be cost prohibitive to use AM, is often used for cycles of 50+ parts. However, AM can still be used to produce a tool to mold composites for 1–10 parts typically. One advanced research engineer summarizes how AM tooling has benefitted composite manufacturing.

Most SMEs do not spend large amounts of money on metallic tooling for producing a handful of parts; the use of ALM allows not just the F1 teams to make low-volume–high-quality parts but also allows anyone that has access to the materials and processes for producing composite components. The benefits of using ALM over other manufacturing methods are the designer/toolmakers can reduce the overall amount of material in the tool, making it easier to handle and lay up plus the heat soak of the material when curing in the oven/autoclave is reduced as there is less material to heat up, hence the overall curing time can be reduced. Until recently, the limiting factor has been finding suitable materials that can withstand the heat and pressures of autoclave curing and be large enough for industrial applications. The development partners have developed AM processing, this time with a ceramic SLA tooling product that can withstand operating temperatures of up to 200 °C.

We have seen other development partners starting with SLA go on to have a wide variety of different processes including recent expansion into metal additive manufacturing. At present, aluminum powder is being used, but the aim is to process using stainless steel, titanium, or Inconel in the near future. Some of the benefits and flexibility of using AM are explained in the following sections by a colleague.

This technology opens up many areas for designers to consider, from adding rads or fillets to parts to relieve stress while building orientation of build and where support needs to be added. This technology has huge advantages such as minimizing material waste, speed of build vs machining, cost of build vs machining, being able to fit a number of components on one build platform, and relative design freedom, although this can sometimes be a false statement as there will always be some kind of restraint depending on the geometry or complexity of the part, which, in turn, may have to be tweaked a little to maximize the build process if possible.

17.6 Medical Industry Usage and Materials Development

Opportunities to use AM for medical applications can come from partnerships. We have seen great progress by working collaboratively with other specialists. An example of this is 3D printed scaffolds as shown in Figure 17.11. Being able to adjust the design and produce small batches with varying pore sizes make this a more development-friendly technical solution as opposed to injection molding, where batches in thousands would be required because of tooling costs.

Progress is being made in using blended materials that could be used for making other medical devices, and this in combination with making them on desktop 3D printers can lower the cost of producing personalized medical devices, making them more accessible for patients. *Gartner* indicates that 10% of people in the

(a) (b)

Figure 17.11 3D scaffold prints in PLA material.

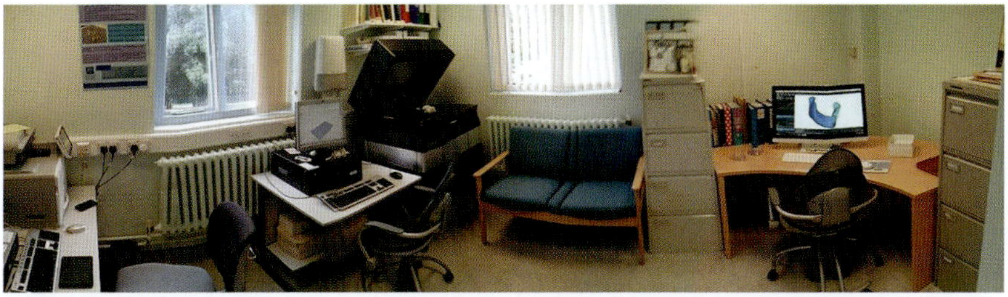

Figure 17.12 Digital manufacturing suite at North Manchester General Hospital.

Figure 17.13 The new Objet 30 3D printer in action.

developing world will be living with 3D printed parts on or in their bodies by 2019 [13].

Our recent R&D endeavors and interest in the development of FDM materials and processes for medical devices brought us into contact with Oliver Burley, a reconstructive scientist/maxillofacial laboratory manager at the North Manchester General Hospital where they have recently invested in a Stratasys Objet 30 3D printer and Materialize 3D surgical planning software. This complete in-house, specialist "design for printing" and manufacturing room (as shown in Figures 17.12 and 17.13) will be used to directly treat patients. This marks a clear step forward and an opportunity for AM technologies to enter into the National Health Service (NHS) with other hospitals in Bath and Bristol following suit.

Oliver further explains how the new suite will be used: "The 3D printing service will be run by the team of reconstructive scientists and dental technicians. Its main purpose is to assist the oral and maxillofacial consultants in treating and restoring patients who have undergone resection for the management of disease (predominantly cancer), suffered facial trauma, or have congenital anomalies. This requires presurgical planning to facilitate accurate plates and bars that are designed specifically for individual patients by matching their unique anatomy. Previously, we would have had to create a model of the device by using external companies that typically took in excess of two weeks. Now, the 3D printer is based in-house; we are able to design and manufacture a variety of medical devices within hours."

It is great to see AM being used for the treatment of patients, but until material ranges expand and regulations are in place, it will be mostly used in the NHS for external uses. Currently, PEEK and titanium are the go to options for medical grade materials but have high material and processing costs. An example below shows the characteristics required for Maxillofacial Deep Buried Facial Implants, provided by Oliver again and highlights the challenges in-store for AM companies and the medical sector as a whole.

- Biocompatible, radiopaque, be chemically inactive, withstand fibrous proliferation
- Fully functional so not impede, limit patient, maintain corrosion resistance, withstand external stress
- Ability to sterilize using hospital-based processes – must be able to sterilize without damaging the implant and must have microorganisms equal to or less than 1×10^{-6}.
- Ability to fixate in position to bone – titanium fixation plates and screws are common; however, screws would only be advantageous.
- Readily available within a hospital setting – to be attainable for clinicians within the NHS in terms of time, manufacturability, and cost.
- Be designable using medical-based software applications.
- If a material solution can be provided with all these properties, it can be fully regulated for use and be affordable that it would be a serious option for use in the NHS.

Another rapidly growing market for personalized 3D printing parts is the dental industry. There are some very good examples of the tailored solutions with materials and machinery being developed specifically for dental applications. Objet has 3D printers that can be configured purely for dental use, and in the last 2 years, Formlabs has introduced a range of specialist resins to be used for dental requirements. One of our customers has developed a novel 3D scanning solution for inside the mouth with software that generates a stl file ready for printing. The aim for the dental technician is to complete this process all via a mobile phone application and then the resulting stl file can then be used to produce a personalized mouthguard (Figure 17.14). The material properties are very close to the desired end properties for the mouthguard.

More softness, elasticity, and the material (or surface treatment) to be medical grade/approved for intended use will be required for these products to be sold in the market and available to the consumers. These areas, where on-demand material solutions are required, are the focus of our materials development at 3D printing engineering as we enter exciting times with the continued expansion of materials and processes within additive manufacturing.

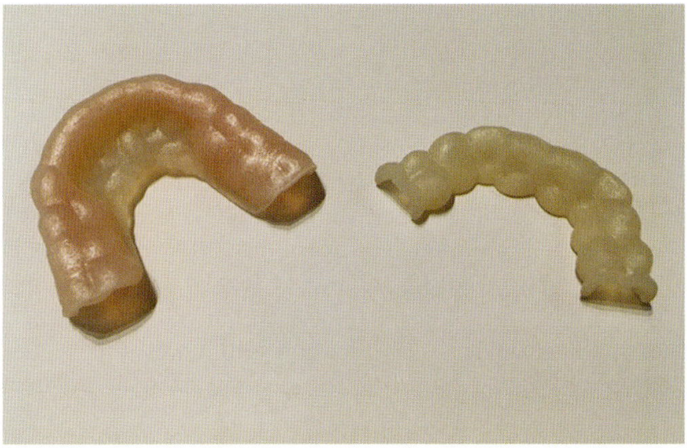

Figure 17.14 3D printed mouthguard.

17.7 3D Printing of Medical Devices: FDA's Perspectives

A number of additive manufacturing processes are currently used in industry and research. One of the most fascinating additive manufacturing processes is a 3D printing, allowing fast and efficient production of customized products at the point of need. Often, 3D printing and additional manufacturing terms are used reciprocally; therefore, here, we will refer to both concepts as 3D printing. 3D printing is a creation of three-dimensional objects by adding sequential layers of corresponding material(s). Objects (shapes) are designed using magnetic resonance image (MRI) or CAD software. Next, the digital design is sent to a 3D printer, which produces the object by adding new successive layers to the previous one until target shape is achieved [14–19].

Versatility and a potential of 3D printing is hard to overestimate, especially in regard to tailored (or customized) medicine. Flexibility of 3D printing allows on-site manufacturing of medical devices and implants tailored to patient's anatomy and individual features. Objects with highly complex internal structure and advanced shapes can be easily manufactured as well. Such medical devices and parts include, but not limited to, orthopedic, cranial, and dental replacements, tissues with blood vessels, surgical instruments, prosthetic parts, and, even, organs. Not only medical devices can be manufactured by 3D technology but also automotive parts, food products, household items, beauty products, and many others are can be printed using the technology. Flexibility of 3D printing allows a reasonably quick and easy way to adjust the design without need of additional tools or equipment. Such versatility generated a massive interest in 3D printing and its application.

17.7.1 FDA's Role in 3D Printing of Materials

The role of FDA's Center for Devices and Radiological Health (CRDH) is to regulate companies that manufacture, relabel, repackage, and/or import any medical devices sold in the USA [16].

Medical devices manufactured by 3D printing technology, such as medical objects produced by any other technology, are subject to regulation by the FDA. Some medical devices are subject to regulatory requirements before their marketing (premarket requirements), and other medical devices are subject to regulatory requirements after their marketing (postmarket requirements). Medical devices are classified as Class I, Class II, or Class III, and the regulatory control intensifies from Class I to Class III. According to the FDA, most Class I medical devices are an exempt of Premarket Notification 510(k); most Class II devices require Premarket Notification 510(k); and most Class III devices require Premarket Approval (PMA).

Detailed description of medical devices classification and a web link to the product classification database can be found at the classification of medical devices [20].

In 2016, the FDA published a draft guidance for manufacturers using 3D printing techniques to advice on Technical Considerations for Additive Manufactured Devices [21]. The guidance is not in effect at a current moment and is not binding; it has been publish as a consultation to obtain public view. Currently, FDA assessed applications for new medical devices printed by 3D technology to determine their safety and effectiveness. The 2016 FDA draft offers a guidance to manufacturers of 3D printed medical devices regarding their design, manufacturing, and testing. The requirement for premarket submission of new 3D-printed product is still determined by its regulatory classification [21].

The draft guidance largely consists of two subject areas:

1. *Fabrication considerations*: This section offers a guidance on technical review of new device production in order to fulfill quality systems (QS) requirements. QS requirements are dictated by the device's regulatory classification (see above), if applicable. The FDA draft guide is not intended to comprehensively address all issues or regulatory requirements for device production, while offering a number of considerations for 3D manufacturing process.
2. *Testing considerations*: According to FDA's guide, this section defines what type of information should be submitted in premarket notification submission (510(k)), PMA application, humanitarian exemption (HDE) applications, *de novo* requests, and investigational device exemption (IDE) for devices printed with 3D technology.

17.7.2 Classifications of Medical Devices from FDA's Viewpoint

Approximately 17 000 different generic types of 3D printed devices have been classified by FDA and grouped into 16 medical specialties, namely panels. Based on the level of control required to ensure the device' safety and efficiency, each of the 17 000 generic types was assigned to one of three regulatory classes [16–21]. The three classes and applicable regulatory requirements are as follows:

FDA's Device Class and Regulatory Controls:

1. Class I General Controls
   - With exemptions
   - Without exemptions

2. Class II General Controls and Special Controls
   - With exemptions
   - Without exemptions
3. Class III General Controls and PMA

   The type of premarketing submission/application necessary for FDA clearance is determined by which regulatory class a 3D printed medical device is assigned to. If the device is classified as Class I or II, and if it is not exempt, a 510(k) application will be required before marketing. According to the FDA, all devices are categorized as Class I or Class II. With exemption are subject to limitations on exemptions. These limitations are covered under 21 CFR xxx.9 (Code of Federal Regulations), where xxx denotes Parts 862–892. For Class III medical device, a PMA application will be necessary, unless the device is a pre-amendment device and did not have PMA called for. The pre-amendment devices are as follows:
   - Have been marketed before Medical Device Amendments approval in 1976 OR
   - Are largely equivalent to a device marketed prior Medical Device Amendments (1976)

Otherwise, Class III medical device will require 510(k) before marketing.

As established, the device classification is dictated by its *intended use* and, as well, by *indications* for its use. FDA offers another example to clarify these terms: "For example, a scalpel's intended use is to cut the tissue." A subset of intended use arises when a more specialized indication is added in the device's labeling such as "for making incisions in the cornea." Device labeling clarifies its intended use but may also be communicated verbally during sale of device. The 510(k) Program: Evaluating Substantial Equivalence in Premarket Notification [510(k)] discusses the meaning of intended use of 3D printed medical devices [22]. Classification of a device is risk-based, which means the risk the device can pose to the patient and/or the user is a major reason for class assignment. Medical devices posing lowest risk are classified in Class I, and devices posing highest risk are classified in Class III.

All classes of devices are subject to General Controls, as discussed above. General Controls apply to all medical devices (Class I–Class III) and are the baseline prerequisite of the Food, Drug and Cosmetic (FD&C) Act [22].

17.7.3 Medical Applications of 3D Printing and FDA's Expectations

Modern 3D printers can be used to produce a great variety of medical and other devices, including objects with highly complex geometry that match patient's specific anatomical features. 3D printed medical devices can be either manufactured from the standard design to produce multiple identical copies or can be patient-matched and feature patients' unique anatomy. The latter objects are created using patients imaging data and are highly customized [14].

Commercial medical devices include, but not are limited to

- Instrumentation (e.g. guides to assist with proper surgical placement of a device)
- Implants (e.g. cranial plates or hip joints), and
- External prostheses (e.g. hands).

There is preliminary research in the 3D printing field to explore manufacturing of complex functional organs, such as liver or heart, but these experiments are on the early stage of development.

Several different technologies are used to manufacture 3D objects. Many factors, including intended use of the final product and ease of 3D printer operation, dictate the choice of technology. The most common 3D printing technology is known as powder bed fusion. This technology is used on routine basis as it can work with a wide variety of materials such as titanium and nylon used in medicine.

The powder bed fusion is known as a technique when a three-dimensional object is built from very fine metal or plastic powder. The powder is applied to a platform and very carefully leveled. Next, the powder layer is melted with high precision by the laser or electron beam, and melted material fuses with the previous layer and the powder around. Once full level is complete, the stage retracts down and a new carefully leveled layer of powder is applied on top of the previous one, and the process is repeated again until the final shape is achieved.

To better evaluate 3D technology for medical device manufacturing and the public health benefit, FDA owns and operates a number of 3D printers. Importantly, FDA uses different printing technologies, including powder bed fusion, to ensure quality and safety of finished medical products by assessing critical parts of manufacturing processes and workflows [14–16].

17.7.4 Person-Specific Devices

Within addition to producing multiple identical replicas of an object from a digital file, 3D printing technology can be used to manufacture exclusive medical devices tailored to each patient. A unique feature (such as individual anatomy) of the patient serves as a basis to produce patient-matched (or patient-specific) devices by creating a template model. The matching is achieved by using patient's imaging data and applying digital techniques such as device scaling. 3D printed medical devices are regulated by the FDA through same pathways as traditional medical devices, which means FDA assesses safety and effectiveness according to the information submitted by the manufacturer of the device. Unlike traditional medical devices that are available in increment sizes, patient-matched 3D printed devices can be produced with predefined minimum and maximum specifications in continuous range of shapes. The FDA evaluates if device will maintain its performance for the intended use by reviewing these specifications. For example, they may define thickness and porosity of the wall or a curvature shape.

"Custom" medical devices are an exempt from FDA review according to a provision in Federal Law, but patient-specific medical devices do no automatically meet all the conditions" [14].

17.7.5 Process of 3D Printing of Various Medical Devices

As described above, there are multiple sequential steps to produce a device by 3D printing technology. The number of steps is defined by a number of factors, such as complexity of the device and its size.

These steps include the following:

Design: Digital models matching patient's specific anatomy or with prespecified measurements are used to create a 3D digital design.
Software workflow: A digital design is translated into the buildable file that can be sent to the 3D printer. The software usually divides the desired shape into many layers, incorporates support material(s) to assist printing, and specifies where the object will be built on the printing platform. Also, 3D printer settings usually need changes, such as material type, design type, and intended use.
Material controls: In order to produce consistent high-quality objects, 3D printing, like all other manufacturing processes, requires high-standard materials that meet consistency specification. In order to meet these requirements, the so-called material controls are established through the supply chain between suppliers, purchasers, and end users. Material controls include specific procedures, requirements, and agreements and must be checked for every batch of the material.
Printing: The 3D object is manufactured (printed) according to the design and specifications encoded in the file.
Post-processing: One or more post-processing steps might be required after a 3D object or a part is complete. These steps can include controlled cooling (so-called annealing), cleaning of the object to remove debris and residual material(s), and/or other required steps such as cutting, drilling, polishing, and, often, sterilization using an appropriate technique.
Process validation and verification: Designed features of a finished 3D device or a component can be verified individually to ensure that they meet specifications and will perform accordingly. This is especially true for specifications that can be assessed quickly and not destroying the object, such as geometric shape. If functional features cannot be checked individually because it is unpractical or because of possibility to destroy the object (for instance, mechanical strength test), the 3D manufacturing process has to be validated before the production. The process validation ensures that, while processing specifications are controlled and monitored, the manufacturing process will result in consistent quality object within specified parameters.
Testing: To ensure that 3D printed devices conform to regulatory requirements, sufficiently safe and effective for their designated use, the device's testing procedures and corresponding results are submitted to the FDA.

A specific set of tests is associated with each 3D printed device or a class of a device. This set of tests may be based on a number of requirements, such as internal process control, FDA guidance policies, or international standards. As mentioned above, 3D printed medical devices follow same regulatory requirements as conventionally manufactured medical devices (Figure 17.15).

17.7.6 Materials Used in 3D Printed Devices Overall

Usually, FDA does not approve specific materials for general use in the manufacturing of medical devices, either traditionally produced or 3D printed. Instead, FDA approves finished medical objects, instruments, etc. For instance, FDA has

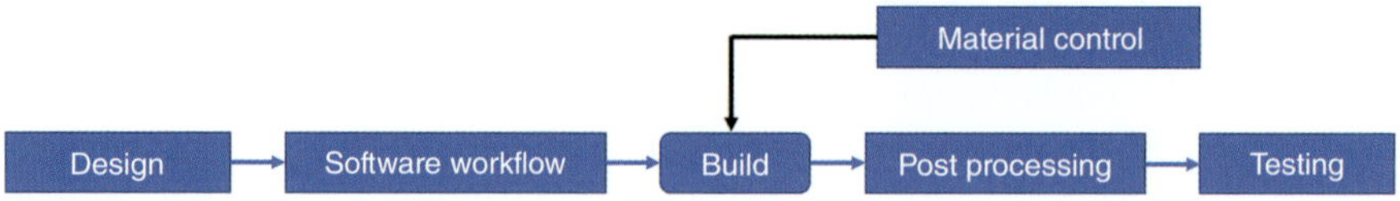

Figure 17.15 Flowchart of 3D printing manufacturing process.

cleared spinal implants produced from titanium alloy; however, FDA does not review or approve usage of titanium in medical objects [14, 16, 23].

FDA evaluates materials used in medical device manufacturing within its context of safety and efficiency of product for its intended use. In detail, the material is evaluated as a part of completed object and its intended use, and FDA assesses if device is reasonably safe and efficient for its intended use, or essentially equivalent to safety and effectiveness of a legally marketed device. If these prerequisitions are fulfilled, the FDA approves or provides clearance (respectively) to particular devices for specified intended use. Ultimately, this does not support an approval or clearance to use the same material in other devices.

Not all devices consisting of new material require FDA's stringent premarket review process (PMA review). Actually, if new material does not invoke a question of safety and/or effectiveness, device containing new material may be cleared through the 510(k) premarket notification process. In such a case, a submission review has to illustrate that the new material is at least as safe and effective as already legally marketed device of same equivalence [23].

17.7.7 Materials Used in Specific Application (Printed Dental Devices)

Some engineered materials for dental devices are considered as finished products that are suitable for use by health care professionals. These materials are patient-matched or fitted at the point of care. FDA clears such engineered materials for a specific indented use as a device. Examples of such materials include prosthetic devices and dental restoratives: dental cements, direct filling resins, denture resins, crowns, bridges, orthodontic retainers, night guards, inlays, and onlays. Usually, performance testing on material in its finished form is necessary for FDA for device clearance. The performance testing has to prove the material's proper physical and performance properties for the intended use [23].

Importantly, "the FDA does not clear materials for unlimited intended uses". Only device materials for that specific use such as "to fabricate a denture base" or "to restore a structural defect in teeth" can be cleared by the FDA. Every engineered material is cleared to make a specific device that has specific physical characteristics and is intended for specific use. For instance, if device material is cleared for "tooth shade resin material" as a specific intended use, FDA will not clear it by default for another purpose, such as an "endosseous dental implant abutment." The FDA will evaluate each specific case if manufacturer wishes to use same engineered material for new purpose. The information provided by manufacturer is used to ensure that material has adequate properties for the new intended use [23].

If a new intended use is classified under different classification regulation, the manufacturer will be obliged to comply with any regulatory requirements for that particular classification regulation.

17.8 Conclusions

Being an additive manufacturing technique, 3D printing continues to enjoy a renaissance in the area of materials fabrication and process engineering across industries. It is not far away when 3D printing will be widely utilized in almost every part of our daily lives. To date, 3D printing has primarily been used in engineering to create engineering prototypes. However, 3D printing has the potential to enable mass customization of medical goods or devices on a large scale also. Various cases have reported that 3D printing has already been used in several medical applications including the manufacture of eyeglasses, custom prosthetic devices, and dental implants. The major advantages of the 3D printing over moldings or paste injection include patient-specific geometries and controlled spatial patterning of materials or polymers within the complex texture. In order to ensure the widespread applications of various 3D printing techniques to fabricate medical products, a suitable set of regulatory guidelines has to be implemented although there seem to have some guidelines been available by the US FDA.

References

1 Tan, D.K., Maniruzzaman, M., and Nokhodchi, A. (2018). Advanced pharmaceutical applications of hot melt extrusion coupled with fused deposition modelling (FDM) 3D printing for personalised drug delivery. *Pharmaceutics* pharmaceutics-361582 in press.
2 Chakroborty, J.P., Boateng, J.S., Nokhodchi, A., and Maniruzzaman, M. (2017). *UK Pharmaceutical Sciences* 149.
3 Goyanes, A., Wang, J., and Buanz, A. (2015). *Molecular Pharmaceutics* 12: 4077–4084.
4 Alhanan, M.A., Okwuosa, T.C., and Sadia, M. (2016). *Pharmaceutical Research* 33: 1817–1832.
5 Maniruzzaman, M. and Nokhodchi, A. (2016). *Critical Reviews in Therapeutic Drug Carrier Systems* 33 (6): 569–589.
6 Moulton, S.E. and Wallace, G.G. (2014). *Controlled Release* 193: 27–34.
7 3D Systems (2018). https://uk.3dsystems.com/our-story (accessed April 2018).
8 Stratasys, USA (2018). http://investors.stratasys.com/news-releases/news-release-details/inventor-fdm-3d-printing-and-co-founder-stratasys-scott-crump (accessed April 2018).
9 3D Printing Trends (2017). https://f.3dhubs.com/yZgXoWzB88BhMHwG9fo3mV.pdf (accessed April 2018).
10 Carbon 3D (2017). https://www.carbon3d.com/ (accessed April 2018).
11 Goproto (2017). http://goproto.com/ (accessed April 2018).

12 Frost and Sullivan (2016). https://ww2.frost.com/event/calendar/enabling-materials-3d-printing/ (accessed April 2018).
13 Gartner (2018). https://www.gartner.com/doc/3834064 (accessed April 2018).
14 US FDA Medical Devices (2018). https://www.fda.gov/MedicalDevices/ProductsandMedicalProcedures/3DPrintingofMedicalDevices/default.htm (accessed April 2018).
15 Emergo (2016). https://www.emergogroup.com/blog/2016/05/us-fda-issues-technical-guidance-medical-devices-using-3-d-printing (accessed April 2018).
16 Medical Devices Regulated by FDA's Center for Devices and Radiological Health (CDRH) (2014). https://www.fda.gov/downloads/MedicalDevices/DeviceRegulationandGuidance/GuidanceDocuments/UCM259172.pdf (accessed April 2018).
17 Biologics Regulated by FDA's Center for Biologics Evaluation and Research (2017). https://www.fda.gov/AboutFDA/CentersOffices/OfficeofMedicalProductsandTobacco/CBER/ (accessed June 2018).
18 Drugs Regulated by FDA's Center for Drug Evaluation and Research (2017). https://www.fda.gov/AboutFDA/CentersOffices/OfficeofMedicalProductsandTobacco/CDER/default.htm (accessed April 2018).
19 US FDA (2018). Manufacturers of medical devices should refer to FDA guidance documents and quality systems regulations for more information on each specific application. https://www.fda.gov/MedicalDevices/DeviceRegulationandGuidance/PostmarketRequirements/QualitySystemsRegulations/default.htm (accessed April 2018).
20 US FDA (2018). https://www.fda.gov/MedicalDevices/DeviceRegulationandGuidance/Overview/ClassifyYourDevice/ucm2005371.htm (accessed April 2018).
21 Technical Considerations for Additive Manufactured Medical Devices (2017). https://www.fda.gov/downloads/MedicalDevices/DeviceRegulationandGuidance/GuidanceDocuments/UCM499809.pdf (accessed April 2018).
22 The 510(k) Program: Evaluating Substantial Equivalence in Premarket Notifications [510(k)] (2014). https://www.fda.gov/downloads/MedicalDevices/DeviceRegulationandGuidance/GuidanceDocuments/UCM284443.pdf (accessed April 2018).
23 US FDA (2018). https://www.fda.gov/MedicalDevices/DeviceRegulationandGuidance/HowtoMarketYourDevice/PremarketSubmissions/PremarketNotification510k/ucm2005718.htm (accessed April 2018).

Index

3D and 4D Printing in Biomedical Applications: Process Engineering and Additive Manufacturing, First Edition. Edited by Mohammed Maniruzzaman.

g

u

x

y

z